IET TRANSPORTATION SERIES 8

Road Pricing

Other related titles:

Volume 1	**Clean Mobility and Intelligent Transport Systems** M. Fiorini and J-C. Lin (Editors)
Volume 2	**Energy Systems for Electric and Hybrid Vehicles** K.T. Chau (Editor)
Volume 5	**Sliding Mode Control of Vehicle Dynamics** Antonella Ferrara (Editor)
Volume 6	**Low Carbon Mobility for Future Cities: Principles and Applications** Hussein Dia (Editor)
Volume 7	**Evaluation of Intelligent Road Transportation Systems: Methods and Results** M. Lu (Editor)
Volume 38	**The Electric Car** M.H. Westbrook
Volume 45	**Propulsion Systems for Hybrid Vehicles** J. Miller
Volume 79	**Vehicle-to-Grid: Linking Electric Vehicles to the Smart Grid** J. Lu and J. Hossain (Editors)

Road Pricing

Technologies, economics and acceptability

Edited by
John Walker

The Institution of Engineering and Technology

Published by The Institution of Engineering and Technology, London, United Kingdom

The Institution of Engineering and Technology is registered as a Charity in England & Wales (no. 211014) and Scotland (no. SC038698).

First published 2018

The Institution of Engineering and Technology
Michael Faraday House
Six Hills Way, Stevenage
Herts, SG1 2AY, United Kingdom

www.theiet.org

British Library Cataloguing in Publication Data
A catalogue record for this product is available from the British Library

ISBN 978-1-78561-205-3 (hardback)
ISBN 978-1-78561-206-0 (PDF)

Typeset in India by MPS Limited

Contents

Author's Preface

Dr. John Walker

The author's preface is defined as dealing with 'the genesis, purpose, limitations, and scope of the book and may include acknowledgements of indebtedness' [1].

Genesis

Why did I initiate the book? Because I had been professionally involved in road pricing for over two decades, I had noted the long history of economic theory, going back to Pigou in 1920 [2], his illustrious successors in the United Kingdom, Reuben Smeed [3] and Alan Walters [4] and the American Monetarist Milton Friedman [5], the apparent inspiration for Ken Livingstone and the London Congestion Charge – see Chapters 2 and 13. I had also been involved in the technology of electronic road pricing, which renders toll plazas obsolete, and, contrary to the beliefs of some, is sufficiently well developed to implement more subtle and comprehensive road pricing schemes, including charging by Time, Distance and Place (TDP), as well as by vehicle emission characteristics.

Road pricing is indeed already being used worldwide, to combat congestion and pollution, to compensate for falling revenues from fuel duty ('gas tax'), to provide an alternative (and fairer) means of assessing road usage-based liabilities than currently provided by the 80-year-old fuel tax, to improve the efficiency of the existing transport infrastructure and to fund new transport infrastructure.

However, it could and should be used far more widely. Eddington's 2006 report for the British government said, 'the potential for benefits from a well-designed, large-scale road pricing scheme is unrivalled by any other intervention' [6]. However, governments, both national and local, worldwide have been reluctant to adopt road-pricing policies in the *mistaken* beliefs that (a) the technology is not yet ready and (b) that they are a vote loser and that voters are implacably opposed to such policies.

These beliefs, particularly the latter, are not without foundation. The referenda in Edinburgh and Manchester and the petition on the Prime Minister's website in the United Kingdom, and the referendum in Gothenburg in Sweden, as described in a number of chapters in the book, show that some people feel strongly about the subject and vote against it. However, the evidence is that the public, once they have experienced the benefits, will vote for road pricing, directly (as they did in Stockholm and Milan) or indirectly (based on political manifestos, as in London) – especially if other charges and taxes are reduced to compensate, or if any additional revenues are reinvested in transport; see Chapters 5, 14 and 19.

There are important lessons to be learned from Manchester and Edinburgh. One can often learn more from failures than from successes. It would seem that Manchester did not learn the lessons of Edinburgh. And of course, one can also learn from successes, as Chapter 5 demonstrates in Minnesota and Chapter 19 in Milan.

Another myth is that the press is always opposed to road pricing – see for example the section on New York in Chapter 8. The *British Daily Mail* newspaper, often regarded as anti-road pricing, designated as 'A hero for our time', a businessman who built a temporary toll road around some long-term roadworks that were impeding his commute. He 'should be knighted in the next honours' list. He is an inspiration' [7]. See also an article in the *Seattle Times* reviewing various road pricing schemes and headlined 'Time for Seattle-area drivers to pay up? Tolling doesn't get much love, but it eases gridlock' [8].

History is also important. Toll roads go back thousands of years: there were toll charges on the Susa–Babylon highway in the seventh century BC and in the Holy Roman Empire in Europe in the fourteenth century AD. They facilitated the Industrial Revolution in England in the seventeenth century via the turnpike roads from 1663 onwards, when there was more transport demand, leading to an expansion in highways, better road maintenance and increased economic activity. (Note that the railways came much later – from the 1820s onwards.) So in a sense, we are living in a period of historical anomaly, where roads are free at the point of use, in most cases and in most countries. Of course, roads are not free – we mostly pay for them through our taxes, which are often regressive – the poor pay proportionately more, and effectively subsidize the rich. It would seem only fair to accept the 'user pays' principle.

Of course, some would argue that motorists pay a lot more in their taxes than they receive back in road construction and maintenance; but if the costs of congestion, pollution and health and safety are taken into account, the scales are more evenly balanced.

Purpose, scope and limitations of the book

So, the purpose of the book is to outline some of the economic theory, indicate the different kinds of road charging schemes that are possible, describe the electronic technology being used, show that it is available and already in operational use in many countries, address how public acceptability can be achieved and demonstrate that people will accept road pricing if they understand the reasons for using it and above all, if they have experienced it in use.

Inevitably, there is some repetition in the book; a number of chapters talk about London and Stockholm, or the Edinburgh and Manchester referenda. I have chosen not to edit out such repetition, partly because different chapter authors may have somewhat different perceptions; but mainly because such references are integral to the way the chapter is developed, and deleting them would make a chapter less coherent and less easy to read standalone.

There are very few engineering-oriented books in this field, or books aimed at transport planners – published books are mainly by transport economists. This book aims to fill that gap. A book is needed to inform engineers and planners how to prepare for and to implement road pricing schemes, which technologies to use, and which technologies are in use successfully already throughout the world. The book is aimed at engineers working in the equipment supply and end-user transport industry, central and local government officials tasked with improving their transportation system and students on ITS and transportation-oriented academic courses. Another important readership is politicians and policy advisors, to show them what has been successfully achieved and what is possible now and in the very near future.

Of course, it is impossible to cover all road pricing schemes worldwide, or all interests in the subject at national, regional or city government levels. But a representative sample of schemes and technologies are described in the following chapters, chosen for their opinion-leading influence and the lessons that can and should be learned from them

I see road pricing as inevitable – because it is a means of addressing the problem of declining fuel duty ('gas tax') revenues as well as congestion and pollution. Roads are congested, especially at certain times and locations (rush hours in towns and cities, motorways at Bank Holiday weekends). Congestion has been estimated to cost £13 billion in 2013 in the United Kingdom, increasing to £21.4 billion in 2030 [9]. The figures would be ten times this in Europe overall; in the United States in 2014 it was estimated to be \$160 billion, with 6.9 billion hours of travel delay and 3.1 billion gallons of wasted fuel [10]. Such estimates depend on the assumptions made, especially when projected into the future; but the costs are enormous and will increase unless something is done.

Air pollution, especially NO_x and particulates, to which the biggest contributor is road traffic, causes 50,000 premature deaths per year in the United Kingdom – far more than the number of people killed on the roads, annually. (There were 1,732 reported road deaths in Great Britain in 2015 [11].) In fact, air pollution is now the world's largest single environmental health risk – the impact is highest in Asia but significant in Europe and the United States. In 2012, 7 million people died as a result of air pollution exposure – one in eight of all deaths worldwide, compared with 1.3 million global road accident fatalities [12].

Road traffic emissions also contribute to global warming. Of the 568.3 million tonnes of CO_2 equivalent – the total net UK emissions from all sources – 21% were from transport, up from 15% in 1990, of which 93% were from road transport [13] (DfT 2015).

Furthermore, revenues from vehicle taxation are declining. The Institute for Fiscal Studies, using UK government statistics, projects that revenue from motoring taxation will drop from £38 billion per year in 2010 to £25 billion by 2029, due to improvements in the fuel efficiency of vehicles and adoption of alternative fuels [14]. This is also driving the distance-based road pricing pilot trials in some US states, include Oregon, California, Colorado, Delaware, Hawaii, Oregon and Washington (Chapter 18).

Acknowledgements of indebtedness

I am indebted to colleagues too numerous to mention, but they certainly include the members of the ITS (United Kingdom) Road User Charging Interest Group, of which I am proud to be Honorary Secretary, and my IET colleagues who have helped me to put together this book for publication, as well as my wife for her forbearance. But above all, I am grateful to the contributors to this book, who have provided authoritative and readable chapters on their areas of expertise, their experience, the technology and the road pricing implementations they are involved with; if they can't convince you that 'the potential for benefits from a well-designed, large-scale road pricing scheme is unrivalled by any other intervention', then no one can.

John Walker
Reading, UK
June 2017

Biographies of the authors and editor

Foreword. Professor Phil Blythe

Professor Phil Blythe CEng, FIET, is Chief Scientific Advisor for the Department of Transport and Professor of Intelligent Transport Systems (ITS) at Newcastle University. In his role at the DfT, he provides a challenge function to the Department on the use of science and engineering evidence in policy making and also ensuring the Department is best informed on new innovations and technologies that may impact on the delivery of transport schemes. Through the CSA network, he also ensures that there is significant cross-government cooperation on science, engineering and technology issues.

Phil's original research area was road user charging and he was credited with developing the first multi-lane automated toll-collection system in the world in the EU-funded PAMELA and ADEPT projects, the design and implementation of the first road pricing demonstrator, the Cambridge Congestion Metering Scheme in the mid-1990s, and was a member of the feasibility team of ROCOL (Review of Charging Options for London) which made the recommendations for the technology to be used for the London Congestion Charging System in 2000. Since then, Phil's research has evolved into work on smart ticketing, pervasive wireless sensor networks and most recently into electric vehicles, recharging networks, connected and autonomous vehicles and smart cities. Phil has published widely and co-authored a book on Road pricing technology with Andrew Pickford in 2006 where John Walker was the series editor. ('Road User Charging and Electronic Toll Collection', Andrew TW Pickford & Philip T Blythe, Artech House (Boston & London), 2006)

For the IET, Phil chairs the IET's Transport Policy Panel, is a member of the Prestigious Lectures Committee and supports the IET in the United Kingdom and abroad in areas such as intelligent transport and more recently also in the connected and autonomous vehicles, electro-mobility and smart cities/big data/IoT agendas.

Chapter 1: "Introduction to road pricing". Dr John Walker.

The editor is Dr John Walker MA (Oxon), MSc, PhD, CPhys, FIET. Following a first degree in Physics at Oxford University, an MSc in Applied Solid State Physics at Brighton University, a PhD in the physics of diamonds at Reading University and diamond research at the University of Paris VII John joined Racal (now Thales) as Technical Manager in 1980. He managed collaborative projects in Artificial Intelligence, Software Engineering, Intelligent Transport Systems, Traffic Information Broadcasting and road pricing, funded by the UK DTI, Technology Strategy Board, DfT and European Commission. He led the security work-package of the DfT-funded DIRECTS Road Pricing demonstration project. His role as Chief Technical Consultant before he retired from Thales in February 2010 included business development and project management in road-pricing and ITS. John was an Expert Advisor to Transport for London in their Congestion Charging technology trials. He is a member of the Executive Team of the Institution of Engineering and Technology's Automotive and Road Transport Technical and Professional Network, and of its Berkshire Network; he has organized and chaired five international seminars on Road Pricing and two on Managed Motorways for the IET. He is the editor of two books: 'Mobile Information Systems' (1990) and 'Advances in Mobile Information Systems' (1998), and was formerly the Series Editor of two Artech House book Series, 'ITS' and 'Mobile Communications'. John is currently an independent consultant, and the Editor-in-Chief of the IET series of books on Transportation. He is also a Visiting Senior Research Fellow in the Transportation Research Group at Southampton University, and Honorary Secretary of the Road User Charging Interest Group of ITS (United Kingdom). His professional interests have included solid state physics, software engineering, artificial intelligence and mobile communications, but are now centred on Intelligent Transport Systems, especially road pricing (including technology, deployment and public acceptability).

Chapter 2: "The Smeed Report at 50: will road pricing always be 10 years away?". Stephen Glaister.

Stephen Glaister was Professor of Transport and Infrastructure at the Department of Civil & Environmental Engineering, Imperial College London from 1998 until 2008, having spent 29 years in the Economics and then Geography departments of the London School of Economics. From 2008 to 2015, he was Director of the RAC Foundation and he is now Chair of the UK Office of Rail and Road. Stephen has been an advisor to many transport-related organizations and committees. He served on the boards of London Regional Transport and Transport for London and as a member of a number of official bodies including the Independent Advisory Committee on Trunk Road Assessment, the Road Pricing Feasibility Review and the Eddington Study. He has advised government and the Transport Select Committee. Stephen has written on a wide range of transport subjects including roads, railways and the London Underground, as well as regulation in the telecommunications, water and gas industries.

Stephen Glaister CBE FICE FTRF FCGI, Chair, Office of Rail and Road, Professor Emeritus of Transport and Infrastructure, Department of Civil and Environmental Engineering, Imperial College, London, SW7 2BU, UK.

Chapter 3: "Types of road pricing, and measuring scheme cost and performance". Dr John Walker & Andrew Pickford.

Andrew Pickford B.Sc. C.Eng. MBA FIET FHKIE

Andrew is one of the world's leading management consultants focused on the design and practical application of Intelligent Transport Systems (ITS) to road infrastructure and Smart City applications in the built environment to help deliver optimal policy outcomes. His competence includes investment appraisal; policy & regulatory stress testing; policy road mapping; business process design; procurement strategy development; scoping funding & financing mechanisms; development of performance management regimes; programme management and long-term implementation support. Andrew has been providing a full spectrum of management consultancy services to government agencies, banks and private sector organizations internationally since 1990, including United Kingdom, Switzerland, the Netherlands, Australia, Chile, the United States, South Africa, Asia, Singapore, Indonesia, Hong Kong, Taiwan and mainland China.

Chapter 4: "We can't get there from here: ecofiscal policies to address traffic congestion in Canadian cities". Nancy Olewiler.

Nancy Olewiler is an economist and Professor in the School of Public Policy at Simon Fraser University. She was the Director of the School from its origin in 2003 until 2014. Prior to coming to the Economics department at SFU in 1990, she was a professor in the Economics Department at Queen's University. Her PhD is in economics from the University of British Columbia. Nancy's areas of research include natural resource and environmental economics with a focus on public policy, including studies on energy and climate policy, natural capital and ecosystem services, transportation and federal tax policy. She has published in academic journals, edited books, has written two widely used textbooks (The Economics of Natural Resource Use and Environmental Economics) and produced numerous reports for the Canadian federal and provincial governments. From 1990 to 1995 she was Managing Editor of Canadian Public Policy. Nancy is a commissioner on Canada's Ecofiscal Commission and was a member of BC's Climate Leadership Team. She has served on the Board of Directors of BC Hydro, Powertech and TransLink, serving as their Chair from 2010 to 2013. She is a member of the National Statistics Council and Chair of the Macroeconomic Accounts Advisory Committee of Statistics Canada.

Chapter 5: "The public acceptability of road pricing – a US case study". Lee W. Munnich, Jr, Frank Douma & Joe Loveland.

Lee Munnich is a senior fellow at the University of Minnesota's Hubert H. Humphrey School of Public Affairs. He served as director of the Humphrey School's State and Local Policy Program from 1991 to 2015. Mr. Munnich's research focuses on transportation policy and regional economic development strategies. Mr. Munnich had over 23 years of experience with state and local government before joining the Humphrey School, including serving as deputy commissioner of the Minnesota Department of Trade and Economic Development and being elected twice to the Minneapolis City Council during the 1970s. Mr. Munnich began studying political and institutional issues in congestion pricing in the 1990s. His research and outreach activities led to the adoption of legislation authorizing high-occupancy toll (HOT) lanes in Minnesota in 2003 and the subsequent development of the MnPASS/managed lane system in the Minneapolis-St. Paul region. Mr. Munnich co-chaired the Transportation Research Board (TRB) Congestion Pricing Committee from 2011 to 2015 and chaired the TRB Pricing Outreach Subcommittee from 2002 to 2011.

Frank Douma is Director of the State and Local Policy Program and Coordinator of the Masters of Urban and Regional Planning Degree Program at the Humphrey School of Public Affairs, and a Research Scholar at the Center for Transportation Studies, located at the University of Minnesota. Mr. Douma's work related to innovative finance includes managing a parking pricing demonstration project for the USDOT, evaluating the qualitative results of introducing MnPASS lanes on I-394 and producing a report on the opportunities and challenges of introducing MnPASS on the I-35E corridor in the northeast Twin Cities metropolitan area. He is currently starting a new project examining the possible role of pricing in an environment with an increasing number of self-driving cars. In addition to working at the Humphrey School, Mr. Douma has a wealth of experience in transportation, having worked for the Canadian Pacific Railway, the Metropolitan Airports Commission and the Minnesota Department of Transportation. Mr. Douma has a Masters' Degree in Public Affairs and a Law Degree from the University of Minnesota, and a Bachelor's Degree in Political Science from Grinnell College.

Joe Loveland, Principal, Loveland Communications. As a public relations professional for over 30 years, Joe Loveland has served a wide variety organizations engaged in transportation, public health, health care, environment, human services and early education issues. Over the past 15 years, Mr. Loveland has provided strategic communications and public policy counsel and services related to congestion pricing and mileage-based user fees for the University of Minnesota's Humphrey School of Public Affairs, including leading the communications initiative described in Chapter 5.

Chapter 6: "How road pricing was implemented in Singapore, and planned technology augmentations". Dr Kian Keong CHIN.

An engineer by training, **Dr Chin Kian Keong** has been involved in the planning, designing, implementing, constructing and managing of various road and traffic management projects in Singapore, including the DSRC-based Electronic Road Pricing (ERP) system, which started full-scale operations in 1998, and various Intelligent Transport Systems (ITS) to better manage traffic. Currently, he also oversees the project to upgrade the ERP system to a satellite-based one that would not require physical charging gantries on the roads. He is now the Group Director, Transportation & Road Operations and concurrently the Chief Engineer, Transportation at Singapore's Land Transport Authority. He has presented and published papers at many conferences and technical journals over the past years, and had lectured at the local universities' Masters programme in transportation.

Chapter 7: "Communication and governance challenges in Greater Manchester's 'congestion charge' referendum". Dr Graeme Sherriff.

Dr Graeme Sherriff is a Research Fellow at the University of Salford and Associate Director of the Sustainable Housing and Urban Studies Unit (SHUSU). He leads the unit's Sustainability and Society theme. With a background in urban planning, his research focuses on the intersection of environmental sustainability and social justice with a view to informing policy and critically engaging with debates on the relationship between environment policy, inequality and social inclusion. He has published in the fields of fuel poverty, energy-efficient retrofit, sustainable transport, urban food and environmental justice. He works extensively with the local government, housing, voluntary and community sectors.

Graeme Sherriff, University of Salford, g.sherriff@salford.ac.uk

Chapter 8: "Case studies of communication and consultation strategies for road pricing". Andrew Pickford & Dr John Walker.

See biographies for Chapters 1 and 3.

Chapter 9: "Road pricing standardisation". Jan Kersten, Trafficteq, Germany & Jasja Tijink, Kapsch TrafficCom.

Jan Kersten holds a diploma in electrical engineering. Since 1993, he has been working in ITS business – mainly in the area of Electronic Fee Collection (EFC) – in different roles and positions in market leading companies (like Bosch, Kapsch TrafficCom and Toll Collect). During his professional career, he gained experiences in concepts, architecture, design and implementation of devices and systems in large-scale commercial EFC projects of different technologies. He is also familiar with business processes, operational and commercial aspects in the EFC systems, processes and services area. For almost 10 years now, Jan Kersten has been successfully running his own consultancy business – Trafficteq – in the field of ITS. In most of the projects and activities, he supported his clients in medium and large scale projects in which he took over responsibility and leadership roles in tasks like overall system and service architecture, business process analysis and technical project management. Over the last 20 years he contributed to several European funded research projects and international standardization. Currently, he is actively involved in the work of CEN/TC 278 WG1 (Standardization in the field of Electronic Fee Collection). Jan Kersten was project team leader of the project team PT136 'Guidelines for EFC applications based on in-vehicle ITS

Stations – CEN/TR 16690', and currently he is project team leader of project team PT140 'Conformance evaluation specifications for equipment based on CEN/TS 16986 (Interoperable profile for ISO EN 12855)'.

Jasja Tijink holds a master's degree in electronic engineering. He has been working in the field of ITS since 2002, especially in the areas of Electronic Fee Collection (EFC) and Cooperative ITS. He is employed with Kapsch TrafficCom AG since 2006 and is responsible for the coordination of international standardization activities. Over the last 10 years, he has personally contributed to standardization in ITS and EFC through participation in numerous standardization development organization's working groups, funded project teams and research projects. His focus lies on system architectures and information security.

Chapter 10: "The European Electronic Toll Service – EETS – and the REETS Project". Mike Hayward (Transport Logic) and Hubert Resch (ASFINAG).

Mike Hayward CEng MIET FIHT is managing director of Transport Logic Limited, an independent provider of technical and management consultancy focused on the deployment of intelligent transport systems. He has advised on many electronic toll collection projects worldwide, including the heavy vehicle truck tolling system in Poland, and acted as a consultant during the REETS project. Mike graduated from the University of Hertfordshire in 1986 with an honours degree in electrical and electronic engineering subsequently joining Jaguar Cars product development team. He qualified as a Chartered Engineer in 1993 and has since held senior roles in engineering-based consultancy companies Atkins, Carl Bro and Hyder Consulting. Mike's project delivery experience covers all aspects of transport systems research, development and deployment including traffic control centres, electronic toll collection, communications systems, electronic ticketing, passenger information and driver assistance systems. He established Transport Logic in 2008 and has since worked for clients in the highways, transit and automotive sectors worldwide, including as a short-term advisor on electronic toll collection projects for the World Bank.

Dr. Hubert Resch is managing director of ASFINAG European Toll Service GmbH, subsidiary of ASFINAG, Austria's motorway operator. In addition, he is secretary to the European Initiative EETS Facilitation Platform since 2015. Hubert was born in Austria in 1969. He graduated from the University of Vienna (Law) in 1995 and was granted the Doctor in Law in 2002. He was assistant professor at the Department of Public Law at the University of Business Administration in Vienna. Hubert worked as an associate at Freshfields; he was admitted to the Austrian Bar in 2006. In 2008, he became the Head of the Legal Department of ASFINAG and since 2009 he is managing director at ASFINAG.

Chapter 11: "Standardisation and implementation of ANPR – a practical guide". Peter Vermaat.

Peter Vermaat is a Principal ITS Consultant at the Transport Research Laboratory (TRL Ltd) in the United Kingdom. He has over 30 years' experience in the electronics, image processing, communications and intelligent transport businesses, including research and development, product development, systems engineering, integration, test and operations. He was part of the team which developed one of the UK's first ANPR systems in the mid-1990s, and went on to develop applications for ANPR in both the commercial and security markets. He was also actively involved in the development of ANPR-specific cameras.

Chapter 12: "Engineering interoperability in the US: video tolling & multiprotocol tags & readers". JJ Eden.

James (JJ) Eden is an industry leader in the field of traffic technology, with a specialization in tolling technology and infrastructure. Currently, a Vice President and Director of Tolling at global engineering firm AECOM, JJ's career spans 40+ years in the business of helping America's highways operate with safety and efficiency. He previously served as President, CEO and co-founder of The Alliance for Toll Interoperability, COO of the North Carolina Turnpike Authority and Director of ETC at ACS/Lockheed Martin, IMS. He spent many years as a contractor and employee of the Pennsylvania Turnpike Commission, where he directed installation of the Turnpike's first computer systems and designed the world's first weigh-in-motion toll barrier. In 1988, JJ was a founding member of the E-ZPass (electronic toll system collection) Network, which has grown to become a $15B inter-agency group with over 27 million customers. JJ's contributions to the industry have evolved to focus on consulting, where he supports executives in the areas of effective operations management, systems integration and assembling highly talented teams to take on the future challenges of tolling design and operations. A graduate of Lincoln University/Ryder Technical Institute, JJ resides in Raleigh-Durham, North Carolina with his wife of 43 years, Winnie.

JJ Eden, President, Alliance for Toll Interoperability, VP AECOM.

Chapter 13: "London congestion charging – a personal account". Nick Patchett & Jeremy Evans.

Nick Patchett is a partner in Pillar Strategy. Nick was first involved in road pricing in 1993 exploring technology options for 'tags' on the Birmingham North Relief Road (now M6 Toll). He has a background in systems development, telecommunications, complex procurement and outsourcing but has spent the last 17 years applying the same principles in transport. He became the systems team lead for the London Congestion Charging scheme in 2000 and managed TfL's road pricing technology trials 2003–2008. He has also advised on over 30 other road pricing, congestion charging and tolling projects in the United Kingdom and around the world.

Jeremy Evans had 8 years' experience in the Congestion Charging Directorate in Transport for London as part of the Senior Management Team responsible for the design, procurement, implementation and operation of the London Congestion Charging Scheme and the London Low Emission Zone. In this role, he worked closely with the TfL Legal Team and Leading Counsel to provide evidence in response to a legal challenge to the original scheme. He did the same and appeared as the principal witness at a planning inquiry after planning enforcement action was taken against the installation of enforcement infrastructure in Kensington and Chelsea. Jeremy was responsible for the traffic management measures and enforcement infrastructure for the original scheme (2003), the upgraded and expanded scheme (2007) and the London Low Emission Zone (2008). He was responsible for the Transport for London technology trials of new technology for road user charging (GNSS satellite positioning, GSM mobile positioning, DSRC tag and beacon). Jeremy represented TfL in work with the UK Department for Transport (DfT), the Dartford-Thurrock River Crossing Operator and the M6 Toll Road Operator to develop interoperability between back offices, irrespective of the detection and enforcement infrastructure or on board vehicle devices. As a Director of Harpur Evans Consultancy Ltd, he worked as a subcontractor to PA Consulting, advising the UK Department for Transport (DfT) on the requirements for the Dartford Free-Flow Project, converting a barrier tolled river crossing of the Thames to free flow toll collection by video and DSRC data collection. In 2015, Jeremy provided advice and training to the Istanbul Metropolitan Municipality as they were designing a congestion charging scheme for the historic part of Istanbul. In 2016, Jeremy advised a leading toll road operator on the prospective tolled tunnel under the Thames at Silvertown, being planned by Transport for London.

Chapter 14: "The Swedish congestion charges – lessons learnt". Dr Ida Kristoffersson & Dr Maria Börjesson.

Dr. Ida Kristoffersson is a researcher in the Traffic Analysis and Logistics division at VTI – Swedish Road and Transport Research Institute. She has evaluated the effects of congestion charging in both Stockholm and Gothenburg, regarding traffic effects, adaptation strategies, socio-economic benefit and equity. She has also developed new transportation models that can evaluate new policy strategies such as congestion charging with higher accuracy. Her research interests include transport policy and pricing, cost–benefit analysis, transport demand modelling, travel behaviour, sustainable cities and transport system effects of digitalization and autonomous vehicles.

Dr. Maria Börjesson is Associate Professor of Transport Systems Analysis at KTH Royal Institute of Technology in Stockholm, Sweden. She is also externally affiliated to the Choice Modelling Centre at ITS Leeds and secretary of the board of International Transportation Economics Association (ITEA). She has been involved in the design process of the Gothenburg congestion charges and evaluated the charges in both Stockholm and Gothenburg. Her research interests include transport cost–benefit analysis, appraisal and sustainability, transport policy and pricing, public opinions, transport modelling, travel behaviour, gender differences, active mobility and cycling, the impact of the transport system on employment and GDP, survey design, choice experiments and econometrics for valuation of non-market goods such as travel time, reliability and security.

Chapter 15: "Moving from conventional tolling installations to Open Road Tolling". Bjarne Olav Tveit.

Bjarne Olav Tveit. BSc (Hons), University of Newcastle-upon-Tyne (1975)

Independent ITS advisor with over 25 years of international experience on different applications of toll collection systems – traditional exit barrier design and multi-lane free flow design. His range of services cover: feasibility studies, cost benefit analysis, process flow and scenario descriptions, analysis and capture of functional and non-functional system requirements, tendering and contract negotiations, project planning and management, quality planning and management and finally planning and management of customer's final acceptance tests and commissioning. Among his customers have been:

Trafikverket (the Swedish Transport Administration), Sweden
Helsinki Regional Transport (HSL), Finland
Øresundsbro Konsortiet (the Øresund bridge toll collection system), Denmark/Sweden
JYW Consulting/West Connex Delivery Authority, Australia
Transportstyrelsen (the Swedish Transport Agency), Sweden
Queensland Motorways Ltd, Australia
World Bank, US Aids and the Road Administration of the Republic of Makedonija
Dornier Consulting/Gauteng Provincial Government, Republic of South Africa
European Commission (DG TREN), Belgium

Chapter 16: "GNSS tolling: standards and implementations". Norbert Schindler, Global Sales Manager, Siemens, Erich Erker, Siemens.

Norbert Schindler was born in the United States to Austrian parents. After he completed his undergraduate studies in Computer Science at Carnegie Mellon University, he moved to Europe. Parallel to his work in Software Engineering, Norbert completed his Master's degree at the Technical University of Vienna in 1997, with a focus on Artificial Intelligence. He then moved to the Telecommunication industry, leading projects in air traffic control in Europe, North America, and East Asia. Later, he led a prize-winning team in the rollout of third generation mobile technology (UMTS) at Ericsson before he started with Electronic Tolling at Kapsch in 2003. Norbert then joined the Electronic Tolling Headquarters at Siemens in Vienna, Austria, becoming Bid Manager and Global Sales Manager for GNSS-based tolling solutions. After nearly 13 years with Siemens, Norbert established his own company in 2017, GNSS-Consulting, supporting industrial leaders in the acquisition of new projects in Electronic Tolling and acting as an Expert for the European GNSS Agency of the European Commission in the evaluation of proposals for the development and deployment of solutions using GALILEO satellite positioning technology. Norbert has three children and is married to a remarkable businesswoman from Belarus with whom he loves travelling and taking long hikes in the Austrian Alps.

Erich Erker was born and raised in Vienna, Austria. After graduating from the Technical University of Vienna, Austria (Diploma Engineer in Communications Engineering, 1997), he continued his studies at the University of California, Santa Barbara (USA), working as research assistant focusing on nonlinear millimeter wave circuits, meanwhile earning a degree in Wave Electronics (Master of Science, 2000). Since his subsequent return to Europe, Erich Erker works for Siemens AG Austria, first leading small R&D projects in RF Engineering, then contributing to the Siemens activities in Electronic Tolling Systems covering multiple roles ranging from project management, system architecture, internal and external technical consulting (e.g. at the EU Commission) to standardization work, IPR issues and others. After a short interlude at Siemens Rail Automation sector in the second half of 2016, he transferred to Kapsch TrafficCom, working as a solution engineer on business processes and system architecture topics for Electronic Tolling Systems. He is husband to a wonderful, supportive wife, proud father of four great children, slow but dedicated Ultra-Trail runner and lover of music, science and literature.

Chapter 17: "HU-GO: the Hungarian distance based electronic toll system". Zoltán Varga, CEO, Toll Service PLC, Hungary.

Zoltán Varga joined the National Toll Payment Services (NTPS) PLC – Hungary in 2011 as director of road charging. He is currently the CEO of Toll Service PLC. Both companies are 100% state owned and they are the Road User Charging and Tolling competence centre of Hungary. Zoltan was in charge for all related operational tasks to be executed of the successful implementation of the HU-GO distance-based commercial vehicle electronic toll collection system in 2013, on more than 6,500 km toll road network of Hungary. As a director of road charging, he was reorganizing the service side of NTPS in line with the development of the new and the already applied processes and work-flows. This created a solid basis to develop and extend the organization by making it ready and available for the implementation of the HU-GO system. HU-GO is the only single platform GNSS tolling system that is using already existing tracking technology for tolling purposes and creates a multi-utilization environment for integrated ITS solutions. Prior to 2011, Zoltán spent 17 years in the vehicle financing industry with companies like General Motors, BMW and Scania.

Chapter 18: "West Coast distance charge programs: an open market as the gateway to implementation in the United States". James M. Whitty.

James M. Whitty, J.D. is a partner in D'Artagnan Consulting LLP. Mr. Whitty is a thought leader on transportation funding and a pioneer on distance charge policies and systems. Mr. Whitty has provided consulting expertise on distance charge projects in the United States, Canada and Australia. Mr. Whitty led Oregon's distance charge efforts from commencement in 2001 to 2016. During this period, Mr. Whitty administered the Road User Fee Task Force, Oregon's key policy body on alternative road funding, to identify distance charging as the principal broad-based alternative to the fuel tax as a road funding mechanism, lead policy development of a distance charge for light vehicles as a revenue mechanism, structure the fundamentals of distance charge systems and guide legislative action on distance charging for light vehicles, including giving testimony 45 times before the Oregon Legislative Assembly, six times before other US state legislatures and three times before US Congress. Mr. Whitty presented and negotiated the nation's first, operational distance charge programme for light vehicles through the Oregon Legislative Assembly in 2013. During his 14 years leading Oregon's distance charge efforts, Mr. Whitty led all strategic development for distance charge policy and systems, including crafting and orchestrating two successful distance charge trials and the United States' first operational distance charge programme known as OReGO. Mr. Whitty is co-founder of the Western Road Usage Charge Consortium, now branded RUC West, a distance charge research entity consisting of 14 US states, participating on RUC West's steering committee from conception and chairing from October 2014 through January 2016. Mr. Whitty has written several publications and reports on distance charging and has spoken on the topic in dozens of US states and in Canada, Australia, Singapore and Europe.

Chapter 19: "Four years of Milan's road charge: effectiveness, acceptability and impacts". Paolo BERIAa, Politecnico di Milano, Milan, Italy, Luca TOSI & Davide NUCCIO, AMAT (Agenzia Mobilità Ambiente e Territorio), Milan, Italy.

Paolo Beria holds an MSc in Civil Engineering and a PhD in Urban Projects and Policies (Politecnico di Milano, 2003, 2008). He is an Associate Professor of Transport Economics and director of the TRASPOL Research Centre on Transport Policy in the Department of Architecture and Urban Studies at the Politecnico di Milano. He is a member of the management board of the Special Interest Group on Transport Economics and Regulation (SIG-E3) of the World Conference on Transport Research. His fields of research are transport planning and geography, assessment of transport projects and policies and regulation. He published more than 70 scientific products, of which 16 papers in indexed journals.

Luca Tosi currently is the Managing Director of the Public Transport Agency of Milan, Monza, Lodi and Pavia. From 2002 to 2016 CIO IT at Agenzia Mobilità Ambiente e Territorio (AMAT, agency for mobility, environment and land use) of the city of Milan, he was responsible for monitoring and control of local public transport service and for transportation modelling systems. He was Technical Director of the Traffic Control Centre of the Milan Municipality, from 2015 to 2016.

Davide Nuccio is an environmental engineer and GNU/Linux enthusiast. Since 2011 he worked for the Agenzia Mobilità Ambiente e Territorio (AMAT, agency for mobility, environment and land use) of the city of Milan analysing the traffic data and pollution/congestion charge. The daily tools are postgreSQL and python, but he also loves to develop user interfaces for data visualization.

Chapter 20: "Optimising use – using incentives to address traffic congestion". Rob Mouris (Arcadis), Jorrit Nijhuis (Rijkswaterstaat), Colin Black (Arcadis).

Ing. Rob Mouris is senior consultant in the area of traffic demand management at Arcadis, the leading global Design & Consultancy firm for natural and built assets. On behalf of Rijkswaterstaat (the Dutch national agency for road and water management of the Ministry of Infrastructure and Environment), he has worked as a project manager responsible for several rush-hour avoidance projects in the Netherlands. Because of his specific knowledge about the rush-hour avoidance projects he has been asked by the Dutch Ministry of Infrastructure and Environment to support the Optimising Use Programme. Within this programme, he advises all the rush hour avoidance projects (Spitsmijden) on behalf of the Ministry. Prior to his career at Arcadis, he has worked for XTNT, a consultancy bureau in Utrecht and he studied traffic engineering at the NHTV, an international higher education institution in Breda.

Dr Jorrit Nijhuis works as a senior consultant on mobility management (or TDM) at the Dutch national agency for road and water management of the Ministry of Infrastructure and Environment (Rijkswaterstaat). He is also involved in the Programme Optimising Use, a public-private investment programme containing over 300 local measures (mobility management, ITS/smart mobility, road improvements) aimed to reduce congestion and improve accessibility. Starting point within this programme is optimizing the use of existing roads by offering travellers more travel choices and better services. Within the Optimising Use Programme, Jorrit works as theme coordinator for projects on rush hour avoidances (spitsmijden), Park+Ride and parking. Prior to his career at the Ministry of Infrastructure and the environment, he worked at the Environmental Policy Group at Wageningen University. Here he wrote a PhD thesis on the active involvement of citizen-consumer groups in transitions towards sustainable mobility.

Dr Colin Black is UK Transportation Managing Director for global consulting firm Arcadis. He has worked internationally as a consultant to governments at all levels, specializing in Transportation Demand Management, Sustainable Transport and Mobility Management. Over 25 years, he has constantly broadened and refined expertise to develop sustainable transport policy and implementation strategy through understanding the complexities of cultural, institutional and legislative frameworks.

References

[1] Writers and Editors (2017) "What is the difference between a preface, a foreword, and an introduction?", Available at http://www.writersandeditors.com/preface__foreword__or_introduction__57375.htm. [Accessed 16 October 2017].

[2] Wikipedia (2017) "Road Pricing", Available at https://en.wikipedia.org/wiki/Road_pricing [Accessed 16 October 2017].

[3] Smeed, R.J., "Road pricing: the economic and technical possibilities". HMSO, 1964.
[4] Walters, A. A., 'The theory and measurement of private and social cost of highway congestion' *Econometrica*, 676–699, 1961.
[5] Friedman M. and Boorstin D. J. "How to Plan and Pay for the Safe and Adequate Highways We Need" in Roth, G., Roads in a market economy. Aldershot, England, 1996.
[6] Eddington, R., "The Eddington Transport Study. The case for action". Her Majesty's Stationary Office, 2006.
[7] "A hero for our time", *Daily Mail*, Tuesday, August 5, 2014.
[8] Seattle Times (2017) "Time for Seattle-area drivers to pay up? Tolling doesn't get much love, but it eases gridlock", Available at http://www.seattletimes.com/seattle-news/transportation/time-for-seattlearea-drivers-to-pay-up-tolling-doesnt-get-much-love-but-it-eases-gridlock/. [Accessed 30 June 2017].
[9] Inrix (2014) Joint study with the Centre for Economics and Business Research. http://inrix.com/press-releases/traffic-congestion-to-cost-the-uk-economy-more-than-300-billion-over-the-next-16-years/.
[10] Texas A&M Transportation Institute & Inrix (2015), "2015 Urban Mobility Scorecard", August 2015.
[11] DfT (2016) "Reported road casualties in Great Britain: main results 2015", UK Department for Transport Statistical Release, 30 June 2016.
[12] Jackson N (2016) "Future mobility & energy demands – the role for hybrid and electric vehicles". Keynote presentation at IET 6th Hybrid and Electric Vehicles Conference, London, 2–3 November 2016.
[13] UK DfT (2015) "Transport Statistics Great Britain 2015".
[14] Johnson P., Leicester A. & Stoye G. (2012) "Fuel for Thought: The what, why and how of motoring taxation'. Institute for Fiscal Studies & RAC Foundation, May 2012.

Chapter 1
Introduction to road pricing

John Walker and Andrew Pickford***

1.1 Introduction to the book

Each chapter is written by people eminent in the field. Unfortunately, it is impossible to cover all existing road pricing schemes worldwide in one book; but the schemes and topics addressed are chosen to be representative of leading-edge activity in road pricing, with wider lessons and implications than the particular schemes described.

And the emphasis is on electronic systems, not cash or paper-based schemes (vignettes[1]) [1], often described as multi-lane free-flow – there are no toll barriers, and no need for vehicles to stop in order to pay; stopping is undesirable since it results in congestion, pollution and accidents.

1.2 Terminology

Although there are no generally-accepted definitions, we should explain the difference between tolling, which is a payment for using (usually new) infrastructure such as a road, a bridge or a tunnel, and road user charging, which involves a payment for driving on an existing road, within an existing corridor or in a particular area. The latter is more controversial, because people believe that they are being asked to pay again to use roads which they have already paid for – though of course, roads are never fully paid for since they need to be maintained – and maintenance costs may be greater than the original construction costs.

*Transportation Research Group, University of Southampton, UK
**MVA Hong Kong Ltd., Hong Kong

[1] Vignettes are coloured paper stickers attached to windscreens, and have been used to implement road pricing for many years. They usually have a fixed period of validity such as a month or a year. They are cheap to implement, but manually intensive to enforce, and inflexible in use, and are gradually being phased out in many places in favour of electronic road pricing. See Walker [1]. In some cases, the number-plate is used as a vignette and is read by Automatic Number-Plate Recognition – see Chapters 3 and 11.

Table 1.1 Terminology

Term	Meaning and usage
Road tolling	Charging for use of a (usually new) road
Road pricing	Used by transport economists and by the UK Department for Transport
Road user charging	Often synonymous with 'road tolling' or 'road pricing', but may apply to charges on existing roads. The implication is that the payment is a user charge rather than a tax (see Section 1.6). Used by the UK Department for Transport
Congestion charging	The payment is limited to certain times of the day, or days of the week, when there is congestion. See Chapter 13
Electronic fee collection	As used by the European Commission; it emphasises electronic as opposed to paper-based schemes for charging in general
Automatic debiting systems	An older term used by the European Commission
Road use charging	An attempt to emphasise that the payment is for use of the road, switching the focus from payment by the driver. Used in the United States and Europe – but not much
Open road tolling	Used particularly in the United States and South Africa
Value pricing	Used particularly in the United States, to emphasise that there are benefits to the payments
Mobility pricing	Used in Canada, especially Toronto
Electronic road pricing	ERP – used particularly in Singapore

Other terms used are indicated in Table 1.1 – they are sometimes used interchangeably but strictly speaking they are more specific, or are used in different countries or domains.

1.3 Context: congestion, pollution, taxation

Current motoring taxation is a blunt instrument, with limited scope to influence driver behaviour. Furthermore, the improved fuel efficiency of vehicles and the switch to alternative, untaxed fuels such as electricity or hydrogen reduces the fuel duty collected; so fuel taxes should be replaced by 'demand-based' and usage-based charges.

Road traffic congestion and pollution were alluded to in the Preface; annual congestion costs in Europe and the United States are in hundreds of billions of euros and dollars; Chapter 4 details the costs in Canada. In 2012, 7 million people died as a result of air pollution – one in eight of all deaths worldwide – compared with 1.3 million road accident fatalities.

Other attempts to combat these problems haven't worked so far; they can't be cured by building new roads or by improving public transport[2] [2], as was also demonstrated in the Stockholm trials (Chapter 14). Road space is a scarce commodity and has a value. Congested road space imposes costs ('externalities') on other drivers (delay and pollution) which should be paid for ('internalised'). And potentially, charges create an investment stream for transport improvement – though there is a tension between that and 'revenue-neutrality' where road user charges simply replace existing motoring taxes and drivers on average pay no more.

The strengths of road pricing are that it targets congestion and pollution effectively, it is flexible and adaptable and it complements other 'urban demand management strategies' such as access control and parking restraint[3]. Its weaknesses are the (often lack of) political and public acceptability[4] [1], the visual impact on the landscape or streetscape (depending on the technology used – see Chapter 3 and elsewhere in the book) and perceived equity issues (Section 1.4). See also Table 1.2.

1.4 What are not the solutions to congestion and pollution

We have just seen above that the problems of congestion and pollution can't be cured by building new roads or by improving public transport[5] [3]. Other ways have been tried, some of which are outlined in Chapter 3 – see especially Tables 3.1

[2] 'We investigate the relationship between interstate highways and highway vehicle kilometers traveled (vkt) in US cities. We find that vkt increases proportionately to highways and identify three important sources for this extra vkt: an increase in driving by current residents; an increase in transportation intensive production activity and an inflow of new residents. The provision of public transportation has no impact on vkt. We also estimate the aggregate city level demand for vkt and find it to be very elastic. We conclude that an increased provision of roads or public transit is unlikely to relieve congestion and that the current provision of roads exceeds the optimum given the absence of congestion pricing'. Duranton and Turner [2].

[3] Smart parking is addressed in Chapters 3 and 4. It isn't Road User Charging but it is relevant – it does reduce the number of drivers cruising around looking for parking space, and hence minimises congestion and pollution.

[4] 'We are nourishing a monster of great potential destructiveness. And yet we love him dearly ... the motor car is clearly a menace that can spoil our civilisation. But translated into terms of the particular vehicle that stands outside the door, we regard it as one of our most treasured possessions or dearest ambitions, an immense convenience, an expander of the dimensions of life, an instrument of emancipation, a symbol of the modern age. To refuse to accept the challenge it presents would be an act of defeatism ... we must meet it without confusion of purpose, without timidity over means, and above all without delay'. Sir David Crowther, in his foreword to the UK Buchanan Report, 'Traffic in Towns', 1963.

[5] As Ellis [3] points out, in the United Kingdom, it is argued that the answer to road congestion is to improve public transport, build more roads and increase cycling. In comparison, the Netherlands have more roads per vehicle (especially motorways), more extensive and more frequent train and tram networks that are less crowded and cheaper than in the United Kingdom. Dutch usage of bicycles is at levels undreamt of in the United Kingdom. Yet despite all this, congestion problems remain – which is why the Dutch have had several attempts to implement road pricing.

Table 1.2 Ten things Transport Ministers should know about road pricing

1. Road pricing is not new. Toll roads are 2,700 years old; tolls were paid by travellers on the Susa–Babylon highway in the reign of Ashurbanipal in the seventh century BC. Aristotle and Pliny refer to tolls in Arabia and other parts of Asia
2. At its peak, in the 1830s, over 1,000 Turnpike Trusts administered 30,000 miles of turnpike roads in England and Wales, with 8,000 toll-gates. Although the turnpikes permitted an expansion of trade, they were not always popular – there were the famous 'Rebecca riots' in South Wales between 1839 and 1843 in which men dressed as women (because rioting was a capital offence for men but not for women) attacked toll gates, as a protest not just about tolls but about other taxes as well. The toll gates were simply the most visible form of an oppressive taxation regime
3. The United Kingdom exported the toll road concept to the United States, where the name 'turnpike' is still in common use, and to many other countries
4. England has a number of tolled facilities, including the M6 Toll road, the Durham and London Congestion Charges, estuarial crossings such as Dartford–Thurrock River Crossing, the M4 Severn River Crossing, the Mersey and Tyne Tunnels, the Humber and Tamar Bridges and the Eynsham and Whitchurch toll bridges across the river Thames
5. Contrary to popular belief, roads, bridges and tunnels are NOT free – we pay for them through our taxes. And they are never paid for; they always need ongoing maintenance, and the maintenance costs often outweigh the original construction costs
6. No one likes to pay what they perceive as extra cash to travel on roads. But when people understand why tolls and congestion charges are being imposed, and particularly when they have experience of them, they will accept the charges, if only as a necessary evil. There is evidence for this worldwide, including from the United Kingdom, Norway, Sweden, Southern Europe, the United States, Singapore, South America and Australia
7. Truck tolling is now widespread in Europe, with the United Kingdom following on after various European countries including Switzerland, Austria, Germany, the Czech Republic, Slovakia, Hungary and Poland
8. Technology is no longer a problem. Several technologies are used in different countries, often in combination. The United States uses radio frequency or microwave tags combined with video tolling using Automatic Number Plate Recognition, and is successfully experimenting with satellite-based and smart-phone technology. Germany, Hungary and Slovakia use Global Positioning System (GPS) backed up by ANPR. Austria, the Czech Republic, Singapore (currently) and Poland use microwave tags plus ANPR. London, Stockholm and Gothenburg use ANPR. Singapore is moving towards a distance-based charging method, to be employed from 2021 onwards
9. Road pricing can address pollution as well as congestion and revenue-raising. Charges in the German truck-tolling scheme depend on the emission characteristics of the truck, resulting in a cleaner vehicle fleet. The London Low Emission Zone uses exactly the same technology as the congestion charge, and could be extended to private cars at minimal extra cost
10. It is generally accepted amongst transport professionals that, as Eddington put it in his report to the British Government, 'the potential for benefits from a well-designed, large-scale road pricing scheme is unrivalled by any other intervention' [19]. That view was endorsed by the Department for Transport in 'Towards a Sustainable Transport System' where it states: 'The Government accepts the Eddington analysis regarding the exceptional case for exploring the potential of road pricing'

See thinkinghighways.com, A supplement to Thinking Highways, Vol 8 No 4, for an earlier (and illustrated) version of this table.

and 3.2. Whilst they – and the methods below – have been effective to some degree, they have not solved the problems.

1.4.1 Improvements in public transport

Improvements in the quality of public transport include new infrastructure, integrated ticketing, new routes, improved capacity and geographic expansion but this could also relate to improved timeliness and the ambience of waiting areas and vehicles themselves. Whilst such improvements are often regarded as an integral part of investment packages that also include congestion charging (e.g. for schemes in Gothenburg, London, Singapore and Stockholm – and proposals in New York, Edinburgh and Jakarta), it is arguable as to whether such improvements can mitigate against congestion without pricing or other access conditions. For example, a light rail line that was implemented to relieve congestion on Interstate 10 from Santa Monica to downtown Los Angeles had no measurable effect on congestion on Interstate 10 despite exceeding ridership forecasts [4]. Similarly, sustained investments in the extension of the mass transit network in Hong Kong between 2002 and 2011, during a period in which the population of cars increased by 25% and mass transit line length of 83%, it resulted in virtually no change in modal split between public and private transport usage [5,6], and travel times for surface transport have continued to increase. It may be concluded that whilst improvements in capacity and quality of public transport are justifiable on several grounds, including supporting economic sustainability with long-term population growth, they appear to be insufficient alone to mitigate congestion.

1.4.2 Car-sharing

Since many cars have only one occupant, particularly during rush hours, attempts have been made to encourage car-sharing. But there is evidence from the United States [7] that High-Occupancy Vehicle lanes (dedicated lanes on highways that can be used only by vehicles with more than two or three occupants – see Chapter 3) do not significantly encourage car-sharing; they are often used by vehicles which already had several occupants, such as a family going on holiday. And as described in Chapter 6, the intention of the Singapore's original Area Licensing Scheme (ALS) was to reduce traffic on the roads, and initially vehicles were exempt from charges if the cars had four or more occupants. This exemption was removed in 1989 because the additional occupants were observed to come mostly from commuters waiting near bus stops, and it was no longer achieving the car-pooling objective.

1.4.3 Park-and-ride schemes

Another attempt to reduce car traffic into Singapore (Chapter 6) was a park-and-ride (P&R) scheme; drivers were encouraged to park in several new car parks outside the city and take a bus to the centre. But this was not popular, and the car parks were eventually used for other developments. Such schemes do seem to be successful in other places, such as Oxford and Reading in the United Kingdom, but according to Wikipedia [8], 'There is not much research on the pros and cons of park and ride schemes'.

1.4.4 Autonomous vehicles

According to the Chief Executive Officer of the Toyota Research Institute [9], it will be decades before autonomous vehicles become widespread enough to tackle the problems of congestion and pollution – assuming they can. In fact, unless resource sharing, restrictive access schemes or price differentiation relative to other modes is employed, increases in vehicle activity from Autonomous Vehicles could worsen congestion; they may initially decrease the effective road capacity, especially in high-speed, high-flow situations such as the UK Strategic Road Network [10]. They are also claimed to be able to reduce road traffic accidents, most of which are due to driver error; but this is as yet an unproven assertion although early results show that the accident rate could be better than vehicles that do not have the capacity to operate in automated modes [11]. Governments are right to invest in Autonomous Vehicles. They have great potential, and will have substantial spin-off benefits whether or not the final goal of full autonomy is achieved. But they will not make roads better, safer, cleaner and more reliable in the short-term.

1.4.5 Workplace Parking Levy

A Workplace Parking Levy (WPL) scheme applies a mandatory charge (e.g. an annual charge) as a licence for all parking spaces (usually above a specified number) in order to incentivise reduction of such spaces (and thereby related trips) and raise funds for complementary measures. After 10 years planning, the City of Nottingham implemented WPL [12] which demonstrated that any direct congestion reduction impact was inconclusive but that benefits from revenue raised enabled investments in complementary measures. Other destination charging schemes have been implemented in Australia and Canada. Chapter 3 makes further references to WPL.

1.4.6 Smarter travel

So-called smarter choices such as workplace travel plans and personal travel plans do not seem to have had a significant long-term impact on congestion. Similarly, promotion of active travel (walking and cycling), whilst desirable on health grounds, and for the reduction of short journeys by car, cannot substitute for long journeys, or journeys with heavy loads, and the provision of cycle-ways has contributed to congestion where they have taken road space away from cars.

1.5 Equity

One argument against road pricing is that less-well-off motorists may be priced off the roads, or at least financially disadvantaged. Cars are important to low-income as well as higher income households. But on balance, low-income car-owning households will not be disproportionately affected by road pricing if it is revenue-neutral; that is, if other taxes or charges are reduced to compensate, though there may be individual winners and losers. In a study in Oregon in the USA, Zhang *et al.* [13] found that a $0.12 per mile flat vehicle mileage tax fee (as an alternative to the 'gas tax') was not significant in either the short-term or long-term, i.e. it did not place a significantly greater burden on either lower income or rural populations.

It also depends on how any surplus revenue is spent; it may be hypothecated (that is, 'ring-fenced') into transport improvements, especially public transport, thereby supporting lower income groups as well as potentially being environmentally friendly.

Third, as Schweitzer and Taylor [14] pointed out, road pricing may be regressive (lower income groups pay a larger share of their income on transport than do the rich); but other methods of taxation and of financing roads may be equally or more regressive. And road pricing has benefits that other transportation finance mechanisms do not, such as reducing journey times, fuel consumption and vehicle emissions. Furthermore, a transparent payment mechanism is a good thing. ('Knowing that resources have a cost is essential to using those resources judiciously, and our road network will function better when drivers pay the costs of their travel'.) We can still be concerned about the burden tolls placed on the poor, but 'we should not subsidise all drivers (and charge all consumers) to help the small number of poor travellers who use congested freeways in the peak hours and peak directions. Rather we should help those who are less fortunate, and see to it that the rest of us pay our own way on the roads'.

Road pricing also offers the potential to signal the cost of driving and apply the principles of 'user pays' on a more fine-grained basis, potentially to individual road segments [as used by the German truck tolling scheme and to be applied by all vehicles in the planned update to the Singapore Electronic Road Pricing (ERP) scheme].

Chapter 4 distinguishes two kinds of equity: vertical equity where a policy does not disproportionately affect lower income households, and horizontal equity, where drivers or groups of drivers are treated equally and receive benefits commensurate with the fees they pay. See also Chapter 18.

1.6 Public acceptability of road pricing

Equity is one of the issues in making road pricing acceptable to the voting public – and hence to politicians, who usually hold the purse strings. A particularly interesting and extensive study of acceptability was carried out by Owen *et al.* in [15] for the UK Department for Transport (Table 1.3).

Some of these themes are addressed in subsequent chapters, especially 4, 5, 7, 8, 18 and 19. We will also return to the theme in Chapter 21.

A more detailed and comprehensive treatment of the acceptability of road pricing will be found in Walker [1].

Table 1.3 Factors for the acceptability of road pricing. Based on Owen et al. [15]

Factors
A recognition that there is a transport problem such as congestion or pollution that requires a solution like road pricing and that road pricing is seen to work to reduce the problem
A road pricing scheme needs to be part of an overall traffic management plan within a consistent transport strategy, including public transport improvements, green travel plans, park-and-ride schemes and car park levies
Travel alternatives need to be available to facilitate choice, including free-of-charge alternative routes and good public transport
Revenue should be identified and used appropriately; support is increased when revenue benefits the motorist (i.e. reduction in road tax and fuel duty), the local transport network (especially public transport) or is used on expanded road capacity and maintenance
The scheme must be simple to understand
The scheme must be fair; it must take account of vehicle size, people's income, health and disability needs and proximity of residence or business to the charging zone
Information on the scheme, and on the success of road pricing elsewhere, must be disseminated via education, publicity and marketing; lack of knowledge leads to lower acceptance
A leader/champion is important
The local authority or the agency operating the scheme must be trusted to deliver, manage and run the scheme in an efficient and cost-effective way. People do not want to see an undue proportion of their payments going into the scheme operator's pockets
The technology used must be reliable and easy to understand; it should minimise evasion; and the implementation cost should not be borne 'up front' by the motorist
Concerns over privacy and data protection must be addressed

1.7 Low emission zones

There are environmental as well as social and time-saving benefits from the use of road pricing. Charges can be related to vehicle emissions class – as they are for example in the German truck tolling scheme (Chapter 16), which has seen a dramatic increase in the proportion of modern low-emission Heavy Goods Vehicles since it was introduced. And by congestion reduction and hence minimisation of stationary traffic and stop-go driving, road pricing reduces emissions from vehicles, since engines are more polluting and less fuel-efficient at low speeds. And specifically, Low Emission Zones – pricing by harmful emissions rather than congestion – can be established using the same technology as conventional road pricing, as for example in London, and may be a useful (and publicly acceptable) stepping stone to it – see also Chapter 19.

As alluded to in the Preface, environmental issues are becoming more prominent, and are driving low emission measures, including banning of diesel vehicles in some European cities, and the proposed Ultra-Low Emission Zone and other measures in London from 2019 [16].

1.8 Taxes or charges?

In Stockholm and Gothenburg, the congestion fee is a tax, not a charge. There was much discussion about this by Swedish legal experts [17]; basically, it was implemented as a national tax because in Sweden, a municipality cannot tax or charge citizens in other municipalities unless they are given a service in return, and in Sweden road space does not count as a 'service'.

In most other places, the payment is a charge, so is subjected to civil rather than criminal penalties for non-payment, with greater flexibility to vary the charges. It is also more consistent with the 'user pays' principle.

The distinctions between charges (user fees) and taxes are indicated by Cassidy and O'Reilly [18] related to whether toll roads represent a form of taxation. US courts find that tolls are user fees, not taxes – important because a 'tax' is a mandatory payment imposed by governments and is likely to encounter significant political resistance. User fees are voluntary payments for access to facilities or services, and are generally viewed as more equitable, being non-compulsory. Courts have applied three questions to distinguish between user fees and taxes; if the answer to each question is yes, the charge is a user fee, not a tax.

1. Is the government service for which the fee is collected, something that benefits the party paying the fee in a manner that is not shared by other members of society?
2. Is the fee paid by choice?
3. Is the purpose of the fee to compensate the governmental entity providing the services for its expenses (and not simply to raise revenue for general purposes)?

1.9 An outline of the book

As indicated in the Preface, Eddington's 2006 report for the British Government [18] said that 'the potential for benefits from a well-designed, large-scale road pricing scheme is unrivalled by any other intervention'. But Governments have not adopted road-pricing policies in the mistaken beliefs that (a) the technology is not yet ready and (b) that they are a vote loser and that voters are implacably opposed to such policies (probably because they are unaware of the benefits described in the following chapters). This book presents the evidence that the technology for road pricing is indeed ready and that it is not necessarily a vote loser.

The chapters are written in different styles, with some of them (especially 12 and 13) being rather personalised accounts of the congestion charging in London and interoperable tolling in the United States, respectively. There is also some repetition – some topics are treated in more than one chapter – but this has deliberately not been edited out because different authors have difference perspectives – and also so that each chapter can be read in isolation if necessary.

We begin in **Chapter 2** with transport economics. Whilst this chapter uses data predominantly from the United Kingdom, the lessons are generic and apply worldwide – similar points are made about the United States in Chapter 18. Chapter 2 also deals with misunderstandings that have led some people to discount road pricing: it is not dependent on privatisation; it is more efficient than fuel duty; it has both congestion-mitigation and revenue-raising effects. The history of road pricing in England is also outlined, as is the issue of 'fairness'. The chapter concludes with some potential future scenarios.

It is clear from many chapters that technology is no longer an issue in road pricing, given the number of real implementations worldwide – it can all be done now – though of course technology developments will continue, especially given economies of scale as road pricing is more widely adopted. But **Chapter 3** describes the technologies available, as well as some technology-independent issues. It begins with the policy context, scheme design issues and the functional requirements. It then describes the three major technology options: radio frequency or microwave tag and beacon [aka Radio Frequency IDentification or Dedicated Short-Range Communication (DSRC)]; Global Navigation Satellite System/Cellular Networks (GNSS/CN) and Automatic Number Plate Recognition (ANPR)[6] – used for enforcement of the previous two technologies as well as being the primary charging/enforcement mechanism in some cases such as London and Stockholm. See in particular Chapters 6, 9, 10 and 14 for more details on DSRC use, 6 and 16–18 for GNSS, and 11–13 for ANPR. Chapter 3 also addresses the measurement of scheme cost and performance.

Chapter 4 is an updated and condensed version of Canada's Ecofiscal Commission's report, 'We Can't Get There From Here: Why Pricing Traffic Congestion is Critical to Beating It' [20] by one of its authors. The chapter starts with the economic and social costs of congestion in Canada; similar costs apply in other countries, of course. It stresses that increasing transportation infrastructure reduces congestion in the short-term but encourages additional vehicle use and the return of congestion. Pricing is the solution – but different kinds of pricing are needed in different situations; the options are: single-entity pricing; high occupancy toll (HOT) lanes; zone-based pricing; distance-travelled pricing; parking pricing. In Vancouver, the recommended option is single-entity pricing – applying variable pricing to the region's bridges and tunnels – with a comprehensive distance-based charge as a long-term option. Calgary's low population density and lack of obvious natural cordon, plus low public and provincial support for road pricing, make HOT lanes the preferred option. The Greater Toronto Area has the country's biggest congestion problem; it is also polycentric, with multiple hubs of activity and traffic; the proposed solution is HOT lanes [partly by conversion of existing High Occupancy Vehicle (HOV) lanes] and tolling on major freeways. The natural cordon formed by the Island of Montreal facilitates congestion pricing in that metropolitan region. Canada's smaller cities could also benefit from pricing policies tailored to their individual circumstances. Finally, the chapter makes recommendations for Canadian cities and provincial and Federal Governments – recommendations that apply equally well in other countries.

Chapter 5 is a US case study of the public acceptability of road pricing. In the mid-1990s, in Minnesota's Twin Cities Metropolitan Area, road pricing was suggested to manage the region's growing traffic congestion problem. But public acceptance was a problem; a public–private toll road was vetoed by a municipal

[6] Known in the United States particularly as ALPR – Automatic Licence Plate Recognition.

government; a proposal to convert the I-394 HOV lane to a HOT lane was withdrawn after it failed to gain public support and was opposed by the soon-to-be State Governor. Yet by 2003, the Minnesota Legislature authorised the Minnesota Department of Transportation to convert HOV lanes into HOT lanes, with remarkably little opposition. By 2005, 65% of citizens adjacent to Minnesota's first HOT lane thought the project was a 'good idea'. HOT lanes were subsequently added to both I-35W and I-35E, and the region's transportation plan calls for HOT lanes to be added to other corridors in the future. How things changed so rapidly and dramatically in terms of public acceptance, and the lessons that we can learn are the focus of this chapter.

Chapter 6 addresses how road pricing has been implemented in Singapore, and outlines the planned technology augmentations. Singapore was a pioneer in urban road pricing; it operated a manual paper-vignette-based ALS from June 1975. It was an effective traffic management measure, but was labour-intensive and inflexible, and was replaced by an ERP scheme in 1998. This worked well and was acceptable to the motoring public once certain concerns had been addressed; indeed motorists would ask for more features for their convenience. However, the ERP gantries are expensive to install, the equipment needs updating every few years, and point-based charging is less equitable than a distance-based pricing scheme. Consequently, the Singapore Land Transport Authority investigated the option of a GNSS-based system. There were problems to be overcome, including inadequate GNSS constellations – just one constellation wasn't good enough – and telecommunications networks, the 'urban canyon' effect and cyber security; but these have effectively been overcome and the GNSS-based road pricing system will come into operation from 2021 onwards.

In December 2008, Greater Manchester (United Kingdom) voted to reject a £3 billion funding package of investment in the conurbation's transport infrastructure, partially funded by congestion charging. **Chapter 7**, written by someone who was personally involved, focuses on the ways in which the plans were communicated to the electorate and the context in which they were debated. It draws lessons on how to pursue controversial policy developments, and the implications for road pricing; what does the experience mean for transport policy in cities, and for our understanding of car dependency?

Chapter 8 outlines case studies of communication and consultation strategies for road pricing schemes. It points out that consultation on any Travel Demand Management (TDM) scheme, including road pricing, cannot be independent of the context of the intended scheme, and these broader policies should be the basis of stakeholder consultation, rather than just the TDM and road pricing measures. The experiences described in Chapter 7 fully confirm that viewpoint. Furthermore, consultation should not be a one-off but must cover all scheme phases to maintain a critical mass of support.

Chapter 9 covers standardisation in road pricing, aka Electronic Fee Collection (EFC), with the emphasis on Europe, addressing both DSRC at 5.8 GHz and GNSS/CN, as specified in the European Interoperability Directive 2004/52/EC. Standards ensure the safety, reliability and quality of products and services. They reduce costs, increase productivity and facilitate access to markets. But they are not easy to produce; the process is still going on after more than 20 years.

Following on naturally, **Chapter 10** describes the European Electronic Toll Service (EETS) – the vision of a driver travelling across Europe using a single electronic on-board unit and a single account. The vision is yet to be achieved, but significant progress has been made. Key steps include the concepts of Toll Charger and Service Provider introduced by the Common EFC System for a Road Tolling European System project, which became the basis for the EETS concept. The chapter then details the Regional EETS project, which has resulted in a number of organisations being registered as EETS Providers.

Chapter 11 is a practical guide to ANPR, the automatic reading of vehicle registration plates, its standardisation and implementation. The first practical ANPR system was trialled on the UK M1 motorway in 1984. An early use of ANPR in tolling was on the 407 ETR Express Toll Route in Ontario, Canada, in 1997, since when its use has expanded significantly, including vehicle identification of occasional users. It is the primary charging mechanism in London, Stockholm, Gothenburg (Chapters 13–14) and elsewhere, and is an enforcement system in practically all other road pricing schemes.

In contrast to Chapters 9 and 10, **Chapter 12** describes the US approach to interoperability. It is also a personal account of the development of road tolling in the United States, from the first turnpike (Pennsylvania in the 1930s), through the adoption of electronic tolling, to the latest in multiprotocol tags and readers. It covers the setting up of E-ZPass and of the Alliance for Toll Interoperability, and the development of multiprotocol tags and readers. ANPR is used as a basis for interoperability between regions where there are different tag standards.

Chapter 13 focuses on London, which has been particularly influential in the domain of city congestion charging, having inspired other attempts such as Stockholm (Chapter 14), and unsuccessful ones such as New York, Edinburgh and Manchester (Chapters 7 and 8). Like Chapter 12, it is a personal account – by two people intimately involved in the development of the London Congestion Charging Scheme from initial proposals to final implementation. They indicate lessons learned, the four main business processes and evasion of charges using 'cloned' vehicles. The 'zone of influence' of a scheme is much larger than the Congestion Charging Zone itself. Finally, they make some suggestions for implementing national road pricing.

Chapter 14 is a forward-looking summary of what we have learned from the Swedish Congestion Charges in Stockholm and Gothenburg, 10 years after toll cordons were introduced around the former in 2006, and the latter in 2013. Stockholm's objective was to reduce congestion and improve the local environment; Gothenburg's was to finance a large infrastructure package. A comparison between the two helps to understand the transferability of the congestion charge effects between cities with different features and of different sizes. The Stockholm scheme was controversial when first proposed but currently has high political and public support. The design of a congestion charging system is not easy; it depends on local conditions, such as bottlenecks and route choice possibilities. In both

Stockholm and Gothenburg, some suggested options were rejected because transport modelling indicated that they would create problems. Cities interested in implementing congestion charging must model proposed schemes to evaluate different designs. Other lessons are also listed.

Chapter 15 advises on moving from conventional barrier-controlled tolling installations to Open Road Tolling (ORT), resulting in more efficient solutions: increased traffic throughput, reduction in traffic incidents and less pollution and fuel wastage from vehicles queueing at toll booths. However, attention is needed to various issues and design features in order to enable the full potential of ORT and to avoid negative impacts.

Chapter 16 covers standards and implementations of tolling using GNSS technology, which is being used in Europe for truck tolling, will soon be used in Singapore for city congestion charging (Chapter 6) and is being trialled in the United States as a replacement for the 'gas tax' (Chapter 18). Schemes in Switzerland, Germany, Slovakia, Hungary, Russia and Belgium are described, as well as failed but instructive schemes in the United Kingdom, the Netherlands and France. Technical challenges of using GNSS in tolling are described, including accuracy, availability, signal reception and implementation issues, as well as deliberate interference by 'jamming' and 'spoofing'. But the future is bright.

Chapter 17 describes a particular instantiation of GNSS-based tolling, the Hungarian distance-based electronic toll system, branded as 'HU-GO'. It is noteworthy for the very short time of implementation – less than 3 months – and low cost, by using the existing fleet management and telematics service providers and their on-board units as Toll Declaration Operators. A centrally-maintained road database is the basis of the digital roadmap of the country, which identifies the tolled sections. All elements of the system existed previously but no one had considered linking them in the way the HU-GO system developers did.

Chapter 18 describes the West Coast Distance Charge Trial Programs in the United States. The Pacific Coast states – Oregon, California and Washington – are trialling a privately operated distance charge system that could replace volume-based fuel excise taxes as the primary road funding mechanism. A second reason for distance charging is equity; some light vehicles pay a lot per-mile for road use while others, such as hybrid and electric vehicles, pay very little or nothing. However, it seems that putting road charging and congestion pricing policies together in one road pricing legislative proposal is too politically challenging for the first quarter of the twenty-first century, at least in the United States. Nonetheless, most contentious issues have been essentially resolved or substantially calmed as a result of these and other pilots, and it seems that widespread, mandated distance charging in the United States is not that far away.

Chapter 19 describes Milan's cordon pricing, introduced in 2008, originally as a pollution charge, but converted in 2012 to a genuine congestion charge, called 'Area C', following a referendum in 2011. It summarises the outcomes in terms of traffic reduction, emissions and impact on user groups, concluding with a discussion of the **very unexpected** public and political acceptability of the policy – 79% voted 'yes' on a higher than usual 48% turnout – and how the experience may be extended more widely.

Chapter 20 describes a particularly interesting alternative technique to address traffic congestion: motorists, rather than paying to drive, are paid not to drive, at certain times (especially peak hours) and locations (such as roadworks). Used in only the Netherlands so far, 'Spitsmijden' ('avoiding peak traffic') projects are an effective and cost-effective congestion reduction measure which deserves serious investigation from other countries. Especially during large road construction works, Spitsmijden projects can significantly mitigate the negative effects of capacity reduction and increased travel time, and the effects may persist long afterwards. They can also be used to alleviate recurrent congestion on roads, either as an alternative to, or in advance of, road charging schemes. Spitsmijden seems to work well for such localised areas; whether it would work equally well on a larger scale is yet to be determined.

Chapter 21 summarises the book and attempts to draw some conclusions about the future of road pricing worldwide, and how best to present it to sceptical politicians, the media and the public. Anecdotal evidence suggests that many politicians are persuaded of the virtues of road pricing, but are concerned about the public reaction – understandably so given events in Edinburgh, Manchester, New York and elsewhere.

Finally, as an Afterword, **chapter 22** reminds us that the problems and challenges of the motor car have been recognised for 55 years, and have not yet been solved. Road pricing is the nearest thing we have to a "silver bullet" to deal with congestion, pollution, greenhouse gases, climate change and energy shortages in the transport field.

References

[1] Walker J. *The Acceptability of Road Pricing*. RAC Foundation; 2011. http://www.racfoundation.org/research/economics/road-pricing-acceptability [Accessed 13 October 2017].

[2] Duranton G, and Turner MA. The fundamental law of road congestion: Evidence from US cities. *American Economic Review*: 2011. NBER Working Paper 15376.

[3] Ellis T. Dutch courage, English lessons. *Traffic Technology International*: August/September 2010: 044–047. Available from: http://viewer.zmags.com/publication/e8eddeaa#/e8eddeaa/1v [Accessed 22 February 2017].

[4] Mernit JL. For Whom the Road Tolls: Is It Fair to Put a Price on Freeways? *Capital & Main* 19 July 2017. Available from: https://capitalandmain.com/for-whom-the-road-tolls-is-it-fair-to-put-a-price-on-freeways-0719 [Accessed 27 July 2017].

[5] Transport Department. *Travel Characteristic Survey 2002 Final Report*. 2002. Available from: http://www.td.gov.hk/en/publications_and_press_releases/publications/free_publications/travel_characteristics_survey_2002_final_report/index.html [Accessed 24 July 2017].

[6] Transport Department. *Travel Characteristic Survey 2011 Final Report*. 2012. Available from: http://www.td.gov.hk/filemanager/en/content_4652/tcs2011_eng.pdf [Accessed 27 July 2017].

[7] Gilroy L, and Pelletier A. HOT Lanes: Frequently Asked Questions, *Policy Brief No. 59*, Reason Foundation. Available from: www.reason.org [Accessed 01 March 2007].

[8] Wikipedia. Park and Ride. 2017. Available from: https://en.wikipedia.org/wiki/Park_and_ride and https://en.wikipedia.org/wiki/Park_and_ride_bus_services_in_the_United_Kingdom [Accessed 27 July 2017].

[9] Pratt G. *Fully Autonomous Cars 'Not Even Close'*. ITSNow.org. 2017. Available from: http://www.itsnow.org/news20170113.html [Accessed 27 July 2017].

[10] UK Department for Transport/Atkins. Research on the Impacts of Connected and Autonomous Vehicles (CAVs) on Traffic Flow, Summary Report, UK Department for Transport; May 2016.

[11] Blanco M, Atwood J, Russell S, Trimble T, McClafferty J, and Perez M. *Automated Vehicle Crash Rate Comparison using Naturalistic Data*. Virginia Tech Transportation Institute; 2016. Available from: http://www.vtti.vt.edu/featured/?p=422 [Accessed 30 July 2017].

[12] Burchell J, Ison SG, and Enoch MP. *Managing Congestion: Evaluating the Nottingham Workplace Parking Levy*. 93rd TRB Annual Meeting, 12–16 January 2014. http://docs.trb.org/prp/14-1004.pdf [Accessed 30 July 2017].

[13] Zhang L, McMullen B, Valluri D. and Nakahara K. Vehicle mileage fee on income and spatial equity: short- and long-run impacts. *Transportation Research Record*; 2009; 2115: 110–118.

[14] Schweitzer L, and Taylor BD. Just road pricing. *ACCESS (magazine of the University of California Transportation Center)*, 36: 2–7; 2010.

[15] Owen R, Sweeting A, Clegg S, Musselwhite C, and Lyons G. *Public Acceptability of Road Pricing. Final Report for Department for Transport*; 2008.

[16] Transport for London. *Mayor's Transport Strategy: Draft for Public Consultation*. June 2017. Available from: https://consultations.tfl.gov.uk/policy/mayors-transport-strategy/user_uploads/pub16_001_mts_online-2.pdf [Accessed 25 June 2017].

[17] Gullberg A, and Isaksson K. (eds.) *Congestion Taxes in City Traffic: Lessons Learnt from the Stockholm Trial*. Stockholm: Nordic Academic Press; 2009. See especially note 36 on p.171

[18] Cassidy CB, and O'Reilly B. Examining transportation infrastructure funding, *Texas Bar Journal*; July 2016; 79,(7): 528–529.

[19] Eddington R. "The Eddington Transport Study. The case for action". Her Majesty's Stationary Office; 2006.

[20] Ecofiscal Commission. *We Can't Get There From Here: Why Pricing Traffic Congestion is Critical to Beating It*. Ottawa: Canada's Ecofiscal Commission; 2015. Available from: https://ecofiscal.ca/reports/traffic/ [Accessed 7 Dec 2016].

Chapter 2

The Smeed Report at 50: will road pricing always be 10 years away?[1]

*Stephen Glaister**

2.1 Introduction

It is now more than 50 years since the publication of the UK Smeed Report [1], a seminal work that first proposed road pricing as practical policy for dealing with congested roads. This paper is a review of what has happened since and what road pricing has to offer in the future. Although much of the material is UK-centric, the lessons apply world-wide.

For the purposes of this discussion, I use 'road pricing', 'road user charging', 'congestion charging', 'pay-as-you-go' and similar terms interchangeably – though each has a special meaning in specific circumstances. I am referring to any scheme that has an element of payment by the user, the amount depending on how much they use at a unit price that depends on location or time of day.

I give almost no attention to the technologies involved. I take the technological problems as solved – as is evident from other chapters.

2.1.1 The economists versus the rest

To most economists road pricing is an obvious solution. To many others, it is a theoretical irrelevance.

It has an immaculate pedigree. Dupuit [2], a canal engineer and the inventor of cost–benefit analysis, had the solution in 1844 when he analysed question of the correct toll for an uncongested bridge. Distinguished contributors since have included Friedman [3], William Vickrey and James M. Buchanan. Walters [4] set out the best-known theoretical analysis in his famous *Econometrica* paper, and he was an influential member of the Smeed Panel. Other great names had essentially the same solution in a variety of contexts: Hotelling (pricing freight railroads), Ramsey (commodity taxation) and Baumol (public utility pricing).

*Department of Civil and Environmental Engineering, Imperial College, UK

[1] This chapter is based on the Smeed Memorial Lecture, University College London, 7 October 2014, http://www.ucl.ac.uk/transport-institute/pdfs/UCL-smeed-memorial.

The fundamental principle relies on a crucial distinction between 'price' and 'cost'. Price is what the user pays. Cost is what it costs to supply the user.

There are crucial subtleties concerning the definition of 'cost'. Generally, an aim in public policy is to account for all the costs that will be incurred by all parties if the user is provided with each extra unit. That will include noise and air pollution costs suffered by bystanders as well as costs to other users – such as additional delay and unreliability, as well as costs to the enterprise that has to supply the extra unit: the marginal social cost.

To have a price set above marginal social cost is undesirable because the user is paying more than their decision to consume the extra unit (travel an extra kilometre) would cost. With such a price, there would be users willing to pay an amount that would exceed the marginal social cost of supplying it, but less than the price asked. Since they are not willing to pay the full price, they are denied the benefit their use would bring, even though that benefit would exceed the cost. There is a body of public interest regulation directed towards preventing this situation. Examples include monopoly pricing, price discrimination and abuse of dominant position.

Conversely, if price is set below marginal social cost, the user is not being required to cover the costs incurred by other people caused by the decision to consume the extra unit. So the value to the user is less than the cost of supplying them.

The point where price paid and marginal social costs are in balance is one where the user is just covering all the costs they impose. There is no net gain to be had by changing the price or volumes of use. This is said to be the 'efficient' price.

Under certain important assumptions, competitive market forces for general commodities will establish this equilibrium point: then the price guides consumers (who neither know nor care about the costs to others) to the level of use where the value to them is the same as the cost to others of supply. This is the 'invisible hand' expounded by Marshall and Pigou and completely codified in Debreu's [5] definitive *The Theory of Value*.

But the congested, under-priced road is an example of failure of these market forces: every user is imposing delay and pollution costs on others which are not fully reflected in the price they are asked to pay.

This analysis is silent on who would gain and who would lose from a move from inefficient, under-priced, congested roads to efficient pricing. Since the total benefits of such a move exceed the costs there must, in principle, be a way to tax the gainers and use the revenue to compensate the losers and leave everybody at least as well off. This begs the question of what compensation will be paid in practice and by what mechanism. It also neglects the important issue of the implementation costs. Any proposal for public consideration must be explicit about those issues.

In our context, there may be considerable revenues from the new charge which is available to pay compensation or to use for other purposes for the public benefit.

2.1.2 Unintended consequences of under-pricing

Examples of the undesirable and unintended consequences of prices being set too low include price regulated food (empty shelves), rent control (homelessness) and health services (waiting lists).

Economists will typically point out that a move to rationing by price and – crucially – using revenues to compensate those made worse off, can make everybody better off. Many others will reject such a solution, arguing that it is wrong or unfair to use price as a mechanism for allocating an 'essential service'.

So it is with a congested, under-priced road. Of course, administering charges has its own costs so to that extent it might be better to put up with inefficient charges. That is a subject for proper cost–benefit analysis. And there may well be institutional obstacles to using revenues to compensate losers. That is a subject for analysis of equity impacts.

Figure 2.1 illustrates the situation for the English road network. Counting the price people pay for fuel as the price for using the road, the figure suggests that well over 40 per cent of vehicle miles are paying too much, but over 20 per cent are paying too little.

Note, one important implication of Figure 1.1 is that serious congestion is by no means universal. This is confirmed in the British Social Attitudes Survey [7] which found that only 28 per cent of Britons thought that congestion on motorways was a problem for them, and less than half (45 per cent) of respondents considered traffic congestion in towns to be a problem. So it is important that any solution proposed is well targeted and does not impose unnecessary cost and inconvenience in circumstances where there is no problem. A significant proportion of the United Kingdom's road congestion is in London.

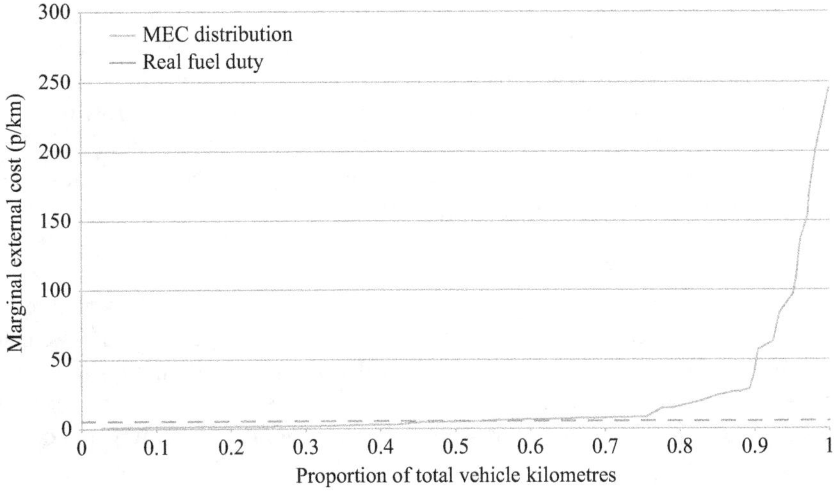

Figure 2.1 Distribution of the marginal external costs of motoring
Source: Institute for Fiscal Studies, RAC Foundation [6]

Table 2.1 Traffic and measures of delay on the English road network: 2010 to 2014

Road type	**2010–14 change**			**% of traffic in very congested conditions**
	Total traffic (%)	**Congestion (lost s/m) (%)**	**Vehicle speed (%)**	
Strategic Road Network (SRN)	46	114	–8	15
Non-SRN	41	56	–9	14
All	43	61	–9	15

Source: UK Department for Transport [8].

2.2 What's new?

Road pricing has been considered several times at the highest level in British (and other) Government and rejected each time. So why keep returning to it?

There are several new considerations in Britain, which are also relevant in other countries.

2.2.1 Forecasts of road congestion

Long-term forecasting is always difficult and controversial, but the official road traffic forecasts for England, shown in Table 2.1, are plausible. They are largely driven by views of demographic change, population growth and economic growth. The source document shows ranges driven by alternative assumptions.

The averages disguise the fact that some places will suffer significantly worsening congestion for a lot of the time. The implications for the level of service on the roads are not attractive.

The 2010 and 2015 UK Governments responded by planning to increase the rate of investment in strategic road-network capacity. But nobody pretends that sufficient new capacity will be provided to deal with the traffic growth in full. In any case, the strategic road network is only about one-third by traffic volume (2 per cent by length), and the rest is the responsibility of local authorities where funding for maintenance and enhancement is very short.

2.2.2 The shortage of capital and the national debt

The combination of historic under investment in many kinds of infrastructure and growing demands because of growing wealth and increasing population has created an investment funding gap that can only be filled by increased charges to users or taxpayer funding. Politicians are reluctant to be seen to sanction price increases and the Exchequer is under immense pressure to restrain public expenditure and contain the national debt.

Between 2015/16 and 2020/21, the annual capital investment budget for Highways England is to increase by a factor of 2.5, and by the end of this period, it nearly reaches the level of spend planned for Network Rail (though they are both topped by the high-speed rail scheme, HS2). The £15 billion over 6 years represents an annual average spend of £2.5 billion each year. Echoing the 1989 fanfare with the White Paper, *Roads for Prosperity* (DoT [9]), the 2013 White Paper *Action for Roads* (UK Department for Transport [10]) talks of 'the largest investment in roads for half a century'.

Any increased expenditure on roads must be funded entirely from either new charges or existing local or national taxation. The competition for desperately scarce public capital will remain fierce. Against this background, delivery of the greatly increased capital programme for roads entirely from public expenditure will be hard work. This will be the reality for the calls on public capital from all sectors.

Road pricing could be configured to provide a new source of funding that would protect the roads investment programme. But any such proposition must be seen against the history that since the mid-1970s, the total of central and local spending on UK roads has fallen far short of the contribution to the Exchequer from Vehicle Excise Duty (the annual fixed charge for owning a vehicle) and fuel duty. Roundly, for every £4 of revenue from Vehicle Excise Duty and fuel duty, £1 is spent on roads; the difference goes to fund general public expenditure.

The general public is acutely aware of this gap, and many argue that the Exchequer should spend more on roads before seeking to raise money from distance-based charges.

2.2.3 Decarbonising and tax yields

Receipts from fuel duty and Vehicle Excise Duty are one of the biggest sources of funds for Her Majesty's Treasury, contributing about £38 billion in 2010, some 7 per cent of the Exchequer's income (Institute for Fiscal Studies [6]). The government uses more than one way of classifying taxes but motoring taxes account for 85 per cent of all those classified as 'green taxes'.

Despite a projected growth in traffic the Institute for Fiscal Studies [6] notes that revenue from motoring taxation is set to drop by £13 billion a year by 2029 (to £25 billion). This is simply due to the improvement in the fuel efficiency of vehicles.

The Institute for Fiscal Studies estimates that to preserve the current level of fuel duty revenue, the duty rate per litre would have to be increased by over 50 per cent above the present 58p/litre[2]. In the longer term, government will have to redesign the system of motoring taxation to be fairer and more transparent across the spectrum of users of different technologies, and to raise the necessary revenue for the Exchequer.

It is sensible to include within conventional fuel duty a charge that reflects the carbon cost of burning a litre of fuel. This suggests a contribution to the duty rate of less than 20p/litre at official carbon prices. But note that other sectors burning hydrocarbons and producing carbon dioxide pay significantly lower rates of duty than road users: agriculture, railways and bus companies. Most notably, domestic householders pay no duty on gas or oil for heating, and 5 per cent Value Added

[2] Plus Value Added Tax at 20 per cent, as at December 2016.

Tax. 13 per cent of all greenhouse gas emissions come from this source, as against 19 per cent from roads.

The advent of the pure electric car charged from the domestic electricity supply creates a further problem. Electric vehicles can only compete on ownership cost with conventional vehicles because they are excused duty on their fuel which is imposed on petrol and diesel.

The Institute for Fiscal Studies is driven to the conclusion that if the 'paying the external costs' argument is to be sustained then a significant portion of current fuel duty must be replaced by some form of 'pay-as-you-go' charge based on distance driven in congested conditions.

Congested roads cause fuel to be burned inefficiently. To the extent that road pricing reduces traffic and controls congestion it will contribute to reduction in greenhouse gas emissions. Banks, Bayliss and Glaister [11] estimated about a 10 per cent reduction in national Carbon Dioxide emissions.

2.2.4 Air quality

On top of the component for carbon, one can reasonably add fuel duty elements corresponding to external costs such as air pollution.

Air quality damages from road transport are now being more widely recognised (see, for example, Ricardo AEA [12]). Like traffic congestion it is highly geographically specific – because the damage depends upon the population exposed – and it is highly correlated with it. Road pricing designed solely to deal with congestion is likely to have beneficial effects on air quality. Further, air quality effects can be added to congestion effects in computing road prices. Since the exposure of people to noxious emissions varies greatly by time and place, this would imply strongly differentiated charges on air quality grounds (see Banks, Bayliss and Glaister, [11] and Glaister, Lytton and Bayliss [13] where air quality damages are included).

2.2.5 Cost of technology

The cost for the relevant charging technologies is falling rapidly. The costs remain an issue that cannot be ignored: in particular the compliance costs imposed on individuals. But there is no longer any doubt that there are several distinct technologies available that can be relied upon. Walker [14] gives a survey. See also other chapters.

2.3 Misunderstandings

There are misunderstandings that have led some people to discount the road user charging proposition.

2.3.1 Privatisation

Some of the 'free market' economists who have advocated pricing have argued that roads should be owned and provided by private enterprise as an important component of a reform – like most other goods and services. But there is a risk of conflation of road user charging with 'privatisation'. It is neither necessary nor sufficient to have charging in order to involve private ownership.

Whatever the merits or demerits of private ownership, sensible charging is an issue: this applies if a facility is owned and provided by the public sector. And one can involve the private sector in provision without charging users, providing some other source of payment is provided. This is illustrated by the fourteen private finance initiative, shadow tolled roads in the United Kingdom.

Pricing is about efficiency, not ownership. However, governance is absolutely crucial – who has ownership and control of the revenues?

2.3.2 Fuel duty is a better mechanism?

Some people claim that fuel duty is a 'charge' for using the road. However, Figure 2.1 illustrates that the *efficient* prices would vary enormously by time of day and location. However, fuel consumption and hence duty on fuel does not vary in the same way. Most of road taxation is best viewed as a non-specific contribution to the Exchequer.

2.3.3 Road tolls 'do not work' in England

It is sometimes claimed that the experience of the Birmingham North Relief Road (the 'M6 Toll'), the United Kingdom's one major charged length of road, demonstrates that tolled roads will not work (see, for instance, the Campaign for Better Transport [15]).

The M6 Toll road certainly 'works' in the sense that road users have an additional, reliable, high-quality choice available to them and there has been no cost to the UK taxpayer. What the M6 Toll road does illustrate is that an agreement in which there are no restrictions on what state-funded developments there might be on competing facilities in return for complete lack of regulation of charges is unlikely to produce a satisfactory outcome.

It also illustrates the difficulty in the English geography, with its high densities, of competition for an isolated toll road. So one must be cautious about generalising from the copious international experience. To the extent that the objective of charging is to raise funding, it may be better to deal with a network rather than an isolated road.

One cannot dismiss charged roads on the grounds that they are technologically or politically infeasible. Walker [14] and the US Department of Transportation [16] survey a number of schemes from around the world that are now in successful operation. They include: the tolled motorways in France, Spain, Portugal and Italy; the urban road charging schemes in London, Singapore and Stockholm; the several schemes in Norway; the lorry-charging schemes in Germany, Austria and the Czech Republic and the privately provided highways in Australia.

In any case much of the English road network was originally funded through tolls. The public trust (or 'public benefit corporation' in US parlance) has a long tradition (see Bayliss [17] for more detail). In the eighteenth and nineteenth centuries, the Turnpike Trusts were largely instrumental in creating and maintaining the main roads in Great Britain, using revenues from tolls.

2.3.4 Road charging is only about congestion or only about raising money

It is not sensible to regard road pricing solely as a congestion-controlling measure, or a revenue raising measure. It is both.

There is a tendency to discuss road pricing in isolation and a reluctance to admit that it would generate revenues. However, as with any infrastructure, pricing, level of service, physical investment and funding are inextricably linked and should be considered as a whole. Price determines volume of demand and revenue. Volume of demand determines quality of service given the capacity of the infrastructure. Revenue provides the funds to operate the infrastructure and, if justified, to service the capital cost of increasing capacity. The funds generated may exceed what is required, or may fall short. (Archer and Glaister [18] give an exposition.)

When electricity, gas and water were nationalised industries in the United Kingdom, there was a tendency for government to treat pricing and capacity as separate – as now with roads. Creation of independent, privatised industries has closed this loop. This is achieved under the umbrella of private ownership subject to independent, public interest regulation. But it would have been possible to achieve something similar under public ownership. This is a fundamental change in governance of the industries, which addresses a problem that bedevils roads policy.

2.4 Reactions to past UK proposals

The bookshelves are replete with studies on road pricing.

2.4.1 The Smeed report[3]

When Smeed undertook his study he was deputy to the director at the Road Research Laboratory which belonged not to the Ministry of Transport, but the non-ministerial Department of Scientific and Industrial Research. Smeed and the Road Research Laboratory had not appreciated the need or even the advisability of getting Minister of Transport, Ernest Marples's permission to publish. Marples was furious when he realised the report was complete and on its way. He was himself up to bold initiatives. But the idea to be pursued had to be his. Therefore his reaction was to get the government not to allow the Road Research Laboratory to release it.

It was eventually published by the Ministry of Transport but only after the 1964 General Election. General Elections were fought in 1964 and 1966 with transport as a major election issue. But the political will atrophied in the United Kingdom as the minister requested more feasibility reports, until, in 1970, the government changed and the scheme effectively died.

[3] Sir Christopher Foster in correspondence and Wikipedia.

2.4.2 The London congestion charge

Central London has had a successful scheme in operation since 2003, the Congestion Charge (see Chapter 13).

The London congestion charge would have been very much less likely to have come to fruition had it not been for the coincidence of several unusual circumstances.

In 1974, a joint Steering Group including the Greater London Council (GLC) worked up in detail Smeed's suggestion ([1] paragraphs 5.3.1 to 5.3.13) for a paper-based, daily, area licence or permit: 'Supplementary Licensing'. This was approved by the GLC for consultation. Several public meetings were held but the reactions at these were strongly negative. As a result it was decided to drop the scheme. In 1979 GLC officials produced the less ambitious 'Area Control: a scheme for reducing car traffic through Central London to limit congestion and improve the environment'. This study was approved by the Chairman of the Transport Committee. However, an article appeared prematurely in a London newspaper and the majority party leader, Horace Cutler, 'got very cross and demanded to have the offender's "head on a pole". That was the end of that'.[4]

In 1986, Mrs Thatcher's Conservative national government abolished the GLC, the local authority and took much of the administration of London back into Whitehall. The leader of the GLC at the time was Ken Livingstone, then a radical Labour Party member.

The national government changed to Labour, under Mr. Blair, in 1997, with a manifesto commitment to restore a city government for London. The Greater London Act 1999 enables any mayor to introduce a congestion charging scheme and the revenues must be mandated for transport purposes in the Greater London area. The decision to insist on the 'hypothecation' of the net revenues in this manner was highly unusual in UK governance: the Treasury was persuaded that it was a necessary condition for the political acceptability of a scheme. Also, unlike most of the world's great cities, London government would have little other income over which it would have direct control.

Two years before completion of the legislation, the national government set up a study group to propose outline designs. At that stage, it was not known who the candidates for mayor might be. The group duly researched the matter and published a report [19].

Ken Livingstone had wanted to stand as the official Labour Party candidate but he was rejected as candidate and excluded from the Labour Party. So he stood as an independent. He saw value in a radical innovation and included a manifesto commitment to introduce the recommended congestion charging scheme. Livingstone was a popular figure and always likely to win. Once elected he had a clear electoral mandate for a specific scheme so it was immune to the inevitable attempts to stop it through judicial review.

The scheme was easily explained and readily understood. Being based on automatic number plate recognition it required no modification to any vehicle.

[4] David Bayliss, private communication.

The need to implement the scheme well before the end of the 4-year term of office and to minimise the technological, administrative, legal and political risk explains the unadventurous and expensive technical design that was selected. Automatic number plate recognition (Chapter 11) is now proven and has substantial advantages in the right applications, but the analogue implementation and the back-office systems used then can now be achieved much less expensively.

The London congestion charging scheme went 'live' in February 2003, worked well and was inspected with considerable interest from round the world.

The factors enabling successful implementation were: time spent on careful preliminary research into a specific, practical and reliable scheme design; easy understandability; and a radical, independent politician with sufficient personal support to win the election and willing to take risks. Crucially the scheme was introduced as part of the creation of a completely new administrative system, rather than being added to an existing one.

There is a common perception that the scheme has 'stopped working'. Congestion levels have returned to their old levels and yet traffic volumes are lower than they were. But this is because the two mayors who held office between 2000 and 2016 took the opportunity offered by the Congestion Charge to take road capacity away from conventional traffic in favour of pedestrians, cyclists and other uses (Transport for London, [20]). If the scheme were to be lifted now, there would be a significant worsening of congestion on London's roads. London First [21] has given an evaluation of the London congestion charging scheme.

For the future London faces the same issues in respect of road traffic as the rest of the nation, much amplified. Mayor Johnson's statutory Transport Strategy [22, Figure 42] indicates that with no intervention vehicle delay would increase by up to 20 per cent by 2031. Even with all the interventions he was then proposing a residual worsening of up to 14 per cent was forecast. It is hard to see how this situation can be resolved without the extension of some form of road charging to a much bigger area of London. That would produce significant revenues for London Government at a time when the funding generosity from central government has been reduced.

Road pricing will always carry political risks. One way in which the opposition from the London public might be mitigated would be if charging for the use of the roads were integrated into the popular universal payment system that is now being implemented for all public transport modes. Then the citizen would be paying for their travel through a common system no matter what mode they chose to use (Glaister [23] proposed this in 1998).

2.4.3 The 2004 road pricing feasibility study

Alistair Darling, was appointed Secretary of State for Transport by Prime Minister Blair in 2002 and the success of London congestion charging encouraged him to take the idea seriously at a national level. He commissioned the Road Pricing

Feasibility Study (Department for Transport [25]) and national road pricing formed the core of the *Transport White Paper* of July 2004 [25]. The Prime Minister wrote a foreword: he was leading on this matter, securing consistency across Whitehall and delegating policy development to the relevant departments.

Then, in early 2007, an 'e-petition' appeared on the official No. 10 Downing Street website: 'We the undersigned petition the Prime Minister to scrap the planned vehicle tracking and road pricing policy'. It closed with over 1.8 million signatories. Within a day the Prime Minister abandoned the idea: he sent a personal email response to every signatory, and he created on the No. 10 website a podcast interview with a motoring journalist and TV personality.

The episode laid bare the extent to which the government had failed to address some of the fundamental issues: in particular, what would happen to the revenues raised. It certainly had not explained a complex proposition. It had sought to imply that it would be possible to spend the revenues on offsetting other motoring taxes and spend them on improving other modes of transport as well as defraying the costs of the scheme itself. This was not plausible.

A significant issue contributing the adverse public reaction to road pricing propositions was the perception that they would create a new, open-ended item for everybody's household budget. For instance deep in the text of the Road Pricing Feasibility Study it was noted that the highest rate of charge considered had been 80 pence per vehicle kilometre and that this would apply to 0.5 per cent of the traffic. The press reported the Study with a front page headline 'Motorists to face £1.34-a-mile toll' (*Daily Mail,* 21 July 2004).

Whatever kind of pricing proposition may be envisaged, a published, pre-determined cap on liability should be incorporated. While this may not be strictly 'correct' from an economic standpoint it is helpful from a psychological and emotional point of view. Well-chosen, it will only relate to a small portion of the traffic.

2.4.4 The 2006 Eddington Transport review

Sir Rod Eddington was commissioned jointly by the Chancellor and the Secretary of State for Transport in 2005 to carry out an independent review of the United Kingdom's transport policy. One of its principal conclusions was (Eddington [26], Paragraph 1.108) 'Introducing markets (pricing) where none exist can have a very powerful and positive economic effect in any sector. ... the potential for benefits from a well-designed, large-scale road pricing scheme is unrivalled by any other intervention'.

He pointed out that there is strong case for investment in new road capacity with or without road pricing, but that the quantum of physical investment required would be much reduced if road pricing were used to manage the peak demands.

Although the government of the day formally accepted the recommendations of the review, little happened to implement it. It was largely forgotten during the 2008 financial crisis and the advent of the coalition government in 2010.

2.4.5 The 2010 coalition government roads policy review

The 2010 coalition government soon realised that it had a problem with capacity on the strategic roads. Prime Minister Cameron himself became publicly involved. In a speech he said (Cameron [27]), 'There's nothing green about a traffic jam – and gridlock holds the economy back .... We need to look urgently at options for getting large scale private investment into the national roads network – from sovereign wealth funds, pension funds and other investors'.

It is significant that, while recognising the need, he was not making a commitment that more conventional, Exchequer funding would be forthcoming. Rather, the Prime Minister was hoping that institutional infrastructure investors would provide the capital.

Cameron indicated that he had the United Kingdom water industry in mind. In principle, that is a good analogy: since privatisation, a large capital investment programme has been *financed* by institutional investors. But it has been *funded* by charges to users. The same could be achieved for roads, but only if some form of charging for use were implemented, to create a defined cash flow, or if some other source of funds were to be dedicated to servicing the interest and repayment of capital. However, in his speech the Prime Minister emphasised that: 'we are only considering [road tolling] for new, not existing, capacity ...'. Since most schemes are incremental improvements to existing roads, this stricture largely precludes the water industry solution – of servicing private capital investment out of charges.

2.5 Charges to replace existing taxes

Road traffic is a robust 'market': notwithstanding the current high rates of taxation it is possible to further increase (reduce) revenues by increasing (reducing) charges or taxes.

In an approximately revenue-neutral scenario Glaister, Lytton and Bayliss [12] postulate a 5 pence per vehicle mile charge for cars on motorways and rural trunk roads at all times of day. Heavy goods vehicles pay three times the charge for cars. In this scenario, Vehicle Excise Duty for all vehicle owners would be halved. Fuel duty for all users in the country would be reduced by 10 per cent. For car users of those roads with the distance based charge, the cost per mile would rise from 9 pence per vehicle mile to 13 pence per vehicle mile. For all other car users, the cost would fall to 8 pence per vehicle mile because of the cheaper fuel.

Traffic volumes would reduce on the charged roads and increase on the others. Overall there would be a reduction of about 1 per cent in national car and light goods vehicle traffic and 3.0 per cent in heavy goods vehicle traffic, and therefore a small reduction in total carbon emissions of 1 per cent. Car traffic on the motorways would fall by 7 per cent.

In practice this policy could be considerably refined by applying the charges only during the congested times – say, during the working day.

As some alternate scenarios indicate, it is possible to allow Vehicle Excise Duty or fuel duty rates to decline and still raise of the order of £1 billion a year extra. There is, of course, a range of alternatives available.

2.6 Fairness

The public acceptability of any road pricing scheme will depend on who would gain and who would lose and, in particular, the extent to which it is perceived to be 'fair'.

This can only be analysed in the context of a specific scheme:

- who will pay what and at what times and locations,
- what will happen to existing road taxation and
- who will enjoy the benefits of any net revenues?

What follows is simply intended to illustrate that analysis is possible once a scheme has been specified.

2.6.1 Equity and revenue-neutral, fully efficient pricing

A first example relates to a fully efficient national pricing scheme in England, in (predicted) 2010 conditions with the revenues being fully used to reduce fuel duties: 'revenue neutral' (see Glaister and Graham [28]).

An analysis of Census wards suggested that there are concentrations of highly deprived wards both in the big urban areas and in remote rural areas. The former areas would tend to pay more with efficient pricing but the latter would pay less because of the reduction in fuel duty under a revenue neutral policy.

Figure 2.2 illustrates the relationship between the degree of deprivation of census wards in England and the change in traffic induced by the introduction of a revenue-neutral package of road pricing and road tax reduction. High values of the deprivation indicate greater deprivation.

The wards fall broadly into two groups. One group has a traffic increase and it tends towards the less deprived end of the scale. These are relatively non-deprived areas enjoying a fall in the cost of using their roads. The other group has a traffic

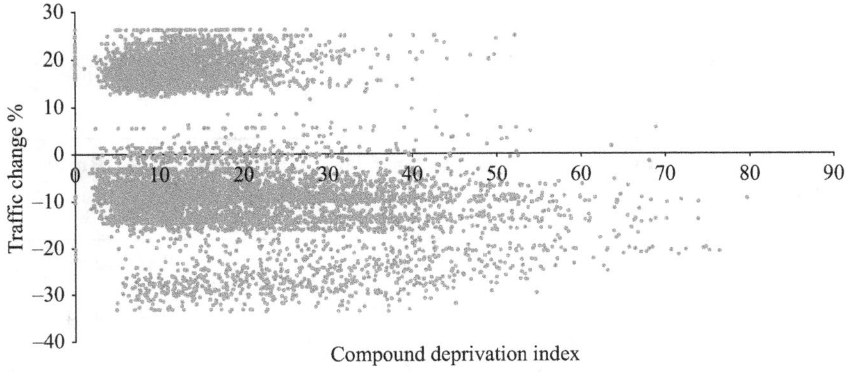

Figure 2.2 Per cent traffic change and compound deprivation index, England. Revenue neutral
Source: Glaister and Graham [28]

reduction because of an increase in road charges and it spreads across the scale of deprivation. There is no obvious, systematic relationship between deprivation and the degree of traffic reduction. Some deprived, rural areas would benefit. Some deprived, urban areas would face higher road charges. But note that in the latter case car ownership and use is likely to be relatively low and residents are more likely to be bus users who will benefit from more freely flowing roads.

2.6.2 Road pricing and household income

In a second example, an analysis by Glaister, Lytton and Bayliss [12] focuses on the financial impacts. These are not revenue neutral changes: Vehicle Excise Duty and fuel duty are replaced by fully efficient charges reflecting congestion, accidents, air pollution, noise and climate change. The net effect is to increase overall revenues and no account is taken of any benefits from the use of the revenues – for instance in improving public transport or roads (Table 2.2).

There would be a surplus of £3.6 billion per year from which the costs of operating the scheme and additional improvements to the road network could be funded. This scenario is one of many and it would be possible to increase or reduce the surplus by varying the tariff structure.

Whilst road users would pay more, in return they would benefit from both reduced congestion and better roads in the longer term. The Eddington study [26] estimated that marginal social cost pricing would lead to welfare benefits of £28 billion per year in 2025; the study also predicted that it would raise Gross Domestic Product by around £14–15 billion in 2025 and that congestion could be reduced by 50 per cent. If this 50 per cent is applied to the estimated £20 billion annual cost of road congestion [12, p43], then the road users' additional payment would be more than compensated for by more freely flowing traffic.

The greatest increase in costs would be in the large towns and cities, but it is here that the congestion benefits would also be greatest. People living in small towns and rural areas would pay less than at present and receive some benefits from easing of localised peak congestion.

Table 2.2 Ratios of pay-as-you-go charges to road taxes by type of area and income range for car-owning households

Area type by population income	London	Metropolitan areas	>250k	>25k	>10	>3k	Rural	All
Lowest quintile	1.84	1.13	1.08	1.00	1.02	0.96	0.83	1.13
Second quintile	1.89	1.22	1.11	1.04	0.97	0.91	0.78	1.12
Third quintile	1.87	1.21	1.11	1.02	0.94	0.86	0.74	1.11
Fourth quintile	1.61	1.07	1.02	0.98	0.93	0.81	0.68	1.03
Highest quintile	1.30	0.93	0.90	0.76	0.80	0.76	0.63	0.96
All incomes	1.72	1.15	1.09	1.00	0.93	0.84	0.70	1.10

Source: Glaister, Lytton and Bayliss [12].

2.7 Governance is the key

The public are acutely aware that whatever the primary aim of road pricing, it raises revenue. There is some evidence that many people feel it is reasonable to pay for things as they use them and that applies to roads. But they care a great deal about what happens to the revenues (see, for example, an Ipsos-MORI survey [29], RAC Foundation [30]).

This issue has to be dealt with explicitly. The acceptability of the London scheme depended on statutory hypothecation of the revenues to transport purposes in the London area and there is sufficient transparency for the mayor to be held to account. This was a crucial difference from the other conurbations, such as Manchester and Birmingham at the time they contemplated and rejected their own schemes – an important lesson for other cities contemplating congestion charges. Although the 2007 New York scheme did not go ahead for 'political' reasons, it had majority public (and newspaper) support because the revenues would have been devoted to transportation improvements [14].

This ring fencing and accountability is missing at the national level for roads.

2.7.1 A ring fenced fund

Wadsworth ([30], paragraphs 3.1 to 4.7) argues that, in the context of the strategic road network, 'Creating a free-standing, dedicated fund for roads is *by far* the most important single thing the government could do to transform the management of our road network and its contribution to the wider economy. A new road fund could be established in the wider public sector, sufficiently distanced from central government to take most of the electoral politics (and associated risk) out of roads planning and delivery'.

A public corporation or public trust could set policy, allocate budgets, make investment decisions and execute those decisions. High-level objectives would be enshrined in the enabling legislation and subject to variation by parliament. The overall budget would be set by government and funded by grant.

It could be given powers to make charges to users and to issue debt. There is a long and continuing tradition of using this structure, which still operates for a number of English ports and airports. The London Passenger Transport Board was a trust between 1933 and nationalisation in 1948, funded by charges to users. Another precedent might be the North American and Australasian trusts set up by statute to deliver public services, including Nav Canada, the Canadian body delivering air traffic control services.

The July 2013 Strategic Highways reforms are significant and welcome. Arguably, the separation of the Highways Agency by creating Highways England is not, in itself, the most fundamental reform possible. However, under a future government it could become a change in which ownership transfers to a body such as a statutory public trust, mutual or private shareholder company. This could accompany a change in the charging regime from the present road tax system to, in full or in part, using a distance-based charging scheme.

2.8 Conclusions

It has become a standing joke that no UK politician is willing to implement road pricing 'within the next 10 years' – by which time they will be long gone. (London, of course, is an exception.)

If it is the case that the level of public understanding of the problems is insufficient for anybody to risk proposing the implementation of a 'big bang' national scheme then there is scope for considering a staged transition.

2.8.1 One problem or five?

The Road Pricing Feasibility Study [24] and Eddington Study [26] assumed a single national scheme, and therefore that the same 'solution' would apply in all circumstances, differences being reflected in rates of charge. Arguably there are several distinct problems as is apparent from Table 2.3.

We might identify five kinds of situations:

1. London
2. Motorways and dual carriage-ways
3. Single carriage-way main roads
4. Large urban areas
5. Rural areas and small townships

The case is clear in (1). There is already a successful scheme that is generally accepted. The congestion case for some kind of extension is more pressing than ever and London needs the money. Technical and administrative solutions are available. The implementation of new forms of electronic payment for bus and rail services in London offer a natural opportunity to include road use.

For (2), the Motorways and some of the other strategic roads in the Highways England portfolio, the case depends upon the likelihood of future governments being able to find sufficient from the Exchequer to fund the maintenance and investment that is required. The historical precedent is not encouraging on this.

Table 2.3 Charge rates for cars by type of area and type of road (pence per vehicle km)

Road type	London	Metropolitan areas	>250k	>25k	>10k	>3k	Rural	All
Motorway	5.1	5.0	4.1	3.2	2.3	1.8	1.6	4.1
Dual trunk			3.5	3.2	2.9	2.7	2.5	2.6
Dual principal			3.3	3.2	2.9	2.7	2.5	2.1
Single trunk	30.3	9.9	8.5	7.1	5.7	4.2	3.9	10.2
Single principal	21.5	11.5	10.2	8.9	7.5	6.2	5.7	8.0
B&C	23.8	10.6	9.0	7.4	5.8	4.2	3.8	7.7
Minor	12.1	9.2	7.9	6.6	5.3	3.9	3.6	5.7
All	19.0	8.7	7.5	6.3	5.1	4.0	3.6	6.1

Source: Bayliss in Glaister, Lytton and Bayliss [12].

A simple two-level licensing scheme seems like a good place to start for raising new funding (see Wadsworth [31] and it was proposed in the Smeed Report [1], Section 5.3). But it would only gain acceptance if a dedicated fund were created to receive the payments and proper governance were created. There would also have to be a degree of offsetting change in existing road taxation. This could create a strategic highway system comparable with the long-established and successful ones in continental Europe.

In the Summer Budget of 2015 ([32], paragraph 1.251) the Chancellor (Finance Minister) announced that 'From 2020 to 2021 the government guarantees that all revenue raised from Vehicle Excise Duty in England will be allocated to a new Roads Fund and invested directly back into the strategic road network'. This could be a beginning.

For (3 and 4) it is likely that in Manchester and one or two other conurbations sufficiently strong local governance will soon be created to allow replication of the London experience. In practice, memories of the failure of an earlier scheme in Manchester will make this harder.

Apart from these, successive government moves towards devolution of powers and funding to Local Enterprise Partnerships and Local Transport Bodies has created much confusion on accountabilities. There seems little hope of moving to road pricing in these areas, unless it were a part of a national scheme. So, absent national road pricing, the alternatives for a large portion of the instances of serious shortages of road capacity are

1. Tolerate the ever-declining service quality
2. Attempt to use alternative traffic reduction measures
3. Provide more physical capacity

Alternatives (2) and (3) will be sensible and productive in particular situations but they are expensive and of limited efficacy.

Category (5) accounts for the greater part of the road network by length. Any congestion problems are best dealt with, as now, by parking and other traffic management measures. Having said that one must not neglect the problem of traffic diverting from newly-charged roads onto local networks.

2.8.2 Lessons from experience

It is near-impossible to win a direct vote specifically on a proposal to introduce road pricing. The arguments in the abstract are too difficult to convey to the general public. The repeated experience is that, once in place the public will understand the benefits of a well-designed scheme and vote to keep it (Walker [14], US Department of Transportation Federal Highways Administration [16]).

Any public discussion must have a properly-researched, specific scheme. People have to know who will pay what, when and what the effects are estimated to be. Only then can the crucial issue of 'fairness' be debated.

Not infrequently proponents of road pricing have allowed themselves to be portrayed as requiring motorists to pay for what they already get 'for free'. Road pricing should never be discussed in isolation: a proposed scheme must be clear about what the net revenues will be and the extent to which they will be used to offset existing road taxation and to improve the service offered to those paying the charges. The costs of the scheme must be presented and they must explicitly recognise the compliance costs to the users.

For whatever reason the English public are resistant to the notion that there would be any element of privatisation. Proposals need to be clear that the organisational reforms proposed do not need to include privatisation.

Trust must be established that the promises will be delivered: in particular that there is transparent accountability for the net revenues. That implies being explicit about the relevant governance.

The merits of road pricing stand on their own. But the public are attracted by the argument that public transport, as an alternative to the car, would benefit. But it is important to be realistic about this last argument. Revenues spent on public transport cannot also be spent reducing conventional road taxation. Also, without the especially high densities to be found in Central London where public transport is already good, the cost of improving public transport sufficiently can be prohibitive.

Any scheme should put simplicity above 'accuracy'. Successful schemes can be rough-and-ready in that they should not require absolute geographical precision or complete compliance – which are both expensive – just as we do not pretend to achieve perfection in other areas of transport charging.

Confidence in a clearly declared cap on charge rates allays fears of an open-ended liability. This would help mitigate the destructive effect of alarmist publicity such as accompanied the Road Pricing Feasibility Study and the Manchester proposals (Chapter 7).

Schemes should minimise the extent of physical modification to vehicles. Automatic number plate recognition has been in successful operation for 10 years in London and could be used easily on motorways – as indeed it is being used in the United States – see Chapter 12.

It is important not to inflict the cost and inconvenience of road pricing when and where there is not a sufficient problem to justify it: and that is most times and places. Having said that, developments for other purposes are making sophisticated electronic charging systems more feasible. The ubiquity of geographic location systems for navigation, eCall and pay-as-you-go motor insurance are creating new possibilities. See for example Chapter 18.

It is crucial that a proper assessment is presented of what will happen if nothing is done and what road pricing has to offer as an alternative. If charges vary by time of day they also will shift commercial vehicles and some other journeys into the off-peak. They will reduce payments in much of the countryside and other uncongested areas. They will have benefits for carbon dioxide emissions and air quality.

2.8.3 What next in England?

Road pricing is not 10 years away. It is here now in London and a number of places overseas.

It is hard to see how London's special problems can be dealt with unless the London scheme is extended. An attractive way to do this would be to include charging for use of the roads on the standard charge methods currently used for all other modes. This may now be possible using the electronic technologies.

National government could consider introducing a sensible, simple system to deal with the pressing funding and financing issues for the motorways and some other major roads, developing the recent reforms to the strategic roads and delivering the proposal to create a ring-fenced roads fund to receive and be accountable for the revenues; and developing a realistic, long-term policy on sources of adequate cash flows. This would have to be accompanied by an explicit policy on the future of conventional road taxation.

That leaves the bulk: non-London local authority roads and the single carriageway strategic roads. The majority of these do not have a sufficient problem to justify intervention. But there are many locations with pressing congestion problems. These will have to be dealt with by managing demand and making local improvements to the road network. It is hard to see how they can be addressed by road pricing except as part of a single national scheme.

A *national* scheme does, indeed, appear to remain 10 years away at the least. But if the targets on carbon dioxide emission are to be met, current fuel duty revenues will have to be replaced and reform will have to come sooner or later.

The move towards replicating the London government model in other major conurbations with more powers and more autonomy over local taxation creates the opportunity to reconsider the matter in the other conurbations.

Whilst this piecemeal approach does not have the appeal of the comprehensive 'solution' envisaged by Reuben Smeed, Sir Rod Eddington and many distinguished others, it is more likely to come about within a decade and it could achieve the pricing of a significant portion of the road congestion in England.

References

[1] Smeed, R. J., *Road Pricing: The Economic and Technical Possibilities.* London: HMSO, 1964.

[2] Dupuit, J., "De la mesure de l'utilité des travaux publics", *Annales des ponts et chaussées,* 8 (2 sem), 332–375, 1844.

[3] Friedman, M. and Boorstin, D. J., "How to Plan and Pay for the Safe and Adequate Highways We Need" in Roth, G., *Roads in a market economy.* Aldershot, England: Ashgate Publishing Ltd., 1996.

[4] Walters, A. A., "The theory and measurement of private and social cost of highway congestion", *Econometrica*, Vol. 29, No. 4, 676–699, 1961.

[5] Debreu, G., *Theory of Value: An Axiomatic Analysis of Economic Equilibrium.* New Haven: Yale University Press, 1987.

[6] Institute for Fiscal Studies, *Fuel for Thought: The What, Why and How of Motoring Taxation.* London: RAC Foundation, 2012.

[7] UK Department for Transport, *British Social Attitudes Survey 2013: Public Attitudes Towards Transport.* July 2014.

[8] UK Department for Transport, *Road Transport Forecasts.* London: Department for Transport, 2013.

[9] UK Department of Transport, *Roads for Prosperity.* London: HMSO, 1989.

[10] Department for Transport, *Action for Roads: A Network for the 21st Century.* London: The Stationery Office Limited, 2013.

[11] Banks, N., Bayliss, D. and Glaister, S. in association with ARUP, *Motoring Towards 2050: Roads and Reality*. London: RAC Foundation, 2007.
[12] Hitchcock, G., Conlan, B., Kay, D., Brannigan, C. and Newman, D., *Air Quality and Road Transport: Impacts and Solutions*. RAC Foundation, 2014.
[13] Glaister, S., Lytton, L. and Bayliss, D., *Funding Strategic Roads*. London: RAC Foundation, 2011.
[14] Walker, J., *The Acceptability of Road Pricing*. London: RAC Foundation, 2011.
[15] Campaign for Better Transport, http://bettertransport.org.uk/media/19-mar-toll-roads-dont-make-sense, 19 March 2012.
[16] Federal Highways Administration, *Reducing Congestion and Funding Transportation Using Road Pricing in Europe and Singapore*. Washington, DC: US Department of Transportation, 2010.
[17] Bayliss, D., *What Went Wrong? British Highway Development before Motorways*. Motoring Towards 2050 – Roads and Reality. *Background Paper No. 1*. London: RAC Foundation, 2008.
[18] Archer, C. and Glaister, S., *Investing in Roads: Pricing, Costs and New Capacity*. London: Independent Transport Commission, 2006.
[19] Government Office for London, RoCOL Working Group, *Road Charging Options for London: A Technical Assessment*. London: Her Majesty's Stationary Office, 2000.
[20] Transport for London, *Drivers of Demand for Transport in London*. London: Transport for London, September 2014.
[21] London First, *Jams today, Jams Tomorrow: How Smarter Congestion Charging Can Keep London Moving*. London: London First, 2014.
[22] Mayor's Transport Strategy, *Mayor's Transport Strategy*. London: GLA, 2010.
[23] Glaister, S., *Virtue out of Necessity: Practical Pricing of Traffic in Towns*. London: Social Market Foundation, 11 March 1998.
[24] Department for Transport, *Feasibility Study of Road Pricing in the UK*. London: HMSO, 2004.
[25] Department for Transport, *The Future of Transport*. CM6234. London: HMSO, 2004.
[26] Eddington, R., *The Eddington Transport Study. The case for action*. Her Majesty's Stationary Office, 2006.
[27] Cameron, D., *Speech on Infrastructure*. London: Institution of Civil Engineering, 19 March 2012.
[28] Glaister, S. and Graham, D. J., *Road Pricing in Great Britain: Winners and Losers*. London: Independent Transport Commission, 2006.
[29] Ipsos-Mori, *Survey on Motorway Use and Charging*. London: RAC Foundation, June 2103.
[30] Wadsworth, B., *The Road Ahead*. London: Road Ahead Group, 2014.
[31] Wadsworth, B., *Moving On: Fairer Motoring Taxes and Investment for Growth and Jobs*. London: RAC Foundation, 2011.
[32] HM Treasury, *Summer Budget* 2015, HC 264 2015–16. London: HMSO.

Chapter 3

Types of road pricing, and measuring scheme cost and performance

John Walker and Andrew Pickford***

3.1 Introduction

Different types of road pricing schemes (point, cordon, area, distance) will be addressed, including how we pay at the moment, how technology might enable different types of payment, examples of different types of payment and how we might pay differently to use roads in the future. The chapter will also introduce the technologies used in road pricing. It is an update of an Intelligent Transport Systems (UK) White Paper [1]. New approaches to measuring the cost and performance of road user charging schemes will also be covered, based on Pickford [2]. We aim to address many of the misconceptions that arise from comparing the operating cost and performance amongst schemes that have different policy objectives and operating characteristics.

Given the authors' background, and the White Paper on which it is based, the chapter is written mainly from a UK perspective – but the issues are generic. We generalise and update any conclusions to ensure application in an international context and reinforce this with international references.

3.2 Policy context

Management of travel and traffic demand is necessary because of the finite capacity of the road network, and the threat to sustained mobility and the environment caused by worsening congestion. Potentially, price-based Travel Demand Management (TDM) in the form of road user charging is cheaper, more effective and more environment friendly than building extra road capacity. However, it cannot be used to compensate for inefficient road operating practices, such as poor traffic management and weak compliance with local traffic regulations.

Road pricing should be viewed in conjunction with other complementary measures and options for road building, improved traffic management, incident management, enforcement and the opportunities for ITS to enhance these

*Transportation Research Group, University of Southampton, UK
**MVA Hong Kong Ltd., Hong Kong

operations. Road pricing should become a routine part of future travel activity, incorporating the 'user pays' principle to reflect a means of paying for a local, visible benefit such as reduced journey time variability and reduced levels of traffic. Public acceptance will be improved via these benefits, enabled by the use and operation of the appropriate technologies, funded by the charges and offset by operating costs. Road User Charging revenues can also be recycled into socially desirable projects which could be unaffordable otherwise – unless a revenue-neutral approach is adopted (in other words, the revenues raised by the road pricing scheme are compensated by reductions in other motoring charges and taxes, so that on average motorists don't pay any more to travel on roads). RUC is also a candidate to replace fuel duty, or to reduce the taxes paid to acquire and operate a vehicle, that have been considered in Europe, the United States, Australasia and elsewhere – see Chapters 2 and 18.

Sections 3.2.1–3.2.4 are based on DETR [3], but we have generalised it beyond urban charging schemes to include interurban and national schemes, and hence national and regional highway and other authorities as well as existing tolled roads and crossings.

3.2.1 Policy options

3.2.1.1 Aims of road use charging

Road user charging schemes have three main aims:

- Congestion reduction
- Mitigating the impact of harmful emissions
- Revenue raising

More recently, the policy aims of road use charging may also overlap with the policy aims of taxing vehicle ownership (e.g. first registration tax and annual taxes) or use (e.g. fuel duty and levies). In this case, road use charges may be used as a full ('revenue neutral') or partial replacement for these taxes or levies.

Congestion reduction can be achieved by using price signals to encourage changes in travel behaviour that lead to reduced traffic levels at defined times and locations. Desirable short-term responses include:

- Use of alternative modes (public transport, walking, cycling, car sharing)
- Changing time of travel to avoid the most congested periods
- Changing route to avoid the most congested areas of the road network
- Reducing numbers of trips
- 'Trip-chaining': combining several activities and destinations into one trip
- 'Car-pooling': users share the same vehicle for common parts of the trip

A less desirable response, at least from a local economic perspective, is when road users change trip destination or decide not to travel at all. RUC schemes that reduce traffic levels will also reduce vehicle emissions, energy consumption, vehicle noise and visual intrusion.

Revenue raising enables a local authority to generate funding for other transport improvements, including public transport, cycling and walking facilities, as well as improvements to the road network and subsidies in public transport, and this is generally acceptable. However, political and public acceptability may be somewhat harder to achieve in this case than in revenue-neutral schemes.

3.2.1.2 Scheme objectives for urban and wider area-based schemes

Once a highway authority has set its overall aims for charging, the scheme objectives need to be defined. They are important because the system design is based on them, so they need to be specific, detailed and defined with a broad understanding of the capital expenditure and operating costs. The objectives should include:

- The geographic area where traffic reduction or environmental improvement is sought, often defined in relation to visible, physical boundaries.
- The times of day and days of the week where improvement is sought, such as periods of congestion.
- Any quantified targets for improvements in traffic conditions, reductions of harmful emissions and revenues generated net of costs.
- Recognition of user groups which should be targeted by the scheme (e.g. commuters or through traffic).
- Recognition of other stakeholders that may require special consideration (e.g. local traders, people with accessibility challenges, potentially reflected in differential discounts or exemptions). (Though in general, the fewer exemptions the better, because it simplifies scheme design and enforcement.)

3.2.1.3 Measures complementary to road user charging

Before designing a scheme, the authorities must consider complementary and alternative measures – road pricing is not a panacea and it should not be used in isolation. There are a number of road network management and demand management measures, with the necessary enablers such as interoperable payment systems, that encourage the public to change their travel habits (Table 3.1). See DETR (2000) [3]. Section 4.5 for further guidance on potential complementary measures.

Table 3.1 Examples of non-restraint alternative measures

Road network management measures	Non-restraint demand management measures
• Advanced traffic control • Improved traveller information • Parking guidance and improved parking enforcement • Route guidance to drivers • Transport network infrastructure improvements • Regulations that improve traffic flow such as no-stopping zones	• Employer-based green transport plans, including car-sharing and tele-working • Implementation of high-quality, accessible public transport services to encourage modal shift • Flexible working hours and telecommuting, where possible • Personal Travel Planning (PTP) to inform travellers

However, where congestion or environmental problems are visible and severe, most studies have shown that restraint-based demand management, supported by complementary measures, is required. Prior to London implementing congestion charging, the road network was being managed efficiently; effective enforcement and behaviour change programmes had all been implemented with declining benefits [4]. Restraint-based demand management has much larger potential impact than non-restraint measures, but greater barriers, including political and public acceptance. Restraint measures can be physical, fiscal or regulatory; examples are shown in Table 3.2, together with some key issues.

3.2.2 Major scheme design issues

Many issues must be considered within the design of a charging scheme, and many of them interact, so that design decisions in one area may influence choices in another (see e.g. Chapter 14). The basic options include:

- Point-based charging
- Area pricing and entry permit schemes
- Cordon charging schemes
- Multi-cordon/zone-based charging schemes
- Distance-based charging schemes (according to distance travelled or the sum of the length of road segments on which a vehicle travels)
- Time-based (in area) charging schemes
- Time-based (driven time) charging schemes
- Closed tolling
- National Road Pricing

The principal charging scheme options must be articulated in the initial design stages, taking account of local objectives and priorities, political and public acceptability (Chapters 4, 5, 8 and 9) and the advantages and disadvantages of the restraint-based demand management options identified above. Their characteristics are described below. Other advantages and disadvantages may be associated with the technologies for implementing the schemes – these are covered in other chapters.

3.2.2.1 Point-based charging: tolling

Point-based tolling, also known as 'open' tolling is an access fee to specified road infrastructure, and until recently in modern times, only tunnels and bridges were tolled. The larger schemes have moved partially or entirely to electronic tolling mechanisms, including the Dartford-Thurrock Crossing (London) and the Humber bridge (see also Chapter 13) in the United Kingdom, 407 ETR (Electronic Toll Road) in Toronto (Canada), Melbourne City Link (Australia), N1 in Pretoria (South Africa), the National Freeway No. 1 (Taiwan) and Costanera Norte in Santiago de Chile (Chile) amongst many others. Ideally, the means of identifying a vehicle and its means of payment, such as by reading a tag or the Vehicle Registration Mark (VRM) used for these schemes, should be acceptable and interoperable with other

Table 3.2 Examples of restraint-based demand management measures

Restraint-based demand management measure	**Description and key issues**
Reallocation of road space	Bus, High Occupancy Vehicle (HOV) and High Occupancy Toll (HOT) lanes, pedestrianisation, cycle lanes, extending the duration of the time provided to crossing pedestrians, prioritisation of public transport at selected intersections
Creation of traffic cells and corridor management schemes	Barriers to prevent road traffic between sectors (cells) of a city centre. Aimed at through traffic, but may inconvenience others
Parking restraint	Parking charges and control of parking space supply. Doesn't affect through traffic, or travellers with private non-residential (PNR) parking. Supply restraints could result in increased traffic as motorists search for parking spaces; pricing restraints can ensure that there is space available. Price control may be limited to publicly owned parking capacity which means that potential effects of restraint are limited by the proportion of all capacity that is publicly owned
Workplace parking restrictions and levies	Charge on employers for each parking space available to employees. Restraint effects are uncertain (employers may absorb costs); there are practical issues of definition and enforcement. No effect on through traffic or traffic not related to employer-provided parking
Access restrictions	Regulatory restrictions on access to defined areas, by vehicle type, weight or other basis (e.g. time restrictions on commercial vehicles). Difficult to achieve specific target traffic reductions. Can be complex and costly to implement and enforce
Low Emission Zones or Low Emission Strategy	A Low Emission Strategy may apply to designated areas, corridors or routes. Greater London is designated a Low Emission Zone (LEZ), in order to achieve air quality objectives and to improve the health and quality of life of Londoners. The most polluting vehicles (older, diesel-engined lorries, buses and coaches – but not cars) are discouraged from driving in Greater London. See https://tfl.gov.uk/modes/driving/low-emission-zone (accessed 31 December 2016) From 2020, an Ultra-Low Emission Zone (ULEZ) will apply to a wider range of vehicles to encourage compliance with exhaust emission standards (ULEZ standards) https://tfl.gov.uk/modes/driving/ultra-low-emission-zone (accessed 31 Dec 2016). See also 'Draft Mayor's Transport Strategy 2017' [5]
Road user charging/pricing	Charging directly for road use in a defined area at defined times for defined vehicles. Covered in this chapter

Figure 3.1 The windshield of a toll-compliant truck today. © Siemens AG 2014. Image provided courtesy of Siemens Electronic Tolling. All rights reserved

regional schemes. The alternative to interoperability is unattractive, to say the least (Figure 3.1).

3.2.2.2 Area licencing and entry permit schemes

Area licencing schemes require that road users who wish to use (or keep) their vehicles within a defined area during a defined time period, purchase and display a supplementary licence. In the past, this took the form of a paper licence displayed in the vehicle but currently the VRM – aka number-plate or licence-plate (see Chapter 11) – may be used and listed on a computer database to identify an eligible vehicle.

With entry permit schemes, vehicles need to have a valid supplementary licence to operate a vehicle within a defined area (the restricted zone). These schemes therefore apply to vehicles crossing entry points on a defined boundary, or otherwise operating or parked within the zone. The paper-based Singapore 'Area Licensing Scheme' (ALS) that operated from 1975 until 1998 was actually an entry permit scheme. (Area licencing schemes enforce on the basis of vehicle presence (moving or stationary) within the zone; entry permit schemes control the vehicle's entry to the regulated area.) The London Congestion Charging Scheme (Chapter 13) is a more recent example of an area licencing scheme based on road user declaration of the vehicle's VRM, and offline or online payment enforced through the capture of image-based evidence of the presence of vehicles on public highways and public off-street parking within the zone. Berlin's Low Emission Zone (LEZ) is one of over 30 German permit schemes that depend on vehicles displaying a supplementary licence.

The legislation under which area licencing and entry permit schemes are set up and operate can be framed to apply charges in a number of ways. Hence, the enforcement methods that are applicable depend on the regulatory regime adopted.

The main attraction of area licencing or entry permit charging as implemented is that they are simple to understand and straightforward to implement. However, a significant disadvantage is that they are a 'blunt instrument' for influencing travel demand compared to other options, because charges are typically levied on a period basis such as a day, month or year, rather than a 'per trip' basis, and there is no incentive to restrict the number of trips made once the licence has been purchased either by road users or by the presence of vehicles being detected automatically[1]. They also provide limited flexibility and geographic scalability. There are also practical limits on the number of licence variants (e.g. charging zones, time periods, vehicle types) that can be accommodated within a scheme, to keep the range of licence types understandable to road users.

3.2.2.3 Cordon charging schemes

Cordon charging schemes are the most commonly used form of electronic road pricing (ERP), and have been implemented in cities such as Bergen, Oslo and Trondheim in Norway, in the Singapore ERP scheme since 1998 (Chapter 6), Milan (2012 – Chapter 19), Stockholm (2007) and Gothenburg (2013) (Chapter 14). These involve setting up a cordon of charging points around a defined area of a town. Road users are then charged (usually electronically) each time they cross the cordon and on selected additional routes such as in Singapore.

Cordon charging has the significant advantage over licencing and permits schemes that every trip made into the defined area is subject to a charge, and is therefore the subject of a time-sensitive choice decision influenced by the charge. Charging regimes can also be more sophisticated, with variations by time-of-day, vehicle type and point of entry. Simple cordon charging schemes are, however, likely to have 'boundary effects'. These may include increased parking just outside the boundary, local difficulties related to trip origins or destinations located just inside or outside the cordon and redistribution of traffic onto roads outside the cordon. There was some evidence of 'loitering' by road users in Singapore just outside the cordon to take advantage of an imminent reduction of time-varying charges but otherwise these secondary impacts are largely manageable.

3.2.2.4 Multi-cordon and zone-based charging schemes

Multi-cordon and zone-based charging schemes are similar to cordon charging. The former typically have two or more concentric cordons, while the latter levy charges for travelling across defined zone boundaries, which may intercept circumferential movements as well as radial ones. Such schemes can give a finer level of influence over travel patterns since the charging points can more closely reflect problem traffic movements, and boundary problems can also be reduced. However, they may be more expensive to implement and more complex for the public to understand.

[1] An account-based system is now implemented in London.

There do not seem to be any extant examples of such schemes. The proposed (and rejected) Edinburgh and Manchester schemes in the United Kingdom (Chapter 8) were based on dual cordons. This complexity seems to have been a factor in the Edinburgh rejection, and suggests that for public acceptability it is better to implement a simple scheme initially, even if it is sub-optimal [6,7]. The original scheme for the Stockholm Congestion tax was based on an outer cordon and single diametric inner cordon. This was rejected in favour of a single outer cordon to simplify scheme operation. The current Singapore ERP scheme combines a cordon scheme (with different pricing regime for each point of entry) enclosing its Central Business District (CBD), plus charge points on road segments on the strategic road network. Singapore's next generation ERP scheme to be implemented (Section 3.2.2.5 and Chapter 6) will retain this approach but with the potential to allocate a price to each road segment to offer a more fine-grained approach to TDM accompanied by an increase in complexity faced by road users.

3.2.2.5 Distance-based charging schemes

Distance-based charging schemes levy charges directly on the basis of distance travelled. Such schemes are used on toll roads (where the distance between toll plazas, electronic gantries or the length of chargeable road segments is known) and in national truck-tolling schemes (Chapters 16 and 17) but have not yet been implemented in urban areas, though Transport for London has trialled the technology in the past and Singapore plans to switch from a microwave gantry-based scheme to a GNSS (Global Navigation Satellite Systems)-based distance-charging scheme in 2019 or 2020 (Chapter 6). Pricing individual road segments can also be used to calculate a total cost for a trip on priced routes although this has currently been employed only on high quality roads (e.g. motorways) in Germany and the Czech Republic for heavy goods vehicles. Distance-based charging has attractions in that it charges directly for actual distance travelled on priced routes, and hence can be related to road damage and polluting emissions, and has routinely being employed in many tolling schemes worldwide such as in France, Taiwan [8], Malaysia and South Africa. In these schemes, the cost of tolls is proportional to the length of the road between intersections, although the multiplier can vary between urban routes and rural routes (e.g. Taiwan).

Distance-based charging may be the logical end-point in a process of creating denser and denser networks of charging zones, in addition to local-specific policies to effect TDM or as a form of tolls (e.g. to fund the construction of specific bridges or tunnels, or to protect natural habitats). Overall, the policy of distance-based charging also reflects a usage-based approach more accurately than other policies.

This policy may also reflect the end point in evolution of road use charging for all vehicles, regardless of method of propulsion. It levels the playing field since all vehicles would be subject to the charge and enables alternative fuel vehicles to be promoted without conflicting with pricing for restraint. Recent pilots in Oregon, Seattle and California have demonstrated the technical and policy options that are now available as a fuel tax replacement mechanism (Chapter 18).

3.2.2.6 Time-based charging schemes

Time-based charging schemes levy charges based on time spent in a defined area. However, this does not directly relate either to distance travelled (and hence to wear and tear on roads), or to congestion – a vehicle may be parked.

A variant on this is 'time-based (driven time) charging'. For example, time-in-congestion-based schemes levy charges based on time spent in congested conditions, as defined by some combination of parameters (e.g. persistent low speed conditions). This approach was put forward [9] as having some theoretical advantages in addressing travel reduction or congestion reduction directly. A demonstration of congestion-based charging in Cambridge was a part of the Automatic Debiting and Electronic Payment for Transport (ADEPT) project [10]. However, research into these options has suggested that they would induce drivers to drive less safely [11] to minimise time spent on driving and has raised difficulties over public acceptance due to the unpredictability of charges in advance, including journey times that could increase due to congestion.

These are interesting aspects of road user charging and the closest analogy would be to extend traditional time-based parking as a means of destination pricing for its potential TDM effects. Nottingham introduced a Workplace Parking Levy (WPL) [12] as a charge for off-street parking which aims to reduce the amount of free workplace car parking in order to reduce car journeys and promote more use of public transport [13]. (WPL is a charge; it also provides a source of revenue for funding public transport – an important consideration in achieving public acceptability.)

However, time-based road user charging policies are not considered further in this chapter since there are no schemes of this type currently in operation.

Another related topic is the 'Pricing Parking by Demand' approach adopted by Shoup *et al.* [14,15]. Shoup is mainly interested in optimising parking charges to ensure optimal utilisation rather than using such charges to influence demand on a road network – but his approach does reduce the number of drivers cruising around to find an available parking space, and hence minimises congestion and pollution. Mobile phone apps enabled by on-street parking space occupancy detection can partially mitigate this, however. See also Chapter 4 Section 4.3 (Pricing Congestion: A Basket of Ecofiscal Policies).

3.2.2.7 Closed tolling

The most common form of interurban highway tolling is closed tolling, in which the toll is related to the distance the vehicle travels on the toll road and so this is a form of distance-related charging. The charge is measured by registering when and where the vehicle enters and leaves the toll road network, hence the term 'closed'. Thus, there is a need for a series of entry and exit points, with toll plazas or (increasingly) electronic gantries on the toll road network.

There are many examples of closed toll facilities worldwide, including much of the French and Italian motorway networks, Melbourne CityLink (Australia), the US turnpikes, the 407 ETR in Canada and the tolled network facilities that form the Santiago Urban Concessions tolling regime in Chile. The distinction between open and closed is blurred when tolled routes charge electronically on each segment travelled, rather than calculating the charge from entry to exit.

3.2.2.8 National road pricing

National Road Pricing introduces the principle of a usage-based (time, distance, place) charging regime applicable to most or all vehicles on most or all roads. It has been off and on the political agendas at various times in the United Kingdom, the Netherlands, Hong Kong, New Zealand, Indonesia and elsewhere, but is not being actively considered anywhere at the time of writing – though 'state-wide distance-based charging schemes are under consideration in a number of American states as a potential replacement for fuel tax' (Chapter 18).

3.2.3 Major scheme design issues: other charging options

There are a number of other options in addition to the ones outlined above. They are

- Corridor charging: charging on strategic and interurban routes (an urban motorway, e.g. may be subject to an urban or to an interurban charging policy).
- High Occupancy Vehicle (HOV)/High Occupancy Toll (HOT) lanes that incentivise vehicle occupancy above a specified threshold by offering discounted or exemption from tolls that would otherwise be paid in dedicated tolled lanes in parallel with general purpose (untolled) lanes.

These are described below.

3.2.3.1 Strategic and interurban routes: corridor charging

Certain routes or transport 'corridors', whether urban or interurban, can be tolled. One example is the M6 Toll in the United Kingdom, built to relieve congestion on the 41-km-long M6 motorway in the Birmingham area of the West Midlands. Another is the Melbourne CityLink (Australia) which is urban and was built to connect three of the major arterial roads that formerly ended at the city fringe, and which were bottlenecks that created traffic gridlock during peak hours. A third is 407 ETR in Toronto (Canada) – effectively a bypass around the north of Toronto – which could not have been built in the timescale without tolling as a means of financing – and hence a particularly interesting example given that Canada, like the United Kingdom, does not currently toll most of its highways (though see Chapter 4). At the time of writing, the city of Jakarta is considering a corridor-based charging scheme to incentivise a modal shift to public transport, but otherwise this policy is relatively under-utilised.

As advocated above, the means of identifying a vehicle and a valid means of payment for any such 'corridor' schemes should ideally be acceptable and interoperable with any other local, regional or national schemes – but unfortunately, this is seldom the case. Notably, the tolled portion of the Gauteng Freeway Improvement Project (GFIP) that is largely employed within the collective Johannesburg, Ekurhuleni and Tshwane economic area in South Africa is part of a nationally interoperable Electronic Toll Collection scheme.

3.2.3.2 High Occupancy Vehicle and High Occupancy Toll (HOV/HOT) Lanes

In the United States in particular, HOV and HOT lanes are a widely used and interesting variant of road charging and are used as a traffic demand management measure [16–21]. HOV lanes remain the most prevalent form of managed lane in the United States, with lane-miles in service doubling from 1,500 in 1995 to over 3,000 in 2005, and estimated at 3,300 miles in 2010 [22]. California alone had 1,592 lane-miles in 2014 [23]. See also Chapters 4 and 5.

The difference between the two approaches is

- HOV – the lane can be used only by vehicles with more than a prescribed number of occupants – usually two or three.
- HOT – the lane can be used free of charge by eligible high occupancy vehicles but single occupancy vehicles are also allowed if they pay a charge.

US policy encouraged the design and implementation of HOV lanes for use by car-poolers – though the evidence is that very little car-pooling took place – most users of these lanes already had several occupants in vehicles – for example a family going on holiday [24]. Such lanes are usually separated from the general purpose lanes to allow eligible vehicles to avoid congestion and to travel at a higher average speed than traffic in the normal lanes. Some HOV lanes are under-utilised while the adjacent regular freeway lanes are congested for many hours of the peak period, so many HOV lanes have been converted to HOT to improve their capacity utilisation. Whilst there have been some concerns about equity, studies have shown that in reality HOT lanes are used by all social classes and income groups [25]. However, it is clear that public acceptability is important and that outreach and (continuing) education programmes are necessary, preferably with an emphasis on congestion relief rather than revenue raising [26]. There are a few HOV lanes in other countries, including Canada, Western Europe and Asia-Pacific. See Chapter 5 for the acceptability and implementation of HOT lanes in Minnesota.

3.2.4 Practical considerations in defining charging schemes

A charged area (or corridor) is the area within which non-exempt or non-zero-rated vehicles are subject to a charge. In the case of entry permit or cordon-based schemes, vehicle users become liable to pay when they enter the area. For area-licence-based schemes, they become liable to pay when inside the area, whether being driven or parked on a public highway as in London. The geographic extent of the charged area is a key element in the scheme design because it has a direct effect on its impact, and is also likely to raise concerns in public consultation. The following factors need to be taken into account; but some of them are conflicting and the optimum solution will therefore require careful balancing (and modelling – see Chapter 14).

- Extent of congestion or pollution
- Simplicity
- Impact on through traffic

- Pricing policy for multiple crossings/entries
- Availability and perceived quality of alternative transport modes
- Equity and land use issues
- Increased activity at (and around) the boundary
- Costs and revenues
- Exemptions, discounts and concessions
- Streetscape and aesthetics
- Vehicle classification and relation to charges.

Each of these is described in Sections 3.2.4.1–3.2.4.11.

3.2.4.1 Extent of congestion or pollution

As far as possible, the charged area should include only areas of congestion or areas of measurable levels of harmful emissions although current operating schemes target one or the other as a single policy target – but not usually both, due to the potential for 'policy conflict'. For a congestion charging scheme, when uncongested areas are included, the effectiveness and acceptability of the scheme are diminished. For example, reducing traffic in uncongested areas may produce no benefits as there will be no speed improvements, or reductions in the variability of journey times, and indeed the impact may be adverse. Strategies targeted at congestion are much more likely to find favour with the public than those that impact more generally on road usage. The ability to target congested areas is often compromised by the need to address other factors, as discussed below.

3.2.4.2 Simplicity

A scheme that is simple in concept and design will be more acceptable, better understood and easier to enforce. Ideally, the boundary of the charging area should be describable in familiar terms to the user, for example 'within the inner ring road', or 'between junctions 9 and 13', or defined by natural features such as rivers. Stockholm is a good example (Höök [27] and Chapter 14), since the city centre is virtually an island. A boundary that minimises the number of charging points with a clear tariff structure also has attractions from the acceptability and cost viewpoints. Such a scheme will need less education and publicity, has a reduced burden of enquires and complaints and will reduce the number of unintentional violations and therefore the enforcement burden. In the longer term, as users become familiar with the scheme and can see its benefits, it can be expanded and refined. In the most extreme case, the policy could be changed, such as Milan that switched from its 'Ecopass' environmental charging scheme to 'Area C' – a congestion charge [28] in 2012 – see Chapter 19.

3.2.4.3 Impact on through traffic

The effect on through traffic requires particular care in the design of the charged area; it is essential that quality data on trip patterns is available and is used in the analysis. Adequate capacity on by-pass routes needs to be provided for traffic diverted from the charged area, for example by improving the signing and improving traffic management. Through traffic should not need to take wide detours to avoid the charged area, as this will result in an overall increase in vehicle kilometres; alternatively, traffic will be forced to use inappropriate roads or pass though sensitive areas. The location of the area boundary should therefore allow convenient by-pass routes with adequate capacity. (Or as in Stockholm, an exemption is given for residents of the suburb of Lidingö that spend less than 30 min within the cordon area, since the residents have no other access than through the charged area.)

3.2.4.4 Pricing policy for multiple crossings/entries

The area needs to be carefully designed to minimise or eliminate the need for multiple boundary crossings. Travellers of metro systems are already familiar with zone-based pricing schemes but this has not yet been considered for charging schemes and should not be until the concept of road pricing has been much more widely accepted.

3.2.4.5 Availability and perceived quality of alternative transport modes

The boundary of the charged area should be located to take advantage of, or allow introduction of, modal interchanges. For example, park & ride facilities located on the periphery of the charged area would provide an alternative means of access. In reality however, facilities requiring large amounts of space will be located away from a city centre where land costs are high. But where a choice exists for a car park or rail/bus station/terminal, the facility should be left outside of the charged area, with measures taken to ensure that it has sufficient capacity to handle additional demands. Note that users may consider the cost of parking and alternative modes in their decision on modal choice, alongside the impact, potential delay times and convenience of modal transfer. The efficiency of transfers is therefore important.

3.2.4.6 Equity and land use issues

Any scheme, especially one based on charging for entry, will raise concerns over equity. Residents or businesses located just outside the charged area making regular trips into it will feel unfairly affected by the scheme, as well as residents within the scheme who may be obliged to pay the charge. Boundaries should not divide residential or commercial areas, although acceptability may be increased by providing discounts to some of the interested parties located within the zone, provided that the economic viability of the scheme operation is not compromised. Distance-based schemes, or those with several, lower charge, zones will minimise this problem although this may compromise any aim to manage demand within a small well-defined area. Other non-price restrictions such as traffic calming or weight limits may help discourage diversion onto inappropriate routes.

3.2.4.7 Increased activity at the boundary

Increased parking and interchange activity can be expected at the boundary with travellers changing modes to cheaper alternatives. Car passengers may be dropped off at bus stops or terminals, taxi passengers may alight close to the boundary and continue on foot (if passengers are liable for charges applied to taxis). Traffic approaching a charged area is increasingly likely to adopt an alternative route around the zone and this behavioural response needs to be expected during the scheme design phase. The volume of this additional activity needs to be estimated and accommodated in a safe and organised manner, for example by discounting charges applied to taxis, improved traffic management, revised signing and extensive multiphased public education campaigns to enable all travellers to make informed choices on the time and mode of travel.

3.2.4.8 Costs and revenues

The location of the boundary may have significant effects on the costs and revenues of a scheme, and these need to be considered in the design. The positioning of the boundary to minimise the number of entry points can provide substantial cost savings through reduced equipment and enforcement requirements. Schemes covering larger areas will normally generate greater revenues (since higher volumes of vehicle users become liable to a charge). This may be attractive for the financing of complementary measures, but care should be taken that the larger scheme does not encroach into uncongested areas, as discussed above.

3.2.4.9 Exemptions, discounts and concessions

Apart from the equity issues outlined above, there are also issues of who or what should be exempt from the charge, or pay at a reduced rate. See Figure 3.2.

The London Congestion charge – Discounts & Exemptions

Discounts

To benefit from a discount, applicants must have successfully registered with Transport for London (TfL). The cost is £10 annually per person or vehicle.

Residents discount: People living in the charging zone or in a designated area next to the zone may be eligible for a 90% discount.
Blue Badge holders: (parking permit for disabled drivers and passengers) in the European Economic Area are eligible to register for a 100% discount.
Accredited breakdown vehicles: A 100% discount applies to such vehicles operated by organisations in the European Economic Area that are accredited to BS EN ISO9001:2008 (and in accordance with the specification for applying that standard to the industry).
Vehicles with nine or more seats: eligible for 100% discount of the Congestion Charge.
Ultra Low Emission Discount (ULED): cars or vans (not exceeding 3.5 t gross vehicle weight) which emit 75 g/km or less of CO_2 and that meet the Euro 5 standard for air quality qualify for a 100% discount on the Congestion Charge. This also applies to 'electric' vehicles and some 'plug-in hybrid' vehicles.

Figure 3.2 (Continued)

Motor tricycles: vehicles 1 m or less wide, and 2 m or less long are eligible for a 100% discount.
Roadside recovery vehicles: if registered with the DVLA (UK Driver & Vehicle Licensing Agency) and have a taxation class of 'recovery vehicle' will be entitled to a 100% discount.

Exemptions
It is not necessary to pay or register for the Congestion Charge if the vehicle is recorded at the DVLA in one of these categories:

- Two-wheeled motorbikes (and sidecars) and mopeds
- Emergency service vehicles, such as ambulances and fire engines, which have a taxation class of 'ambulance' or 'fire engine' on the date of travel
- NHS vehicles that are exempt from vehicle tax
- Vehicles used by disabled people that are exempt from vehicle tax and have a 'disabled' taxation class
- Vehicles for more than one disabled person (e.g. Dial-a-Ride) that are exempt from vehicle tax and have a 'disabled' taxation class

Taxi and Private Hire Vehicles
Taxis and Private Hire Vehicles are exempt from paying the Congestion Charge when actively licenced with London Taxi and Private Hire (TPH); the exemption for Private Hire Vehicles only applies when undertaking private hire bookings.
Other exemptions
Certain vehicles, including buses, registered in European Economic Area member states, must also be registered with TfL to qualify for an exemption. Other categories of exemption include:

- HM Coastguard and Port Authorities
- Certain operational vehicles used by the London boroughs
- The armed forces
- Royal Parks Agency
- Breakdown organisations

See https://tfl.gov.uk/modes/driving/congestion-charge/discounts-and-exemptions (14 July 2016)

Figure 3.2 The London Congestion charge – discounts and exemptions

The rejected Edinburgh congestion-charging scheme would have exempted residents' vehicles, motorcycles, public transport and public service vehicles, taxis, recovery/breakdown vehicles, pay-as-you-drive vehicles and disabled persons' vehicles. Of course, the larger the number of exemptions, the more difficult it will be to confirm if vehicles and their users are compliant, and users are more likely to misunderstand the exemption structure – all of which increases the costs of operations and the burden on users. If user (rather than vehicle) discounts are applied, then some means of checking the compliance at the time of charging needs to be defined, potentially based on associating the user with the vehicle and then regarding the vehicle as a proxy for the user to facilitate compliance checking. Overall the system of user declaration and association should ensure that deterrents to abuse are adequate.

3.2.4.10 Streetscape and aesthetics

The intrusiveness of vehicle detection infrastructure on the street is a key consideration for a scheme. Rural areas, conservation areas, areas of historic interest and streets into which a theme has been designed may be disrupted by the location of roadside infrastructure such as poles, cameras and Dedicated Short-Range Communication transceivers. To date, the majority of charging has occurred on motorways, on bridges and expressways; these areas are typically not impacted by the structures needed for charging. In cities, however, the impact can be substantial – and expensive – see especially Chapters 6 and 14.

In Figure 3.3, one example is provided of streetscape impact of roadside infrastructure, used for vehicle detection, enforcement and charging on a trial basis in London.

Figure 3.3 Streetscape impact of detection infrastructure – London Bridge (trial system). Courtesy Andrew Pickford/Transport Technology Consultants

These same structures are unlikely to be accepted near historical buildings, squares or areas of natural beauty. In mitigation, since the primary policy aim for an urban scheme is likely to be to manage demand rather than, as in tolling, to collect all revenue due at the point of detection, the performance of roadside infrastructure for urban charging schemes can be lower, since vehicle speeds are lower and the objective of behaviour change is emphasised more than revenue maximisation – unlike a tolling scheme where performance is likely to be more critical.

3.2.4.11 Vehicle classification and relation to charges

Charging by vehicle classification would seem sensible but brings with it a number of technical and practical difficulties. For example, charging could be based on vehicle dimensions; but especially in free flow traffic, it is difficult to measure such parameters in a reliable and consistent way for all types of vehicles and under all environmental conditions. And generally, the use of a vehicle's dimensions as a charge differentiator, for example, does not map neatly onto taxation classes, which are based on maximum gross weight bands, numbers of axles or purpose – metrics that can only be revealed by records (paper or electronic) relating to the vehicle itself. Hence, if vehicles need to be registered for a scheme, this provides an opportunity to confirm compliance.

Some classification sensors need to be mounted on gantries and therefore have a negative environmental impact. Buried sensors overcome this issue but are less reliable in the harsh environment of the roadway and are likely to have higher maintenance costs. Above-ground and sideways-facing sensors such as scanning lasers or stereoscopic cameras overcome this disadvantage and are able to provide volumetric classification, and at a greater cost, axle counting capability also. The choice of sensors will also consider their ability to detect and separate vehicles, track vehicles and trigger enforcement cameras at the appropriate location to ensure usable evidential-quality images.

Generally speaking, if vehicle classification is deemed to be advantageous to a charging policy, then very simple classification rules should be applied ideally based upon measurable characteristics of vehicles, especially if road-way or road-side vehicle classification sensors are to be used.

If the vehicle class or charging band is declared by an on-board device and interrogated as the vehicle passes the charge point, or may be easily confirmed by checking vehicle registration records against a local database, then reliable charging is possible, at least where cross-border interoperability is of limited concern. However, appropriate (and potentially expensive) enforcement must be applied to prevent fraudulent use [e.g. a car On-Board Equipment (OBE) in a Heavy Goods Vehicle (HGV)]. The introduction of discounts for low emission vehicles or vehicles in use by a disabled person, for example, means that enforcement strategy needs to consider the complexity of vehicle- and user-specific variations in charges and the process for vehicle and user registration and updates to such information.

In summary, classification is difficult to implement, and the measurable quantities do not align with taxation classes. Overall, a charging regime that is not complex, has fewer (rather than more) vehicle classes, and that supports high levels of compliance, contributes to lower operating costs.

3.2.5 Security, privacy and fraud

Security, privacy and fraud are important policy issues. Although it is clear that many people are unconcerned about privacy issues, especially as regards location identification (otherwise they would not carry a mobile phone), schemes should respect people's privacy and anonymity, at least while they comply with scheme requirements and pay the charges. See for example Chapter 18 for some approaches being used in US trials.

Scheme security is also crucially important, since large amounts of money are involved, inevitably leading to attempts at fraud. As a revenue collection system, best practice in financial security should be observed, preferably based on banking standards applicable within the jurisdiction and, if regional interoperability is beneficial, adjacent jurisdictions also.

The European Electronic Toll Service (EETS – see Chapters 9 and 10) enables the separation of the roles of levying the charge for road use and account management. For compliant users, the chosen account manager does not need details of the location at which charges were applied, if the charge records are provided by the 'toll charger' as a trusted party. Similarly, the toll charger would not need the account details of each road user if the payment was guaranteed by an account manager. In this case, an interoperable on-board unit (OBU) would usually be the most effective means of communicating details of the vehicle and the relevant account authority to enable payment to be secured.

The integrity of the evidential regime, from collection of images through debt collection to the imposition of fines or penalties, depends on being able to trust the evidence itself based on the use of legally enforceable data protection principles. For example, images may only be retained until the purposes for which they were collected have been concluded and no longer. Also, such images must satisfy any locally applicable 'evidential test' required by the courts, enabled by ensuring physical security of the cameras used, watermarking and encryption technologies throughout as well as the means of image transmission (often over public networks) and storage.

The charging and enforcement system itself, together with its interfaces to other government agencies or third parties, must also be sufficiently secure against intentional cyber-attacks that may be aimed at deleting or revising critical data, at interfering with the operations and negative impact on the reputation of the scheme and its operator. Simulated attacks and maintaining compliance with prevailing security standards are necessary but not always sufficient; implementing a robust design initially is politically more acceptable than repairing systems and reputational damage later.

3.2.6 Enforcement

Another important issue is enforcement. A charging scheme will inevitably be subject to evasion, so there must be some alternative means of deterring or detecting and identifying drivers who try to evade the charge. Enforcement is the way in which a Road Pricing System Operator maintains high levels of compliance, safeguards the primary revenue stream and ensures that the charges are applied to users in a fair and equitable way according to the applicable tariffs. The choice of enforcement policy and related enforcement technologies play a large role in optimising the operating costs of the system.

With no enforcement the revenue could be severely eroded. At the other extreme, 100% enforcement will guarantee that 100% of the revenue is collected, but the cost will be prohibitive and net returns would be reduced. Accordingly, the enforcement rate should be set at the point of diminishing returns, at least to ensure that net revenues collected meet the intended economic targets and that the primary policy objectives are met. This is also dependent on the acceptability of evading payment; for example a tax might be expected to have a higher enforcement rate and penalty than a civil charge; see Chapter 1 for the distinctions between taxes and charges.

From a user perspective, the equation for whether it is 'worth' attempting to defraud the system is: whether the 'Likelihood of Enforcement' times the 'Cost of Penalty' is less than the 'Cost of Normal Use' or

LoE * CoP < CONU

The Likelihood of Enforcement is a user perception of the probability of being detected and targeted for recovery of unpaid funds, increased through effective marketing that explains how non-compliant behaviour will be detected. The Cost of Penalty (including the effort incurred in paying) may offer a discount for early settlement, also marketed widely. Increasing the perception and cost, reinforced by other deterrents such as inability to renew annual vehicle licences (e.g. as used in the United States and South Africa) can improve compliance. Depending on jurisdiction, the income from enforcement operations may not be recognised as a 'revenue' and therefore may not be applied within the economic model for scheme operations – a technical as well as a political matter.

3.2.6.1 Criminalised vs. decriminalised enforcement

There is an issue about whether evasion is a civil or a criminal matter, which is related to the question of whether the fee collected is a charge for use of the road network, or a tax. The London scheme (Chapter 13) is a civil charging scheme, whereas in Stockholm (Chapter 14) it is a tax, and hence a criminal offence not to pay it, at least for Swedish citizens. (Originally, foreign-registered vehicles were exempt, but in 2015, this exemption was removed and non-payment of debts could lead to a vehicle being banned from Sweden's roads[2].) Much of the precedent in place for enforcement in the United Kingdom is in the civil arena and has used legislation similar to that for parking, which has been largely decriminalised.

Enforcement typically requires images of vehicles – in the United Kingdom the Home Office Select Committee on admissibility of evidence has generally determined that images are required to determine that a vehicle is in a given place

[2] See https://www.epass24.com (accessed 31 Dec 2016).

at a given time so that enforcement can occur. However, in Melbourne, the City-Link Act provided for tag transactions being used as evidence for enforcement if necessary.

There are two charging paradigms used in different parts of the world, as follows:

- Paradigm 1: Users are compliant whether they are equipped or unequipped. Equipped Users pay by Pre/Post Payment Account or an On-Board Account. Non-Transactors are billed post event, based on data captured at the charge point sites (e.g. images). In this case, evasion is a civil matter.
- Paradigm 2: Users are compliant if they are equipped but are non-compliant if unequipped. Equipped Users pay by Pre/Post Payment Account or in On-Board Account. Non-Transactors may either be billed post event, based on data captured at the charge point sites (e.g. images) and/or pursued as Violators using this data. In this case, evasion may be a criminal matter.

In practice, the enforcement regime may initially regard unpaid revenue as a debt which may be collected through an invoice, possibly including an additional administration fee (e.g. as used in 407 ETR, Melbourne's CityLink, all New Zealand toll schemes and the South African GFIP scheme). The debtor would then be given a short period to pay, after which the offence is determined as failure to pay an invoice (that may include records of several charges) rather than failure to pay a charge, which would then be subject to civil penalties or criminal fines. A commercially minded approach to enforcement aims to reduce expensive and time-consuming arbitration or court processes by facilitating payment and, for first time users, recognising the invoicing process as an opportunity to convert the debtor into an account-holder.

3.2.6.2 What is being enforced?

There are three main aspects of enforcement:

1. Paid/unpaid tolls or charges
2. Compliant/Non-compliant vehicles:
   - (i) (Some) qualifications for access to specific services or areas
   - (ii) Vehicle type
   - (iii) Permit or permission
   - (iv) Driving regulations
3. Registered or declared details match actual details:
   - (i) Registration number
   - (ii) Vehicle characteristics
   - (iii) Axle count
   - (iv) Weight
   - (v) Trailer
   - (vi) Emissions
   - (vii) Number of occupants

The type of evidence required to produce a successful prosecution includes:

- Evidence of non-compliant activity
- Adequate supporting information to ensure prosecution, such as accurate details of vehicle owner
- Identification of vehicle, including its make and colour
- Location
- Time and Date
- Vehicle characteristics that are aligned with tariff structure

The location of enforcement equipment can be either 'fixed' or mobile. Fixed infrastructure is easier to equip with electrical power and communications and requires little direct intervention from the system operator and is suited to high traffic volumes; however, the location of such points becomes well known and evaders would consider avoiding these areas. Mobile enforcement is often vehicle based and provides the ability to set up a temporary enforcement point anywhere on the tolled road network, informed by 'intelligence-based' decisions on historic behaviour, likelihood of detecting non-compliant vehicles, etc.

Most evidential requirements need photographic evidence of the wrongdoing so that the vehicle, the felony, the location and time of the incident can be proved to the satisfaction of the courts, even if the aim is not to use the courts for the majority of events. In the London Congestion Charging Scheme and the Belgian National Truck Tolling Scheme (Figure 3.4), this takes the form of a colour, contextual overview image showing the vehicle and the number of axles, and separate (front and/or rear) licence plate image capable of being read automatically. Where an OBU is present, information captured from the unit can also be included in the record.

Figure 3.4 Evidential record from the Belgian National Truck Tolling Scheme. © 2017 EFKON AG, www.efkon.com

3.2.6.3 Fixed infrastructure enforcement

Automatic number plate recognition

In a tolling environment where only non-payment is enforced, for example on a toll bridge, an automatic number plate recognition (ANPR) process (aka Automatic Licence Plate Recognition – ALPR) combined with overview camera(s) can be used to determine the licence plate number to improve the efficiency of enforcement operations since ANPR itself is not used to improve evidential integrity of the system or images. Where the enforcement criteria are more complex (see below), and vehicles are charged according to specific categories or classifications, then both ANPR and additional subsystems will be required. For more details of ANPR and its implementation, see Chapter 11.

Classification

Many systems vary their charges by some characteristic of the vehicle for example:

- Vehicle type – motorcycle, private light goods vehicle, public bus, long-distance coach, HGV
- Fuel type – petrol, diesel, electric, hybrid, liquid petroleum gas, compressed natural gas
- Vehicle dimensions
- Maximum laden weight
- Type of body
- Presence of trailer, caravan, etc.
- Emissions, including CO_2, particulates (PM_{10} and $PM_{2.5}$) and nitrogen dioxide (NO_2)
- Jurisdiction of registration
- Occupancy

For some systems, these characteristics will have been declared during the registration process and be included in the personalisation records, regardless of whether the vehicle is associated with an OBU or only its VRM if non-OBU (e.g. video tolling) is offered. Where the account type requires an OBU, the enforcement system must be capable of communication with the OBU and comparing the declared and observed vehicle characteristics. For non-OBU schemes, the enforcement equipment must be able to compare the observed characteristics with those stored in the customer's record in the central system that is addressed using the VRM.

Most vehicle characteristics can be classified automatically but equipment will have limitations such as how fast vehicles can pass, whether overtaking is permissible and so on. Tables 3.3 and 3.4 identify some of the more common technologies and indicate which characteristics they can measure and what their vulnerabilities are.

We would promote the concept of 'exception handling' (Section 3.3.3.1) rather than immediately calling it enforcement. Individual subsystems are not perfect and can make mistakes. Any anomaly (non-detected OBU, misclassification, ANPR error, etc.) will trigger an exception process which collects image data and all other available data into a single package which is forwarded for compliance checking.

Table 3.3 What can be measured by different technologies

	VRM	Vehicle type	No. of axles	Height	Length	Width	Weight	Trailer	Cost
ANPR camera	X	Manual	Manual					Manual	££
Loop/treadles		*f*	*X*					X	£
Light curtain				X	*f*			X	£
Scanner		*f*	*f X*	X	*f*	X	*f*	X	££
Stereoscopic camera		*f*	*f*	X	X	X	*f*	X	£££
Weigh-in-motion		*f*					X		££

f = can be derived from the information gathered, X = can measure, Manual = can be determined by manual observation and £ = relative cost.

Table 3.4 Vulnerabilities of different technologies

	Congestion	Weather	Reliability	Maintainability	High speed traffic	Cost
ANPR camera	X	X				££
Loop/treadles			X	X	X	££
Light curtain			X	X	X	£
Scanner	X					££
Stereoscopic camera						£££
Weigh-in-motion				X	X	££

The checking process should be governed by business rules (e.g. 'do we check every one?' or 'do we only check those with low confidence results?', etc.). Finally, for exceptions that meet the business rules for potential enforcement, an authorised enforcement officer (depending on scheme regulations) manually checks the data and confirms or refutes the anomaly leading to the continuation of an enforcement process or a 'normal' transaction in which the user is not charged any more than normal. All efforts should be made to reduce errors that result in road users being incorrectly targeted – this could be enshrined as a Key Performance Indicator.

3.3 Functional requirements

3.3.1 Principal functions

The principal functions required to determine road usage include:

- Time of day
- Position of vehicle and road segment
- Communication with the vehicle

- Identification of the vehicle
- Measurement of the distance travelled by the vehicle
- Charge calculations

Each of the above is described in Sections 3.3.1.1–3.3.1.6.

3.3.1.1 Time of day

Time is a key variable in Road User Charging. Congestion is highly time-dependent and one of the principal aspects of demand management is influencing drivers to change both the parts of the road network they use and the times at which they use them. By applying time as a charging factor, different rates can be levied at different times of day, on different days of the week or during particular events. Decisions on the appropriate rate of charging can be made in order to influence driver behaviour, or through analysis of historical demand and congestion data.

The application of time-based charging also has a spatial element. For example, a time-based charge can be applied within an area, such as in the case of the London or Durham (United Kingdom) congestion schemes. Alternatively, time-based charges can be applied to specific roads or road corridors such as in Singapore where every entry route to its CBD employs locally specific time-based charges.

However, when using time as a charging variable, a standardised measure of time will be required. Charging at time boundaries needs to be applied flexibly (e.g. 'shoulder pricing' – charges that are higher than non-peak times but lower than at the peak) to manage possible congestion caused by traffic waiting outside a time-based zone shortly before the tariff is removed.

3.3.1.2 Position of vehicle and road segment

Vehicle position is the primary factor in generating a billing event in Road User Charging, irrespective of the type of charging technology implemented. ISO Standard 17575 (Chapter 9) defines vehicle position in terms of three measurements: zone, corridor and virtual gantry. However, the choice of technology is relevant. For example, microwave DSRC tags (Chapters 9 and 10) can identify vehicle location only at set points using roadside poles or gantries, whilst Global Navigation Satellite System (Chapters 16–18) can provide a continuous feed of location data where there is visibility of a sufficient number of satellites.

3.3.1.3 Communication with the vehicle

The generation of charging events depends on two-way communication between vehicles and the back office, in order to generate billing events. Note that with DSRC microwave tags, communication is a two-stage process (between (a) the vehicle and the roadside equipment and (b) the roadside equipment to the back-office), whereas GPS charging is typically a one-stage process using a mobile data link to submit batches of data to generate charging events in the back office.

Even where ANPR (Chapter 13) is used as the charging technology such as in video tolling, whilst there is no communication between the vehicle and roadside infrastructure, communication to a charging back office is still required. The availability of appropriate communication methods and communication architecture (which will depend on the regional interoperability strategy, if any) is therefore critical to the successful operation of all charging schemes.

3.3.1.4 Identification of the vehicle

In order to issue correct charges and to avoid issuing incorrect penalty notices, vehicles must be correctly identified. Historically, the VRM (aka licence plate or number-plate) has been used. Billing is then carried out against a particular vehicle, or an OBU within a particular vehicle. Ideally, the on-board equipment must therefore uniquely identify the vehicle or user and communicate this information to the charging infrastructure. This will ensure that the vehicle is correctly identified in all interactions with the charging system, including by the bill issuers, the enforcement organisation and enforcement authorities. In Singapore, an OBU (locally known as an In-Vehicle Unit) is mandatory for locally registered vehicles and must be fitted by authorised service centres ensuring confidence between the OBU and vehicle. In other tolling schemes where the OBU is installed by users, if the OBU is detected, the related account will be charged as well as a consistency check with measured vehicle characteristics – at least the classification and sometimes against the VRM also.

The basis of charging may depend on static (vehicle type, emissions, owner status) and dynamic (occupancy, purpose of trip, etc.) characteristics. It is necessary to confirm declarations and compare with what can be measured to enable effective enforcement.

3.3.1.5 Measurement of the distance travelled by the vehicle

Distance is an effective measurement of road usage, providing greater charging flexibility and probably perceived as a fairer basis for charging. Like a common definition of time, a common measurement of distance is required. There are a number of mechanisms that enable charging for distance. First, based on road link lengths derived from cartographic databases, which can be identified using satellite positioning data and combined with map-matching. Second, using on-board measurements such as an odometer or tachograph, or GNSS. Or third, some combination of these measurements.

It is important to note that the cartographic databases may be more of a problem than the position accuracy of GNSS. Roads may be modified or closed, and new roads may be built. Furthermore, the class of road needs to be identified so that differential charging can be used. Indeed, an even more fine-grained approach may be needed, since charging could be used to discourage travel along roads with schools or hospitals or that pass through residential areas. Consideration also needs to be given to whether charges are being applied only to journeys on public roads, or to all distance travelled (even off-road) as in the case of the Swiss lorry charging scheme. Location-specific charging also allows roads that have a higher capacity to be priced more favourably than roads of lower capacity, the exact approach depending on scheme objectives and acceptability.

Extensions to distance-based charging combine distance measurement with time-of-day and position – hence the term Time Distance Place (TDP) charging.

3.3.1.6 Charge calculations

This technical accuracy underpins the charging accuracy that is based on road segment identification. Although the technical accuracy is important to ensure operational integrity, the scheme operator and road user are more interested in the billing accuracy, which depends on all road segments being correctly reported from the point when the vehicle enters the charged road network. The accuracy requirement for missed or incorrectly reported road segments creates a requirement on three parts of the operation:

- A short time to first fix (TTFF) to ensure that positioning accuracy requirements are met as soon as possible after the trip commences
- The positioning accuracy (relative to the correct chargeable road segment)
- The accuracy of the charges actually levied on the road user as shown in Figure 3.5.

If the positioning accuracy is not sufficient to correctly identify the road segment (e.g. two parallel roads having different tariffs), then the final bill will be wrong.

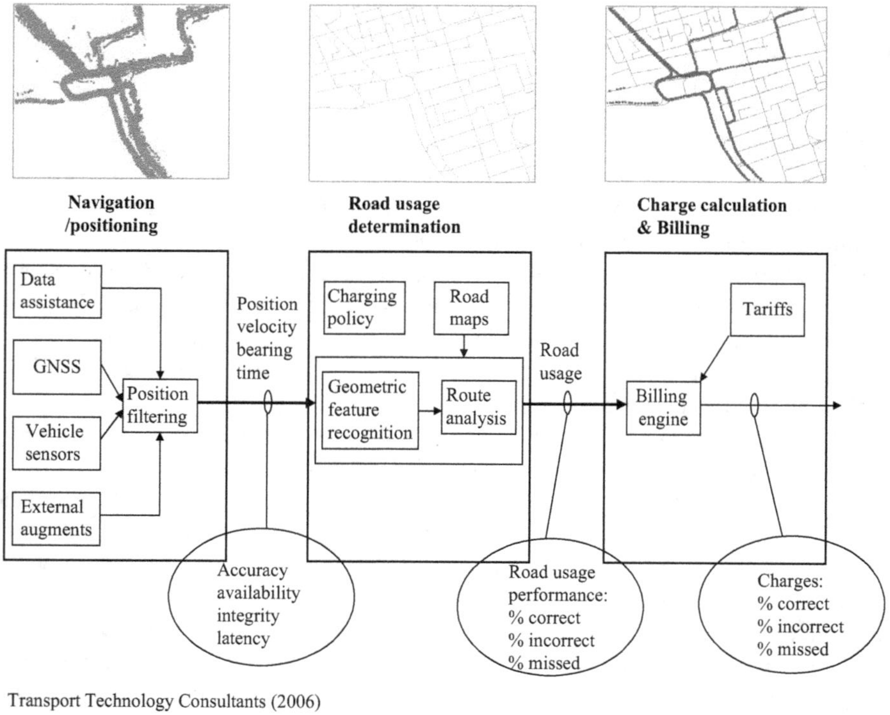

Figure 3.5 Relationship between positioning accuracy and billing accuracy [29]

This may be mitigated by several methods, such as providing additional local augmentation at difficult locations on the road network (e.g. the German truck tolling scheme uses infrared beacons to broadcast the identity of some road segments when satellite visibility is reduced, when there are diversions due to construction and maintenance or where there are close parallel routes); auditing a vehicle journey to identify apparently missed or inconsistent road segments or using the integrity information to decide whether or not to use a position estimate [29]. The distance-based charging trials conducted by Transport for London of GNSS-based OBUs in 2006 identified the potentially high probability of segments being misidentified in an urban environment and the System Evaluation Test in Singapore conducted from 2010 onwards and similar trials in Beijing in 2015–16 focused on addressing this in similar environments. Trials in the United States have employed a variety of automated and manual distance-based recording methods [30].

3.3.2 Security

3.3.2.1 Introduction

Security can mean different things to different stakeholders within a Road User Charging scheme. The major ones are:

- Physical security of equipment at the roadside and within vehicles;
- Physical security of buildings to prevent unauthorised access to private data;
- Security of data for RUC transactions (including evidential images and their meta data), processing and transfer within the end-to-end system.

3.3.2.2 Requirements for physical security of equipment at the roadside and within vehicles

This is one of the most vulnerable parts of a RUC scheme since equipment, especially at the roadside, is unattended. In-vehicle equipment may be damaged or stolen.

- Provide adequate physical protection to roadside equipment by means of strong cabinets and good locking facilities.
- Prevent climbing of any structures such as cantilevers and gantries where RUC equipment is mounted.
- Hide all cables within structures or underground.
- Ensure any cryptographic security keys cannot be accessed by physical entry into either roadside equipment or in-vehicle devices (tamper-proofing techniques ensure such keys are destroyed if physical access (or attempt at access) is detected).
- Minimise the risk of physical attack which would reduce or prevent the collection of transaction data (and hence revenue) and loss of confidence in evidential data collected.
- Reduce casual interest in the equipment by using silent cooling methods and ideally eliminate any obvious means of ventilation.

3.3.2.3 Physical security of buildings to prevent unauthorised access to private data

Any building in which personal or transaction data is received, stored, processed or transmitted should only be accessible by authorised personnel:

- Secure access to the buildings;
- Only to data relevant to their job;
- Limited access to any room where sensitive data or archive material is stored.

3.3.2.4 Security of data for RUC transactions, processing and transfer

Security of data can take a number of forms depending on the requirements. A receiving organisation must be able to trust that the data it is being sent has actually come from the claimed source and has not been tampered with (authentication and integrity). If data is deemed to be sensitive then it should be encrypted to protect it from any unauthorised users. These processes require encryption keys (provided by a trusted third party) and standard algorithms. In addition, error protection methods should be used for transmitted data.

- Secure storage and limited access to any personal account data in accordance with local Data Protection legislation;
- Proof of origin and authenticity of any data collected from in-vehicle equipment;
- Proof of identity of any sub-system which requests data from another;
- Proof of authenticity of any data transferred between any two subsystems so that it can be trusted;
- Encryption of any sensitive data which is deemed to be private or could in some way be used maliciously if accessed by unauthorised personnel;
- Secure transfer and storage of encryption keys used in the transfer and processing of data.

3.3.2.5 Access security to equipment to data to prevent possible fraudulent activity

There should be access control (using e.g. password protection or authentication devices) to any workstation involved in data processing or system maintenance. It should be possible to grant and revoke access remotely; data access should be logged and changes or additions to data should also be recorded.

The development of cloud-based computing and the availability of related cloud-based services increase the dependency on maintaining data integrity even when in the hands of third-party service providers that may be located outside of the jurisdiction when road use charges were incurred.

3.3.3 Enforceability and enforcement

3.3.3.1 Introduction

As indicated in Section 3.2.6, any Road User Charging scheme will require some enforcement, otherwise the RUC operator has no means of identifying users who do not pay or attempt to underpay for road use. Historically, and still in common use today, charging was performed at a toll plaza where each vehicle is stopped, either to pay a human operator or to use an Automatic Coin Machine or card facilities (see Chapter 12). A barrier acts as a deterrent to passing through the toll lane without paying. Even attempted fraudulent use of OBUs at barrier-controlled automatic toll lanes can be monitored and recorded for later pursuit. Enforcement is therefore very practical and effective in plaza-based environments.

With open road free flow traffic, where enforcement is more difficult, the investment in enforcement equipment and processes will depend on the particular business rules of the road operator. The objective is to detect vehicles which are using the road without paying for it, to collect sufficient evidence (usually image-based) to prove vehicle presence with sufficient additional information to confirm any infringement and to use this evidence to identify the entity (company or person) who is liable for the payment. Whilst it is important to detect deliberate or systematic fraud attempts, it is also important that any equipment failures must not falsely accuse users of an infringement. As indicated in Section 3.2.6, the overall process of detecting possible fraud or system failures is called 'Exception Handling'. A detected exception does not necessarily mean that an infringement has taken place and that enforcement is necessary. Methods for checking the collected evidence are required which usually involve human intervention. Only when an exception is confirmed and no other means of payment is available will enforcement action be taken. In any case, a person or company that has not paid may be initially regarded as a debtor or 'potential account holder' rather than a violator but the requirements on evidential integrity remain unchanged.

The reasons for detecting exceptions are diverse, but the main ones which may lead to enforcement are:

- Detection of a vehicle with no (or an invalid) form of identification;
- Detection of declared vehicle details which do not match with corresponding measurements e.g. vehicle class, axle count, weight, trailer;
- Detection of an invalid method of payment (invalid or expired contract or insufficient on-board funds);
- Detection of a vehicle which has been 'hot listed' due to previous infringements;
- Insufficient details collected to allow the vehicle to be charged.

In all cases, it is necessary to collect additional evidence to support any enforcement action and the claim for payment in case this is challenged by the road user. This usually involves the capture of one or more images of the target vehicle along with associated data which clearly and unambiguously positions the vehicle at the correct time and place. Ideally, two different images are captured (see Figure 3.4):

- a narrow field of view (either one or more lane widths depending on camera resolution and ANPR process requirements) monochrome image showing the front of the vehicle (and hence the make, model and VRM) – and sometimes the rear for a second image of the number-plate to reduce the number of false positives from the ANPR process; and

- a wider field of view colour image showing the vehicle colour and its position on the road to resolve potential detection ambiguities.

An 'evidence pack' will typically comprise the following:

- A unique transaction identifier
- A charge point identifier
- One or more images from the monochrome camera
- An ANPR result from the monochrome images
- A monochrome camera identifier
- A time and date stamp for the image captured by the monochrome camera
- One or more images from the colour overview camera
- A colour camera identifier
- A time and date stamp for the image captured by the colour camera
- Any other data collected during the passage (e.g. valid declared information from an OBU)

To avoid any doubt that the data as collected above all relates to the same event, some of the camera-based data is usually embedded in the image itself (e.g. the camera identifier, location and time and date stamp), either as visible text at one corner of the image or as data which can be read electronically. It is also important that the primary evidence (namely the images) is cryptographically protected so that it can be proved to be complete and unmodified, either deliberately or by accident, between capture and the point of use. Everything else is 'meta data' and refers to (or accompanies) the evidence through the evidential process.

The purpose of the evidential data is to help demonstrate that a particular vehicle has failed to pay the appropriate fee for the use of a charged road. If the road charge is a tax, this would be tax evasion which is handled in a criminal court of law, where the burden of proof is on the Prosecutor to prove the evasion 'beyond reasonable doubt' for example, assuming use of common law, or else other confidence-based requirements would be applied. If the road charge is not a tax, evasion is handled in a Civil Court and the burden of proof is on the Claimant (probably the provider of the road service) to prove the evasion 'on balance of probability'. Correct storage, handling and disposal of evidential data for criminal prosecution are very important and various recommendations and guidelines have been drawn up by organisations such as the UK Home Office Centre for Applied Science and Technology and the then UK Association of Chief Police Officers[3] [31–33]. Some of these processes (which can be expensive to operate) may be relaxed when pursuit of unpaid charges is to take place in a Civil Court.

At its simplest, a road user charging scheme is identified on the streets by its enforcement infrastructure, using only the capture of images and the reading of VRMs – as in the original London Congestion Charging Scheme (Chapter 13).

[3] ACPO was replaced in 2015 by a new body, the National Police Chiefs' Council.

If the VRM is found on a prepaid list of subscribers (or within a grace period) then the evidence may be deleted immediately. If it is not, then the evidence is retained until the methods of payment open to the road user have been exhausted or a deadline for payment is exceeded after which a Penalty Charge Notice (as it is known in London) or an Infringement Notice (as used by the Victoria Police in Melbourne) will be issued.

All the requirements for capture of evidence may be applied to fixed infrastructure at the roadside or to a mobile unit. It is clearly optimised for detection of a vehicle at a specific location and is therefore most relevant to enforcement at entry points to charged zones, or detection within a charged zone. Proof of situations where enforcement is necessary for distance-based charging is a lot harder to obtain and would be designed according to the specific requirements. This evidence may depend on archive trip information held by a vehicle's OBU, or by the charging scheme operator, depending on charging policy, operations policy and interoperability strategy.

A jurisdiction may define position information recorded by an authorised device such as an OBU installed by the operator, as being sufficient to meet evidential requirements. Enforcement image data captured externally must be of a minimum quality to enable a human observer to clearly identify the vehicle and to be certain of the time and location of image capture.

3.3.4 Privacy

3.3.4.1 Introduction

Any personal data stored for the purposes of maintaining a User's Road User Charging account is covered in the United Kingdom by the Data Protection Act (and similar privacy laws in other countries) and may not be divulged to any other party. It is retained to ensure payment or invoicing. Privacy also means that no unauthorised party can eavesdrop on data transferred over communication links. The use of data encryption ensures that only those parties authorised to do so are able to acquire the data in 'clear' format.

3.3.4.2 The requirements

The following requirements relate to retaining User privacy:

- All personal data must be stored, processed and when required destroyed in a secure fashion in line with the Data Protection Act.
- Only those elements of personal data required for the purposes of subscribing to a RUC scheme should be requested by the RUC operator.
- As soon as personal data is not needed (and compliant with any other regulations relating to retention of auditable information) it should be destroyed.
- Only authorised personnel should have access to personal data.
- If personal data is transferred between subsystems of a RUC scheme then it should be protected from eavesdropping.

A further extension of privacy relating to Road Charging may be the provision of anonymous methods for payment. As long as the Service provider is guaranteed to receive the appropriate fee for road use, then the service provider does not need to know any details of the vehicle or the driver. As an example, assume that a vehicle owner can purchase over the counter a prepaid device which can be fitted into a vehicle and communicate with the service provider. This device will have a certain amount of cleared 'value' stored in an electronic purse and will also have a method for confirming the device issuer as guaranteeing to pay any road use fee levied for road use. When the vehicle, equipped with this on-board device, transacts with the road service provider, then a cleared fee is debited from the purse. The service provider knows where to get his fee from. At no time does the service provider need to know the identity of the vehicle or occupant. Only if there is some exception which requires enforcement (e.g. insufficient funds to pay for road use), will the driver's anonymity be lost and evidential data captured for pursuit via the appropriate Agency (in the United Kingdom the Driver and Vehicle Licencing Agency – DVLA). Even so, any personal data collected should conform to the Data Protection Act.

The separation of roles between the Toll Charger and EETS providers, as defined by the EETS, also afford some protection to users since the former does not need to have details of accounts and the latter does not need to have details of the location at which the charge liability was incurred (Chapter 9).

3.3.5 Environmental issues

Road usage imposes external effects on other users and on the environment, including noise, accidents, delay due to congestion, pollution and visual intrusion on the landscape. The report 'Towards Fair and Efficient Pricing in Transport — Policy Options for Internalising the External Cost of Transport in the European Union' [34] attempted to quantify these external costs as a road user charge. This process is known as 'internalising the externalities' of road usage.

One cost which is clearly higher for HGVs than for the general vehicle fleet is the physical damage caused by vehicles to the road infrastructure. (This is because damage is related to the fourth power of the weight; if a truck is ten times as heavy as a private car it will cause 10,000 times as much damage.) Various mechanisms for 'internalising' these costs are currently used worldwide in the form of annual taxes and user fees. However, the extent to which they are able to reflect the actual cost of damage has come under scrutiny as part of the move towards fair and efficient pricing in transport, to ensure that each vehicle's contribution to road damage is estimated accurately and charged for appropriately. The relative amount of damage caused by individual HGVs varies widely depending on number of vehicle axles, road surface/construction and climatic characteristics [29].

3.3.5.1 Improved network management

Many UK authorities focus on the management and operation of the existing transportation system and the related development of new tools and approaches, including interactive traveller information. In recent times, the number of new road miles constructed in the United Kingdom has increased by 2%, whereas the Vehicle Miles Travelled have increased more than 50% [35]. From 1996 to 2016, VMT increased by 18.1% and between 2013 and 2015, growth in car traffic in England outstripped population growth [36]. Overall, congestion in the United Kingdom remains a significant (and increasing) social, economic and environmental issue.

One of the benefits of applying ETC to existing toll facilities (such as estuarial crossings) has been to keep vehicles moving rather than to stop and pay a charge manually. The positive impacts of ETC include the reduction of noise from the brakes and engine, reduction of fuel wastage and of localised harmful emissions. Vehicles equipped with OBUs for charging can also act as anonymous traffic probes, informing transportation operators of road status, as has been used in Singapore, Sweden, Germany and the United Kingdom [29].

3.3.5.2 Travel demand management

Road networks that are subject to pricing to modify traffic demand can provide additional environmental benefits. TfL's 4th Annual Monitoring Report in 2006 [37] highlighted the following effects of its Congestion Charging system:

- The combined effect of charging and improved vehicle technology is that NO_x emissions within the charging zone fell by 13% and total harmful PM10 emissions by 15%, comparing annual average values for 2002 and 2003.
- Measured concentrations of PM10 within the charging zone declined somewhat, but this could be various reasons as well as congestion charging. Concentrations of NO_x have declined slowly, but this has not been matched by corresponding decreases in CO_2.
- Limited sample surveys of ambient noise in and around the charging zone find no detectable congestion charging impact.

Thus, the effects range from positive to neutral. Other schemes, including Stockholm and Singapore, have observed similar favourable reductions in harmful emissions within a congestion charging area and its surroundings.

3.3.5.3 Low emission zones

Charging methods and technologies can also be used to monitor and control LEZs – as in London for example. A LEZ covering the whole of Greater London was introduced in 2008 to exclude the worst-polluting lorries, coaches, buses and taxis. Unlike the congestion charge, the LEZ applies 24/7/365. An Ultra-LEZ will come into force in April 2019 in central London, expanding London-wide in 2020 for buses, coaches and trucks, and to the North & South Circular roads for cars, vans, minibuses and motorbikes in 2021 [38].

3.3.6 Back-office processing – the central system

3.3.6.1 Background

A central system is a bundle of functions and administrative processes that follow prescribed business rules to create predetermined outputs meeting quality of service expectations [29]. The Central System is defined as the Information Technology and core services on which charging, enforcement and all external interfaces depend. For road user charging or electronic tolling, the scale of the underlying infrastructure depends on the quantity of events that need to be managed, which in turn depends on factors such as the number of road users, variety of accounts, payment transactions, charging transactions and target level of compliance. In addition to the detection technology and charging regime, the operating infrastructure needs to collect charging data, manage customer accounts, interact with the customer and take payments and generate bills or statements, operate enforcement, assure data quality, deal with third parties and other schemes and enable the management and sourcing of this enterprise in a cost effective manner.

Small barrier-controlled toll plazas may need only a handful of staff, a single server and a couple of workstations. In contrast, a nationwide lorry road charging scheme or city-wide pricing scheme would typically be based on distributed functionality, proven technology components and operational processes, adherence to internationally recognised standards (necessary to varying degrees for public procurements), a robust operations & maintenance structure, secure authenticated interface to third-party service providers, 24/7 operations, quality management, disaster recovery provisions, management information systems and reporting. These large-scale schemes would employ many staff and a complex IT system architecture. A regional interoperability regime could mean that a central system required by a scheme operator could be smaller since it wouldn't need to handle individual accounts, although enforcement and applications for exemptions and discounts may remain localised. In general, the trend is towards customer relationship (CRM) being handled by third parties, in exchange for transaction and account-related fees – for example as seen in Australia, many of the EETS-compliant operators in Ireland, plus regional schemes in the United States and the nationally centralised approach used in South Africa.

The principal factors which drive the design of the 'back office' are

- Accessibility and economics of customer payment channels – to encourage customers to use the lowest cost payment means
- The quality of data needed for billing and enforcement and the need for automated or manual quality checks
- The means and cost of gathering usage data: roadside communications links, wireless, smart card or on-board systems and their security
- Statutory and business requirements to ensure that the customer has a good experience and that the scheme operates fairly
- The opportunity for interoperability or consolidation of functions between schemes to reduce costs and to allow customer accounts to be portable between schemes
- The design of the IT systems which automate the majority of the back office processes.

The back office is generally one of the largest costs at scheme set-up unless this function depends on a third party. It is therefore important to develop its architecture, business processes and systems to accommodate the potential customer contact volumes, data quality improvement measures and charging policies which may need deployment within the life of that design.

Back office is usually also the most substantial element of scheme operating cost and therefore overall cost of collection – as a result its economics tend to drive its structure, rather than any well-defined architectural or technical standard. For example, the cost of a call centre interaction is typically of the order of £1.50–£2.50 in the United Kingdom – this is a very substantial portion of a charge for a scheme – of say £3.50–£8.00 – and incentivises the use of machine-based Instant Messaging and automated enquiry handling systems.

3.3.6.2 Standards for back office

There have been attempts to establish standards for road pricing back-office functions, for example the UK DfT's OMISS specifications [39], and specifications exist in all of the multi-operator interoperability schemes globally. In addition, ISO TC204/CEN TC 278 defines ISO 12855:2015 that specifies the interfaces between electronic fee-collection systems for vehicle-related transport services (see Chapter 9) although its adoption had been very limited in favour of simpler locally defined standards.

A back office whether established by a charge scheme operator or a third-party specialised organisation, is largely similar to any major customer-facing and account payment organisation, with a few elements which are road-pricing related. Technology options are therefore:

- transaction or event processing
- the types of payment channel and the extent to which those channels can be automated to minimise operating costs or enable volume customer event throughput
- enforcement processes (optional, will be provided by a scheme operator or third-party back-office services provider)
- the manner in which data quality is ensured
- the format and duration for which transaction and enforcement data is stored.

3.3.6.3 Transaction and event processing

A transaction or event is a record of a vehicle passage across a point or cordon, along a corridor or into a charged area. It may also include a record of a vehicle travelling on a predefined road segment or a length of journey encompassing several individual segments. Transaction processing can take place in the OBU, roadside reader (if applicable) or camera or central system. Determination of where each part of the transaction process occurs would depend on the scheme type, charging policy, operational requirements, selected technology, timeliness of required updates and where data quality assurance or validation takes place.

Section 3.3.1 defines the functional requirements and information collected to establish the 'liability' for the charge and potentially to support the enforcement process.

3.3.6.4 Payment channels and accounts

There are essentially three types of event related to a road pricing payment:

- Customer initiated payment (e.g. retail, call centre, web) for use of a scheme
- Vehicle passage or journey-related payment or transactions: detection-based payment where a liability is created through the vehicle passing a point, located on a defined road segment, within an area or at a barrier
- Account bill payment where a statement is presented and then paid using direct debit, or credit card.

Payment channels need to accommodate schemes which use on-board accounts and those which do not. There are typically three types of accounts:

- Anonymous
- Prepaid
- Post-paid

Anonymity (Section 3.3.4) can be preserved for road users that are believed to be complying with scheme rules, including making correct declarations of vehicle type, carrying a recognised and locally acceptable means of payment or a link to a valid account.

Prepaid accounts require the account holder to maintain a credit balance in their account. When no credit remains, it would be possible to enforce payment immediately at a toll plaza – but not in open road environment, so violation charges would be incurred as set by the operator depending on operational practices and scheme rules.

Post-paid accounts allow the holder to accumulate charges against the account. For each billing period, the holder receives a statement indicating the amount owing. If the charges exceed a pre-determined maximum (as defined by the account balance thresholds), the OBUs associated with the account are not permitted to proceed through the toll lane (if barrier-controlled) – or, as above, are subject to a violation charge. If post-paid accounts are set up, then there should be sufficient information on the vehicle to trace the driver in case of non-payment.

In general, prepaid accounts afford a lower financial risk to an operator, and to further mitigate risk of non-payment, may be associated with a credit card, effectively transferring non-payment risk to the credit card service provider instead.

3.3.6.5 Enforcement processes

Enforcement is a generic term for what happens when a customer does not follow scheme rules. What is regarded as a minor infringement rather than an offence is the responsibility of the scheme operator or charging authority. In many cases, it is better value to notify the customer of the infringement through a low-cost communications channel rather than initiate a full (and expensive) enforcement process.

Typically enforcement involves the following stages, described in Table 3.5, in a public authority charging scheme based on civil (rather than criminal) process, each of which has technology options.

Table 3.5 Stages of a typical enforcement process

No.	Stage	Technology vs. manual process options
1.	Exception identification	Automated processes for example using number-plate reading system and OBU transaction validation techniques with business rules
2.	Exception validation	Secondary automated processes with human inspection
3.	Customer identification	Retrieval of customer data from database or from the applicable vehicle registration authority using VRM
4.	Choice of warning or enforcement action	Automated process potentially using business rules with human intervention for validation
5.	Issue of penalty charge or equivalent or addition to account	Automated process with human intervention to validate that image of vehicle correctly relates to the VRM concerned
6.	Correspondence concerning enforcement	Typically customers can make a 'Representation' (an official declaration that someone else is liable for charges, taxes, penalties, etc.) if they disagree with the enforcement notice. The charging authority decides whether to accept the representation – largely a human decision, but correspondence can be automated to some extent
7.	Representation process	Largely a human process involving retrieval of the notice and a decision on whether to accept the representation. Business rules can be automated to ensure consistency of decisions
8.	Appeal process	Usually customers have the option of appealing if they disagree with a rejected representation; the appeal has to be processed into a suitable pack for consideration by an adjudicator or equivalent
9.	Adjudication	The adjudication process typically has no well-defined technical standards. There are emerging developments in enabling electronic submission of appeal packs, but they may also need to be processable by the appellant or their legal adviser
10.	Debt registration	There are standard interfaces typically to national debt registration systems
11.	Bailiff Action	Again there are no clear standards for automated systems using bailiffs – communication is usually best enabled using standard communications methods (post, e-mail, etc.). Typically, the penalty fee to be retrieved is sufficiently high to justify the resources used

3.3.6.6 Data quality

Assuring data quality is a key cost for the back office. Charging accuracy has no well-defined standards hence billing accuracy, often described and over- and under-charging accuracy, is typically scheme specific. Depending on the charging policy and detection means, the size of data error can vary substantially. Significant errors can bring schemes into disrepute and fuel challenges from customers for what are otherwise fair charges. Typically, errors can be generated through:

- Number-plate reading rate being of the order of 95%–98% (depending on lighting, levels of congestion, whether front and rear cameras are employed and the number of Optical Character Recognition (OCR) engines employed sequentially and with the option of manual correction of misinterpreted reads[4]
- Tag customer error (tag not fitted, mislaid, not mounted) typically of the order of 1%
- Tag detection error – typically through obscuration – of the order of 0.05%–0.1% depending on free-flow conditions and number of high-sided vehicles
- Journey distance error – typically 1%–5% due to mapping errors, satellite obscuration and long 'TTFF' by the OBU.

The level of data quality needed is therefore very scheme- and detection technology-dependent. The back office can apply a wide range of tools to improve the error in the raw data:

- Automated checking of number-plate data using syntax rules, registration system lookup and filters. This can raise the effective detection rate to 95% or more and reduce the error rate to <1%[4]. However, manual inspection has a cost and workflow to use this resource needs to be optimised.
- Encouraging customers to fit tags or OBUs correctly – this can be through communication with the customer following a failure. Failures can be detected by comparing tag transactions with vehicle number-plate captures and using a tag-VRM lookup table identifying which vehicles should have had a tag transaction related to their passage.
- Obscuration can be improved typically through better mounting of detection equipment so that high-sided vehicles are less likely to obscure a tag in an adjacent vehicle.
- Map-matching processes for GPS-based OBU charging, though these mechanisms are much more complicated.

3.4 Technology options

At present, there are only three fundamentally different roadside technologies that can be used for Road User Charging; they are

- Video + ANPR/OCR
- Microwave tags and readers, including DSRC and RFID (see Chapter 12 for information on multiprotocol tags and readers)
- GNSS/CN (Cellular Networks)

[4] Number-plate read accuracy is a complex issue – see Chapter 11.

Strictly speaking, a fourth option is the tachograph as used in Switzerland for HGV charging [40], but this is only applicable to commercial vehicles which are obliged to use tachographs. In New Zealand, either an approved specialised hub odometer or an approved electronic distance recorder is required to comply with the requirements of the Road User Charging Act for light diesel vehicles and diesel-powered heavy vehicles.

Note that both DSRC and GNSS/CN technologies may in addition use video cameras and ANPR for enforcement of unequipped (or wrongly equipped) vehicles.

In fact, several technology approaches are loosely termed 'DSRC'. This includes:

- CEN-compliant microwave (short range) 5.8 GHz communication commonly utilised in European highway tolling schemes.
- Active Infrared communications in a limited number of tolling applications.

Each of these approaches may have value in a given scheme, but it is important to differentiate between them. For example, the German scheme utilises active infrared DSRC enforcement to meet high-speed data requirements for mobile enforcement on German roads, and is also equipped with microwave DSRC for wider European interoperability. RFID (Radio Frequency IDentification) is employed extensively in Taiwan (with at least 95% of all registered vehicles being equipped), and increasingly in India and Malaysia, amongst other countries.

Some of the issues and options discussed below are covered in more detail in Chapter 9 on 'Standardisation' and Chapter 10 on 'The European Electronic Toll Service – EETS and REETS'.

3.4.1 Option 1: DSRC

In this approach, a small transponder (OBU) in the vehicle, usually battery-powered and mounted on the windscreen, communicates with roadside equipment, usually mounted on a pole or a gantry, to detect the passage of the vehicle (Figure 3.6). There are two charging options available, either prepay where the OBU contains electronic cash which is debited during the passage, or post-pay where the OBU account number is validated and sent to the back office; the vehicle owner is then sent a bill for using the road.

The communications link between roadside and in-vehicle equipment is usually microwave at 5.8, 5.9 or (for older schemes) 2.45 GHz (or 915 MHz in the United States and elsewhere, with variants in Taiwan, Malaysia and India), or infrared. If the vehicle is not equipped with a transponder, a digital image is taken of the vehicle and its number-plate, as described below, and the owner is thereby identified and charged (or fined if the system mandates OBUs). The Melbourne System for example operates both DSRC and a Day-Pass scheme which enables occasional users to be charged via ANPR.

Many DSRC charging schemes are in commercial operation worldwide – on both National Motorway Systems and specific infrastructure installations such as bridges, tunnels and toll roads. However, DSRC is increasingly being used in urban areas, including the United States, and at 2.45 GHz in Hong Kong and Singapore (Chapter 6, though soon to be upgraded to GNSS/CN).

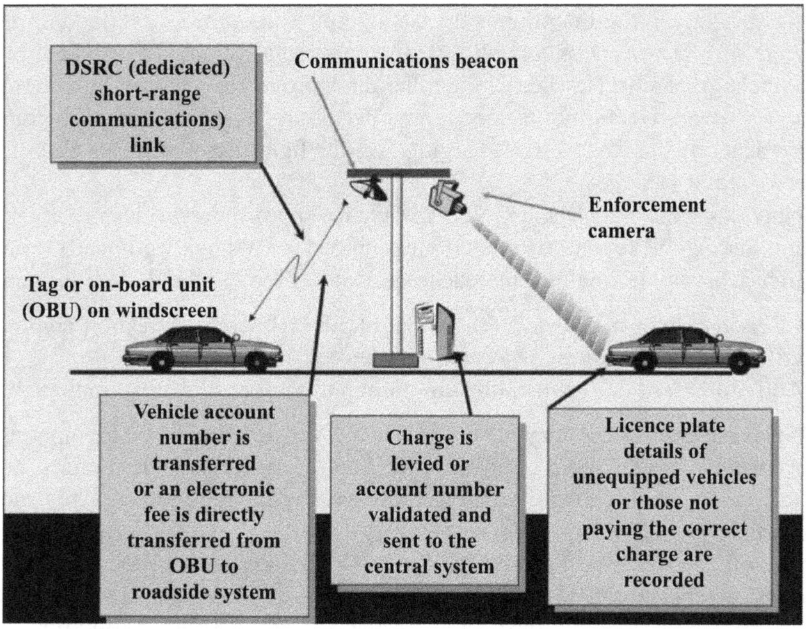

Figure 3.6 Microwave DSRC free-flow tolling. Courtesy of the Transport Operations Research Group, University of Newcastle

DSRC OBUs can communicate with the roadside either via a microwave or infrared protocol. To ensure interoperability, microwave or radio frequency communications systems need to comply with locally applicable technical standards, generally split into the following (see also Chapters 9 and 10):

- CEN DSRC Standards (Europe, South Africa, Australia, Chile and South Africa) and related interoperability specifications
- ISO 18000-6C (e.g. United States, Taiwan, India, Turkey, Argentina, Philippines and Malaysia)
- Others, including GB/T 20851-2007 (China), 18000-6B, Title-21 and E-ZPass (United States), HiPass (South Korea)

The functionality of an OBU is system-dependent and units can be tailored to allow interaction with the user; for example:

- an audible warning when credit is running low
- a display
- a simple means of declaring the presence of a trailer
- a button for axle declaration (Figure 3.7).

When users register with a scheme, an OBU is personalised with their details as required, for example, the number of the account to which it is linked, programmed with vehicle classification or other details. How much or how little information is stored on the unit is dependent on the wishes of the system operator.

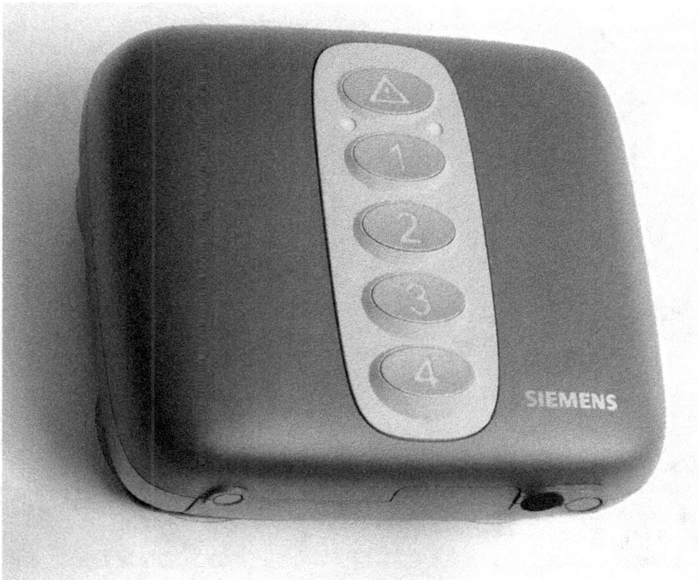

Figure 3.7 Example DSRC OBU with audible warning and button for axle declaration. Courtesy Siemens AG

The charging policy defines the technology choice – and influences the specification of the in-vehicle equipment. Batteryless Radio Frequency IDentification tags (such as those that comply with ISO1600-6C) look like thin but flexible stickers with an area no larger than that of a business card, some providing an NFC (Near Field Communications) connectivity, although lack a user interface. More complex RFID or DSRC devices may have an audible indicator, simple visual indicator, keypad or integral card reader.

3.4.2 Option 2: GNSS/CN

Mounting any DSRC roadside equipment requires gantries (portal or cantilever) which must be compliant with the appropriate Highways Regulations and may be large structures, acceptable on motorways but less so in towns, and still less in country areas or National Parks. Also, they may be expensive to deploy and maintain – see Chapter 6. An alternative method to determine road usage is a system that depends on satellite infrastructure (Section 3.3.1.5), although it must be emphasised that fixed infrastructure for enforcement may still be required, particularly at locations with a high volume of traffic that would be difficult (and uneconomic) to enforce with mobile or portable equipment (Figure 3.8).

3.4.2.1 System operation of GNSS-based charging

GNSS is now a mass-market technology with a much-reduced per-vehicle cost. A GNSS-based charging solution uses an OBU which combines a GNSS location system (augmented with inertial and other sensors if appropriate) and a communications link (usually but not necessarily a cellular radio network – CN), with a

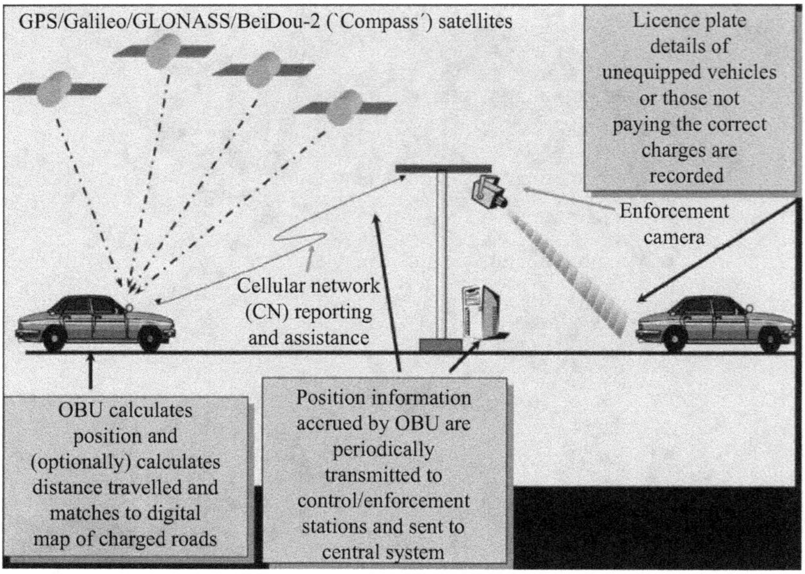

Figure 3.8 Wide area charging systems. Courtesy of the Transport Operations Research Group, University of Newcastle

digital map either on-board or in the back office. The digital map is configured with the location system to define the route travelled. Increasingly, GNSS chipsets within an OBU would include multichannel receivers compliant with multiple satellite types, including GPS, GLONASS (GLobal Orbiting Navigation Satellite System), Galileo (from 2019) and BeiDou-2 (from 2020).

GNSS is particularly flexible in that the geographic coordinates (geo-objects) can be stretches of roads, or zones, permitting distance-based charging along a road, or charging within an area. In the urban environment where satellite visibility may be limited, augmentation may be required to maintain positioning accuracy, or as a minimum the correct identification of the road segment on which the vehicle is located, discussed further below. Any changes to the charging regime may be implemented by updating the digital map either in the back office or over the air for an on-board system[5]. A number of factors can then be attributed, either individually or combined, to the route in order to calculate a charge – see Table 3.6.

The performance of 'pure' GNSS/CN-based schemes, especially the problems encountered due to loss of the satellite signal, can be improved by using 'map-matching' (because in general vehicles can travel only along roads), infill beacons to provide the fix of a known location to the unit (e.g. as in Germany and considered in Singapore) and inputs from sensors such as accelerometers and the odometer. The availability of additional satellite constellations (Galileo,

[5] In 2014, the Slovakian tolled road network was extended from 2,400 km to 17,700 km of motorways, expressways, 1st, 2nd and 3rd class roads – seven times the original size – in just a few months – see Chapter 16.

Table 3.6 Possible charging parameters

Social parameters	Scheme parameters	Vehicle parameters
• Residents	• Distance	• Private cars
• 'Tourist'	• Time of day	• Company cars
• Disabled	• Day of week	• Bus/coach
• Senior citizen	• Public holidays	• Taxis
• Income related	• Events	• Lorries
• Key workers	• Festivals	• Motorcycles
	• Road type	• Emergency services
		• Utilities vehicles
		• Emission class
		• Fuel type
		• Hire vehicles
		• Foreign-registered
		• Exempt & specialist

GLONASS, BeiDou-2, etc.) improves matters by making more satellites visible to a ground-based receiver, reducing the need for support sensors[6]. The disadvantages of a satellite system are

- the satellites may not be visible to the OBU if the vehicle is in an 'urban canyon'; under these conditions a GNSS-based system would revert to other sensors to complete the route data. The digital map is designed to predict and assist in these events.
- ground-based fixed infrastructure may be needed for enforcement; additionally, enforcement can be carried out by mobile enforcement teams.

There are a number of operational GNSS/CN charging schemes currently. The first was the German HGV scheme, which began operations in January 2005, covering 12,000 km of motorway, deploying over 600,000 OBUs and billing 26 billion km. At the time, this represented an increase of 50% in the total number of European roads covered by tolls. Since that time, it has raised over 44 billion euros for the German Government, which has been invested in the transport infrastructure (roads, rail and waterways) [41]. For enforcement, microwave and, in Germany, infrared DSRC may also be integrated within GNSS/CN OBUs for communication between the OBUs and the optional fixed gantries as well as with mobile and hand-held enforcement equipment. The roadside infrastructure uses camera systems and ANPR to detect the vehicle status and validate payment. It is technically interoperable with CEN-compliant DSRC systems such as the Austrian HGV RUC system. See Chapters 6 and 16–18 for some planned, operational and trial GNSS-based schemes.

[6] The OBU for the Belgian Viapass truck tolling scheme (Chapter 16) must be able to use GPS, GLONASS, Galileo and EGNOS.

3.4.2.2 GNSS OBUs

In order to charge a vehicle, an OBU is required to generate information concerning a vehicle's journey. There are primarily two types of GNSS OBU depending upon where the 'intelligence' is (i.e. the ability to match the GNSS information to the physical road network). These are generally referred to as 'intelligent' (or 'thick') and 'thin' client units: a thick client OBU performs the map-matching in the OBU, a thin client in the back office. In practice, though the difference may be academic and dependent on the capability of the charging infrastructure and if intermediary service providers are employed that can support all types of clients to ensure a common presentation of charging records to account service providers.

The complexity of the OBU is a significant driver of the cost of the unit and so thick client units are more expensive than thin ones although this terminology is increasingly used to describe the OBU capability rather than its type (e.g. a thick client OBU can also perform thin client actions plus other value-added services). But the cost of an OBU is primarily dependent on the volume of production – the more units manufactured the lower the cost – as we have seen in the mobile phone market. Arguably, the availability of GNSS-enabled mobile phones has the potential to improve the convenience of location-based charging; but see also Chapters 16 and 18. And as we have seen in the German truck-tolling scheme, it is possible to combine the technologies so that an OBU could contain both DSRC and GNSS/CN technology – the so-called Universal OBU.

3.4.3 Video-ANPR

In a video-ANPR-based system such as London (Chapter 13) or Stockholm (Chapter 14), as vehicles enter the charging zone, images of their licence plates are recorded, and transmitted to the back office (Figure 3.9). Most of the licence plates are read automatically by ANPR and compared to the list of registered users and exempt vehicles. The more difficult images, or rather those that result in a lower confidence interpretation, may be submitted to secondary and tertiary OCR engines (e.g. Stockholm employs three independent OCR steps). Those that are rejected by the process are read manually, for example those which may have been obscured or which are 'non-standard'. Images and licence plate details of registered vehicles are discarded; those of non-registered vehicles are recorded.

Because ANPR is not 100% accurate (several parameters define a camera's accuracy – see Chapter 11), it is advisable to have multiple cameras so that additional images can be captured at the same or different locations making it unlikely that a vehicle would completely evade capture. At a single toll gantry, this could be front and rear cameras (to help reduce false positive errors for individual vehicles) or along a route (to enable vehicle 'signatures' to be matched between different locations). Multiple cameras are necessary in an area-based scheme such as London so that drivers who drive within the area but never cross the boundary can be charged. (Note that, strictly speaking, ANPR is not used for charging in London, but for enforcement. ANPR alone is therefore best suited to zonal charging although event-based charging using front/rear cameras as described above at a

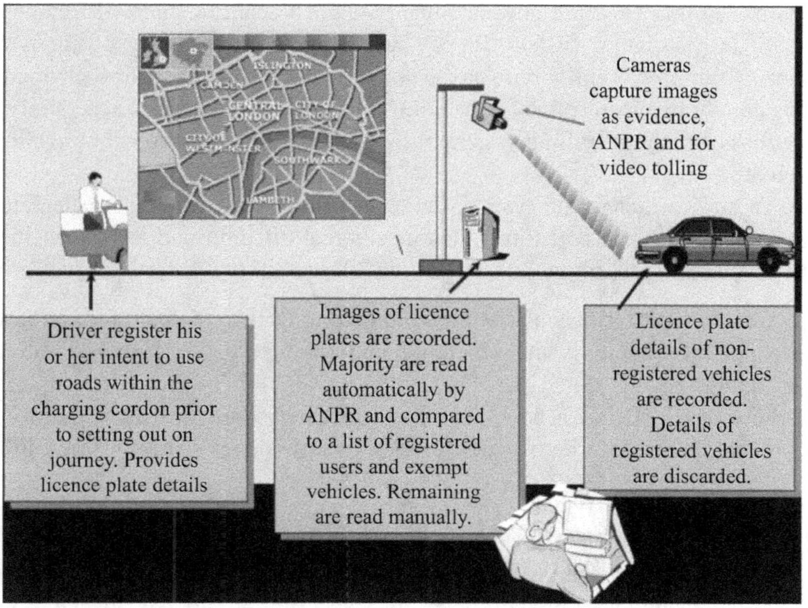

Figure 3.9 Video-ANPR-based charging technology. Courtesy of the Transport Operations Research Group, University of Newcastle

single location has been employed in Stockholm, Australia and 407 ETR in Toronto). The use of ANPR for charging is known as 'video tolling' and images collected can be used for both charging and enforcement. (e.g. Melbourne City Link.) See also Chapter 12, where video tolling in the United States is one way of achieving interoperability between regions where there are different tag standards.

ANPR is less suitable for transport corridors or cordons where there is only one opportunity to capture the vehicle (or at best two opportunities if the cordon monitors outgoing as well as incoming traffic). ANPR is also particularly well suited to schemes with a high number of occasional[7] users who would be difficult or uneconomic to equip with OBUs.

One important drawback of an ANPR approach is that it may encourage number-plate fraud, especially in an uncontrolled number-plate manufacturing market. Of course, the use of ANPR for enforcement will also highlight existing number plate fraud – see Chapter 13, Section 13.3.3.

3.4.4 Impact of charging policy on OBU requirements

3.4.4.1 Prepay and post-pay

With a prepay system the user's account has a credit balance which is debited as tolling transactions occur. This credit balance may be held on the OBU itself, as in

[7] A rule of thumb is that video tolling is economically feasible if a vehicle is seen less that 5–20 times/year.

on-board charging (e.g. the current Singapore ERP scheme, planned for replacement), or in the central system. Prepay facilitates the availability of anonymous accounts if required. With a post-pay account, users are charged for their road use retrospectively by means of a regular bill. In this instance, the OBU acts purely as a device for notifying the billing system of the route, distance or segment travelled by the vehicle.

In an on-board charging system, the charge point must be able to calculate the fee for the passages in real time. This means that all information needed for the operation must be present in the charge point. There are two major reasons for this.

- Transaction performance. The time required to exchange this information with the central system is unpredictable and may exceed the transaction time available.
- Where the OBU has a potential cash value, measures must be introduced so that stolen units can be black-listed (hot-listed) and cannot be 'cashed in'.

3.4.4.2 Occasional users

It is a policy decision whether to mandate use of OBEs in a scheme or not. If the OBE is not mandatory, then provision must be made for charging the non-equipped vehicles such as buying a prepaid ticket for a specific journey (self-declaration) as in Germany or an ANPR-based day-pass as used in Melbourne CityLink.

Where an OBU is mandatory, such as in Austria for heavy goods vehicles, the device must be so designed that it can be fitted quickly at the border crossing point. Accordingly, the OBU for occasional or visiting users cannot require extensive installation (e.g. requiring a connection to the tachograph or odometer), as this would cause the driver to incur cost and time at the border crossings and for goods vehicles this could be considered a barrier to trade. This may necessitate an OBU which can easily be installed by the driver.

Where more than one approach to charging is in operation, it is necessary to ensure fairness between the types of user and between domestic and foreign vehicles.

3.4.4.3 OBU operational issues

Policy decisions must be made with regard to who owns the OBU – the driver or the scheme operator – and accordingly whether a deposit is required. Internal DSRC OBU batteries will require replacing on average every 5–7 years (if feasible, otherwise the OBU is replaced) and a decision should also be made as to what happens to the unit then; should it be recycled or simply disposed of and by whom? For EU recycling directives see http://ec.europa.eu/environment/waste/ (accessed 02 Jan 2017).

3.5 Measuring scheme cost and performance

The political or public acceptability of a scheme also depends on the ability of the scheme to meet its policy and economic objectives, noting that the operating costs must be predictable, ideally based on operating practices proven elsewhere or based on contractual commitments if all or part of the operations are subcontracted or outsourced.

Although charging schemes are policy led, they are enabled by technologies for vehicle detection, communications and automated methods for the collection of evidence to support enforcement processes. The use of such technologies means that charging and enforcement processes can be highly automated although the complexities detailed above mean that manual support is invariably needed, as an integral part of back-office processes. An internal technical or operational view is only part of the story though since the policy aims will include improvements in road network performance, network utilisation, measures for vulnerable users such as cyclists and pedestrians and reducing harmful emissions. In addition, a focus on costs should not ignore the total costs imposed on road users, including proxy measures such as the time to set up an account or pay a charge, detailed further below.

As pointed out by Pickford [2], schemes are often compared by their relative operating cost – a process that is *'fraught with difficulty, prone to large errors, can lead to unfair comparisons and in the worst case could lead to the wrong charging policy being chosen'*. We consider the traditional and more enlightened approaches below, the sources of cost, the rationale to allocate some of the operating costs (and related risks) to contractors or services providers and the benefits of interoperability.

3.5.1 Measures of cost

Sources of cost are known as 'cost drivers' and may include the following:

- Scheme area – a larger scheme invariably means more stakeholders and a higher cost of operations since there are more unique road users, zones (residential and commercial), potential road users, applications for exemptions and discounts, enquiries and complaints.
- Transaction volumes – more transactions mean more payment events, more payment channel options, more account types, more event-based enquiries and more opportunities for errors that impact road users (such as over-charging).
- The proportion of services provided internally (where costs are known) and externally by subcontractors or service providers where the costs to the charging authority are defined in contracts, potentially incentivising improved performance over time.
- The nature of the investments to achieve high or moderate levels of compliance, such as the intensity of any marketing campaign and the introduction of electronic court processes.
- Accounting policies such the amortisation of the costs of scheme development and whether the scope includes the funding of complementary measures, and the cost of enforcement
- The system of law under which enforcement is conducted, generally either civil or common law regimes.

Furthermore, comparing an area pricing scheme such as London to the German LKW (LastKraftWagen Maut – literally 'HGV Toll') truck tolling scheme based on simple cost/revenue measures is inappropriate. Costs are defined by the chosen

Table 3.7 Cost to the user and operator for a payment transaction

Traditional View	New View
Cost to a user = the value of the payment i.e. the user pays \$20 into an account using a retail outlet. The cost to the user is assumed to be \$20. The cost to the operator is the credit card company fee (i.e. 1.5% of the payment value)	Cost to an account holder = the value of the payment plus the value of the time taken to make the payment, and cost to provide payment channel (mobile phone or computer). The cost to the operator is the credit card company fee (i.e. 1.5% of the payment)

operations model and charges are defined by policy and hence the relationship between these is weak at best and not recommended – or otherwise limited to a cost comparison only. Transport for London aggregates all revenues and costs whereas the public reports for the German truck tolling scheme and Stockholm Congestion Charging Scheme do not include any figures relating to enforcement. On the other hand, measures of comparison on the revenue side include the charging policy (i.e. high charges vs. low charges); the purpose of collecting charges (to pay for infrastructure build/operations or to encourage change in road user behaviour) and the demand and willingness to pay charges for services received (elasticity of demand).

Costs incurred by users are often oversimplified. Table 3.7 compares the traditional understanding of costs to a user to make a payment, alongside a more informed 'total-cost' approach.

In this example, the 'new view' could result in easy-to-use payment channels provided via mobile phones, a simple account setup (potentially acquiring an OBU that is ready-to-use, even before an account is set up). This approach would provide a more useful starting point to help prioritise investments in CRM processes, since it incentivises making a charging regime more accessible to users.

There are many more internal process costs that are often ignored unintentionally. Investors, scheme operators, technology vendors and the public may wish to know how well road user charging schemes are being procured and operated. Setting the correct expectations and ensuring that a fair comparison is made to other comparable schemes is critical to ensure the correct statement and interpretation of performance levels.

3.5.2 Interoperability

The move towards regional or national interoperability, enabled through technical standards and operating models such as EETS (Chapters 9 and 10), can have a significant impact on the operating strategy for road user charging – particularly if some functions are centralised to ensure that small operators or new market entrants can benefit from the economies of scale of a large centralised, common back office. For example, the South African model provided centralised account operations to all ETC schemes and an (optional) enforcement services operation [42]. But ETC scheme operators often integrate all activities in-house rather than inviting competitive tenders for outsourced services as demonstrated by the national tolling regime established in Ireland, and the similar functional structures exemplified by the EETS. The enforcement regime and evidential requirements are invariably defined by the system of law in which the charging scheme is located, thereby offering limited opportunity to outsource enforcement operations across borders, although Transport for London and Transport Infrastructure Ireland (operator of the M50, Dublin) require their respective operators to outsource debt collection for EU non-payment offences.

Overall, interoperability encourages specialisation, and specialisation accrues the benefits of economies of scale and cost savings that may be shared between the service provider, scheme operator and stakeholders. The Norwegian national tolling strategy that includes the Oslo Toll Ring is underpinned by a nationally applicable business model, based on proven principles of operation and technical standards, which reduces design costs and development risks for new projects.

3.5.3 Performance management regimes

Performance management aims to reward a service provider if it meets agreed service levels, usually defined by quantitative and qualitative metrics that are traceable to the strategic objectives of the scheme owner, avoiding the usual transactional and potentially adversarial relationship between scheme owner and its contractors [43]. These objectives may include the reduction in traffic entering a defined area, vehicle classification accuracy, capacity of the call centre, email enquiry response times and availability of web-based account management services. Quantitative measures that could be captured and reported automatically through 'dashboards' (high-level reports) may be supplemented by *qualitative* measures that are captured through periodic surveys, such as the satisfaction of account holders or the condition of buildings owned by the scheme owner but maintained by the service provider.

By offering an additional payment if services levels are exceeded, there is an opportunity to create a more collaborative relationship with a service provider, for example to incentivise innovation and continuous performance improvement, hence the term 'incentive-based performance management', typically paying a nominal small percentage above the agreed monthly payments for services provided.

3.6 Summary and conclusions

We have covered different types of road pricing schemes, how payment can be made, the technologies used in road pricing and pitfalls in comparing the cost and performance of schemes of different types with different policy objectives and operating characteristics The relationship between the charging policy and the enabling technologies to assess the charges or taxes due, complemented by compliance checking and enforcement, preserves the business model for road user charging in general. From the examples cited, we would also conclude that policy development in road user charging has been enabled by technologies but overall, policy leads and technology enables rather than allowing technology to dictate policy.

On the subject of costs, the *charging policy* and *scale of operations* are the two most important measures and cost drivers of any charging scheme. At a first level of approximation, a comparison of schemes with similar charging policies and scale will yield a more useful comparison than between schemes underpinned by a different policy and/or those that operate at a different scale. Interoperability enables specialisation and outsourcing of some services such as account management and (potentially) enforcement. In turn, specialisation allows economies of scale to be exploited, and the emergence of dedicated service providers, as described in Chapters 9 and 10.

Subsequent chapters describe proposed and real schemes, based on the principles and technologies described here, the standards that apply, how public acceptability of those schemes was achieved – or not, and lessons learned, in some cases.

References

[1] ITS (UK) White Paper, "Technology Options for Road User Charging in the UK", December 2007.

[2] Pickford A, "New approaches to measuring the cost and performance of road user charging schemes", *ITS World Congress*, Orlando, USA, October 2011, pp. 1–6.

[3] DETR (2000). "Urban Road User Charging Scheme Design Principles: A Preliminary Scheme Design Guide", Prepared for the Department of the Environment, Transport and the Regions by Transport & Travel Research Ltd in association with Ian Catling Consultancy & Wilbur Smith Associates. Version 3.0, February 2000.

[4] London Assembly, "Alternatives to Congestion Charging" (Summary of seminar) (30 January 2002).

[5] https://consultations.tfl.gov.uk/policy/mayors-transport-strategy/?cid=mayors-transport-strategy. *Accessed* 30 June2017.

[6] Gaunt M, Rye T and Allen S, "Public acceptability of road user charging: the case of Edinburgh and the 2005 referendum", *Transport Reviews* 2007; **27**(1):85–102.

[7] Allen S, Gaunt M and Rye T "An investigation into the reasons for the rejection of congestion charging by the citizens of Edinburgh", *European Transport\Trasporti Europei* 2006; **32**:95–113.

[8] Ho F-S, "Going the Distance", *Thinking Highways* 2014; **9**(1):38–43.

[9] Oldridge B, "Electronic Road Pricing: An Answer to Traffic Congestion?". *Proceedings of Conference on Information Technology and Traffic Management*, NEDO, London, 1990.

[10] Clark DJ, Blythe PT, Thorpe N and Rourke A, "The ADEPT Project: 3. Congestion Metering – The Cambridge Trial", *Traffic Engineering and Control* April 1994; **35**(4):256–363.

[11] May AD, Bonsall PW, Hills PJ and Thorpe N, "Evaluating Driver Response to Road User Charging Systems", *Final Report to EPSRC* (Road Transport Information and Control, 1998).

[12] "Workplace Parking Levy: Nottingham", Nottingham City Council, UK. (Undated).

[13] "Funding and Financing Public Transport Infrastructure", April 2016. Steer Davies Gleave & the Campaign for Better Transport. http://www.steerdaviesgleave.com/casestudies/opportunities-improve-funding-and-financing-public-transport-infrastructure (accessed 12 October 2017).

[14] Pierce G and Shoup D, "SFpark: Pricing Parking by Demand", *ACCESS Magazine No.* 43 2013; **Fall**:20–28.

[15] Shoup D, "The High Cost of Free Parking", Chicago, IL: APA Planners Press; Updated edition (1 April 2011).

[16] Parkany E, "Can HOT lanes encourage car pooling?", UCI-ITS-WP-98-5, *Institute of Transportation Studies*, University of California, Irvine, USA (1998).

[17] "High-occupancy vehicle lane", https://en.wikipedia.org/wiki/High-occupancy_vehicle_lane (accessed 31 December 2016).

[18] "A Guide for HOT lane development", Parsons Brinkerhof, the Texas Transportation Institute & the USDoT Federal Highway Administration", 2003. https://ntl.bts.gov/lib/jpodocs/repts_te/13668.html (accessed 12 October 2017).

[19] "Considerations for High Occupancy Vehicle (HOV) Lane to High Occupancy Toll (HOT) Lane Conversions Guidebook", HNTB & Booz Allen Hamilton for the USDoT Federal Highway Administration, June 2007.

[20] "Managed Lanes: A Popular and Effective Urban Solution", Ed Regan, in "2014 IBTTA Global Summit: Innovations & Technologies for Sustainable Mobility, Environment and Road Safety", Prague, Czech Republic, October 19–21, 2014.

[21] "Contemporary Approaches in Congestion Pricing: Lessons Learned from the National Evaluation of Congestion Pricing Strategies at Six Sites", USDoT, Federal Highway Administration, FHWA-JPO-2015-217, August 2015. https://ntl.bts.gov/lib/55000/55600/55668/UPA_2015_Final_9-17-15.pdf (accessed 12 October 2017).

[22] "Frequently Asked HOV Questions", http://ops.fhwa.dot.gov/freewaymgmt/faq.htm (accessed 31 December 2016). USDoT Federal Highway Administration. Last modified: January 12, 2015.

[23] California Department of Transportation (2013), "California High Occupancy Vehicle Lane Degradation Determination Report", 12 December 2014.

[24] Gilroy L and Pelletier A, "HOT Lanes: Frequently Asked Questions", Policy Brief No. 59, Reason Foundation, www.reason.org, 01 March 2007.

[25] Poole RW and Orski CK (2003) "HOT networks: a new plan for congestion relief and better transit", Reason Public Policy Institute, Policy Study 305, February 2003. http://www.reason.org/.

[26] "A Domestic Scan of Congestion Pricing and Managed Lanes". DKS Associates, PBSJ & Jack Faucett Associates for the USDoT Federal Highway Administration, April 2009. https://ops.fhwa.dot.gov/publications/fhwahep09044/index.htm (Retrieved 12 October 2017).

[27] Höök B, "Stockholm: Influencing public acceptability". Paper presented at EU RUC 2007 Conference, London, 22–23 January 2007.

[28] Area C in Milan: from pollution charge to congestion charge (Italy) (Eltis, European Commission, 2105) http://www.eltis.org/discover/case-studies/area-c-milan-pollution-charge-congestion-charge-italy (accessed on 7 May 2017).
[29] Pickford ATW and Blythe PT, "Road User Charging and Electronic Toll Collection", Artech House; 2006.
[30] California Department of Transportation, "California Road Charge Pilot", January 2016.
[31] Lewis SR, "Outline Requirements and Specification for Automated Traffic Enforcement Systems", *Home Office Scientific Development Branch, UK, publication* 3/96; 1996.
[32] Lewis SR, "A Guide to Type Approval Procedures for Speedmeters Used for Road Traffic Law Enforcement in Great Britain", The Speedmeter Handbook, (Fourth Edition), *Home Office Scientific Development Branch, UK, publication* 15/05; 2005.
[33] Lewis SR, "A Guide to Type-Approval Procedures for Automatic Distance/Time Speedmeters used for Road Traffic Law Enforcement in Great Britain", *Home Office Scientific Development Branch, UK, Publication No.* 02/06; 2006.
[34] "Towards Fair and Efficient Pricing in Transport — Policy Options for Internalising the External Cost of Transport in the European Union", European Commission, DG VII, COM(975)691, 1995. http:/europa.eu/documents/comm/green_papers/pdf/com95_691_en.pdf (Retrieved 12 October 2017).
[35] DoDoo NA and Thorpe N, "A Pavement Damage-Based System for Charging HGVs for their Use of the Road Infrastructure," *Highways: Cost and Regulation in Europe*, Bergamo, Italy; 2004.
[36] Department for Transport, "Annual Road Traffic Estimates: Great Britain 2016", 27 April 2017.
[37] Transport for London, "Central London Congestion charging: Impacts monitoring", Fourth Annual Report, June 2006. http:/content.tfl.gov.uk/fourthannualreportfinal.pdf (accessed 12 October 2017).
[38] https://tfl.gov.uk/info-for/media/press-releases/2017/april/gla—mayor-plans-to-introduce-ulez-in-april-2019 (accessed 03 July 2017).
[39] UK DfT, "Open Minimum Interoperability Specification Suite" (OMISS) Volume 3 – Requirements For OBU To RSE Transactions Using 5.8GHz DSRC. Test Specification for Volume 3 – Evaluation of Conformity of OBU and RSE to the Transaction Requirements using 5.8 GHz DSRC; 2006.
[40] ARE-DETEC, "Fair and efficient: The Distance-related Heavy Vehicle Fee (HVF) in Switzerland", Federal Office for Spatial Development (ARE) & Department of the Environment, Transport, Energy and Communications (DETEC); 2004.
[41] https://www.toll-collect.de/en/toll_collect/presse/pressemitteilungen/detail-seite_press_5120.html (accessed 02 January 2017).
[42] PIARC, "Gauteng Freeway Improvement Programme: Open Road Tolling", ITS handbook 2015 (case study). See also www.nra.co.za/content/GFIP_and_ORT.pdf (accessed 12 October 2017).
[43] Pickford A. and Abou-Rahme N., "Made to Measure", *Thinking Highways* 2009; **4**(2):8–12.

Chapter 4

We can't get there from here: ecofiscal policies to address traffic congestion in Canadian cities[1]

*Nancy Olewiler**

4.1 Introduction: congestion costs—an economic and social loss

Traffic congestion costs the economies in Canadian cities conservatively between $5.5 and $8.6 billion per year [2]. These costs represent real losses to the economy and reductions in our quality of life. When congestion exists, people working for hourly wages may incur income losses. Every driver stuck in traffic spends more on fuel and loses time that could have been spent more productively and enjoyably. This lost time has an economic value called its opportunity cost. This is the value of what we would be doing with the time had it not been lost in congestion—getting home to families sooner, taking the children to the park, not having to work overtime because we arrived late, and so on. Between 79% and 94% of the total costs of congestion for passenger vehicles in Canada are attributable to time delays, varying across urban areas and depending on details of traffic flows [3].[2] Additional, longer term costs to urban areas also may be present. For example, businesses may relocate to reduce the commuting costs of their employees and shipments of their goods. People may take a lower wage job that takes less travel time to get to. A recent study estimates these additional congestion costs at between $0.5 and $1.2 billion annually for Vancouver and $1.5–$5.0 billion for Toronto, amounts at least as large as the direct costs from delay [5,6].

A large share of the costs of congestion represents the higher freight cost from delays in getting goods delivered to market. Approximately, 90% of the goods Canadians consume daily come by truck [7]. These costs can translate into lost income, particularly for small businesses that depend on vehicles for work

*School of Public Policy, Simon Fraser University, Canada

[1] This chapter is a modified and condensed version of Canada's Ecofiscal Commission's report, "We Can't Get There From Here: Why Pricing Traffic Congestion is Critical to Beating It" [1]. I was one of the authors of that study.

[2] Although the Transport Canada study did not include many of the environmental and health costs, research in the United States suggests that these costs are significantly smaller than the economic costs of congestion [4].

(e.g., construction workers, skilled trades that need to drive to jobs). The timing of freight movement is typically less flexible than for drivers of passenger vehicles, making it harder to avoid congestion by changing where and when the trucks travel [8]. Congestion costs to the trucking sector can amount to as much as $200 per hour [9]. Estimates for greater Vancouver alone put the costs of congestion for goods movement at between $417 million and $2 billion in 2013 [10]. There are no comprehensive estimates for all of Canada's major cities. Congestion also creates large social costs in the form of increased air pollution, noise, and greenhouse gas emissions. Our cities become less liveable when traffic snarls city streets. Drivers report increased stress and anxiety; traffic accidents rise with attendant costs of injury and property damage. It becomes challenging to plan for routine trips when travel times to destinations are unreliable [11]. Drivers and transit users must then build more time into their schedules to avoid being late. Congestion is thus wasteful and harmful. The direct economic costs are the value of lost time, lower productivity, higher costs of goods and services, more accidents and resultant damage, and harder to quantify costs—greater stress, more environmental degradation, and reduced quality of life.

Canadian municipal governments are increasingly paying attention. In 2015, Toronto Mayor John Tory, Vancouver Mayor Gregor Robertson, and Gatineau Mayor Maxime Pedneaud-Jobin noted that about 80% of Canadians now live in cities that face crippling congestion and failing infrastructure [12].

This chapter examines the economic rationale for pricing congestion and provides examples of how congestion pricing can be implemented in a sample of Canadian cities.

4.2 The economic rationale for congestion pricing

Increasing transportation infrastructure—more roads and public transit—may reduce congestion in the short term but is not a solution over time. Congestion may temporarily abate, but any gains will be short lived. The benefits from new and better roads allow more vehicle travel more quickly, but additional vehicle use ultimately results in a re-emergence of congestion [11]. It may seem paradoxical that expanding public transit may also not be a long-term solution. The reason is that increased transit use frees up road space, making driving more appealing and hence, can increase the number of vehicles on the road over time [13]. In general, any action that creates more road capacity for vehicles provides an incentive for more people to drive. The problem is, as various chapters explain, there is no price attached to the cost of congestion. One's use of the road and hence, the costs to other drivers is not priced. This does not mean that increasing the overall capacity of the transportation system in response to population and economic growth is not desirable. Without some form of congestion pricing, there is no incentive for anyone to adjust their road use to make more space for others. Reducing traffic congestion significantly and in a sustained manner requires some form of pricing to correct the incentives associated with driving and can reduce traffic, improve the transport of people and goods, increase incomes and productivity, and enhance the liveability of our cities.

4.3 Pricing congestion: a basket of ecofiscal policies

Congestion pricing is not a one-size-fits-all policy solution. Different policy tools can price traffic congestion in different ways and need to reflect the specific circumstances of each jurisdiction. Ecofiscal policy tools are ones that correct market prices or introduce prices where none previously existed to reduce harmful activities, congestion in this case, and encourage positive economic activities such as job creation, innovation, and investment. Potential ecofiscal tools fall into five broad categories that are covered briefly: (1) single-entity pricing; (2) high occupancy toll (HOT) lanes; (3) zone-based pricing; (4) distance-traveled pricing; (5) parking pricing. Examples are given of each type with lessons learned about the benefits and drawbacks of each type.

Single-entity pricing instruments are the most common form of congestion pricing. It takes the form of a fee to use a specific piece of infrastructure, typically in the form of a toll for a road, tunnel, or bridge [14]. Canadian examples include Ontario's Highway 407 in Metro Toronto, bridges on Quebec Autoroutes 25 and 30 in Metro Montreal, two bridge crossings over the Halifax harbor, and Metro Vancouver's Port Mann Bridge. The toll road, Highway 407 (called the 407 ETR—electronic toll road) serves as an example of single-entry pricing.

Highway 407 was built by the Province of Ontario and sold as a concession soon after its opening. Tolling began in October 1997, making the 407 ETR the first open-access, all-electronic toll highway in the world. The private operator sets tolls, collects toll revenues, and is responsible for the operation, maintenance, and expansion of the highway. Tolls are paid through transponders placed in customers' vehicles or through cameras and license-plate recognition. The innovation compared to a number of tolled highways (and virtually all other tolled highways in Canada) is the use of variable "time-of-day" pricing that is also distance based. The per-kilometer charge for light vehicles during weekdays, for example, varies from 24.43 ¢/km during off-peak hours to 42.42 ¢/km (including provincial tax) during the morning peak hours and also differ within that range by whether one is traveling in the light or regular zone [15]. There are annual and monthly fixed charges for the use of a transponder. Those without a transponder pay a flat rate per trip of $4.10 for license plate recognition. Everyone pays a $1.00 per trip charge. The Ontario government has extended the 407 route further to the east with Highways 407 and 412 with further extensions underway. Tolling of Highways 407 and 412 began February 1, 2017 with rates varying from .19 ¢/km off-peak to .29 ¢/km at peak times [16]. Drivers will receive one invoice for use of both the private and public sections of the route.

Lessons learned. Commuters using the 407 ETR are found to reduce their total commute time by 18%–36% compared to nontolled routes, a time saving of about 20 min/day [17]. Drivers rarely experience congestion even during peak travel times and with the increasing use of the 407 over time. While data are not extensive, time-of-day pricing appears to be effective. While there was some initial opposition to the highway, there is growing acceptance of both the toll and the operator. Residential and commercial development has emerged along the corridor. For example, job hubs close to the 407 have generally grown at a faster pace than others in the Greater Toronto Area (GTA) [17]. A disadvantage of private management is that the government does not influence the toll structure, thus inhibiting flexibility in response to traffic congestion in the rest of the Greater Toronto region. With the addition of Highways 407 and 412 and a common invoice, there will be both private and public determination of tolls. How drivers will respond to the two sets of prices is not yet known. Electronic tolling, after initial glitches were resolved, enables the use of pricing improvements in the technology over time and have reduced its costs. A concern is that despite the success of the 407 ETR, other tolled highways in Canada have yet to introduce time-of-day pricing.

High occupancy toll (HOT) lanes impose a price on a *portion* of a road. This policy typically involves converting high occupancy vehicle (HOV) lanes to HOT lanes or building new HOT lane capacity, not reducing existing road capacity. HOV lanes are restricted to buses and private passenger vehicles with multiple occupants. HOT lanes give single-occupancy vehicles access to these separate, faster, and more reliable lanes, usually with prices that vary over time dynamically in response to demand or the time of day. In 2013, the United States had over 470 km of HOT lanes, with 262 km under construction [18]. Greater Toronto became the first Canadian metropolitan area to introduce HOT lanes in September 2016 with a pilot program that started September 15, 2016 and is slated to run 2–4 years [19]. The HOT lane converts an HOV lane on 16.5 km of the Queen Elizabeth Way—a major limited-access highway through the core of Toronto. Access to the HOT lane requires a permit that is obtained by lottery from the Ministry of Transportation. Only 500 permits were issued in the first lottery, but the government may issue up to 1,000 in future lotteries. The permit costs $180 and is good for a 3-month period and renewable for a maximum of two times. A new lottery is held every 3 months for the duration of the pilot period. Electronic tolling is not used. Compliance will be assessed by visual enforcement from the Ontario Provincial Police. Vehicles with two or more drivers can use the HOT lanes without charge as can certain vehicles (e.g., licensed taxis, electric vehicles). Dynamic pricing is not used. Demand has exceeded the fixed supply thus far.

HOT lane users receive a direct benefit from shorter and more reliable commute times, while other road users benefit from less congestion in the general-purpose lanes. And unlike more comprehensive congestion-pricing systems, road users have the choice to pay and use the faster and more reliable HOT lanes or stay in the free, more congested lanes. This flexibility is a key virtue of the HOT-lane approach to reducing traffic congestion. An example of the impact of HOT lanes comes from the Minnesota MnPASS program.[3]

[3] Chapter 5 covers the Minnesota policy in detail, especially the means of achieving public acceptability.

Minnesota's Department of Transportation (DoT) introduced one of the first dynamically priced HOT lanes in 2005, known as the MnPASS system. It converted underused, two person minimum HOV lanes to electronically tolled HOT lanes. Buses, carpools with two or more people (referred to as "2+"), and motorcycles access the HOT lanes for free, while single-occupancy vehicles must pay an electronic fee that ranges from US $0.25–$8.00, depending on traffic levels [20]. The tolls change every 3 min according to the HOT lanes' throughput and speed, and are displayed on electronic signs posted along the route [21]. MnPASS lanes are usually priced only during peak hours. Over 25,000 drivers drive HOT lanes weekly, using leased transponders that deduct tolls from their prepaid accounts [22]. The Minnesota government has expanded the use of HOT lanes since its introduction on one expressway and plans future expansion to cover much of the expressways in the metropolitan area.

Lessons learned. Minnesota's hot lanes have been effective at reducing congestion, optimizing traffic flows, and increasing transit ridership along the tolled routes [21]. In addition to maintaining free-flow speeds of at least 50 mph in the tolled lanes, the MnPASS system increased traffic speeds by 6% in the general-purpose lanes. Traffic flows increased by 5% on HOT lane highway and decreased slightly on other nontolled corridors [23]. The total number of trips using MnPASS HOT lanes nearly doubled between 2009 and 2013 [21]. While the primary objective of the MnPASS system is to reduce congestion and optimize road use, the HOT lanes also are generating close to US $1.5 million in gross annual revenues [24]. These revenues cover operation and enforcement costs, with half of surplus revenues devoted to capital investments for the corridor and the other half to improving public transit [23].

One of the primary factors for success is the extensive network of HOV lanes in place prior to the introduction of the HOT lanes. As a result, construction costs were lower because building new lanes was not required. Preserving the same number of general-purpose lanes also improved public attitudes toward the program [20]. The practicality of HOT lanes depends on several other city-specific factors, such as: plans for additional HOV lanes, population density, congestion levels, employment hubs, other transportation management tools in use, traffic safety, enforcement capacity, HOV utilization, transit ridership, number of people per vehicle, and public perception [25,26]. Each of these factors played a role in the success of the MnPASS system and requires close consideration in the Canadian context.

Another factor in the success of this program was the involvement of multiple levels of government. The Minnesota DoT acted as the primary decision-making body for designing and implementing HOT lanes. This helped in developing a consistent long-term regional plan for the city. In addition, the Minnesota DoT received financial support from the US federal government with a grant of $133 million to help expand the MnPASS system to the I-35W corridor [27].

Zone-based pricing systems charge a fee to use designated roads within a given geographical area. Terminology varies, with zone-based policies often referred to as *cordon pricing* or *area pricing*. Pricing can apply to any movement into, out of, or within a defined perimeter of a zone [28]. In some systems, drivers pay each time they cross a perimeter (e.g., Stockholm), and in others drivers pay once daily (e.g., London, which is an area-based scheme). No zone-based systems currently exist in Canada. Other chapters in this volume cover examples of zone-based pricing; please refer to Chapters 2, 6, 8, 13, 14, and 19.

Distance-traveled charges are typically levied on all roads in a given region and can vary depending on distance, time, direction, and location [28]. By comparison, zone-based pricing involves charging in a given area, but not based on the distance driven. While the 407 ETR toll depends on distance traveled, Canada currently has no *region-wide* distance-based charges. Several countries in Europe have distance-based charges for heavy-duty trucks, depending on vehicle emissions and the number of axles [29,30]. See Chapters 10, 16 and 17 in this volume for details.

Dynamic parking prices can also be designed to reduce traffic congestion by creating incentives for shifting transportation choices across time, mode, or location [28]. While most cities use flat-rate structures for parking meters, a variable pricing structure can reduce congestion and improve parking availability. Calgary has recently implemented demand-based parking pricing and San Francisco recently completed a 2-year pilot project, SFpark. To optimize parking access and reduce the total time spent "circling" while searching for a parking space, the project increased prices on parking spots with high demand and reduced prices on spots with low demand in several pilot areas. The program aimed to capture benefits from reduced congestion, such as a faster operating at grade public transit system, cleaner air, safer conditions for pedestrians and cyclists, and increased economic vitality. Funding for the project came primarily from a US $19.8 million grant from the US DoT, with the remainder covered by the San Francisco Municipal Transportation Agency [31]. Analysis of the pilot project concluded the first-of-its-kind program was a success. The city recently finished upgrading all of its parking meters to the "smart meters" that enable demand-based price changes and plans to expand the pricing program throughout the city [32]. The use of smart meters cuts the costs of the program considerably compared to the wireless in-ground sensors used in the pilot.

The San Francisco project included eight pilot areas and three control areas, covering 7,000 metered spaces (25% of the total available) and 12,250 off-street spaces (75% of the total available) [33]. Rates varied by each city block and were adjusted monthly in an attempt to achieve the target occupancy rate of 60%–80% of all parking spots. Rates were capped at $6/h, with an $18/h cap during special events. The project used wireless in-ground sensors to detect when vehicles occupied specific parking spaces, connected to a city-wide communications network. Customers had access to real-time information on parking availability via a mobile application, through a website, or via text message. New message signs indicated to drivers entering downtown which parking garages have real-time availability [33]. Roadway sensors monitored the impacts on traffic.

Lessons learned. The policy was effective at not only optimizing parking space but helping to reduce congestion. The amount of time that the target occupancy rate was achieved increased by 31% in the trial areas, compared with 6% in the control areas [33]. Moreover, the amount of time that city blocks were too full for drivers to find a parking space (corresponding to 90%–100% occupancy) decreased by 16% in pilot areas and increased by 51% in control areas [33]. In short, scarce parking spaces were more efficiently allocated as a result of the policy. In addition, traffic volumes went down by approximately 8% in areas where parking availability improved (in both pilot and control areas). In areas where parking availability worsened, traffic volumes increased by 5% [33]. Circling for parking—a major contributor to downtown traffic—was estimated to have declined by 50% [34]. Vehicle miles traveled also decreased by 30% in the pilot areas compared to a 6% decrease in the control areas [33]. The San Francisco Metropolitan Transportation Authority (SFMTA) also claims that peak period traffic congestion decreased by providing people a financial incentive to travel during off-peak periods. Parking availability on the street is improved by 22% during peak periods, and off-peak parking garage exits are increased by 15% [33]. Both statistics suggest that the SFpark program helped to reduce peak congestion.

An independent evaluation of the SFpark program concludes that the target occupancy level did not necessarily correspond with increased levels of availability. Some city blocks achieved the average occupancy target of 60%–80%, yet many city blocks were still too full (90%–100% full) for drivers to find parking in high-demand areas. The study found that the price in high-demand areas would need to be higher than the imposed cap of $6/h to sufficiently increase availability rates. Moreover, price levels in pilot areas did not seem to reduce parking duration, or increase vehicle turnover and carpooling [35]. Moreover, the price cap appears to have contributed to a decrease in the overall average price for parking in San Francisco. This finding suggests that a more dynamic pricing framework in response to demand (e.g., higher peak prices) would make the program more effective. However, this raises important equity issues on the affordability of parking. Pricing structures therefore must balance program effectiveness with overall affordability.

Gaining public support for the SFpark program has been challenging, as many residents were skeptical even after its implementation. This was primarily due to the meter expansion into residential neighborhoods and paid metering on Sundays, dampening plans for expansions of the program in the shorter term [36]. Intergovernmental cooperation was also important. The SFMTA benefited through the funding partnership with the US DoT, while the DoT benefited from implementing a trial program in a major city that others can emulate. Importantly, parking policy in the United States is almost exclusively within the purview of municipalities, allowing city governments to take direct action.

Based on the San Francisco model, Calgary implemented a dynamic pricing approach to parking in 2013. Similar to SFpark, rates vary across 27 discrete areas in the city to encourage more parking in underutilized areas. While price adjustments in San Francisco are made monthly, prices in Calgary are adjusted only annually and aim to achieve a target occupancy rate of 50%–80% [37]. The City of Calgary recognizes that more frequent price changes would be more responsive to demand and thus more effective in allocating the scarce parking spots. But the city also aims to provide stable prices and avoid large seasonal fluctuations [38]. In contrast to the significant opposition to the program in San Francisco, the ParkPlus program in Calgary has kept a low profile. This is partly due to the "smart meters" already being in place when the shift to dynamic pricing was made, and also the minor change in parking price. While still too early to determine its overall effectiveness, using demand-responsive pricing to manage parking demand is expected to help achieve target occupancy levels. This will reduce circling for parking spots in high-demand areas and improve overall traffic congestion. Other Canadian cities (e.g., Vancouver, Edmonton) have installed smart meters and express interest in introducing dynamic parking pricing.

4.4 Designing congestion-pricing policy: evaluation of trade-offs

Evaluation of trade-offs in choosing congestion pricing can be aided by first specifying the policy design, and then using criteria to assess the options. This section explores policy trade-offs with three policy design choices that are assessed using four evaluation criteria. The design choices are

- **Coverage** refers to the scope of the congestion price. Different policy tools apply a congestion price more or less broadly across drivers and trips. A policy that has broad coverage means that the type of congestion pricing applies to more roads or bridges, a larger geographic area, more vehicles. Thus, a toll that applies only to a single road such as 407 ETR or bridge such as the Port Mann is relatively narrow (and even narrower if it applies only to individual lanes on that road). Parking policies such as in San Francisco and Calgary are narrow in a different way, given that they affect only the subset of drivers looking to park in a given area, and not those passing through the area toward some other destination, and also because some people have access to private free parking (so may be unaffected by parking pricing policies). Zone-based approaches tend to be broader, applying to driving in specific areas or across given boundaries. Distance-traveled policies can be broadest of all given that they could apply to all eligible drivers in a region based on the distance they travel.
- **Pricing** refers to the variability of rates across time and space. Tolls for a bridge, road, or zone might be the same at all times or could be higher during peak traffic times. Distance-traveled pricing could vary depending on the class of road, current congestion conditions, with higher prices when and where more congestion exists. The extent of price variability is a design choice for all kinds of pricing tools. Toll rates, the price of accessing a congestion zone, or even parking prices can remain flat or can vary across time, both in a predictable, structured way depending on time of day, or in response to changes in demand.
- **Revenue use** How revenue generated by a congestion-pricing policy is used or "recycled" has important implications for the policy's overall performance. Different approaches—for example, reducing taxes, funding infrastructure, or

improving public transit—involve trade-offs associated with the performance and public acceptability of the policy. As the revenues are a scarce resource, policy-makers must make difficult choices about how to use them.

Four main approaches to revenue recycling help highlight the central trade-offs involved:

- Invest in alternative transportation to personal vehicles (e.g., public transit or bike lanes);
- Reduce existing taxes (e.g., fuel taxes, parking taxes, property taxes, sales, or income taxes);
- Invest in infrastructure (e.g., roads and bridges);
- Make direct transfers to households

Local factors—in particular, the state of existing transportation infrastructure—also affect trade-offs between different approaches to revenue recycling. We highlight relevant factors in the assessment below.

For each design choice the four evaluation criteria are:

- **Effectiveness** is the extent to which a policy achieves its stated objective. Congestion-pricing policies might have different—though inter-related—objectives. The focus here is on the goal of reducing congestion, paralleling Althaus *et al.* [39] who argue that keeping this primary objective at the forefront during design and implementation can provide coherence to pricing systems. We examine implications for revenue generation as a secondary objective where appropriate.
- **Net benefits** refers to the extent to which a policy achieves benefits (from lower congestion, more efficient road use, and smart revenue use) that exceed the direct costs for drivers and the costs of administering the policy.
- **Fairness** considers two kinds of equity. *Vertical equity* is the extent to which the policy avoids disproportionately affecting lower income households. *Horizontal equity* is the extent to which drivers or groups of drivers (i.e., passenger vs. commercial) are treated equally and receive benefits commensurate with the fees they pay [40].
- **Practicality** is the extent to which a policy can actually be implemented. It considers technical barriers, legal and jurisdictional constraints, and public acceptability.

Table 4.1 illustrates a summary of the assessment of the three design features using the four criteria. The design choices are shown vertically and each is assessed using the four evaluation criteria. A plus sign indicates a positive impact of a specific design choice on a given evaluation criterion, while a minus sign indicates a negative impact. Where a question mark is shown, multiple effects make the net impact uncertain. A zero indicates no likely impact on the specific criterion. N/A designates that a given design choice is not applicable to the evaluation criterion. The table illustrates the sort of trade-offs that are assessed in the application of this framework to policy design for Canadian cities. For example, a broad policy is

Table 4.1 Evaluating the trade-off in the choice of policy design

		Effectiveness	Net Benefits		Fairness		Practicality	
		Reducing congestion	**Economic efficiency**	**Administrative efficiency**	**Horizontal equity**	**Vertical equity**	**Legal/ Jurisdictional**	**Public acceptability**
Coverage: Broader Pricing		+	+	–	+	–	–	?
Pricing: Greater Price Variability		+	+	–	+	+	–	?
Revenue Use:	Public Transit	+	+	N/A	+	+	0	+
	Tax Cuts	–	+	N/A	–	?	–	+
	Infrastructure	–	+	N/A	+	–	–	0
	Direct Transfers	0	0	N/A	–	+	0	+

Source: Canada's Ecofiscal Commission [1], Table 3, p. 25.

more effective and efficient than a narrow one, but could be much more administratively complex, have challenges being a larger burden on low-income users than on high (the minus sign on vertical equity), run into legal challenges, and be harder to get community buy in. Similar issues arises the more prices vary according to the level of congestion (compared to a uniform price). Using revenues gained to invest in public transit dominates the other uses of funds in that it meets most of the criteria positively and thus has few trade-offs. Before exploring what might be done in Canadian cities to address traffic congestion, I turn to a brief consideration of principles for successful policy implementation.[4]

4.5 Principles for implementation

The design of congestion-pricing policy is most successful when customized to the circumstances of individual cities and regions. Successful policy, however, is not just a matter of *design*, but also *implementation*. How policy is rolled out can play a key role in overcoming barriers to adoption of congestion pricing. This section lays out six principles for policy implementation. Overall, they suggest that implementation is not a simple, one-time policy announcement, but instead a deliberate *process* over time.

Define objectives clearly. Policy should establish clear, explicit, and relevant objectives [39,41]. Objectives serve several purposes and set up the principles that follow. First, clear objectives focus the design choices. Congestion pricing can have multiple objectives and some design choices can support one objective (e.g., reducing congestion) while being a detriment to another (e.g., raising revenue). Second, they establish consistent messaging as to the policy's rationale, enabling clear communication with stakeholders to build acceptance. Third, they provide a benchmark against which to evaluate the policy's success. Such evaluation is essential in order to adjust the policy over time to improve performance.

Engage and communicate with stakeholders. Althaus *et al.* [39, p. 553] note "decision-makers cannot just decide in favor of a congestion charge, announce its implementation, and think reduced congestion will automatically follow. Care and timing must be paid to incorporating the right timing, communication, and consultation strategy for each jurisdiction, circumstance, political will, and community mood." In short, part of the process of implementation is engaging and communicating with those most affected by congestion pricing. See for example Chapter 5.

Interviews with Canadian mayors and transportation authorities emphasized the importance of clearly communicating the benefits of policy for individual drivers or companies in terms of personal and concrete savings. People need to see how and where the revenues are being used to benefit them. They also noted the need to frame the policy's benefits in terms of price and time: communication efforts should help drivers understand they are not paying to be able to go

[4] Many of the other chapters in this volume consider implementation factors in detail.

somewhere, but rather to get there *more quickly*. It is important for people to recognize that society *already* accepts demand-based pricing policies in many areas of the economy (e.g., utilities, movies, hotels, and airlines) and that congestion pricing is a logical extension of such practices. The precise framing used may vary depending on local circumstances such as the nature of the congestion, political factors, and community attitudes.

Phase in policy over time. Timing is generally a critical factor for successful policy implementation. Gradual implementation of a policy can help build public support [41,42]. De Palma *et al.* [43] and Verhoef *et al.* [44] highlight the importance of "implementation paths" as a sequence of policy measures, overcoming barriers to policy in stages. The crucial implication here is that policy design need not be static: it could become broader or more variable, for example, over time. Priorities for revenue recycling might also change over time. A pathway approach might optimize design for practical concerns in the short term to overcome implementation barriers but transition to more effective and cost-effective policy over time. Starting with a demonstration project or trial period is a particularly effective way to build public support, as Chapter 5 indicates. Congestion pricing can generate concrete benefits for drivers in the form of reduced traffic congestion and saved time.

Timing of public engagement also matters. Holding a referendum on congestion-pricing policy before implementation has not led to success (e.g., Manchester and Edinburgh), compared to first operating a pilot program, then holding a referendum on the policy (e.g., Stockholm). The trial period gives voters more complete information on the system by the time of the referendum. Alternatively, make it part of your political manifesto, as Mayor Livingstone did in London.

Coordinate decision-making. Overlapping jurisdictional authority for congestion pricing can create challenges for policy implementation. Coordination between provincial and municipal governments as well as local transit authorities can be complex, though it is almost certainly necessary. Clear objectives can be harder to define. And coherent incentives across different modes of transit and different policy instruments can be hard to coordinate, leading to challenges for effective and cost-effective policy [43]. For example, congestion pricing does not necessarily require price changes for public transit, although the scope for users to substitute between driving and public transit suggests that some coordination on this front may be valuable.

Evaluate and adjust policy over time. The design and implementation of congestion-pricing policy is almost certainly an iterative process. As a result, clear processes for evaluating the performance of policy and adjusting its design over time are essential [41]. Collecting data—both before and after a policy has been implemented—is necessary for evaluating the effects of congestion pricing.

Congestion pricing is best implemented as part of a policy package. Congestion pricing is likely to have its greatest impact when accompanied by complementary nonpricing measures—for example, road and transit improvements that improve alternatives for drivers [44]. The US DoT [45, p. 3] shows how complementary actions and technologies "(1) extend the benefits of congestion-pricing strategies to those directly and indirectly affected, (2) improve public acceptance of congestion-pricing strategies, and (3) improve the value of existing travel options available to individuals." In particular, other measures can complement congestion pricing by affecting driver behavior through information or behavioral cues. For example, they can seek to influence the behavior of travelers en route, via roadside message signs updated with current traffic conditions and suggested routes, and radio traffic reporting. They can influence travelers *before* they travel via web and telephone-based traffic services. Or they can influence vehicle flow directly through responsive variable speed limits, ramp metering (for regulating the flow of traffic that enters highways), adaptive signal control, and automated vehicle location for transit [45].

Several additional complementary policies must be examined when considering congestion-pricing policies. These include changes to transit pricing, encouraging bicycling, staggering of work hours by employers, promoting teleworking, and reforming public and employer-provided parking spaces [46,47].

4.6 Options for congestion pricing in Canada's four largest cities

Policy design and implementation is even more complex than suggested by Table 4.1, given that local context in different cities also affects the trade-offs between design choices. What do the trade-offs mean for a city considering congestion pricing in practical terms? How should a city's local context be translated into policy design that is effective, cost-effective, fair, and practical? Identifying detailed, customized solutions for different cities is outside the scope of this chapter. Yet based on the analysis and principles examined above, and information from interviews with local experts, this section provides some practical policy suggestions for Canada's four largest cities—which will of course apply to many other cities.

To explore differences across Canada, Canada's Ecofiscal Commission interviewed key stakeholders in different parts of the country to gain insight on how to move forward with practical and effective congestion pricing in Canada. This section also draws upon findings from new research gauging public opinion on traffic congestion and the role of congestion pricing in alleviating it [48] and reaction to and events following the Ecofiscal Commission's recommendations in its report [1].

The interviews help to focus on similarities and differences in Canada's four biggest cities: Metro Vancouver, Calgary, Greater Toronto, and Greater Montreal. Lessons also emerge for smaller Canadian cities—and indeed for cities worldwide.

Four main themes emerge that help to define the barriers and opportunities faced by different cities: *geography, infrastructure, governance and institutions* and *attitudes and culture*. While Vancouver, Calgary, Toronto, and Montreal have much in common as they work to combat rising traffic congestion, they also have important differences along these dimensions.

The local context of each city and the design framework developed above provides an outline for a congestion-pricing proposal for each of the four cities. Following Lindsey [49], the focus is on how congestion pricing "might be introduced in the next few years using established technologies." While these proposals follow logically from the analysis of local context and design trade-offs, other options may also be suitable. These are intended only as initial policy proposals to begin more detailed policy conversations in each city.

4.6.1 Metro Vancouver

Metro Vancouver's transportation network is at a turning point. The region suffers from one of the worst congestion problems in the country and has an overburdened public transit system. The 2015 referendum failed to establish new funding mechanisms for road and transit projects, leaving the region's transportation system in an uncertain state. While recent announcements by the region's mayors and transportation authority have supported congestion pricing (known locally as mobility pricing), many barriers remain.

The geography of Metro Vancouver is constrained and defined by northern mountains, the US border, the Pacific Ocean, and many bridges across rivers and inlets as illustrated in Figure 4.1. In terms of land-use and travel patterns, the region is considered *polycentric*, meaning that the network of regional city centers results in "everywhere to everywhere" travel, as opposed to travel directed primarily to and from a single downtown core. Metro Vancouver is also home to the busiest port in

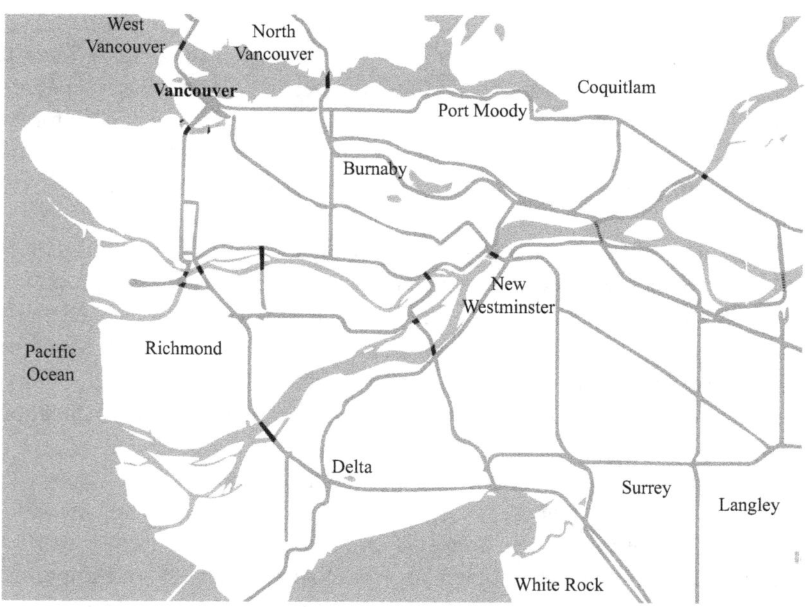

Figure 4.1 Greater Vancouver Bridges. Note: Untolled bridges are indicated in black. Two bridges shown with dashes were tolled until 1 September 2017

Canada with a growing volume of container truck traffic [50]. Growth in container volumes is projected to increase container truck traffic by 300% by 2020 [51].

Congestion in Metro Vancouver is largely a function of population growth and transportation infrastructure. Despite significant improvements and expansions, the capacity of the region's roads, bridges, and public transit has been unable to keep pace with the mobility needs of a growing population, which is expected to increase by approximately one million residents by 2041 [52]. Bridge and tunnel crossings are an essential component of the region's transportation network. A total of 20 bridges and 2 tunnels serve as corridors between municipalities, such as the crossings between Vancouver and North Vancouver, Richmond and Delta, and Burnaby and Surrey. The major crossings are shown in Figure 4.1 with black shading. Only two bridges were tolled—the Port Mann and Golden Ears—but these tolls were abolished on 1st September 2017. As there are few high-volume roads, congestion or an accident will turn large areas, such as North and West Vancouver, into gridlock. The main downtown peninsula is also separated by water bridges, a sea bus crossing, and several roads that access the peninsula from the east.

Metro Vancouver's governance and institutional framework for transportation policy and operations is very complex. The region comprises 21 municipalities, 1 electoral district (encompassing several unincorporated areas), 1 treaty First Nation, and the regional government of Metro Vancouver. But unlike many Canadian cities, the regional transit operator (TransLink) has significant authority over the region's major roads and public transit infrastructure.

Culture and attitudes toward driving in Metro Vancouver are also complex. According to a recent public-opinion survey, 70% of Metro Vancouver residents reported that current traffic congestion hinders their mobility [48]. Yet its residents decidedly voted no in the 2015 plebiscite on whether to increase the sales tax to help fund the mayors' council's $7.5-billion package of infrastructure and transit improvements. The same survey results also suggest that most referendum voters did not mention traffic congestion as a major factor in their decision to vote yes or no.[5] Regional variations are an important factor: while residents of the downtown core of the City of Vancouver have access to multimodal transportation, including walking, cycling, and transit infrastructure; many residents in surrounding municipalities are heavily car dependent and travel is not all periphery to city centers but widely dispersed around the region. While TransLink has a well-developed network with bus/Skytrain connectivity and express runs during peak periods, it is simply too expensive to provide frequent service to low-density regions.

[5] One should interpret the May 2014 referendum results with caution as there were many factors at play in the "no" vote that include dissatisfaction with TransLink, the regional transportation authority, the regional mayors' opposition to holding a referendum, the lack of strong support for the referendum by the provincial government, and history in the province of previous failed referenda. The referendum question was "Do you support a new 0.5% Metro Vancouver Congestion Improvement Tax, to be dedicated to the Mayors' Transportation and Transit Plan?" While the tax was framed as a congestion tax, there was nothing in the transportation plan about congestion other than noting it was a problem, the increase in the sales tax did not link to congestion pricing, and as noted, voters did not find congestion a factor in their decision of how to vote.

Proposed Vancouver policy: variable bridge tolls. Considering the region's constrained geography, polycentric travel patterns, and complex governance structure, applying variable pricing to the region's bridges and tunnels is a practical option to reduce regional congestion. This option has been suggested for over a decade, and ranks highly in a report from the Joint Technical Committee, with representatives from the British Columbia Ministry of Transportation and Infrastructure, TransLink, and the municipalities of Surrey and Vancouver [52,53].

Tolling the region's bridges and tunnels could be an effective approach for several reasons. Due to the high number of water crossings and a lack of land-based alternatives, tolls on the region's water crossings would intercept a large portion of traffic and would be expected to reduce congestion and its subsequent air pollutants and greenhouse gas emissions [49,54]. As reinforced by an interview with members of the BC Ministry of Transportation, the provincial government would need to change its existing policy on tolling, which limits the application of tolls to cover infrastructure costs for projects that increase capacity and have a reasonable non-priced alternative [54,55]. Variable pricing in response to congestion levels has been shown in other cities to be effective in reducing traffic in peak congestion periods and reducing bottlenecks.

Vancouver could pilot a program similar to that proposed for New York City [56]. New York has many similarities to Vancouver geographically and in their projection of a million more people moving into the region. Their current system, like that in Vancouver, has some bridges with tolls (with some tolls rates much higher than those on Vancouver's two tolled bridges); others are free. The same sort of distortion in traffic flows and increased congestion occurs because people divert to untolled bridges to avoid paying a toll. The proposal is to differentially toll bridges as a proxy for a congestion charge and to recognize the extent of public transit. Tolls would vary by bridge (but with a constant price per bridge, i.e., not dynamically priced over the course of the day). Bridges with good public transit alternatives would have higher tolls than those without. A pilot of sufficient length in Vancouver would allow TransLink and the Ministry of Transportation and Infrastructure to see how traffic would flow without the distortions created by the presence of tolled and untolled bridges and adjust tolls as need. A plan of this sort does require, as noted, coordination between the provincial government, TransLink, and the municipalities as different bridges in the region come under the jurisdiction of all three. This too is similar to New York's differing "ownership" of their bridges.

The region's geography, governance structure, and commuting patterns make other approaches to congestion pricing currently impractical. Unlike Toronto and Calgary, Metro Vancouver lacks long stretches of controlled-access highways and has limited HOV lanes. As argued by Lindsey [49], this makes HOT lanes an ineffective option for Metro Vancouver. Ideally, a comprehensive distance-based system is a plausible long-term option for the region. Costs of the technology should decrease over time as more regions adopt the approach. Distance-based pricing would charge all drivers within Metro Vancouver a per-kilometer fee that could vary by day, time, and even the emissions intensity of the vehicle.

Of all available pricing options, a comprehensive distance-based system is viewed as the most equitable, because each driver pays directly for the time and use of the roads, no matter where they live in the region [54]. However, unlike bridge tolling, a comprehensive distance-based option has not yet been implemented anywhere in the world. Singapore plans to implement a system in 2019–20. The existing distance-based trials program in Oregon and California (see Chapter 18) are similar; however, in those pilot programs, drivers have the ability to opt in to the program, and the primary objective is to replace existing gasoline taxes to provide revenue for the road system, rather than to reduce congestion. There are plans to have a system in place that includes Oregon, California, and Washington (see Chapter 18). A comprehensive distance-based system that specifically targets congestion in Metro Vancouver could provide an option similar to that in the US pilots where vehicle owners can choose between distance-based pricing or pay a fixed but high charge annually in lieu of distance-based pricing. A distance-based system requires each vehicle to be fitted with an on-board unit using GPS technology, and would necessitate tracking the time and distance of travel from each vehicle and communicating this information to the pricing authority. Even though people commonly carry electronic devices that collect and store information regarding their location, privacy concerns are expected to be a major point of contention in implementing distance-based pricing. The Oregon pilot showed however that once people are familiar with the system and see how it protects their privacy, their skepticism turned into support, allowing the legislation for a distance-based charge to pass state legislatures (see Chapter 18).

In April 2016, Metro Vancouver's mayors delivered to the province their *10-Year Metro Vancouver Transit & Transportation Plan* [57] that includes a request that the province establish a multistakeholder commission to develop an approach to regional mobility pricing that can be implemented within 5–6 years. Support for mobility has been expressed by BC's Chamber of Commerce, various regional Boards of Trade, and other stakeholders.

4.6.2 Calgary

Road congestion in Calgary is shaped by its unbounded geography and low population density as illustrated in Figure 4.2. Unlike most cities in Canada, the city limits of Calgary are unrestrained by natural or man-made barriers, allowing the city to expand outward. With a large land base, the historical direction of Calgary's development encouraged the expansion of suburbs while fostering a dense downtown core. With such dispersed geography, road and rapid-transit infrastructure stand out in Calgary. Calgary has significantly increased its funding for transportation infrastructure in the past decade, and public transit has received a large portion. Yet even with consistent year-on-year ridership growth, Calgary lags behind Toronto and Montreal in the number of transit trips per capita and the number of residents living within 1 km of rapid transit. Residents rely on the extensive network of freeways and major arteries that crisscross the city. One major network is the Calgary ring road, which is a series of provincially owned highways

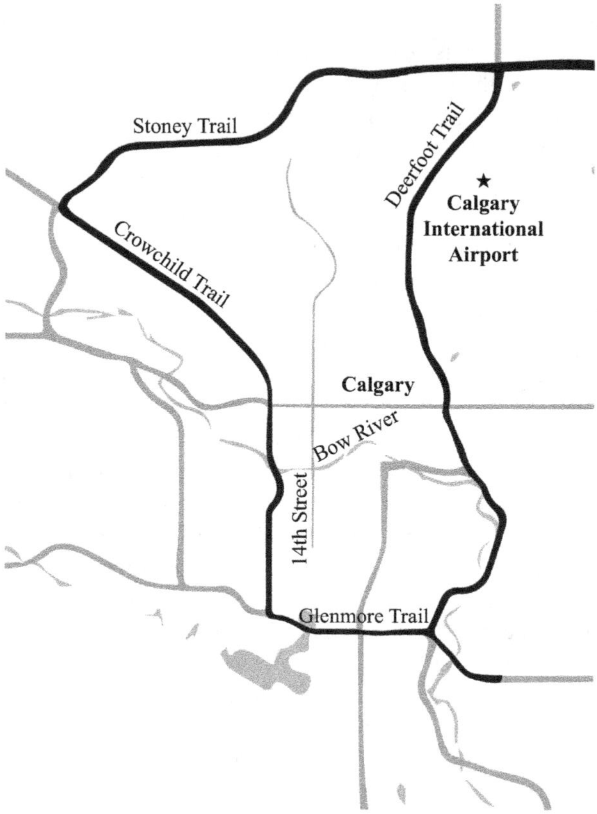

Figure 4.2 Calgary Highways with HOT Lane Potential

that run around the city's edge. The second major network is the municipally owned "skeletal road network," comprising a series of freeways and major arterial roads designed to provide free-flowing transportation through and around the city [58].

In terms of governance and institutions, Calgary, like other Canadian cities, contains overlapping jurisdictions with the provincial and municipal governments. The Alberta provincial government is responsible for planning the region's transportation network and is directly responsible for the maintenance of provincial highways. Previous provincial governments have considered tolling only *new* highways. A new government, elected in 2015, has also rejected the idea of tolling to raise revenue but has not directly addressed congestion issues.

Increasing the capacity for HOVs is a major planning priority for Calgary. Outlined in its 2009 Transportation Plan, the city developed and proposed the Primary HOV Network, which includes plans to build over 440 lane-km of HOV lanes for Calgary's biggest corridors over the next 10–60 years [59]. Interviewees from the region noted that driving is deeply engrained in the culture and attitudes of most Calgarians. Despite ongoing efforts to build multimodal transportation alternatives, driving is still the preferred mode of transportation. Approximately 77% of Calgary's population commute to work by car or truck, compared to roughly 70% in Montreal, Metropolitan Vancouver, Toronto, and Ottawa-Gatineau [60]. And 80% of those commuting car trips are made by single-occupant drivers [61]. In addition, since Albertans have for many years paid the lowest provincial tax rates in the country, Calgarians may be more averse to road pricing than residents of other Canadian cities [62].

Proposed Calgary policy: HOT lanes. Calgary's low density and lack of public and provincial support for road pricing make HOT lanes a very practical policy option. Based on the extensive network of freeways in Calgary, a number of recent reports recommend HOT lanes as the best option to increase time savings for commuters, with relatively low implementation costs [63,64]. A recent report by the city evaluated 16 different revenue-generating options over several criteria (e.g., efficiency, equity, sustainability, etc.) and HOT lanes were recommended as one of the best ways to reduce congestion, generate modest revenue, and increase the efficiency of the transportation network [38]. A similar recent report identifies five major roadways as suitable for HOT lanes: the Crowchild Trail, Glenmore Trail, Deerfoot Trail, 14 Street South, and Stoney Trail [64] as shown in Figure 4.2. Congestion pricing may also complement Calgary's ongoing efforts to expand public-transit capacity, influencing residents to make better use of the growing transit network.

Due to its low density, high levels of car use, and moderate use of public transit, broad congestion pricing may not actually have a significant effect on Calgarians' driving behavior. And because the city does not have a clear natural cordon around the downtown area, establishing a zone-based congestion-pricing system would likely be challenging [64]. The existing legislative context in Alberta prohibits the use of road tolls. In order to implement HOT lanes in Calgary, the provincial government would thus need to coordinate with the City of Calgary (and potentially Edmonton) and alter legislation to allow an effective pricing scheme. Passing legislation in advance of any specific tolling plans might be advisable, learning from the experience cities such as New York, where opposition to congestion tolling proposals allowed a few state legislators to block passage of legislation. Another important consideration is public perception. Not only is driving deeply embedded in Calgary's culture; residents also have very limited experience paying directly for road use. As such, building HOT lanes is the most pragmatic pricing option for the Calgary context, as it maintains personal choice for drivers while still tackling the growing problem of traffic congestion.

4.6.3 Greater Toronto Area

As Canada's largest city, the GTA likely has the country's biggest problem with traffic congestion. In a 2015 survey of Toronto residents, 89% of respondents expressed serious concerns about traffic—the most of any city considered in the survey done for the Ecofiscal Commission [48]. In terms of geography, the GTA is polycentric, with multiple hubs of activity and diffused congestion problems with drivers traveling between the various hubs in different directions. Compared to Metro Vancouver, the GTA is relatively unbounded (except on the south by Lake Ontario). Greater Toronto's transportation system relies heavily on the highly congested 400-series freeway network, which includes Highway 407, the tolled expressway with dynamic pricing discussed above as illustrated in Figure 4.3. Congestion is widespread across all types of roads, but highways are critical arteries for commuting throughout the densely populated region.

Two kinds of infrastructure stand out in Toronto: HOV lanes and transit capacity. A 2007 report from the Government of Ontario laid out plans for a 450-km network of HOV lanes to be constructed by 2031, and to be built with new capacity rather than by converting existing lanes [65]. The same report also included a commitment to convert selected HOV lanes into HOT lanes. Following the implementation of a temporary expanded HOV network during the 2015 Pan-Am Games, the Government of Ontario announced its intention to implement HOT lanes in the GTA, renewing and updating the 2013 provincial budget commitment [66]. As discussed above, the first HOT land pilot program began in the GTA in September 2016.

The GTA's phenomenal growth in population has not been matched by an equivalent expansion in public transit infrastructure, which is uncommon for cities in the developed world. The GTA's public-transit systems are highly over-burdened: the region has the greatest number of annual transit trips per capita but

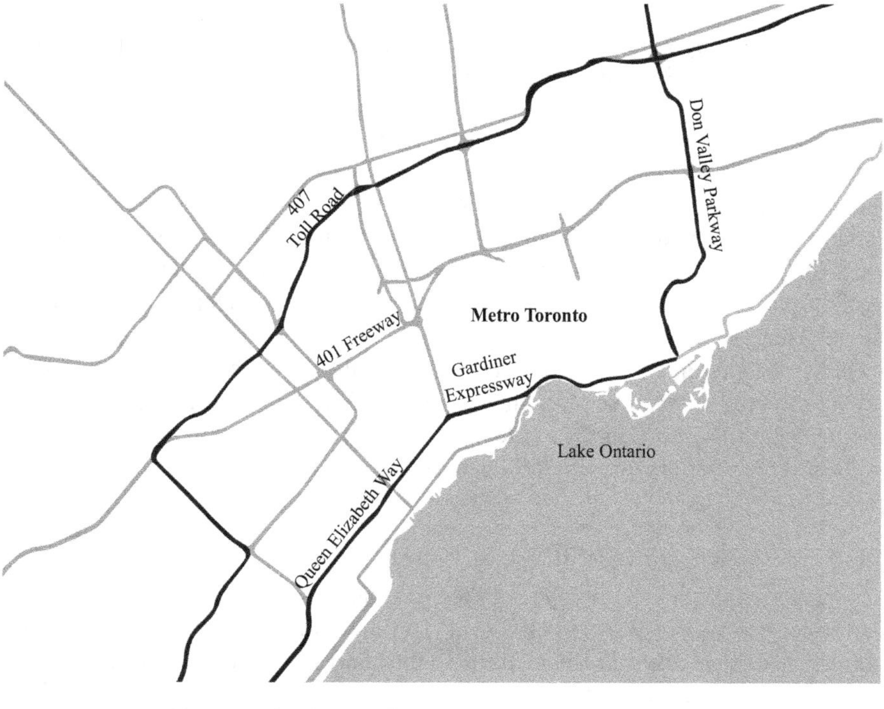

Figure 4.3 Major Expressways in Metro Toronto

the second-lowest capacity per capita (as measured by length of lines), as documented by Singer and Burda [67].

As with other Canadian provinces, overlapping jurisdictions covering transportation policy in Ontario makes achieving coherence and consistency challenging. The Province of Ontario owns the 400-series of highways. Metrolinx, an agency of the provincial government, manages and integrates road transport and public transit in the Greater Toronto and Hamilton Area. While Metrolinx is legally empowered to levy tolls on local roads, it would still require a special regulation to do so, and the provincial government would need to be involved [49]. However, as discussed below, the City of Toronto announced in November 2016 its intention to toll some of the freeways lying fully within its jurisdiction.

As in Calgary, driving is deeply embedded in the attitudes and culture of GTA residents, and concerns about traffic congestion are extremely high. In 2013, a survey on funding tools found moderate support for congestion pricing such as HOT lanes (with 49% support) and central core charges (46%), and wide support for dedicated use of these revenues to fund the GTA's needed transit expansion (92%) [68]. As in Metro Vancouver, suburban drivers in the GTA rely more on driving, with less access to alternative modes of transit. Nevertheless, the existence of the well-used private 407 ETR suggests a rising degree of acceptance for congestion pricing in the region.

Proposed Toronto policy: toll routes and HOT lanes. The Ecofiscal report on which this chapter is based recommended two policies for the GTA: HOT lanes and toll routes. Since its publication, both have been announced and one HOT lane pilot program operates on the Queen Elizabeth Way (see Figure 4.3).[6] The Commission's arguments for HOT lanes were as follows. Converting HOV lanes to HOT lanes or building new HOT lane capacity on the provincially owned 400-series of highways is a practical option to price and reduce congestion in the area. In addition, using a pricing system which responds dynamically based on demand could help to improve vehicle flow on these key transportation corridors, with implications for traffic throughout the city. Lindsey [49] and Dachis [69] both propose the conversion of provincially owned HOV lanes to HOT lanes. This previous research, along with knowledge from our interviews with city experts, suggests that this is a good near-term option for incorporating congestion pricing into a mobility strategy for the GTA region. It can be done as a trial and can be seen as a first step, with more comprehensive pricing introduced as population and congestion continue to grow. This is indeed what has happened, except that the HOT lanes will have flat rates rather than dynamic pricing—a missed opportunity to fully see the impact of the pilot on congestion.

HOT lanes on provincial highways likely provide the most straightforward opportunity to bring congestion pricing into the GTA's congestion-reduction strategy [69]. Given Toronto's geography, a cordon is impractical, and a distance-traveled

[6] Canada's Ecofiscal Commission would like to take credit for seeing its recommendations adopted, but undoubtedly there were other factors influencing the decision beyond the Commission's well publicized report and follow-up opinion pieces.

pricing approach is both technically and politically challenging. Further, if the Province of Ontario continues to expand its network of HOV lanes, in line with its 2007 plans, permitting them to be built as HOT lanes instead would likely be a superior choice, as HOT lanes enable highways to be used by more people [70]. Focusing on provincial roads exclusively could reduce coordination challenges that would come from tolling both local and provincial roads. As well, HOT lanes respond to concerns regarding fairness, as unpriced options remain available.

The Ecofiscal Commission also recommended tolling on the major freeway system as a longer term solution, again using dynamic pricing. In November 2016, the mayor of the City of Toronto announced his support for tolling the two major expressways into Toronto's downtown core—the Don Valley Parkway and Gardiner Expressway (see Figure 4.3). Toronto City Council endorsed the plan in mid-December 2016 and issued a formal request that the Province of Ontario approve the tolling plan. The early design details indicated that the toll would be a flat \$2 per trip with no dynamic pricing [71]. If so, this is a huge missed opportunity (as is not making Toronto's HOT lanes dynamically priced). The flat rate will simply raise revenue, not affect congestion other than perhaps to drive congestion to city streets—a problem seen in cities such as Vancouver when the tolling of the Port Mann Bridge created an incentive for drivers to avoid the toll by taking another bridge (the Pattullo) which only added to its congestion. The toll should also be distance based as are the tolls on the 407 ETR. The mayor's motivation for tolling was however to raise revenue for the city, estimated at \$200 million annually [72], but the tolls may take up to 5 years to be implemented. This again represents missed opportunity if it is the case. A report by the City of Toronto staff in 2015 said electronic tolling could cost up to \$50 million to install and \$30 million per year to operate and a \$3 flat toll that diverts traffic to other roads and transit would produce a modest positive effect on travel on the two tolled expressways [72]. There has been a mixture of support and opposition to the proposal. Environmental groups and the Toronto Board of Trade are supportive, while antitax groups and the Ontario Trucking Association are opposed, as are some of the region's suburban mayors and Toronto Councillors [72]. However on January 27, 2017, Premier Wynne of Ontario rejected the tolling proposal [73]. Tolling thus appears to be in abeyance for now.

4.6.4 Greater Montreal

Congestion is widespread throughout the Island of Montreal and on connectors to surrounding off-island areas, which together make up the Greater Montreal region. Institutions governing regional transportation are in flux, and an aging road network—and especially the upcoming replacement of a major Interchange (the Turcot) and the federally owned Champlain Bridge—create some uncertainty for the future of the regional transportation system, and for moving forward with congestion-pricing policies. For example, when replacement of the Champlain Bridge was announced by the federal government under Stephen Harper, the Prime Minister announced that the bridge would be tolled. The Trudeau government's

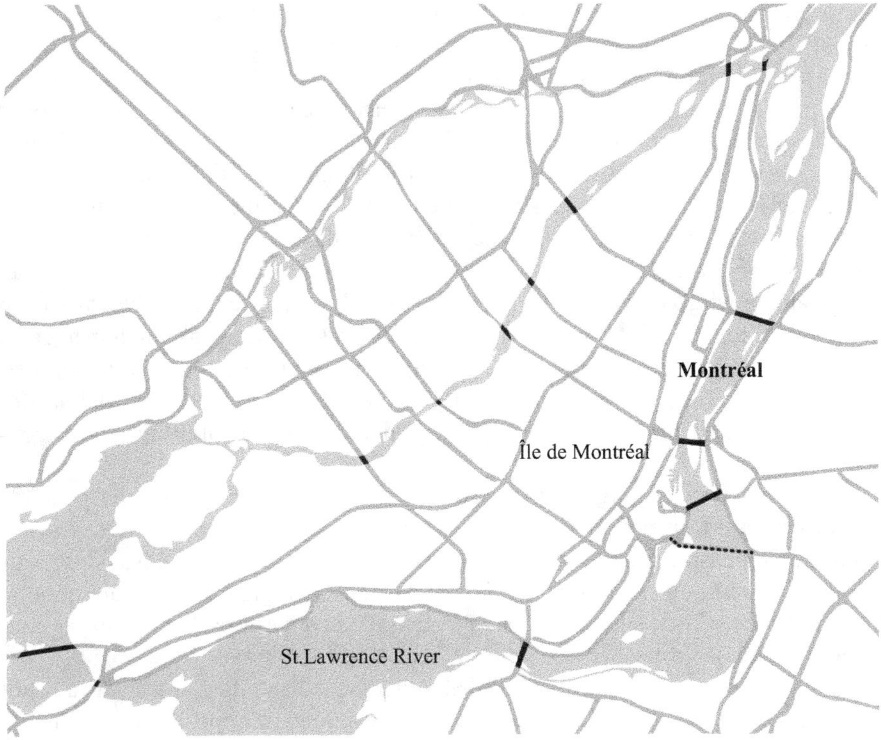

Figure 4.4 Greater Montreal. Note: Major bridges are shown in black. The Champlain bridge is indicated by dashes

campaign pledge in Canada's 2015 election was no tolls for the new bridge. As of July 2016, the federal Infrastructure minister was adhering to the commitment for no tolls [74].

Two aspects of geography are notable in Greater Montreal: the natural cordon formed by the Island of Montreal, and extensive commuting on and off the island as illustrated in Figure 4.4. While there are multiple centers of activity, the island boundary and its constraints have led to development that is less polycentric than in Metro Vancouver or the GTA. The island is connected to the mainland and other surrounding islands by 18 bridges (one of which is partially a tunnel, and some of which include rail lines and pedestrian or bike routes). Of the 18 bridges, 3 are federally owned.

Montreal's transportation infrastructure stands out from other major Canadian cities, with two key factors shaping the regional transportation system and efforts to manage congestion. The public transit system is relatively good, but the urgent need to replace aging road infrastructure has generated major funding challenges, as well as difficult-to-predict congestion due to ongoing repair and maintenance. A member of a government agency in Montreal noted that upgrading the city's aging infrastructure is indeed a high priority. Two bridges in the region are tolled.

Both have private-sector involvement and both aim to provide good vehicle flow, but only one features rates that vary by time of day.

Greater Montreal is considering governance and institutional changes to centralize public transit and urban planning decision-making. Under the proposed plan, the provincial government would abolish the existing Agence métropolitaine de transport, replacing it with two new institutions. The first would integrate several off-island transit agencies into a new group called the Réseau de transports métropolitain. The second, the Agence régionale de transport, would be responsible for long-term planning and would include mayors from the metropolitan region along with other experts [75]. Historically, Montreal's policy has been aligned with its objectives, for example, with the regional gas tax implemented to fund transit infrastructure. But more recently, governance and financing issues, with inconsistencies between governments and a lack of integrated visions for transportation and transit, have exacerbated political and public barriers to congestion policy.

Challenges with aging road infrastructure permeate the culture and people's attitudes toward transportation in Greater Montreal. Interviewees in Montreal, for example, emphasized that the city's deteriorating infrastructure poses a serious problem for transport of both goods and people and is a high priority locally, noting that this concern may take precedence over traffic congestion. Nonrecurrent congestion (which is costly because of time delays and low travel–time reliability) is perhaps more of an issue in Montreal than other Canadian cities, mostly as a result of the persistent and widespread road repairs.[7] Despite local opposition to tolling on the new Champlain Bridge, there are tolls on two new bridges on Autoroutes 25 and 30 (with one using time-varying pricing) and both bridges are well used. Thus with the region's modest history of tolling (some bridge tolls existed until 1990), current acceptance of tolling may be reasonable.[8]

Proposed Montreal policy: variable bridge tolls. The natural cordon formed by the Island of Montreal is an opportunity to relatively easily implement congestion pricing in the metropolitan region. Reducing congestion on the surrounding bridges, which are key choke points for traffic, can lead to reduced congestion on the island, in the dense downtown core, and also in surrounding areas off-island. Targeting the pricing policy to congested areas via the limited number of the island's access points makes area-based pricing more straightforward than in Toronto or Calgary as illustrated in Figure 4.4. Moreover, Montreal is less polycentric than Metro Vancouver, with the downtown core a major hub for business. Significant traffic comes from commuters from off-island driving in for work. Customizing breadth of policy to this geographic context could thus target congestion zones with relatively low administrative costs. As with our proposal for Vancouver,

[7] See Chapter 20 for more on this issue in other cities.

[8] Denis Coderre, Mayor of Montreal, announced in November 2016 that he was not opposed to tolls on roads and bridges if the revenues were used to fund public transit [76]. His opposition to tolls on the Champlain Bridge was because the revenue would be used to pay off the bridge's debt.

tolls that vary by time of day, or even dynamically in response to demand would be most effective for reducing congestion while minimizing the costs of the policy for drivers.

Concerns around the city's aging road infrastructure—and the associated delays of ongoing construction—could have implications for the timing and acceptability of implementing congestion pricing. Residents may be more supportive of congestion pricing if infrastructure problems are also addressed. This could suggest infrastructure investments as an appropriate approach to revenue recycling in the Montreal context.

Recent research on pricing road access for Greater Montreal has led to mixed recommendations. Lindsey [49] similarly proposes a cordon charge around the Island of Montreal, noting how such a zone-based scheme could effectively target the most significant congestion issues. Research conducted as part of a 2014 Quebec fiscal review is also in support of tolling the network of 18 bridges around the Island of Montreal as a medium-term option, indicating that the province should work with the federal government on harmonizing tolls on the three federally owned bridges [77].

On the other hand, Boulenger *et al.* [78] and the Institut de développement urbain du Québec [79] consider a range of direct road-use pricing options and have come out in favor of distance-traveled pricing, which could also effectively target congestion in the region. However, both studies have approached the subject primarily from the point of view of revenue generation for financing infrastructure, as opposed to congestion reduction.

The City of Montreal's 2007 and 2008 transportation plans have expressed support for user-pay charges, and considered the possibility of tolls on bridges around the island, although such plans do not appear to be on current agendas. Still, one of the bridges already has a toll (on Autoroute 25). But as noted, there are currently no plans to toll the new Champlain Bridge (see Figure 4.4).

4.6.5 Congestion pricing could benefit other Canadian cities

Congestion can be a problem in Canada's smaller cities as well as its large ones. In Halifax, for example, regular congestion occurs on the seven roads and bridges that connect the downtown peninsula with the rest of the municipality [39]. The region has one of the lowest population densities in Canada, which results in people traveling significant distances on the region's sprawling network at peak commuting times [80]. Similarly, congestion in Ottawa–Gatineau is characterized largely by travel between its many suburban communities. Highway 417 cuts directly through Ottawa's downtown core and is often congested at peak times, along with bridges linking Ottawa and Gatineau and several major arterial roads [81]. Victoria, Edmonton, Saskatoon, Winnipeg, and Québec City are also medium-sized cities with their own isolated congestion problems that could be alleviated by well-designed congestion-pricing policies. As with the larger cities examined in this paper, each city warrants individual assessment of what policies best address its congestion issues.

4.7 Summary

This chapter assesses the costs of congestion for commuters, the economy, the environment, and for Canadian cities overall. It examined the merits of congestion-pricing policies and explored policy experience in implementing such policies. Trade-offs between different design choices are noted as are principles for smart policy implementation. There are three main, overarching findings.

1. ***Traffic congestion is a growing problem in our cities, imposing significant costs on Canadians***
   Congestion on our roads and freeways leads to wasted time for commuters and goods movement. Given the importance of the movement of goods and people through our cities, this lost time translates into a less efficient economy. Congestion also affects choices about where people live, undercutting the ability of cities to attract businesses, jobs, and workers. Finally, congestion increases air pollution from vehicles with corresponding health implications for Canadians.
2. ***Congestion pricing is an essential—but missing—piece of smart transportation policy***
   Congestion pricing is an ecofiscal policy that prices road use or parking with the aim of reducing costly traffic congestion. A growing body of evidence and policy experience suggests that congestion pricing works, particularly as part of a broader policy package. When designed well, it reduces traffic congestion and creates net economic benefits both for the economy as a whole and for individual drivers.

   Despite the evidence of its potential benefits, Canada has very limited experience with congestion pricing. The traditional approach to dealing with traffic congestion has been to expand public transit and to build more roads. These policies are key parts of the transportation puzzle: they increase the overall capacity of the transportation system and can reduce congestion in the short term. But in the absence of congestion pricing, more drivers will ultimately fill this increased road capacity, and congestion will not be reduced in the long term. Moreover, the building of new road infrastructure to meet growing demand is constrained by land-use policy and increasingly stretched government budgets.

   Congestion pricing is therefore the crucial, missing piece of a broader, coordinated package of policies to create greater mobility for a growing urban population. More public transit, roads, and cycling infrastructure provide drivers with alternatives, making it easier for them to respond to the congestion price by changing their behavior. They are essential complements to congestion pricing. But without addressing the fundamental issue of misaligned incentives around free access to roads, traffic congestion in Canadian cities will only get worse.

   Despite the evidence of its potential benefits, Canada has very limited experience with congestion pricing. Lack of public understanding of the benefits from congestion pricing remains an important barrier to implementing these policies.

3. ***The design details of congestion-pricing policy matter***
Congestion pricing is not a one-size-fits-all policy solution. Different cities face different types of congestion problems, and tailoring policies to local circumstances is critical for success. Policy design includes a range of choices. Should pricing be narrowly targeted or broadly applied? That is, should it price access to some roads, to all roads, to parts of roads, or even to parking? How should the price vary? Should it be higher at times of peak traffic, or even vary dynamically in response to real-time traffic levels? How should revenue from the policy be used? Smart policy design can reduce congestion, improving efficient transportation and travel outcomes for all travelers. It can also ensure that low-income travelers are not disproportionately affected. But the specific details of effective, cost-effective, fair, and practical policy solution will vary from city to city.

4.8 Recommendations

What then, are the next steps for reducing traffic congestion in Canada? Congestion pricing is increasingly needed in Canadian cities. To demonstrate the benefits of congestion pricing and test policy design, we make four recommendations:

- ***Recommendation 1: Major Canadian cities should implement congestion-pricing pilot projects, customized to their local context***
As illustrated by the case studies in this and other chapters in the volume, trial periods for congestion pricing are low-risk policy initiatives. They can be voluntary for drivers, as in Oregon (Chapter 18). They can take place for a limited time, as in Stockholm (Chapter 14), and they can apply to a narrow scope of drivers, as in San Francisco's parking pilot. The benefits of such trials could be huge. If well-designed, they can demonstrate the concrete benefits that congestion pricing can deliver. They can also provide opportunities for learning about how well different policy designs work in different contexts, thus allowing policy design to evolve and improve over time.

  Municipalities best understand their own congestion context and should play a major role in designing pilot projects. They should design their pilot projects according to their unique policy objectives and their local geography, governance, infrastructure, and attitudes. Different trial policies are not only more likely to succeed when customized to local context, but can also provide more information to other Canadian cities regarding what works and what does not.

  The four proposals for congestion-pricing policies for each of Canada's four largest cities outlined in this chapter could form the foundation for finite, time-limited trials in each city. The details of each proposal draw on lessons that emerge from past experience with congestion pricing in other jurisdictions, take into account local context (gauged in part from interviews and polling), and consider key elements of policy design. They are not recommendations in themselves, but instead are intended as initial policy proposals to kick-start more detailed policy conversations in each city:

- **Metro Vancouver** has constrained geography bounded by mountains and ocean, polycentric travel patterns with multiple hubs of activity, and a complex governance structure with involvement from multiple municipalities and the provincial government. Applying variable pricing to each of the region's bridges can price access to key driving arteries to reduce regional congestion. A long-term solution is region-wide road pricing.
- **Calgary** has low density, a lack of familiarity with congestion pricing, and more localized congestion problems. In this context, HOT lanes could be practical to implement, provide unpriced alternatives, and reduce congestion in key locations.
- The GTA has polycentric travel patterns with drivers traveling between multiple hubs in multiple directions and relatively unconstrained geography. Converting HOV lanes to HOT lanes or building new HOT lane capacity on the provincially owned 400-series of highways—a backbone of the regional transportation network featuring the privately operated and variably tolled Highway 407—is a practical approach for reducing congestion in the area. Current pilots on a portion of their highway system will aid in the analysis of the viability of HOT lanes. Dynamic tolling of other highways is a longer term solution, one that may be tested if the tolling of the Don Valley Parkway and Gardiner Expressway occurs, and it is not a flat-rate toll.
- **Greater Montreal** has extensive commuting to and from the central Island of Montreal, relatively widespread congestion, an existing, time-varying toll on the Autoroute 25 bridge connecting the Island, and plans to replace the aging, highly used, and federally owned Champlain Bridge. The natural cordon formed by the Island provides a practical opportunity to implement variable pricing on the full array of surrounding bridges and tunnel, harmonizing tolls and reducing congestion throughout the area.

• ***Recommendation 2: Provincial governments initiate, enable, or facilitate congestion-pricing policies***

Provincial governments can play multiple roles in enabling congestion pricing. First, not all roads are municipally owned and operated. In some situations, it is provincial governments that should directly implement congestion-pricing policies. We considered approaches for Toronto, for example, that would price access to all or some lanes on the provincially owned 400-series freeways. While coordination with the municipal government would be essential, the province should implement the congestion-pricing policy.

Second, provincial governments should play a coordinating role. A key governance challenge in many urban areas (e.g., Metro Vancouver and Greater Montreal) is the diverse collection of municipalities with highly linked and overlapping transportation corridors.

Finally, provincial governments should provide municipalities with explicit authority to implement congestion-pricing policies. The existing legal framework for implementing road pricing in Canadian municipalities is unclear and is complicated by overlapping jurisdictions. Generally, most municipalities are unable to implement broad congestion pricing on their own without changes to provincial policy. Provincial governments should reduce the existing ambiguity and make space for municipal policy by passing explicit legislation allowing municipalities to implement these policies.

- ***Recommendation 3: The federal government should help fund pilot projects***
  Funding for congestion-pricing pilot projects remains a barrier. Physical and digital infrastructure will be required to set up, monitor, and enforce the pricing policy during the trial period. While revenue could be generated, the scale of this revenue is uncertain, and depends on the details of how the policy is implemented. Municipalities have very limited revenue sources and could face significant financial challenges in initiating pilot projects.
  Federal funding to establish pilot projects would generate benefits for Canadians well beyond the individual municipalities involved in the project. As discussed below, evaluation of the pilot would lead to valuable lessons learned about congestion-pricing policy design and implementation that could be applied in other Canadian cities. In addition, the cross-country benefit of efficient goods movement means that the federal government has a direct interest in supporting regional congestion pricing.
  Support from the US federal government has played an important role in helping cities and states pilot and implement pricing policies. For example, Federal support helped enable the parking-pricing trial period in San Francisco and also helped finance the development of Minnesota's HOT lanes.
- ***Recommendation 4: Governments should carefully evaluate the performance of pilot projects, communicate the results broadly, and incorporate lessons learned into future mobility policies***
  The full benefits of pilot projects can only be realized if they are monitored over time, with data from before and after projects are implemented. The projects should be set up so that the impact on congestion, and also the overall administration costs, can be measured and assessed. Communicating these results to stakeholders and to the general public can generate new, city-specific information about the efficacy of congestion pricing. Demonstrating policy success can be a powerful tool for building public support.
  This data-driven evaluation of the policy should be used to inform the next steps. If the policy performs less well than anticipated, its design can be adjusted over time to respond to problems, or the policy can be terminated. If, on the other hand, the policy performs well it can be expanded more broadly. Both the benefits and the costs of the policy should inform subsequent policy decisions.
  Implementing and evaluating pilot projects is only a first step in addressing Canada's congestion problems. Yet as cities grow and congestion problems build, a starting point for smart policy is desperately needed. Demonstrating the effectiveness of congestion pricing on a small scale can create a launching pad for a transportation system that gets prices right—a transportation system

that fosters cleaner air, more liveable cities, and ensures people and goods move efficiently rather than wasting time in traffic.

References

[1] Ecofiscal Commission. (2015). 'We Can't Get There From Here: Why Pricing Traffic Congestion is Critical to Beating It'. Ottawa: Canada's Ecofiscal Commission. Available from https://ecofiscal.ca/reports/traffic/ [retrieved 7 Dec 2016].

[2] Urban Land Institute. (2013). 'When the Road Price is Right: Land Use, Tolls, and Congestion Pricing'. Available from http://www.uli.org/wp-content/uploads/2013/03/When-the-Road-Price-is-Right_web_F.pdf [retrieved 18 January 2017].

[3] Transport Canada. (2006). 'The Cost of Urban Congestion in Canada. iTRANS for Transport Canada'. Ottawa: Transport Canada. Available from http://www.adec-inc.ca/pdf/02-rapport/cong-canada-ang.pdf [retrieved 20 Jan 2017].

[4] Levy, J., Buonocore, L.J., von Stackelberg, K. (2010). 'Evaluation of the Public Health Impacts of Traffic Congestion: A Health Risk Assessment'. *Environmental Health.* 9(65). Available at https://ehjournal.biomedcentral.com/articles/10.1186/1476-069X-9-65 [retrieved 22 Jan 2017].

[5] Dachis, B. (2015). 'Tackling Traffic: The Economic Cost of Congestion in Metro Vancouver'. Toronto: C.D. Howe Institute e-Brief; 2016. Available at www.cdhowe.org/pdf/e-brief_206.pdf [retrieved 19 Jan 2017].

[6] Dachis, B. (2013) 'Cars, Congestion and Costs: A New Approach to Evaluating Government Infrastructure Investment'. Toronto: C.D. Howe Institute Commentary. Available at https://www.cdhowe.org/cars-congestion-and-costs-a-new-approach-to-evaluating-government-infrastructure-investment/22210 [retrieved 18 Jan 2017].

[7] Canadian Trucking Alliance. 'Canada Moves by Truck'. Available from http://member.cantruck.ca/iMISpublic/Content/NavigationMenu2/CTAIndustry/TruckinginCanada/default.htm [retrieved 18 Jan 2017].

[8] Holguín-Veras, J. (2008) 'Necessary Conditions for Off-hour Deliveries and the Effectiveness of Urban Freight Road Pricing and Alternative Financial Policies in Competitive Markets'. *Transportation Research Part A: Policy and Practice*. 42(2): 392–413.

[9] Roelofs, M. and Springer, M. (2007). 'An Investigation of Congestion Pricing Options for Southbound Freight at the Pacific Highway Crossing'. Border Policy Research Institute, Western Washington University. Available at http://www.wwu.edu/bpri/files/2007_Aug_Report_No_4_Congestion_Pricing.pdf [retrieved 18 Jan 2017].

[10] Arnold, J. (2014). 'Caught in Traffic: Road Congestion in Metro Vancouver and its Impact on Commercial Goods Movement'. Burnaby: Simon Fraser

University Masters Project. Available at http://summit.sfu.ca/item/14143 [retrieved 18 Jan 2017].
[11] Small, K. A. and Verhoef, E. T. (2007). *The Economics of Urban Transportation*. New York: Routledge.
[12] Canadian Broadcasting Corporation. (2015, February 6). 'Big City Mayors to Push Federal Parties for 'Robust' Urban Agendas'. Available at http://www.cbc.ca/news/canada/big-city-mayors-to-push-federal-parties-for-robust-urban-agendas-1.2945908 [retrieved 8 Dec 2016].
[13] Duranton, G. and Turner, M. (2011). 'The Fundamental Law of Road Congestion: Evidence from US Cities'. *American Economic Review*. 101(6): 2616–2652.
[14] Litman, T. (2015). 'Online TDM Encyclopedia'. Victoria: Victoria *Transport Policy Institute*. Available at http://www.vtpi.org/tdm [retrieved 8 Dec 2016].
[15] 407 International Inc. (2017). 'Light Vehicle Rate Chart'. 407 ETR. Available at https://www.407etr.com/en/tolls/rate-charts/rate-chart-light.html [retrieved 18 Jan 2017].
[16] Ontario Ministry of Transportation. (2017, January 3). 'Backgrounder, Tolling Rates on Highway 407 and Highway 412'. Ministry of Ontario. Available at https://news.ontario.ca/mto/en/2017/01/tolling-rates-on-highway-407.html [retrieved 18 Jan 2017].
[17] Gill, V. and Knowles, J. (2013). 'The Value of Travel Time and Reliability: Commuting on 407 ETR'. Ottawa: Conference Board of Canada. Available at http://www.conferenceboard.ca/topics/energy-enviro/traveltime.aspx [retrieved 8 Dec 2016].
[18] Urban Land Institute. (2013). 'When the Road Price is Right: Land Use, Tolls, and Congestion Pricing'. Available at http://www.uli.org/wp-content/uploads/2013/03/When-the-Road-Price-is-Right_web_F.pdf [retrieved 8 Dec 2016].
[19] Ontario Ministry of Transportation. 'High-Occupancy Toll (HOT) Lanes'. Available at http://www.mto.gov.on.ca/english/highway-bridges/hot-lanes.shtml [retrieved 8 Dec 2016].
[20] Buckeye, K. (2012). 'Innovations on Managed Lanes in Minnesota'. *Public Works Management and Policy*. 17(2): 152–169.
[21] Government of Minnesota. (2013). 'MnPASS Express Lanes: Use and Performance'. Minnesota Department of Transportation. Available at http://www.dot.state.mn.us/rtmc/pdf/100313mnpassperfactsheet.pdf [retrieved 8 Dec 2016].
[22] University of Minnesota. (2014). 'Study Finds Two Types of MnPASS Lanes Equally Safe'. Centre for Transportation Studies. 2014. Available at http://www.cts.umn.edu/Publications/catalyst/2014/august/mnpass [retrieved 8 Dec 2016].
[23] Cambridge Systematics. (2006). 'I-394 MnPASS Technical Evaluation'. Prepared for the Minnesota Department of Transportation. Available at

http://www.mnpass.org/pdfs/394mnpass_tech_eval.pdf [retrieved 8 Dec 2016].

[24] US Department of Transportation. Federal Highway Administration. 'Minnesota: HOT Lanes on I-394 Minneapolis'. Available at (http://www.ops.fhwa.dot.gov/congestionpricing/value_pricing/projects/involving_tolls/hot_lanes/mn_hotlanes_i394minn.htm) [retrieved 19 Jan 2017].

[25] Eisele, W., Wilner, H.T., Bolin, M.J., et al. (2006). 'Evaluating Criteria for Adapting HOV Lanes to HOT Lanes: Development and Application of HOT START Software Tool' (No. FHWA/TX-06/0-4898-1). College Station, TX: Texas Transportation Institute, the Texas A&M University System.

[26] Transport Canada. (2010). 'High Occupancy Vehicle Lanes in Canada'. Available at http://data.tc.gc.ca/archive/eng/programs/environment-utsp-hovlanescanada-886.htm [retrieved 9 Dec 2016].

[27] Munnich, L. (2008). 'Minnesota's Proposal for Urban Partnership Agreement Case Study of Political and Institutional Issues in Congestion Pricing'. Transportation Research Record. Available at http://www.hhh.umn.edu/centers/slp/transportation/congestion_pricing/pdf/MinnesotasProposalforUrbanPartnershipAgreement.pdf [retrieved 9 Dec 2016].

[28] Samdahl, D., Swisher, M., Symoun, J., and Lisska, W. (2013). 'Congestion Pricing: A Primer on Institutional Issues' *(No. FHWA-HOP-13-034)*. Washington DC, USA: Federal Highway Administration. Available at http://ops.fhwa.dot.gov/publications/fhwahop13034/fhwahop13034.pdf [retrieved 9 Dec 2016].

[29] Toll Collect GmbH. (2015). 'Toll Rates'. Available at https://www.toll-collect.de/en/toll_collect/rund_um_die_maut/maut_tarife/maut_tarife_neu.html [retrieved 9 Dec 2016].

[30] European Commission. (2015). 'Road Infrastructure Charging: Heavy Goods Vehicles'. Mobility and Transport. Available at http://ec.europa.eu/transport/modes/road/road_charging/charging_hgv_en.htm [retrieved 9 Dec 2016].

[31] San Francisco Metropolitan Transportation Authority (SFMTA). (2011). 'SFpark: Putting Theory Into Practice'. Available at http://sfpark.org/wp-content/uploads/2011/09/sfpark_aug2011projsummary_web-2.pdf [retrieved 9 Dec 2016].

[32] Bialick, A. (2015, April 17). 'All Meters Now SFpark-Ready—More Demand-Based Parking Pricing to Come'. Streetsblog, SF. Available at http://sf.streetsblog.org/2015/04/17/all-meters-now-sfpark-ready-more-demand-based-parking-pricing-to-come/ [retrieved 9 Dec 2016].

[33] San Francisco Metropolitan Transportation Authority (SFMTA). (2014). 'SFpark Pilot Project Evaluation Summary'. Available at http://sfpark.org/resources/docs_evalsummary [retrieved 9 Dec 2016].

[34] Millard-Ball, A., Weinberger, R. R., and Hampshire, R. C. (2014). 'Is the Curb 80% Full or 20% Empty? Assessing the Impacts of San Francisco's Parking Pricing Experiment'. *Transportation Research Part A: Policy and Practice*. 63: 76–92.

[35] Chatman, D. and Manville, M. (2014). 'Theory versus Implementation in Congestion-Priced Parking: An Evaluation of SFpark, 2011–2012'. *Research in Transportation Economics*. 44(June): 52–60.

[36] Manville, M. (2014, December). 'Battle Won, War Lost? Parking Reform and its Backlash in San Francisco'. *Transportation Research Board 94th Annual Meeting Compendium of Papers*. Paper #15–1932.

[37] Calgary Parking Authority. (2015). Calgary Parking, Parkplus System. 'Street Parking Rates'. Available at https://www.calgaryparking.com/web/guest/onstreetrate [retrieved 9 Dec 2016].

[38] City of Calgary. (2015). 'Comprehensive Analysis of Shortlisted Funding Mechanisms'. Calgary. Prepared by AECOM.

[39] Althaus, C., Tedds, L. M., and McAvoy, A. (2011). 'The feasibility of Implementing a Congestion Charge on the Halifax Peninsula: Filling the "Missing Link" of Implementation'. *Canadian Public Policy*. 37(4): 541–561.

[40] Litman, T. (2002). 'Evaluating Transportation Equity'. *World Transport Policy and Practice*. 8(2): 50–65.

[41] Eliasson, J. (2010). 'So You're Considering Introducing Congestion Charging? Here's What you Need to Know: a FAQ Based on Stockholm's Experiences'. *ITF/OECD Round Table, 2010*, OECD: Paris (No. 2010-4).

[42] Lindsey, R. (2007). 'Congestion Relief: Assessing the Case for Road Tolls in Canada'. Toronto: C.D. Howe Institute Commentary. Available at http://www.cdhowe.org/pdf/commentary_248.pdf [retrieved 10 Dec 2016].

[43] de Palma, A., Lindsey, R., and Proost, S. (2006). 'Research Challenges in Modelling Urban Road Pricing: an Overview'. *Transport Policy*. 13(2): 97–105.

[44] Verhoef, E. T., Lindsey, C. R., Niskanen, E., *et al.* (2008). 'Implementation Paths for Marginal Cost-Based Pricing in Urban Transport: Theoretical Considerations and Case Study Results'. In Jensen-Butler, C. Sloth, B., Larsen, M.M., Madsen, B. and Nielsen, O.A., eds. *Road Pricing, the Economy and the Environment* (pp. 49–78). Berlin: Springer-Verlag Berlin Heidelberg.

[45] U.S. Department of Transportation. (2008). Technologies that Enable Congestion Pricing. Available at http://ops.fhwa.dot.gov/publications/fhwahop08042/fhwahop08042.pdf [retrieved 10 Dec 2016].

[46] Arnott, R., Rave, T., and Schöb, R. (2005). *Alleviating Urban Traffic Congestion*. Cambridge: MIT Press Books.

[47] Shoup, D. C. (2005). The High Cost of Free Parking (Vol. 206). Chicago: Planners Press.

[48] Abacus Data. (2015). Urban Congestion—Public Opinion on Possible Solutions—Canada's 4 Largest Urban Markets—Conducted for the Ecofiscal Commission.

[49] Lindsey, R. (2008). 'Prospects for Urban Road Pricing in Canada'. In Burtless, G. and Rothenberg Pack, J., eds. *Brookings Wharton Papers on Urban Affairs. Washington: The Brookings Institution*. pp. 235–293.

[50] Port Metro Vancouver. (2013). 'About the Port: Learn About the Port'. Available at http://www.portmetrovancouver.com/about.aspx [retrieved 10 Dec 2016].

[51] Lam, J.K. and Kitasaka, K. (2009). 'Smart Commercial Corridors'. Prepared for the Annual Conference of the Transportation Association of Canada. Vancouver: TransLink and Delcan Corporation. Available at http://conf.tac-atc.ca/english/resourcecentre/readingroom/conference/conf2009/pdf/Kitasaka.pdf [retrieved 20 Jan 2017].

[52] TransLink. (2013). 'Regional Transportation Strategy: Strategic Framework'. Available at http://www.translink.ca/-/media/Documents/plans_and_projects/regional_transportation_strategy/rts_strategic_framework_07_31_2013.pdf [retrieved 20 Jan 2017].

[53] Cayo, D. (2012, February 24) 'Road Tolls Recommended for Metro Vancouver: Extensive Levies Aimed at Raising Millions for TransLink'. Vancouver Sun. Available at http://www.transportfutures.ca/news/vanfunding [retrieved 21 Jan 2017].

[54] Deloitte. (2010). Metro Vancouver Road Pricing Research Study: Final Report. Prepared for TransLink. Available at http://www.translink.ca/~/media/Documents/plans_and_projects/regional_transportation_strategy/Research/Metro_Vancouver_Road_Pricing_Research_Study_Report.ashx [retrieved 10 Dec 2016].

[55] BC Ministry of Transportation and Infrastructure. 'Guidelines for Tolling—April 2003'. Government of BC. Available at http://www.th.gov.bc.ca/Publications/reports_and_studies/tolling/index.htm [retrieved 10 Dec 2016].

[56] Move NY. (2015). The Move NY Fair Plan. Available at http://iheartmoveny.org/wp-content/uploads/2015/02/Move-NY-Fair-Plan-150217v1.pdf [retrieved 19 January 2017].

[57] Mayors' Council. (2016, April). Ten Year Metro Vancouver Transit and Transportation Plan, Funding Strategy Proposal. Available at: http://www.translink.ca/-/media/Documents/about_translink/governance_and_board/council_minutes_and_reports/2016/may/2016_05_26_funding_strategy_submission.pdf [retrieved 20 Jan 2017].

[58] City of Calgary. (2015). Transportation Infrastructure. Transportation. Available at http://www.calgary.ca/Transportation/TI/Pages/Transportation-Infrastructure.aspx [retrieved 10 Dec 2017].

[59] City of Calgary. (2009). Calgary Transportation Plan. Available at http://www.calgary.ca/_layouts/cocis/DirectDownload.aspx?target=http%3a%2f%2fwww.calgary.ca%2fTransportation%2fTP%2fDocuments%2fCTP2009%2fcalgary_transportation_plan.pdf&noredirect=1&sf=1 [retrieved 10 Dec 2016].

[60] Statistics Canada. 'Proportion of Workers Commuting to Work by Car, Truck or Van, by Public Transit, on Foot, or by Bicycle, Census Metropolitan Areas, 2011'. National Household Survey. Available at http://www12.statcan.gc.ca/nhs-enm/2011/as-sa/99-012-x/2011003/tbl/tbl1a-eng.cfm [retrieved 10 Dec 2016].

[61] City of Calgary. (2013). 'Downtown Mode Split and Vehicle Occupancy Methodology'. Mobility Monitor. Available at http://www.calgary.ca/_layouts/cocis/DirectDownload.aspx?target=http%3a%2f%2fwww.calgary.ca%2fTransportation%2fTP%2fDocuments%2fdata%2f2012%2fmobility_monitor_march_2012.pdf&noredirect=1&sf=1 [retrieved 10 Dec 2016].

[62] Zdeb, C. (1993, July 2). 'Klein and Kowalski form tag-team to sell the Alberta Advantage'. *Edmonton Journal.* Available at http://www.edmontonjournal.com/July+1993+Klein+Kowalski+form+team+sell+Alberta+Advantage/11180321/story.html [retrieved 10 Dec 2016].

[63] Graveland, V. (2014, May 6) 'Calgary Think-tank Recommends Toll Lanes to Raise Revenue and Reduce Travel Time'. The Canadian Press. Available at http://www.kelownadailycourier.ca/news/national_news/article_e565cbd1-94a2-59de-a592-6e151a650570.html [retrieved 10 Dec 2016].

[64] Brunnen, B. (2014). 'For Whom the Road Tolls: The Prospect of High Occupancy Toll Lanes in Calgary'. Manning Foundation. Available at http://manningfoundation.org/Docs/road-pricing.pdf [retrieved 10 Dec 2016].

[65] Government of Ontario. (2007). ARCHIVED: High Occupancy Vehicle (HOV) Lanes—Summary 2007 Summary of the Plan for the 400-Series Highways in the Greater Golden Horseshoe—Tackling Congestion in the Greater Golden Horseshoe. Available at http://archive-ca.com/page/4378987/2014-08-11/http://www.mto.gov.on.ca/english/traveller/hov/summary2007.shtml [retrieved 10 Dec 2016].

[66] Moore, O. and Hui. A. (2015, July 22). 'Some Pan Am HOV lanes Could Become Toll Routes, Wynne Says'. The Globe and Mail. Available at www.theglobeanamail.com [retrieved 10 Dec 2016].

[67] Singer, G. and Burda, C. (2014). 'Fast Cities: A Comparison of Rapid Transit in Major Canadian Cities'. Pembina Institute. Available at https://www.pembina.org/reports/fast-cities-report.pdf [retrieved 10 Dec 2016].

[68] Ipsos Reid. (2013, April). Toronto Transportation Investment Survey—Summary Report. Available at http://www1.toronto.ca/City%20Of%20Toronto/Feeling%20Congested/PDFs/CITY-OF-TORONTO-Transportation Investment-Summary-Report-1.pdf [retrieved 10 Dec 2016].

[69] Dachis, B. (2011). 'Congestive Traffic Failure: The Case for High-Occupancy and Express Toll Lanes in Canadian Cities'. CD Howe Institute e-brief. Available at http://www.cdhowe.org/pdf/ebrief_122.pdf [retrieved 10 Dec 2016].

[70] Safirova, E., Gillingham, K., Harrington, W., and Nelson, P. (2003). 'Are Hot Lanes a Hot Deal? The Potential Consequences of Converting HOV to HOT Lanes in Northern Virginia'. *Resources for the Future Issue Brief*, 03-03. Available at http://www.rff.org/files/sharepoint/WorkImages/Download/RFF-IB-03-03.pdf [retrieved 20 Jan 2017].

[71] Dachis, B. (2016, November 24) 'The Time for Tolls is Now, but Toronto must get it Right'. The Globe and Mail. Available at http://www.theglobeandmail.com/report-on-business/the-time-for-tolls-is-nowbut-toronto-must-get-it-right/article33020178/ [retrieved 20 Jan 2017].

[72] Gray, J. and Morrow, A. (2016, November 24) 'Mayor Tory will Require Firm Backing to Impose Road Tolls'. The Globe and Mail. Available at http://www.theglobeandmail.com/news/toronto/as-toronto-mayor-pitches-tolls-city-report-calls-for-more-taxes/article33013689/ [retrieved 20 January 2017].

[73] Radwanski, A. (2017, January 27) 'Wynne has her Gas-plant Moment with About-face on Toronto Tolls'. The Globe and Mail. Available at http://www.theglobeandmail.com/news/national/wynne-has-her-gas-plant-moment-with-about-face-on-toronto-tolls/article33799192/ [retrieved 30 Jan 2017].

[74] Blatchford, A. (2016, July 9). 'Abandoning Tolls on New Champlain Bridge Project Could Save $300M'. Montreal Gazette. Available at http://montrealgazette.com/news/local-news/abandoning-tolls-on-new-champlain-bridge-project-could-save-300m-memo [retrieved 20 January 2017].

[75] Magder, J. (2015, April 27). 'AMT to be Abolished, Planning Transferred to Elected Officials'. Montreal Gazette. Available at http://montrealgazette.com/news/local-news/amt-to-be-abolished-planning-to-be-taken-over-by-elected-officials [retrieved 10 Dec 2016].

[76] Radio Canada. (2016, November 24). 'Denis Coderre ouvert à l′ instauration de péages sur les autoroutes'. Available at http://ici.radio-canada.ca/nouvelle/1002105/denis-coderre-ouvert-instauration-peages-routes [retrieved 20 January 2017].

[77] Gagnon, L., Lefebvre, JF., Théorêt, J., Benoit, R., and Craig, M. (2014). 'Modalités et avantages d'une réforme fiscale écologique pour le Québec: Mythes, réalités, scénarios et obstacles'. Groupe de recherche appliquée en macroécologie (GRAME). Available at http://www.examenfiscalite.gouv.qc.ca/fileadmin/user_upload/etudes/modalites_et_avantages_reforme_fiscale_ecologique.pdf [retrieved 10 Dec 2016].

[78] Boulenger, S., Castonguay, J., and Montmarquette, C. (2014). Étude sur la tarification routière pour la région métropolitaine de Montréal. Available at http://www.cirano.qc.ca/pdf/publication/2014RP-02.pdf [retrieved 10 Dec 2016].

[79] Institut de développement urbain du Québec (IDUQ). (2015). 'Péage dans la région métropolitaine de Montréal'. Available at http://iduquebec.com/peage-dans-la-region-metropolitaine-de-montreal-lidu-recommande-une-taxe-kilometrique/ [retrieved 10 Dec 2016].

[80] Statistics Canada. Population and Dwelling Counts, from Census Metropolitan Areas, 2011 and 2006 Censuses. Available at http://www12.statcan.gc.ca/census-recensement/2011/dp-pd/hlt-fst/pd-pl/Table-Tableau.cfm?T=205&S=3&RPP=50 [retrieved 10 Dec 2016].

[81] Moscato, A. (2015, March, 31) 'Ottawa Third Most Congested City in Canada'. Ottawa Sun. Available at http://www.ottawasun.com/2015/03/31/ottawa-third-most-congested-city-in-canada [retrieved 10 Dec 2016].

Chapter 5

The public acceptability of road pricing—a US case study

Lee Munnich, Frank Douma*, and Joe Loveland***

In the mid-1990s, in Minnesota's Twin Cities Metropolitan Area, there was no significant technical reason why road pricing couldn't be implemented to manage the region's growing problem of traffic congestion. After all, the high-occupancy toll (HOT) lane concept and related technology had already been proven effective on I-15 and SR91 in Southern California, Singapore and elsewhere.

But technical issues aside, public acceptance remained a problem in Minnesota. In 1996, public questions led to a public–private toll road proposal on a new trunk highway 212 being vetoed by a municipal government. In 1997, a proposal to convert the I-394 high-occupancy vehicle (HOV) lane to a HOT lane on a demonstration basis was withdrawn after it failed to gain much public support [1]. In the midst of that debate, current Governor Mark Dayton, then a former State Auditor running for Governor, placed full-page ads in the state's largest newspaper opposing the proposed HOT lanes.

As he withdrew the I-394 proposal from consideration, then-Commissioner of the Minnesota Department of Transportation (MnDOT), James Denn summarized the problem: "I do not believe the proposed I-394 demonstration project enjoys the level of public understanding that is necessary for it to receive the objective analysis and fair consideration we seek" [2]. In terms of public acceptability, the idea had hit rock bottom.

Yet by 2003, the Minnesota Legislature authorized MnDOT to convert HOV lanes into HOT lanes, with remarkably little opposition. For the purposes of marketing to consumers, the HOT lanes were branded "MnPASS" lanes. Shortly thereafter, MnDOT resurrected the I-394 HOT lane project it had withdrawn under public pressure back in 1997. By 2004, the HOV to HOT conversion was underway. By 2005, 65 percent of citizens adjacent to Minnesota's first HOT lane told survey researchers that they thought the project was a "good idea." In 2010, another 26 km (16 mi) HOT lane was added on I-35W, as one component of Minnesota's Urban Partnership Agreement proposal to reduce traffic congestion on

*Humphrey School of Public Affairs, University of Minnesota, United States
**Loveland Communications, United States

I-35W and downtown Minneapolis. By 2016, a HOT lane on I-35E became the third major freeway corridor in the Twin Cities with a HOT lane. Looking forward, the future for HOT lanes looks strong, as the region's transportation plan calls for HOT lanes to be added to several additional corridors in the future [3].

So, what happened? After so much fierce early opposition and failure, how did things change so rapidly and dramatically in terms of public acceptance? Moreover, what lessons can we learn from that experience? Those questions are the focus of this chapter.

5.1 Public acceptance and road pricing

Economists for years have advocated congestion pricing or road pricing as the only real solution to road congestion: unless there is a market-based approach to reducing traffic during peak travel periods in urban areas, roadways reach their capacity and each additional vehicle adds to the delay of all the vehicles causing reduced speeds, more congestion delays for everyone and an inefficient use of road space [4]. Yet, only a few places in the world have adapted full congestion pricing schemes—notably Singapore, London and Stockholm through cordon or area pricing[1]. Once in place, these road-pricing schemes have been supported by the public when they see them work, but the public have inevitably opposed road-pricing plans when they are proposed.

5.1.1 Singapore

Singapore was an early model for road pricing. Singapore adopted road pricing, known as the Area Licensing Scheme, which required vehicles to pay tolls for entering the downtown area during daytime hours. This approach, which required the purchase of window stickers for vehicles and checkpoints at entrances to the downtown area, successfully reduced congestion in the downtown area. The manual system was replaced by an Electronic Road Pricing system in 1998 [5]. Singapore's approach to road pricing, along with a hefty Certificate of Entitlement fees for purchasing vehicles to limit growth of cars on the densely-populated island and investments in transit, have offered a successful model for others in managing congestion, improving the environment and providing mobility and accessibility to residents and visitors. More details of Singapore's history of road pricing and its plans for upgrading the current electronic scheme will be found in Chapter 6.

5.1.2 London and Stockholm lead the way

Beyond Singapore, the most significant adoptions of road pricing as a congestion management approach have been London in 2003 and Stockholm in 2006. The London pricing scheme was proposed by Ken Livingstone when he ran for Mayor of London in 1999. As Mayor, he saw congestion charging in downtown London

[1] Singapore and Stockholm are cordon schemes, London is an area-based scheme.

not only as a way to reduce congestion but also as a way to fund needed improvements in transit, particularly in the bus system [6]. As Mayor, Livingstone initiated the steps to implement congestion charging and the public became more aware of the scheme through media coverage, public opinion turned negative. Many of Mayor Livingstone's political advisors recommended dropping or postponing the congestion charges, but the Mayor stuck doggedly to his campaign promise. When the plan was implemented in February 2003, congestion in downtown London dropped immediately and public opinion polls shifted from strong opposition to strong support. See also Chapters 2 and 13.

A similar phenomenon occurred in Stockholm, Sweden, when they implemented congestion charges in downtown Stockholm. As a compromise between the Social Democratic and Green parties in forming the national coalition government following the 2002 parliamentary election, the Social Democrats agreed to a Green Party proposal to implement congestion charging in downtown Sweden and use the additional revenue to fund transit improvements [7]. Since the Social Democrats had not previously supported the plan, their leadership agreed to support a national tax in the form of a congestion charge on the condition that after implementation for a period, the system would be shut down and a referendum would be held of Stockholm citizens. If the referendum passed, the national government could then vote to reopen the system.

Public opinion polls taken before the referendum indicated Stockholm residents were opposed to the congestion charge. When the congestion charge was implemented in January 2006, congestion immediately dropped and traffic moved smoothly through downtown Stockholm. The system was then shut down as planned in August 2006, and congestion immediately returned to the levels before the congestion charge was implemented. In September 2006, Stockholm residents voted in favor of the congestion charging system. While the conservative party, which had opposed the congestion charging originally, won the national election, they decided to honor the referendum by reinstating the congestion charging system in August 2007 while shifting the use of revenue from transit projects to road projects. For more details see Chapter 14.

5.1.3 Setbacks with Manchester and Edinburgh referenda

Emboldened by the success of London's congestion charging scheme, other cities in Great Britain took steps to explore and implement area pricing in their cities. In particular, Edinburgh, Scotland, in 2005, and Manchester, England, in 2008, developed congestion charging plans and held referenda to determine if there was public support. Both cities undertook outreach and education efforts to explain the plans and the benefits in terms of congestion reduction and investment in transit improvements. However, in each case, the public when given the choice voted against congestion charging—74.4 percent against in Edinburgh [8] and 78.8 percent against in Manchester [9]. See Chapter 7.

A European study in 2008 may shed some light on the peculiar problem of public perceptions toward congestion pricing [10]. Focusing on European cities that either implemented or attempted to implement congestion pricing during the 2000s,

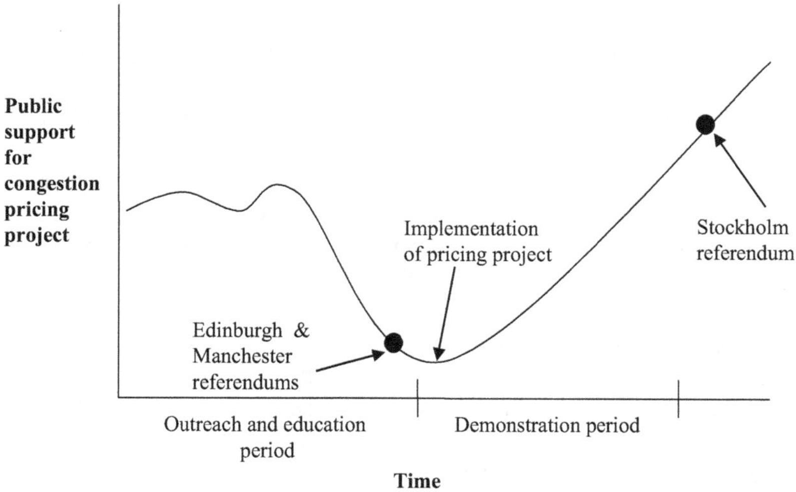

Figure 5.1 Public support for congestion pricing projects before and after implementation
[Source: Institute for Transport Studies, University of Leeds. CURACAO State of the Art Review, 2009]

the study found that the public tends to be open or mildly opposed to congestion pricing when discussed generally but becomes more opposed to congestion pricing as projects move closer to implementation. However, after congestion pricing has been implemented, the public shifts from opposition to support (Figure 5.1).

Projects that were implemented in London and Stockholm in spite of significant public opposition beforehand resulted in high levels of public support after implementation and they are still in operation. However, as indicated above, referenda on congestion pricing in Edinburgh and Manchester before implementation have failed miserably.

While the European study does not attempt to explain the phenomenon an explanation may lie in the fields of psychology and behavioral economics, where loss aversion refers to people's tendency to strongly prefer avoiding losses to acquiring gains. Behavioral studies suggest that losses are twice as powerful, psychologically, as gains (Figure 5.2) [11,12].

When a specific congestion pricing project is proposed in an area or corridor, the public focus is on the potential tolls (loss), triggering the loss aversion reaction among citizens. At the same time, the complexity of congestion pricing results in confusion or doubts about the primary benefit of congestion pricing, time savings (gain), as well as who will benefit from those gains. Media coverage will magnify these negative concerns about potential losses, and positive coverage will be overshadowed.

After a congestion pricing project is implemented, it becomes clear that congestion has been reduced, and the loss aversion affect switches in favor of congestion pricing—the public does not want to lose the time savings benefit, and the congestion tolls are clearly understood and are of less concern.

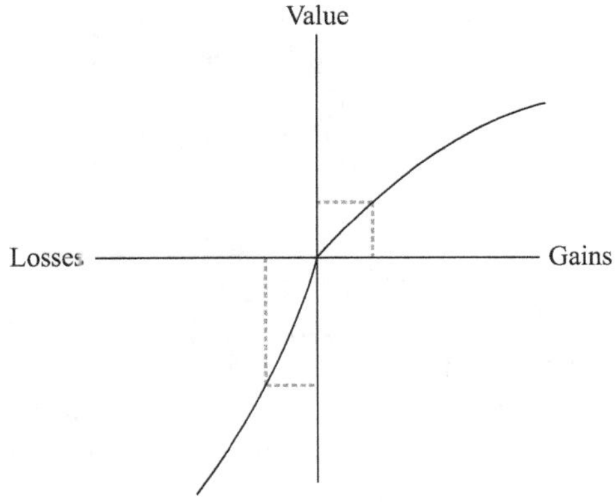

Figure 5.2 Loss aversion affects public perceptions more the potential gains
[Source: Amos Tversky and Daniel Kahneman. The Framing of Decisions and the Psychology of Choice. Science, New Series, Vol. 211, No. 4481. 1981, pp. 453–458. Also, "Thinking, Fast and Slow", by Daniel Kahneman, Penguin (May 10, 2012) – especially Figure 10 and the section on "loss aversion" in Chapter 26]

5.1.4 The US experience with HOT/managed lanes

US cities have not implemented comprehensive congestion pricing approaches of the sort that Singapore, London and Stockholm have done. Mayor Michael Bloomberg attempted to implement congestion pricing in New York City in 2008 but failed to get the support of the New York state legislature. San Francisco has also explored congestion pricing and conducted public outreach but has not attempted to move forward.

On the other hand, US metropolitan areas have had considerable success with a version of congestion pricing known as HOT lanes, managed lanes or priced express lanes. These lanes typically are conversions of HOV lanes built in the 1980s and 1990s to give priority to carpools of two or more (2+) or three or more (3+) occupants and buses. These HOV lanes typically were underutilized. By allowing single-occupant vehicles to use lanes for a price and varying that price up or down in response to changes in demand to assure that traffic flowed without congestion, the lanes could be used more efficiently and the user would have the choice of getting a quicker trip for a fee or using the unpriced lanes which were subject to congestion delays in peak periods.

In 1995, SR91 in Orange County, California, became the first application of peak-period road pricing in the United States offering uncongested trips for a fee that varied by time of day and day of the week. The project initially was run by a private firm but later was turned over to Orange County, which operates the SR91 express lanes today. In 1996, San Diego, California, converted an HOV lane to a HOT lane and dedicated the revenues from the HOT lane to improved transit service in the corridor [13]. Since that time numerous other cities in the United States have adopted HOT lanes, including Minneapolis-St. Paul, Minnesota; Houston, Texas; Denver,

Colorado; Seattle, Washington; Atlanta, Georgia; Miami, Florida; Los Angeles, California; San Francisco, California; and Washington, DC [14].

5.2 The Minnesota experience

In Minnesota, in the mid-1990s, the environment was in many ways becoming more favorable to the adoption of road pricing than it had been in the past.

Importantly, toll collection technology had improved dramatically. Minnesotans had long grumbled about the hassles associated with the traditional toll booths they encountered in Illinois and elsewhere. However, new technology being proven effective and affordable around the world offered the hope of eroding some of that public resistance (see e.g. Chapter 12).

The new technology allowed tolls to be collected electronically at full-speed, with no need to stop and find the correct change. The technology also could use "dynamic pricing," which meant that tolls could be quickly raised or lowered according to the level of congestion in the lane. This offered the promise of getting more use of underutilized lanes, while still preserving free-flow conditions for transit and carpool users.

At the same time, frustration about a carpool lane on I-394 was mounting. Many I-394 general purpose lane drivers resented that the I-394 HOV lane appeared underutilized as they sat stalled in adjacent congested lanes. The federal government would not allow MnDOT to open the underutilized carpool lanes to all traffic, but many believed that the federal government may allow Minnesotans to convert the I-394 HOV lane into a HOT lane, which could get more use of the lanes.

In many ways, the stars seemed to be aligning for a HOT lane demonstration project in the Twin Cities.

5.3 The challenges

So, why did road pricing become such a political punching bag in the 1990s? There are many reasons, but these were some of the driving forces for opposition.

First and foremost, few Minnesotans were aware of the advances in HOT-related technology, and even fewer had experienced or observed them firsthand. Absent that knowledge and experience, skeptical leaders and citizens had scores of unanswered questions about the technology.

There were also deep concerns about equity. That is, Minnesotans questioned whether it was fair to sell premium access to highways, since "freeways" had always been available in Minnesota at no additional cost to all users. Without specific information available to the public about the cost of a typical trip, customer profiles or usage patterns, many assumed that the HOT lane effectively would be only available to the most affluent drivers. Opponents of the approach fed this suspicion by referring to HOT lanes as "Lexus Lanes," referring to a luxury vehicle that is usually driven by relatively wealthy citizens.

Carpoolers, transit users and their advocates were also skeptical. They were concerned that opening up the lane to additional users would lead to congestion of the HOV lane, and a loss of the premium service they enjoyed. HOV advocates were also worried that opening up the lane to solo drivers would weaken the incentive to carpool or use transit.

Then there was concern about "double taxation." For many, tolling was perceived to be about raising revenue, not managing traffic. The notion that dynamic pricing could manage traffic congestion by tapping into market forces was a difficult abstraction to grasp. Without that core understanding, many assumed that the toll was being used only to raise additional revenue. That struck corridor users as unfair "double taxation," since citizens felt they had already paid for the highway through gasoline taxes.

Tolling also struck many as counter to Minnesota's traditional culture. The fact that Minnesota had never had toll roads was a source of pride for Minnesotans, and many were determined to avoid becoming "another Chicago."

Another challenge was that concern about congestion had not yet reached a high level. For instance, a "Citizens Jury" that was convened by the Jefferson Center and the University of Minnesota's Humphrey School of Public Affairs to study the issue rejected the concept. A primary reason given by jury members was that they didn't think such major changes were needed until traffic congestion increased [15,16].

Finally, there was no natural constituency advocating for road pricing. After all, beneficiaries of road pricing didn't yet understand that they would be beneficiaries. The lack of natural advocates caused many public questions to go unanswered. At the same time, even the relatively few leaders and officials who understood the concept well were sometimes reluctant to champion a concept that seemed to have so little organic support in the community.

An October 12, 1997 newspaper account summarizes many of the daunting public acceptability challenges that were facing HOT lanes at that time:

Critics say the $450 million highway was paid for by all taxpayers and every lane, including the carpool "sane lane," should be open to all taxpayers, all the time.

Mark Dayton, a democratic candidate for Minnesota governor, has seized upon the experimental program as a campaign issue, lambasting it in newspaper ads and guest editorials.

He has called for the transportation department to "free the sane lanes."

"So, you no longer have to be 'commuter correct' to use the sane lane, you only have to be rich. That's insane!" editorialized Dayton, who is a multimillionaire heir to the Dayton department store chain.

Elsewhere, critics have attacked congestion pricing as the corruption of one of the few truly democratic places remaining in American culture—the open road where rich and poor drive side-by-side. California state Sen. Tom Hayden has lamented the creation of "Lexus lanes" for the rich, which he feels penalize lower-income drivers.

Taking the "free" out of the freeway is a politically loaded concept. A 2-year study of congestion pricing by a special committee of the Transportation Research

Board, an independent clearinghouse for transportation research and technical information, endorsed the idea in theory but noted it probably would not be embraced by politicians or voters.

"Except for professors of transportation economics and planning—who hardly constitute a potent political force—I can think of few interest groups that would willingly and vigorously fight for the concept," noted the committee chairman Martin Wachs of the University of California. "I don't believe the approach has the slightest chance of being enacted in the coming decade"[17].

5.4 The turning point

So what happened between 1997, when the initial demonstration project was withdrawn by MnDOT, and the I-394 HOT lane launch in 2004? Among other things, a new kind of outreach and education plan was undertaken by a multidisciplinary team assembled by the University of Minnesota Humphrey School's State and Local Policy Program (SLPP) [2].

Starting in 2001, funded by a grant from the Federal Highway Administration's Value Pricing Pilot Program through MnDOT, the Humphrey School used a public outreach strategy that differed from the approach taken in the mid-1990s. First, the Humphrey School convened a Value Pricing Advisory Task Force in 2001, which was made up of a diverse group of community stakeholders and led by a former chair of the Minnesota Legislature's Senate Transportation Committee, Carol Flynn.

The members of the task force were initially skeptical about value pricing. But an intensive year-long learning process allowed members to get their questions thoroughly addressed by local and national road-pricing experts. After examining a number of congestion pricing options, the task force recommended three potential projects for a road-pricing demonstration project: (1) Tolling all lanes on the I-35W/Highway 62 Crosstown Commons during a reconstruction project; (2) converting the existing I-394 HOV lanes, or "sane lanes" as they were sometimes called, into HOT lanes; and (3) tolling all lanes on a bridge crossing the St. Croix River between Minnesota and Wisconsin near Stillwater during a multiyear bridge reconstruction [18].

The first alternative was eventually dropped since MnDOT engineers viewed the approach as too controversial and believed it could disrupt a long-planned project. The second option was pursued and eventually adopted. The third option was considered but never pursued as the second option moved forward.

Beyond the task force, the other major change in approach MnDOT and the Humphrey School project leaders adopted was to bring private sector communications and engineering experts onto their team. First, it retained a communications consultant with experience in the public affairs arena to recommend and implement a proactive public outreach and education strategy. At the same time, an engineering firm with deep knowledge of road pricing and the local transportation landscape was hired. The engineering firm worked closely with the communications consultant to develop strong answers to the many public questions about road pricing and HOT lanes.

5.5 The results

Armed with the task force findings and recommendations, task force members and outreach staff did a number of small group visits with legislators, state and local government officials, interest group leaders and civic groups. The team also commissioned a public opinion survey to learn more about consumers' concerns, and inform messaging strategies. After the initial groundwork was done with high level stakeholders, there was also outreach to the general public. For instance, there were newspaper editorial board meetings, placement of guest newspaper commentaries, convening of several public policy roundtable discussions featuring issue experts and news reporter briefings.

Each of these outreach tactics relied on technical information from the engineering firm, simplified and clarified by the communications consultant. The team's goal was for every public question, concern or criticism to be promptly addressed with a response that was simple and clear enough for the general public to understand. This required close cooperation between the communications and engineering consultants.

The engineering team also developed preliminary details about how the I-394 HOT lanes would work. This helped build stakeholder understanding and confidence that a project was technically feasible.

At the same time, a 2002 public opinion survey of 800 Twin Cities residents commissioned by the team found that 57 percent of a random sample of 800 Twin Cities' residents supported "having the option of paying a fee to use an uncongested freeway when in a hurry," compared with 51 percent in the same survey that supported a gas tax increase. This survey finding surprised many community leaders, and helped convince some of them that the project may actually be politically feasible [19].

With each answer provided by the project team and champions, opposition lessened and skeptics became increasingly open to testing the concept. By the time the state legislation authorizing value pricing was enacted and signed into law by Governor Tim Pawlenty in 2003, there was relatively little opposition. After the legislation was passed, a corridor advisory task force was established to assure that concerns by the cities and citizens in the corridor were addressed. The task force provided a way for legislators and other representatives from the corridor to understand how HOT lanes worked and provide guidance to MnDOT in implementing the new approach. Perhaps more importantly, the task force process allowed a number of task force members to travel to southern California to see for themselves how pricing worked, and how California freeway users of all income levels had embraced priced-express lanes on SR91 in Orange County and HOT lanes on I-15 in San Diego.

During this outreach and education initiative and through the launching of the I-394 HOT lanes, public acceptance grew steadily. Even before the HOT lanes opened in 2005, a fall 2004 panel survey of 750 individuals in the nearby I-35W service area found that 60 percent of respondents believed that allowing single drivers to use carpool lanes by paying a toll was a "good idea." By the spring of 2006, when residents actually had a chance to see and experience the HOT Lane, 65 percent of the respondents thought it was a good idea, while only 22 percent thought it was a "bad idea" [20].

In Minnesota, education and outreach has been more than a 4-year process that ended in 2004; it has been a 15-year process that continues today. Well after the 2001–04 outreach and education effort had concluded, a steady series of public outreach and education initiatives were done in other corridors adopting MnPASS lanes. MnDOT and the Metropolitan Council now have policies requiring the consideration of managed lanes including MnPASS for future highway projects in the Minneapolis-St. Paul metropolitan region. During the lead-up to each new HOT lane project—the I-394 HOV lane to HOT lane conversion launched in 2004, the I-35W Priced Dynamic Shoulder Lane launched in 2010 and an I-35E HOT lane addition launched in 2016—a new outreach effort was organized in the new service area to carefully explain the concept and answer all questions [21].

MnDOT and the Humphrey School, in coordination with the Metro Council, led the subsequent outreach efforts. Operating alongside these efforts was a private marketing agency hired to advertise the newly branded "MnPASS Lanes" to corridor users (Figure 5.3).

5.6 Lessons learned

Although much of Minnesota's experience appears to be the result of continual trial and error, the story follows the pattern that was going on at roughly the same time in London and Stockholm of skepticism through the time of implementation followed by public acceptance. Though each community faces its own unique set of circumstances, some of the individual lessons that have been learned over the past two decades of Minnesota's outreach and education efforts related to MnPASS lanes may be instructive in helping road-pricing champions in other communities follow this same pattern of addressing concerns and achieving eventual success. These lessons are summarized in Table 5.1 and explained below.

5.6.1 Seeing is believing

Perhaps, the most important lesson learned in Minnesota is that support for road pricing grows after leaders and the public observe it in operation. For instance, support on the I-394 MnPASS Advisory Task Force improved significantly following a fact-finding trip to southern California, where some members toured the I-15 HOT lane and SR91 Express Lane facilities and visited with local officials. Similarly, exposing Minnesota leaders to videos of the California HOT lanes was one of the most effective parts of the entire outreach and education effort.

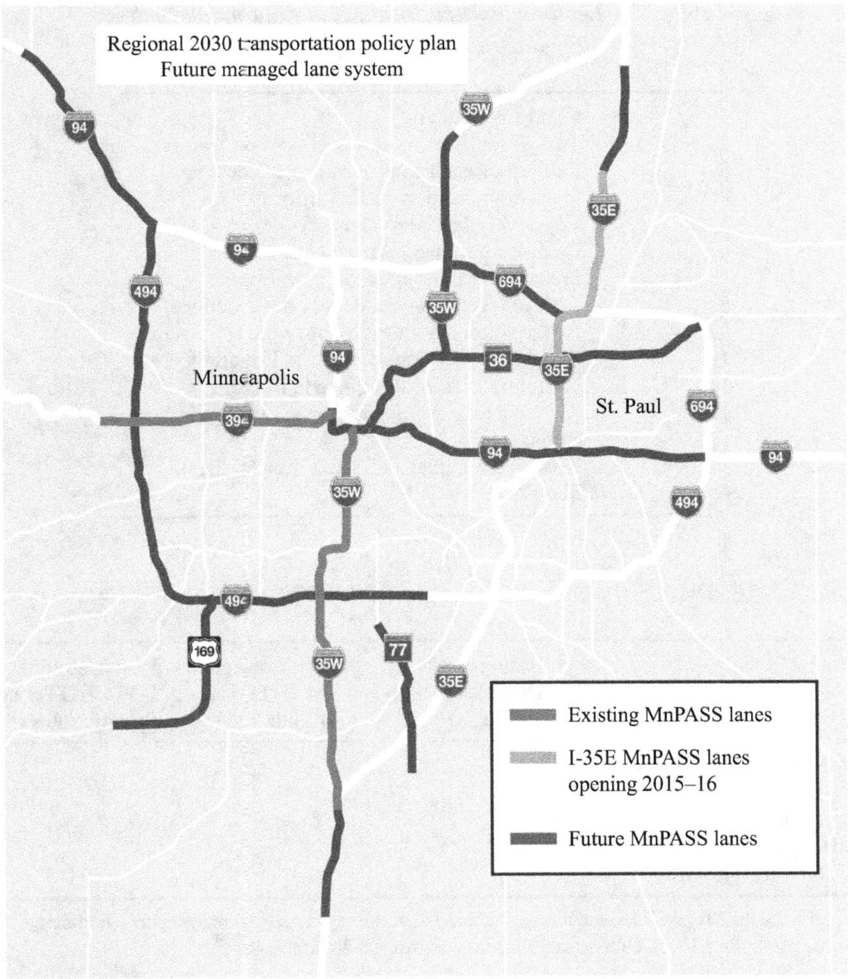

Figure 5.3 The future of Minnesota's HOT lanes now is bright. Note that the existing MnPASS lanes are on I-394 and the southern half of I-35W. The MnPASS lanes on part of I-35E opened in 2016. Future MnPASS lanes or MnPASS system include parts of I-36, I-77, I-94,I-169, I-494, I-694 and northern part of I-35W. "Regional 2030 Transportation Policy Plan: Future Managed Lane System"
[Source: Minnesota Department of Transportation]

Table 5.1 Lessons learned from Minnesota's outreach and education efforts

1.	Seeing is believing
2.	Task force is an in-depth education tool
3.	Public outreach improves project design
4.	Grass-tops support is first priority
5.	Top-level champions are key
6.	Coalition requires constant maintenance
7.	Preparation must precede promotion
8.	Preliminary technical details must be available
9.	No question should go unanswered
10.	Flexibility and opportunism is important
11.	Customized messaging needed
12.	Accentuating the positive pays off
13.	Choice sells
14.	Nongovernmental facilitator can be useful
15.	Show, don't just tell

Table 5.2 Evolution of public acceptance, I-394 HOT lane

	Before I-394 HOT Lane Fall 2004 (%)	**Newly Launched I-394 HOT Lane Fall 2005 (%)**	**Post Launch I-394 HOT Lane Spring 2006 (%)**
I-394 and I-35W service area respondents who believed allowing single drivers to use carpool lanes by passing a toll was a good idea	60	59	65

Source: NuStats. MnPASS Evaluation Attitudinal Panel Survey: Wave 3 Final Report. Humphrey Institute of Public Affairs, University of Minnesota, August 2006, p. 14.

The same "seeing is believing" lesson applies to the general public. Minnesota has seen that after HOT lane projects are launched, public acceptance improves. For instance, a survey done in Minnesota in 2005, *after* area residents had a chance to actually see and use the I-394 HOT lane, found stronger public support for the project than was found by the same researchers with the same respondents in 2003—see Table 5.2.

The ability to see performance data also helped build public confidence. MnDOT did an outstanding job gathering extensive data about the lanes' actual performance, such as average prices, throughput and speeds during peak periods. These data were publicized via the news media and elsewhere, and were very important in helping the public and policymakers understand the truth about the MnPASS lanes, such as the fact that they did not degrade transit service by becoming congested during peak periods.

Whether seeing the project itself or the related performance data, for many citizens, seeing was believing.

This is an important point for road pricing outreach and education practitioners to understand. As an abstraction, road pricing can be difficult for many leaders and citizens to understand and embrace. Many need visual evidence of the concept successfully operating—either in person or via videos, surveys or peer testimonials—before they can be convinced that the approach could be effective and efficient.

5.6.2 Task force is an in-depth education tool

Outreach and education leaders often stress the need to "develop credible champions" for road pricing. But that's often easier said than done. Road pricing is a complex and unfamiliar topic that cannot be quickly explained. For that reason, Minnesota leaders have found community task forces can be an excellent way to facilitate a detailed discussion of the issue amongst key leaders.

The 2001 Value Pricing Advisory Task Force was the first of many such community task forces that have been used in Minnesota. During this effort, the Humphrey School team brought together a diverse group of 30 state and local elected officials, business people and advocates for the environment, transit and carpools to be briefed by experts about value pricing. These proceedings gave the group an opportunity to systematically have their questions thoroughly addressed. Chaired by a respected former legislator, the process was thorough, respectful and transparent, and that helped build trust and understanding.

Obviously, it would be financially and logistically infeasible to offer such an in-depth 1-year briefing session for everyone in the community. But this tactic helped develop an initial core group of champions who could then perform peer-to-peer outreach to incrementally broaden the circle of people who were open to the idea of a HOT lane.

5.6.3 Public outreach improves project design

Not only did the project team's education and outreach efforts influence the public, but the public feedback flowing out of those dialogs also shaped the approach taken by the project team.

For instance, the project team heard loud and clear that area users wanted a guarantee that any excess revenues not needed to pay HOT lane-related costs would be used for transportation and transit improvements within the corridor, and not spent on unrelated purposes. Residents and transit advocates also asked the project team for assurances that transit and carpools would continue to enjoy free-flowing traffic in the new HOT lane. Additionally, in the pre-MnPASS era, residents expressed concern about a proposal to use a monthly pass, causing that approach to be dropped. Finally, many residents also told the outreach team that they would be open to trying a HOT lane if it were structured as a pilot project that would be evaluated for effectiveness and efficiency before being made permanent. While the I-394 MnPASS project was introduced as a permanent project once legislative approval had been obtained, the project became an important pilot demonstrating that the approach worked before introducing HOT lanes in other corridors.

These citizen suggestions gathered during outreach activities ultimately were adopted as part of the I-394 HOT lane project plan, and their adoption helped improve the public palatability of the project.

5.6.4 Grasstops support is first priority

In 1997, following the rejection of an earlier road-pricing proposal, Minnesotans had been told by their political leaders that road pricing was to be rejected because it constituted an unfair "Lexus Lane," used "double taxation" and was simply "too controversial." Given that kind of heated rhetoric coming from Minnesota leaders, changing the opinions of Minnesota's "grasstops" leaders—influential advocates or officials with a high professional or public profile—was an important prerequisite to changing "grassroots" opinion.

Therefore, outreach team members didn't start with a grassroots campaign—taking HOT lane messages directly to ordinary residents. Instead, it started by painstakingly piecing together a coalition of respected leaders. It gave a smaller group of people detailed briefings and answered all of their questions. The time for grassroots messaging would come later, such as when it came time to sell MnPASS toll collection devices to allow residents to use the HOT lane. But early in the process, outreach and education was aimed narrowly at the grasstops in order to ensure community leaders were well informed and not spreading myths.

5.6.5 Top-level champions are keys

In the late 1990s, road-pricing support in Minnesota was primarily limited to a handful of academics, transportation leaders, mid-level government officials and community leaders. Their backing was helpful for building the intellectual case for the concept, but they were not high profile enough to stimulate broad public support.

High-level elected officials are also needed in order to inspire confidence with the general public. Winning the support of Republican Governor Tim Pawlenty for the I-394 HOT lane conversion required the project team to first persuade many of his most trusted advisers, such as Republican Congressman Mark Kennedy and MnDOT leaders. Being able to share a number of Minnesota surveys showing high levels of public support was also likely to be helpful in convincing Governor Pawlenty and other elected officials that public acceptance had improved dramatically since 1997, when the concept had been withdrawn by then-Governor Arne Carlson.

In 2003, when Governor Pawlenty decided to back the I-394 HOT Lane as "a bold and innovative way to reduce congestion for Minnesota drivers," it helped in at least two ways. First, the active support of Governor Pawlenty helped activate his administration's key transportation and transit leaders, as well as his other allies in the public, private and nonprofit sectors. This influx of institutional and individual support instantly broadened and energized the coalition of supporters. Second, the Governor's endorsement led to the Governor using the prominent public messaging platform to persuade holdouts and the general public about the benefits of the approach. This is a messaging platform that only the top elected officials have at their disposal, and, for a relatively low-profile coalition, having this delivering their messages made a huge difference.

5.6.6 Coalition requires constant maintenance

The old adage "politics makes strange bedfellows" applies in the case of HOT lanes. After all, HOT lanes have the potential to appeal to groups of people who often find themselves disagreeing with each other on other issues, such as Republicans and Democrats, solo drivers and transit users, businesspeople and environmentalists, urban interests and suburban interests. This is a source of great strength for any public policy coalition, but it also poses challenges.

Minnesota advocates have found that maintaining coalition and trust and cohesion between such traditional opponents requires continual effort. If coalition dynamics are not closely monitored and managed by a team with strong diplomatic and organizational skills, once-strong networks can quickly weaken.

5.6.7 Preparation must precede promotion

Often, public education campaigns start with tactics involving mass media, such as a news conference bringing messages directly to television, radio, newspaper and social media consumers. However, Minnesota road-pricing advocates have learned from past experience that seeking news coverage too early in the process—before the reporters and community leaders shaping such news stories have been adequately briefed—can backfire. It too often leads to coverage where the primary focus of the story is on consumer costs—"do you want to pay tolls"—and ignores consumer benefits—"do you want to have the option of occasionally paying to bypass gridlock when you're late for an appointment?"

For that reason, news coverage was not sought until a robust set of communications tools were in place, such as simple visuals to explain the HOT lane concept at-a-glance, preliminary project design details, surveys showing that a majority of citizens were open to the idea, well-crafted and supported answers to frequently asked questions and diverse and credible community champions.

Armed with this information, the team was able to convince skeptics that road pricing was a proven traffic management tool, not a risky leap of faith. Once that information was assembled, then, and only then, was news media coverage sought. By that point, many community leaders were well prepared to help shape such news coverage to ensure it was thoughtful and accurate.

5.6.8 Preliminary technical details must be available

The engineering firm retained by the outreach and education team in 2001 was a critical part of the process of developing these communications tools. From past experience, the outreach team knew that stakeholders would have very detailed questions about how HOT lanes work, and whether their concerns about it were justified. In the past, road-pricing advocates had learned that every time they answered a question "we don't know yet," public skepticism grew.

Therefore, the engineering firm coordinated with the communications consultant to identify preliminary answers to all technical questions. For instance, the firm conducted preliminary investigations into the type of technology that could be used, so that stakeholders could see examples of it in action. It also developed preliminary project costs to help address those types of questions. The technical team also gathered information about other projects to build confidence that the concept was technically feasible and could be embraced by the public.

Outreach and public education is about more than mere wordsmithing. It's also about having a strong command of the facts, and that makes having technical expertise embedded into the outreach team extremely important.

5.6.9 No question should go unanswered

Minnesota advocates who had been through the struggles of the late 1990s were all too aware that public opinion was fragile. They learned that an accusation that goes unanswered can quickly become an accusation that is believed.

For this reason, the team formed a rapid response team, similar in some ways to what are sometimes used in political campaigns. The team of communications and engineering professionals continually scanned the environment and met weekly to discuss proactive tactics to preempt criticism and address new questions.

The stream of questions the team addressed was lengthy. The most frequently asked questions had to do with public acceptance, costs of tolls, technical feasibility, equity and preserving HOV service. Armed with technical data and documentation from the engineering firm, the team would simplify the information and work with credible messengers to swiftly provide solid answers to all questions before misunderstandings could evolve into conventional wisdom (Table 5.3).

5.6.10 Flexibility and opportunism is important

The outreach team had a communications plan to guide how and when it would reach out to various audiences. However, when unforeseen opportunities arose, the team frequently supplemented that plan with new tactics developed "on the fly."

For example, when a local survey found a surprising amount of public support for the idea of road pricing, plans were quickly made to ask all road-pricing supporters to share that data with news reporters and others in the community. This was important, because at that time many leaders inaccurately believed that HOT lanes were very controversial with the general public.

Likewise, the outreach team took steps to ensure that key stakeholders were aware of a MnDOT study that documented unused capacity in the I-394 HOV lane. While that finding was not surprising to most observers, the team used the high profile release of the study as a platform for explaining how road pricing could be used to better utilize HOV lane capacity, while still preserving free-flowing service for transit users and carpoolers.

Table 5.3 Examples of frequently asked questions and answers

Q. How much will solo drivers be charged during peak periods, and how is that calculated?
A. "Dynamic" prices automatically increase when congestion gets heavy and decrease when congestion gets light. (Average prices from other facilities also cited.)
Q. Will solo drivers need to stop at a toll booth to pay?
A. No, drivers place small electronic devices in their vehicles, which allow their prepaid account to get charged at highway speeds.
Q. Will only relatively wealthy solo drivers be able to afford to use the lanes?
A. Research from other areas show that many nonwealthy drivers like having the option to use the lane occasionally, such as when late for work or an important appointment.
Q. Will the lanes become congested, thus degrading service for transit users and carpoolers?
A. No, to prevent the lane from getting congested and degrading service for transit and carpools, prices are progressively increased during periods of congestion.
Q. How will the toll revenues raised via the MnPASS lanes be used by state government officials?
A. Revenue amounts that exceed what is needed to operate the MnPASS lane will be dedicated to transportation and transit improvements on and around the highway.
Q. How will authorities stop solo drivers from cheating, or using the lane without paying?
A. Law enforcement will have equipment to easily detect, at full highway speeds, whether a driver has a device and is therefore authorized to be in the lane.
Q. How reliable are the transponders and corresponding billing processes?
A. On other HOT lanes, research shows that errors are extremely rare.
Q. Will the technology perform as intended in Minnesota's severe winter weather conditions?
A. Weather is not expected to be a problem, but if it is, drivers will not be unfairly penalized for weather-related issues.
Q. How will solo drivers obtain a transponder, and what will it cost?
A. Solo drivers can get a transponder via MnDOT online or a facility.
Q. How and where will solo drivers be able to move in and out of the lanes?
A. Drivers can change lanes when there is a break in the double solid lines. (Maps and illustrations provided.)
Q. Won't visitors and tourists get unfairly fined because they aren't familiar with the system?
A. Signs will be clear, and out-of-towners will have their first violation forgiven.

Source: Joe Loveland, Loveland Communications.

In 2002, when a survey showed that residents near a HOT lane in southern California supported it, including residents of all income levels, the outreach and public education team again sprang into action. That information was summarized and provided to area stakeholders and reporters who had long maintained that HOT lanes were inequitable "Lexus lanes" that only benefited the wealthy. The team also shared anecdotes from California, such as nonwealthy parents, who were not every day HOT lane users, but greatly appreciated having the option to use the HOT lane when in a hurry, such as to avoid paying a stiff fine associated with missing a child care pick-up deadline.

These were all unforeseen developments that likely would have gone largely unnoticed by most reporters and stakeholders if there had not been a well connected and flexible outreach team in place to spread the word. It turned out that all of these developments proved to be important "educable moments" for rebutting pervasive myths about road pricing that had been impeding progress.

5.6.11 Customized messaging needed

In the 1990s, Minnesota advocates had long discussed the need for a one-size-fits-all set of "key messages" that would compel Minnesotans to support road pricing. Starting in 2002, it abandoned the idea of universal messages in favor of the use of messages that were customized for each major audience type.

It made no sense to use the same messages for conservatives and liberals, businesspeople and environmentalists or transit users and solo drivers, so different messages were emphasized for each audience type. For instance, when having a discussion with political conservatives, emphasis was placed on the fact that road pricing is a market-based approach and allows for road capacity to be added without the need to raise gas taxes. Similarly, when meeting with political liberals, emphasis was placed on the equity, environmental benefits and the potential to improve transit in the corridor.

Solo drivers were shown the electronic tolling technology, assured that they wouldn't have to stop at toll booths, and reminded that they always had the choice to stay in general purpose lanes and avoid paying tolls. For business groups, advocates stressed the cost of congestion and the fact that HOT lanes improved travel predictability for businesses.

To be sure, some common themes were used for all groups, such as the ability to use road pricing to manage traffic congestion. But a special effort was made to tailor messages to match each group's unique needs, motivations and values. This may seem like an obvious thing to do, but it took a concerted effort to ensure that such customization of messages and materials actually happened.

5.6.12 Accentuating the positive pays off

In past outreach efforts, road-pricing advocates may have inadvertently overemphasized the *cost* portion of the concept. By using terms like "value pricing," "congestion pricing," "road pricing," "tolling," and "high-occupancy toll lane," the cost to consumers is stressed.

On the other hand, terms such as "express lanes" and "MnPASS," which are the terms most often used in public communications today, emphasize the consumer benefit, or the user's ability to access free-flowing premium service. It's not that the cost side of the equation is in any way hidden, it's just that it is no longer overemphasized in the labels chosen for outreach and public education.

This may sound like an esoteric semantic argument, but it does matter. For instance, a survey being planned in Minnesota during the I-394 debate originally was going to ask residents if they would support paying tolls. Project advocates convinced the survey researcher to include both the cost and the benefit side of the HOT lane proposition in the survey question. As a result, the wording of the question became: “Would you support or oppose having an option of paying a fee to use an uncongested freeway lane when in a hurry?” [14]. With that kind of balanced presentation, the survey found more local support than had past surveys, which helped convince political leaders that the idea wasn’t as controversial as they had previously believed.

5.6.13 Choice sells

Most Minnesotans have embraced HOT lanes on I-394, I-35W and I-35E, but they still appear to be wary of road-pricing projects that do not offer users the option of avoiding the fee by using adjacent nontolled lanes. Choice matters.

The 2001 Value Pricing Advisory Task force recommended the I-394 HOT lane, but it also recommended two other projects. One would have charged peak period tolls on *all* lanes in a highly congested construction zone on the I-35W/62 Crosstown Commons reconstruction project. This project received much less community support, in large part because consumers had no way of avoiding the charge. This was not the only reason the Crosstown option was rejected—timing issues were also important—but it was a primary reason.

This experience convinced Minnesota advocates that it is much easier to get people to embrace projects that offer drivers a way to avoid the charge. As a result, the choice aspect of Twin Cities HOT lanes is constantly stressed during outreach and education initiatives. For instance, materials will often describe the concept like this: “With MnPASS express lanes, solo drivers have the option of paying a fee to use an uncongested lane when they are in a hurry.”

As Minnesotans become increasingly accustomed to having tolling in their midst, it may eventually become easier to adopt more comprehensive road-pricing projects. But achieving public acceptance for projects that do not offer user choice has been much more difficult than achieving public acceptance for the Twin Cities’ HOT lanes.

For instance, a 2011 Mileage-based User Fee (MBUF) Policy Task Force, dealing with a concept where users would not be able to avoid the fee, did not recommend moving forward with a comprehensive project. It did, however, recommend “state government conducts MBUF trials, possibly with contiguous states” [22,23]. No such trials have moved forward to date in Minnesota, but MnDOT is still exploring alternative ways of introducing MBUF such as through carsharing and ridesharing. See also Chapter 18 for the trials currently being carried out or planned in other US states.

5.6.14 Nongovernmental facilitator can be useful

In many jurisdictions, a governmental entity takes the lead on outreach and education efforts. There are advantages of such arrangements, such as the ability to tap into government institutions’ outreach expertise and resources. However, there are also disadvantages. Sometimes public institutions must battle against public distrust of government, as well as slow and cautious decision-making that is sometimes characteristic of large and heavily scrutinized institutions.

But beginning in 1997, following MnDOT's withdrawal of the Highway 212 and road-pricing projects in the late 1990s, the University of Minnesota Humphrey School of Public Affairs SLPP took the lead in facilitating public outreach efforts. Using a nongovernmental outreach and public education lead had advantages. To some stakeholders, an academic institution may have seemed more acceptable than a governmental entity. The fact that the Humphrey School was known in the community to be a place for studying public policy questions made it a natural and trusted convener for the task force. Moreover, the fact that the SLPP staff was relatively small and nonhierarchical helped keep the outreach team quick and nimble, particularly when there was a need to respond rapidly to a new development.

Once HOT lane projects officially became state-funded projects, MnDOT took the lead on public outreach and education efforts. But for every new corridor considering a new HOT lane, the Humphrey School has also played a substantial role in initial outreach efforts. The strong partnership between the Humphrey School and MnDOT has been an important part of Minnesota's successes.

5.6.15 Show, don't just tell

Even some stakeholders who had been relatively well briefed on HOT lanes still struggled to fully understand them. It was particularly difficult for them to understand how allowing solo drivers into the lane wouldn't lead to transit and carpools being stuck in gridlock.

As mentioned earlier, for many people road pricing has to be seen to be believed. For many, visuals are more compelling and memorable than verbal descriptions.

When visits to existing projects aren't financially or logistically possible, other types of visuals are the next best thing. For instance, project sponsors provided videotape to news reporters of the underutilized HOV lane, as well as free-flowing HOT lanes from southern California. This helped viewers visualize the nature of the problem, and the feasibility of the solution.

For audiences that had sufficient time and patience, a 13-min videotape was produced by the Humphrey School to show how HOT lanes worked, and how nonwealthy California residents occasionally used the lanes when in a rush [24].

In meetings, project advocates also played "show-and-tell" with the transponder devices that are placed in a vehicle's window to electronically collect tolls. Being able to see and feel these devices made the concept seem more tangible, feasible and intriguing to many people. Similarly, the PowerPoint presentations made to stakeholders were full of simple photographs, charts and renderings explaining value pricing. These types of visuals usually were much more clear and persuasive than verbal and written messages (Figure 5.4).

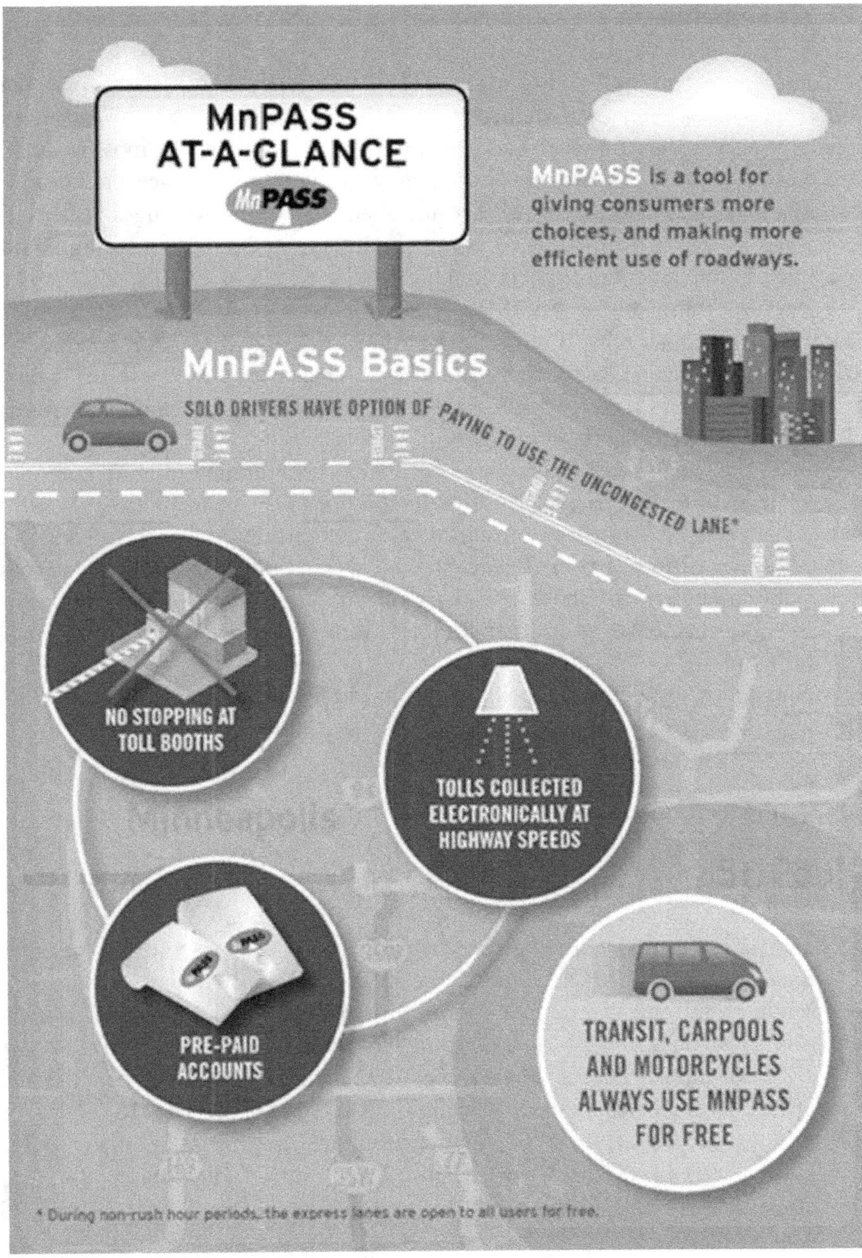

Figure 5.4 *Example of highly simplified visual infographic used in outreach*
[Source: State and Local Policy Program, Humphrey School of Public Affairs, University of Minnesota for I-35E MnPASS outreach project]

5.7 Conclusion

For communities struggling to implement congestion pricing, the Minnesota, London and Stockholm case studies provide hope, and a number of lessons to potentially inform their outreach and education efforts. Perhaps, the most powerful of all these lessons is that citizens are much more likely to embrace pricing after they have a chance to actually observe it, whether it be by touring a facility in another area, viewing a well-crafted video or other type of visual, or seeing a pilot project that is implemented in their community.

For transportation and political leaders, this lesson has important implications. For instance, it suggests that if they take a political risk and implement an effective and efficient HOT lane project on a demonstration basis, they can have confidence that public skepticism likely will dissipate after residents can see and experience a well-implemented HOT lane in operation.

To be clear, however, Minnesota's success has been limited to HOT lanes and not other forms of road pricing. But this lesson does also seem to apply to other types of road pricing as well. For instance, with congestion pricing in London and Stockholm, courageous political leaders moved forward with projects, despite initial public skepticism. Those leaders were subsequently rewarded, as initial public skepticism melted away after projects were implemented. See also Chapter 19 for similar lessons from Milan.

In Singapore, London and Stockholm, as in Minnesota, seeing road pricing was believing.

References

[1] Buckeye K., and Munnich L. 'Value Pricing Education and Outreach Model: MnPASS Community Task Force.' In *Transportation Research Record: Journal of Transportation Research Board*, No. 1960, Transportation Research Board of the National Academies, Washington, D.C., 2006, pp. 80–86.

[2] Munnich L., and Loveland J. 'Value Pricing and Public Outreach: Minnesota's Lessons Learned.' In *Transportation Research Record: Journal of the Transportation Research Board*, No. 1932, Transportation Research Board of the National Academies, Washington, D.C., 2005, pp. 164–168.

[3] Minnesota Department of Transportation. *Minnesota Express Lane System* [online]. 2015. Available from http://www.dot.state.mn.us/mnpass/pdfs/052015MnPASS_One_page_fact_sheet.pdf [Accessed 29 August 2016].

[4] Lindsey R. 'Do Economists Reach a Conclusion on Road Pricing? The Intellectual History of an Idea.' *Econ Journal Watch.* Vol. 3, No. 2, May 2006, pp. 292–379. Available from https://econjwatch.org/issues/volume-3-number-2-may-2006 [Accessed 29 August 2016].

[5] Phang S., and Toh R. 'Road Congestion Pricing in Singapore: 1975 to 2003.' *Transportation Journal.* Vol. 43, No. 2, Spring 2004, pp. 16–25.

[6] Transport for London. *Central London Congestion Charging. Impacts Monitoring – Fifth Annual Report* [online]. June 2007, p. 9. Available from http://content.tfl.gov.uk/fifth-annual-impacts-monitoring-report-2007-07-07.pdf [Accessed 11 August 2016].

[7] Eliasson J. *The Stockholm Congestion Charges: An Overview* [online]. CTS Working Paper 2014:7, KTH Royal Institute of Technology, 2014. Available from http://www.transportportal.se/swopec/CTS2014-7.pdf [Accessed 11 August 2016].

[8] Edinburgh Rejects Congestion Plan. *BBC News* [online]. 22 February 2005. Available from http://news.bbc.co.uk/2/hi/uk_news/scotland/4287145.stm [Accessed 29 August 2016].

[9] Ottewell, D., 'C-charge: A Resounding NO.' *Manchester Evening News* [online]. Updated 13 January 2013. Available from http://www.manchestereveningnews.co.uk/news/greater-manchester-news/c-charge-a-resounding-no-976016 [Accessed 29 August 2016].

[10] Institute for Transport Studies. *CURACAO State of the Art Review* (FINAL). Deliverable D2, Project No. 019788, Coordination of Urban Road User Charging Organisational Issues (CURACAO), University of Leeds, UK, 2009, p. 158.

[11] Tversky A., and Kahneman D. 'The Framing of Decisions and the Psychology of Choice.' *Science, New Series*. Vol. 211, No. 4481, 1981, pp. 453–458.

[12] Kahneman D. *Thinking, Fast and Slow*, Farrar, Straus and Giroux, New York, NY, 2011, pp. 282–283.

[13] Supernak J. *I-15 Congestion Pricing Project: Monitoring and Evaluation Services: Task 13. Phase II Year Three Overall Report*, Department of Civil and Environmental Engineering, San Diego State University, San Diego, CA, 24 September 2001.

[14] Urban Land Institute. *When the Road Price Is Right – Land Use, Tolls, and Congestion Pricing*, 2013. See Figure 2, p. 6. Available from http://www.uli.org/wp-content/uploads/2013/03/When-the-Road-Price-is-Right_web_F.pdf [Accessed 29 August 2016].

[15] Jefferson Center. *Citizens Jury: Report on Traffic Congestion Pricing*, June 1995. Available from http://jefferson-center.org/wp-content/uploads/2012/10/Traffic-Congestion-Pricing.pdf [Accessed 11 August 2016].

[16] State and Local Policy Program. *Buying Time: Political and Institutional Issues of Congestion Pricing, Final Report*. Hubert H. Humphrey Institute of Public Affairs, University of Minnesota, Minneapolis, 1996.

[17] Smith W. 'Minneapolis Drivers Will Be Able to Buy Way into Fast Lane to Ease Congestion: Passes to be Sold for Car-pool Zone.' *Chicago Tribune* [online]. 12 October 1997. Available from http://articles.chicagotribune.com/1997-10-12/news/9710120411_1_congestion-pricing-fast-lane-affluent-western-suburbs [Accessed 11 August 2016].

[18] State and Local Policy Program. *Curbing Congestion: Improving Traffic Flow, Transit, and Transportation Funding Through Value Pricing: Summary of the Work of the Minnesota Value Pricing Advisory Task Force.*

Hubert H. Humphrey Institute of Public Affairs, University of Minnesota, Minneapolis, 2002.

[19] Zmud J., and Arce C. *Compilation of Public Opinion Data on Tolls and Road Pricing: A Synthesis of Highway Practice.* NCHRP Program Synthesis 377, Transportation Research Board, Washington, DC, 2008. Available from http://onlinepubs.trb.org/onlinepubs/nchrp/nchrp_syn_377.pdf [Accessed 11 August 2016].

[20] NuStats. *MnPASS Evaluation Attitudinal Panel Survey: Wave 3 Final Report*, Humphrey Institute of Public Affairs, University of Minnesota, Minneapolis, MN, August 2006, p. 14.

[21] Munnich L., Luedke Karlsson M., and Fure M. 'Designing Minnesota's I-35E MnPASS Managed Lane Extension.' In *Transportation Research Record: Journal of the Transportation Research Board*, No. 2484, Transportation Research Board of the National Academies, Washington, D.C., 2016, pp. 173–181.

[22] Munnich L., Doan J., and Schmit M. *Report of Minnesota's Mileage-Based User Fee Policy Task Force*, Humphrey School of Public Affairs, University of Minnesota, 2011. Available at http://www.dot.state.mn.us/mileagebased userfee/pdf/mbufpolicytaskforcereport.pdf [Accessed 11 August 2016].

[23] Munnich L., Doan J., and Johnson C. *The Minnesota Mileage-based User Fee Policy Task Force*. TRB Paper 13-2753. Presented during the 2013 Transportation Research Board Annual Meeting, Washington, D.C., January 2013, (Unpublished).

[24] State and Local Policy Program and Aurora Pictures. *Buying Time: A New Strategy for Traffic Congestion Relief* [video]. Hubert H. Humphrey Institute of Public Affairs, University of Minnesota, Minneapolis, 1998. Available from https://www.youtube.com/watch?v=IwZgpIDapUk [Accessed 28 November 2016].

Chapter 6

How road pricing was implemented in Singapore, and planned technology augmentations

*Kian Keong Chin**

6.1 The early years of road pricing

Today, drivers in Singapore consider the paying of Electronic Road Pricing (ERP) charges as a normal part of the cost of motoring. Most of them may not be happy paying the ERP charges but will not protest outright, possibly because many had also benefitted from the improved traffic conditions on these priced roads and also drivers in Singapore have been experiencing road pricing since 1975.

1975 was the year that the original road pricing scheme started operating. Known as the Area Licensing Scheme or ALS, it was a manually operated scheme. This involved the purchase and display of pre-paid licences to be stuck onto their vehicles' windscreens as they went through charge points (demarcated by physical ALS gantries that have lights switched on during operational hours) surrounding the Central Business District (CBD). Vehicles without these pre-paid licences would have their vehicles' registration number noted down by enforcement officers stationed at these charge points, and their owners issued with summonses subsequently.

These licences could be pre-purchased at various outlets like post offices, petrol stations, convenience stores and dedicated sale booths located on the major roads leading into the CBD. There were both monthly licences for frequent driving into the CBD and daily licences for the occasional driving into it. Initially, it was only necessary for these licences to be displayed during the morning peak hour, but was extended in 1989 to cover the evening peak period as well. A few years later, licences were also required during the inter-peak period, but this road pricing charge was a reduced amount, in line with the lower congestion on the roads during these off-peak periods. In terms of the type of vehicles covered by the ALS, it was applicable to cars initially but was later extended to include motorcycles and goods vehicles.

Given that the intent was to reduce traffic on the roads, the early years saw exemption from the ALS if the cars had four or more occupants when they crossed

*Land Transport Authority, Singapore

the charge points or ALS gantries, in what is known as the carpooling scheme. This feature was removed in 1989 because the additional occupants to make up the four persons were observed to come mostly from bus commuters waiting at or near bus stops, and it was no longer serving the intent of getting car drivers to pool into one car. The other feature to reduce car traffic into the city was the park-and-ride scheme whereby car drivers were encouraged to park outside the city and hop onto a bus to get into the CBD. Several car-parks on the outskirt of the CBD were built to encourage this, but given it being not popular, the car-parks were progressively taken over for other developments over the years.

The initial reduction in traffic entering the CBD in 1975 was about 44%, but it crept up slowly to give a reduction of about 31% in 1988. It was, nevertheless, an impressive result given that the growth of employment in the CBD went up by about a third over those years, along with a vehicle population increase of 77% during this same period. This drop in traffic came from the decanting of traffic using the CBD roads as bypass routes to reach their destinations on the other side of the CBD, and also partly from drivers shifting their time of travel to enter into the CBD earlier (when the ALS was not operating) so as to avoid paying the ALS charges. The bus services had also improved to some extent over those years, and although the evidence is not strong, there would have been some trips that were now made on the buses instead.

6.2 Shortcomings of the manual area licensing scheme

The ALS, being a manually operated one, relied heavily on manpower resources for both the enforcement needs and the sale of the ALS licences. This was viewed as a significant difficulty for expanding the scheme to cover more roads outside the CBD that had become more congested over the years. Given that the level of congestion varied in the various newly congested roads, and with varied congested hours, an effective road pricing scheme would need varying ALS charges. However, the range of prices permitted by the licences was limited with the manual scheme, as it was necessary to keep it simple and with the differently priced licences easily distinguishable by both the users and the enforcement officers stationed at the ALS gantries.

Another limitation of the ALS was that the licence, once paid for that day's use, allowed for multiple trips into the CBD throughout that day and did not persuade high usage vehicles to reduce their number of trips, and hence, they continued to contribute to the heavy traffic on the roads within the CBD.

The manually issued ALS licence was not meant to be transferable as the vehicle registration plate number was written onto the licences at the point of sale. However, in practice, there had been evidence that some of these licences were transferred from one user to another during the course of the day. There had also been instances of the licences being tampered with to reflect another vehicle registration plate number, or to show a different date for which the licence was to be valid.

6.3 The electronic road pricing scheme

Technology for an ERP system was emerging in the early 1990s, and several technology trials were carried out before the decision was made to replace the manual ALS with an ERP system, culminating with the award of the ERP contract in 1995. A key feature required of this ERP system was that it should not require the slowing down of traffic and that it must work across multiple lanes that are under a multi-lane free-flow condition.

The ERP system comprises three key sub-systems – (a) the electronic device called the In-vehicle Unit (or IU, for short) with its stored-value smart card (Figure 6.1), (b) the ERP gantry (Figure 6.2) with various electronic components to communicate with the IU via Dedicated Short-Range Communication (DSRC) protocol and enforcement cameras and (c) the Control Centre that monitors all the ERP gantries and updates operating hours and level of road pricing charges. At this Control Centre, non-payments and violations are also processed. Non-payments are instances where the driver had forgotten to insert a stored-value smart card into the IU or has an inserted stored-value card with insufficient amount of funds in it. Under these instances, the owners of the vehicles are sent letters to recover the unpaid road pricing charges. Given the additional manual effort needed for this mode of collection, an administrative charge is collected together with the unpaid road pricing charge. However, when a vehicle is detected as having no IU installed when going through the ERP gantry, a summons is issued as this is considered a violation.

The ERP system was designed to be easy to understand and use by motorists, and indeed all that is needed for the motorists to do is to insert the stored-value card into the IU at the start of the trip. The IU is designed to alert the motorists if the

Figure 6.1 The stored-value smart card and the ERP's in-vehicle unit inside a car

Figure 6.2 An ERP gantry

stored-value card is inserted wrongly, has insufficient funds stored in it or if the stored-value card is technically malfunctioning. When the vehicle goes through the ERP gantry, there would be a beep to signify that payment had been made while the display will show the amount deducted, and a few seconds later, the amount of funds left in the stored-value card.

6.4 Effectiveness of the ERP in managing traffic

The ERP system is designed to allow varying levels of road pricing charges for varying periods of time. Hence, while theoretically, it can deduct different level of road pricing charge every minute with an independent pricing scheme at each ERP gantry, it would be too complex for the motorists to understand and respond to. Ultimately, the motorists, being totally confused, will just simply ignore the prevailing road pricing charges and drive on regardless. This will mean that instead of the ERP system being targeted for optimum effectiveness, it will become totally ineffective in managing traffic. Hence, the decision was to have changes in the ERP charge for any ERP gantry only after at least 30 min.

The CBD road pricing scheme remained as a cordon charging scheme, with all the ERP gantries having the same ERP charge for each time period. This was to avoid motorists rat-running on the road network just outside CBD, which could be the case if the cordon-based ERP gantries have different ERP charges. Outside of the CBD, the ERP gantries introduced are based on point pricing and are meant to influence behaviour of motorists on that stretch of road alone.

When the ERP system replaced the ALS scheme for the CBD, there was a reduction of traffic entering the CBD of about 10% – 15%. This was in spite of the

ERP charges being variable (for each 30-min period), ranging from S$0.50[1] and S$2.50, which was lower than the S$3.00 ALS charge during the peak morning and evening periods, and S$2.00 for the inter-peak period. The reason for this drop in spite of the lower road pricing charges was that in the ERP system, each trip is subjected to the road pricing charge, whereas in the ALS, the road pricing charge was on a per-day basis with no additional charge after the first trip into the CBD. It is also possible that with the ERP system, there was a 100% detection of trips (for the purpose of enforcement) which was not the case with the manual ALS system, resulting in almost no one deliberately evading the ERP's camera-based enforcement system.

Foreign vehicles are also subject to the prevailing ERP charges when they come onto the Singapore roads. The number of these foreign vehicles is, fortunately, not as large as Singapore has only land links to its neighbouring Malaysia, across two points. There are only ferry and air connections to its other neighbour, Indonesia. For foreign vehicles that use the ERP-priced roads regularly, they can opt to fit an IU onto their vehicle, just like any other Singapore vehicle. Their drivers would pay ERP charges via a stored-value smartcard that they could buy (and also top-up subsequently) at banks or local outlets such as convenience stores. For the drivers of foreign vehicles who come onto Singapore roads only occasionally, they would probably not find it worth their while to have an IU fitted as they would have to pay for the full cost of the IU at S$150. Instead, they would opt for a scheme that imposes a lump sum of S$5 per day if they were detected using the ERP-priced roads for each day that they were detected. For those who do not travel on ERP-priced roads, they would not be charged at all for that day. The back-end computer system would tally the total ERP charges based on the number of days that the vehicle use ERP-priced roads during their stay in Singapore and have this amount stored and ready to be retrieved for payment at the checkpoints during departure. The mechanism of payment is via their Vehicle Entry Permit (VEP) smartcard, which is a stored-value card issued to each foreign car or motorcycle entering Singapore which can be re-used for subsequent visits to Singapore. This smartcard is akin to a passport for that vehicle. The visitor would top-up their VEP smartcard with the requisite amount when in Singapore, and the total ERP charge payable for that vehicle is deducted from this VEP smartcard at the checkpoints during departure.

6.5 Getting acceptance of the ERP system

A key factor to get the acceptance of the motoring public was that the ERP system must be proven to be completely reliable and easy to use. Hence, intensive testing was necessary to show that the system was completely reliable and indeed, the contract required the ERP system to undergo an intensive system qualification

[1] Currency is in Singapore dollars throughout. US$1 = S$1.39 as at October 2016.

testing before the mass production of the IUs was permitted. A section of an unopened expressway was used as a test track, onto which 12 sets of ERP gantries were implemented. Using a fleet of 250 vehicles fitted with prototype IUs, these vehicles were driven round-and-round this test track, clocking up about 4.8 million ERP transactions. Various test scenarios including different combination of vehicles and formation were also done. It was after this exhaustive test that the ERP system was deemed reliable and the project went on to the next stage.

Publicity was an important part of the programme, with road-shows as well as talk-shows on the radio and TV to answer queries that the public might have. Publicity material was also sent to all vehicle owners to explain how the ERP system worked and how it would benefit them. For example, motorists paying road pricing charges would no longer need to detour to buy a paper licence under the then existing ALS scheme. It was also helpful that the government decided that the first set of IU to be fitted for all vehicles would be at no cost to the vehicle owner. Furthermore, the fitting of the IU was to be non-mandatory as there were some vehicle owners who did not want the modern-looking IU to be fitted on their classic vehicle, or who just did not want an IU because they said that they would not be using priced roads or had concerns about privacy. The privacy issue was critical enough for the government of the day to announce in parliament that all valid ERP transactions would be deleted from the computer system once payments have been settled with the banks managing the stored-value smartcards, and this will typically be within a day. With all the assurance given to deal with the concerns of the motoring community, it was worthwhile to note that at the end of the IU fitting exercise, more than 98% of the vehicles at that time (1998) came and had an IU fitted.

From the publicity programme, there were also concerns raised on health issues relating to the electromagnetic waves coming from the ERP gantries and also about the IU becoming a potential missile if it got dislodged during an accident. Experts were called in and they assured the public that there was no health impact from the electromagnetic waves which in fact were less powerful than the mobile phones that were becoming ubiquitous at that time. On the dislodgment of the IU during an accident, crash tests conducted overseas showed that the fitting was secured and the IU would not become dislodged in crashes.

All in all, the publicity programme proved to be necessary and had been effective in facilitating acceptance of the ERP system by the motoring public and was worth the effort.

6.6 Enhancement of the ERP system and its operations

While the ERP system was viewed as a success when it was launched in 1998, there was always room for improvement. The motoring public was always asking for more features for their convenience, while the transport authority was always seeking for modifications to make it more effective and sometimes, more equitable to various sectors of the motoring community.

Traffic pattern changes over time, often due to changing development in various part of Singapore. Some new developments such as shopping complexes or offices would attract more trips, with the former attracting them later in the day since shops traditionally open for businesses only from the late mornings onwards and the latter (new offices) attracting more trips during the morning peak periods. Hence, there was a need to keep the ERP charges relevant over time so as to keep traffic conditions on the roads acceptable. It was decided that the ERP charges be reviewed every 3 months to reflect the changing traffic patterns. The issue was how to decide on the quantum of change in the ERP charges, if at all? The methodology had to be simple to implement and yet easy for the motoring public to accept.

Eventually, the relationship between speed and traffic flow rate was used to decide on when and how much to change the ERP charges. Based on an extensive traffic survey of free-flowing roads i.e. the expressways and urban roads with traffic signal control, the speed-flow curves for these two categories of roads were determined. Given that the intent of the road pricing scheme was on efficiency, the nearest round-up speed that corresponded with the maximum flow rate was taken to be the minimum acceptable speed. Hence, the ERP charges are increased only when the measured average speed falls below the minimum acceptable speed for these two categories of roads.

The other issue is whether there should be an upper limit to the measured average speed that would dictate that the ERP charges be reduced. The decision was that there should be one, and it was arbitrarily taken as 20 km/h higher than the minimum speed for expressways, and 10 km/h for urban roads. Hence, on expressways, the desired speed range was 45 to 65 km/h and for urban roads, 20 to 30 km/h. Speeds were measured on the roads that had ERP charges or were operating near capacity (that would necessitate ERP) and measured for half-hour periods so that the charges can vary from one half-hour period to the next. These measurements were automated with the use of up to 25,000 Global Navigation Satellite System (GNSS) equipped taxis serving as vehicle probes to measure their speeds on the roads. This was complemented with the many detection cameras along major roads that also measure speeds on the roads.

Besides the quarterly review of the ERP charges, it was also observed that traffic naturally became less heavy during the two extended school holidays in June and in December each year. This was because many parents with their children would take their vacation overseas and this translated into significantly lower traffic on the roads. Hence, ERP charges were also lowered during the school holiday periods but only selectively as many roads would still be operating near capacity in spite of the overall reduced traffic. Roughly about 20% of the half-hourly time periods on ERP-priced roads would have their ERP charges reduced during the school holidays.

Given that the ERP charges can change from one half-hour period to the next, the transition can be significant, and indeed, the difference could be as much as S$2.00. For such instances, it was observed that there would be drivers who would be waiting near the ERP gantries or slowed down on their approach to them, so as to benefit from the lower charged period a few minutes later.

This behaviour could increase traffic congestion, which was clearly not the intent of the ERP system. Hence, it was decided that if the transition of charge from one half-hour period to the next is significant, that is S$1.00 or more, there would be a 5-min transition period to have the change in ERP charge effected on a more gradual scale. For example, if the ERP charge in the first half-hour period was S$2.00 and the next was S$1.00, the 5-min transition period would have an ERP charge of S$1.50. This was to make the change in ERP rate less steep and encourage the motorist to proceed normally through the ERP gantry. This worked to some extent, and in some cases, there had been the need to install enforcement cameras and warning signs to alert motorists that roads in the immediate vicinity of the ERP gantries were under surveillance and that it is an offence to stop before the ERP gantry.

Soon after the launch of the ERP system, the supplier of the system came forward with an alternate use of the IUs that were installed in almost all the cars. This was for the payment of car-park charges based on the time of entry and exit from off-street car-parks. It used to be that motorists on entering these off-street car parks would have to push a button at the car-park entry terminal to get a parking token or a magnetic stripe parking card, which was subsequently used to compute parking charges payable before exiting the car-park. At the exit of these car-parks, the token or magnetic stripe card was used to open the barrier for the vehicle to leave. With the ERP-based parking solution, the IU identity code of each vehicle would be read remotely by a DSRC reader and associated with the time of entry. This information is stored in the car-park central computer, so that on exit, the same IU identity code would be read by another DSRC reader, and the payment computed based on the time of entry is stored in the central computer. The parking charge would be deducted directly from the stored-value card inserted in the IU. If all goes well, the barrier would open to allow the vehicle to leave, without the driver having to do anything other than making sure that a stored-value smartcard was inserted into the IU. This brought significant convenience to motorists as the whole process became automated and without the need to handle an intermediary in the form of a parking token or magnetic stripe card. This was certainly welcomed by motorists and some had mentioned that they would not park in any off-street car-parks that did not employ the ERP's technology.

But this rather convenient solution for off-street car-park was not without its problems. A feature of most commercial car-parks is that the shops would provide complimentary parking coupons if the motorist patronized those shops, usually with spending more than a pre-defined amount. This usually comes in the form of a credit-card size paper-based complimentary ticket with a bar code or some other machine-readable code; the motorist was to scan this bar code (or similar) at the ticket reader near the car-park exit to lift the barrier without payment. This, however, necessitated the removal of any stored-value smartcard inserted into the ERP's IU as failure to do so would mean that the required car-park charge would be deducted from the smartcard well before the driver had the opportunity to wind down his vehicle's window to have the bar code on the provided complimentary ticket scanned. This problem is often dealt with by the displaying notices before the exit of the car-park reminding drivers to remove their smartcards from their ERP's IUs if they have complimentary tickets.

Sometimes, the driver would forget to re-insert their stored-value smartcard into the ERP's IUs. This meant that when the motorist next passed through an ERP gantry, there would be a "non-payment of ERP charge" event occurring and the system would automatically issue a letter to the vehicle owner to demand the unpaid ERP charge and an administrative charge. This can range from $8.00 to $10.00, with the former being applicable with electronic payments via the internet or similar, while the higher than $10 charge when payment is made via cheques or other manually intensive methods. Often motorists would lament that the administrative charge is often higher than the parking charge saved with the use of the complimentary parking ticket.

The payment of the administrative charge for the non-payment of ERP charges due to the absence of a stored-value smartcard inserted into the ERP's IUs could have been due to other reasons but often this is due to the forgetfulness of the driver. Another sizeable number of non-payments of ERP charges is due to insufficient amount of stored value in the smartcard, again primarily due to the forgetfulness of the driver to top up the stored value in their smartcard, even though there would be reminder via audio alerts from the IU when the remaining stored value falls below a pre-defined value which is currently set at $5.00. In addition, when the stored value falls below this pre-defined value, the display on the IU will also show the remaining amount permanently instead of going blank under normal circumstances.

Drivers have been unhappy about this need to pay the administrative charge, even though the trigger for the imposition of the charge is their own forgetfulness. Hence, the authority had worked with the banks to offer an alternative payment mode, relying on the detection of the IU code and have payment of ERP charges settled at the back-end. To prevent the ERP gantry from activating the enforcement cameras as they would for IUs with no smartcards inserted into the IUs, all participants were required to register their vehicles and have the associated IU codes uploaded onto a list stored at each ERP gantry. The banks that administered the scheme had to impose a small fee onto the participants, so as to cover their operating costs. This would be just a few dollars every month, but because of this additional charge, this post-payment scheme was not popular, and the take-up numbers of this scheme remained small.

In recent times, the two stored-value smartcard suppliers got into the act and through their tie-up with banks and credit card companies, started offering these back-end payment schemes to drivers at no cost to the participants. Payment of the accumulated ERP charges was made together with their credit-card bills every month. The business strategy by the two smartcard suppliers appeared to be based on the credit-card companies using this scheme to increase their market share and also to cross-sell other products and services. There had been slight increases in post-payment ERP transactions since the launch of these post-payment schemes, although it was still early to ascertain its popularity at the point of writing this report (September 2016).

6.7 Operating challenges with electronic road pricing

The ERP gantry houses an array of sophisticated equipment, from detection and enforcement cameras to antennas for communicating with the IUs of vehicles passing through them. It is also a relatively heavy piece of infrastructure which needs large foundations. Given that there are many utilities ranging from electrical and telecommunication cables to water and gas pipes under the roads, the location of these ERP gantries had to be chosen carefully to minimize the need to divert these underground utilities, and hence the cost of installing them. This means that the ERP gantries might not be at their optimum pricing location but at a second-best location. Sufficient time is also necessary to build each of these ERP gantries which means that the response time to deal with new congestion locations in the road network can take up to a year or more.

There are also times that an operating ERP gantry needed to be re-located for a number of reasons. For example, this could be due to a road-widening project or to a new development being built next to an ERP gantry. This takes significant amount of time to effect and costs about $2 million for each ERP gantry.

Every 5–7 years, there would also be the need to replace equipment on these ERP gantries due to obsolescence, and again this can be rather expensive.

The gantry-based road pricing system means that the ERP charge is imposed at a point, and there was an issue of inequality because a motorist whose destination is just before the ERP gantry would be seen to be contributing to the traffic congestion but was not required to pay the ERP charge. Likewise, a motorist whose destination is just after the ERP gantry would felt being unfairly treated because she would not be contributing to congestion downstream of the ERP gantry (and her destination) but is required to pay the full ERP charge applicable at that ERP gantry. The argument would be that the actual ERP charge payable ought to be proportional to the actual distance travelled on a congested stretch of road, under what is called a distance-based pricing scheme. Of course, this could be achieved with a number of ERP gantries placed at close intervals along a congested stretch of road but the cost of doing so would be so immense that it would be impractical. The streetscape would also be rather unpleasant especially for urban planners.

6.8 The next step – GNSS-based road pricing system

Arising from the shortcomings mentioned in the preceding section, the authority started looking at the option of a GNSS-based system from the mid-2000s. Since the detection of the location of the vehicle is no longer based on ERP gantries at various locations of the road network, but on the GNSS signals picked up by the GNSS-enabled device installed in the vehicles, there was no longer a need for these physical ERP gantries. The introduction of road pricing points in the road network became more flexible, easier and faster to introduce. All it needed was software to identify the GNSS co-ordinates for which road pricing charges are applicable, and transmit them to the GNSS-enabled devices. The planners would call these road pricing points virtual gantries and the urban planners loved it as it no longer spoiled the streetscape. A quick road pricing solution may be to rely on the ubiquitous smartphones with the appropriate application installed, given that smartphones would already have GNSS positioning capabilities. However, this was assessed as not practical as there is no assurance that drivers will remember to have the smartphone with them when they drive, and even if they have, there is no assurance that the congestion charging application will be activated. Tampering with the application for fraudulent transactions is also possible as access to the smartphone cannot be protected. Hence, the on-board device has to be a customized one that can provide the assured compliance by the driver.

However, translating this ideal GNSS-based concept into reality was not so straightforward. There were issues that had to be dealt with.

First was the accuracy of the GNSS-enabled device to determine the vehicles' location. Initial trials were focused on just one satellite constellation, and it wasn't good enough. Later trials adopted dead-reckoning techniques based on electronic compass and gyroscope fitted into the devices, and then later with map matching techniques to improve the positioning accuracy of the devices. Multiple satellite constellations were later adopted, and this allowed many more satellites to be seen by the GNSS-enabled device, thereby improving the positioning accuracy significantly. There were, however, still difficulties at places with large concentration of tall buildings in the form of the urban canyon effect whereby the satellite signals bounce off these buildings and confuse the GNSS-enabled devices. This had to be overcome with roadside beacons that communicate wirelessly with the GNSS-enabled devices to provide reliable positioning data.

Second was the adequacy of the communications system to update all the GNSS-enabled devices with road pricing locations and charging information. The choice option was to use the ubiquitous 3G or 4G telecommunication network but comes at a cost. Alternatively, these can be achieved via dedicated frequency, and in Singapore, channels in the 5.9-GHz band were reserved for this function.

Third was the issue of cyber security. GNSS jamming and spoofing were of concerns and the GNSS-enabled devices had to be designed to detect and to have mitigating measures to deal with these possibilities.

A series of trials with potential suppliers of the GNSS-based road pricing system were conducted in the early 2010s, and these issues, along with other operating requirements, were tested out. The results were evaluated carefully and it was only when there was sufficient confidence that a GNSS-based road pricing system could work that a tender was called. This was eventually awarded to a consortium comprising Singapore's NCS Pte Ltd and Mitsubishi Heavy Industries in early 2016. This next-generation GNSS-based road pricing system would not only have all the road pricing features of the current gantry-based road pricing system, but will have distance-based road pricing. On top of that, the GNSS-enabled device to be installed in the vehicle, to be simply referred to as the On-Board Unit or OBU, will have a 5-inch display that would be able to provide adequate information of the presence of road pricing points, as well as the applicable road pricing charges in real-time. This would allow the drivers to make informed decisions on the route to take, depending on the prevailing road pricing charge and the driver assessment on the value of paying the road pricing charge.

In addition, the OBUs will also function as a probe vehicle and can transmit back information on the prevailing traffic conditions along the stretch of route that the vehicle is travelling on. Only a selection of vehicles' OBUs is needed to give reasonably accurate traffic information, thereby keeping the communication costs affordable. The traffic information is processed and can be re-transmitted back to the OBUs of vehicles to alert drivers on traffic conditions on the road network, and possibly allows the computation of journey time to their destination. This real-time traffic information will be of value to many road users, especially logistics vehicles as well as emergency vehicles, as the quickest route to their destination could be determined in real-time. Another application of the OBU is the continued use of the device to facilitate the payment of car-park charges, both for off-street and on-street parking.

6.9 Conclusion

Road pricing as a means to manage traffic in Singapore has a long history, starting with a manual system in 1975 and then replaced with a DSRC-based road pricing system that offered improved convenience to motorists and improved effectiveness for the authority. This will be improved further when the GNSS-based road pricing system comes into operation from 2020 onwards and will double up to provide other value-added services such as the monitoring of traffic conditions and disseminating such traffic conditions to drivers on the roads to make the road network operate more effectively. While road pricing has been effective in Singapore for managing traffic, this approach to be adopted elsewhere has to be customized to suit the nature of the urban city as well as the culture and acceptance of the people.

Further reading

[1] Menon A.P.G., and Chin K.K., 'ERP in Singapore – What's Been Learnt from 5 Years of Operation?' Traffic Engineering & Control. 2004; **45**(2): 62–65.

[2] Chin K.K., 'Road Pricing – Singapore's 30 Years of Experience.' CESifo DICE Report, Journal for Institutional Comparisons. 2005; **3**(3): 12–16.

Chapter 7

Communication and governance challenges in Greater Manchester's 'congestion charge' referendum

*Graeme Sherriff**

7.1 Introduction

In December 2008, the Greater Manchester (GM, United Kingdom) electorate voted to reject a £3 billion funding package including wide-ranging investment in the conurbation's transport infrastructure together with, and partially funded by, congestion charging. The vote followed a 2-year period of debating and campaigning, upon which this chapter reflects. The chapter focuses in particular on the ways in which the plans were communicated to the GM public and the governmental, political and media context in which they were debated.

It adds to work by the author and others [1–4], providing a deeper analysis and reflecting on subsequent developments. By investigating the parameters of the debate and exploring the implications of the vote, it makes a contribution to the under-researched field of transport policy development in GM through a highly significant case study in which tensions and political differences over the future of transport in the area are brought to the surface. It therefore contributes to international literature on sustainable transport in general, and road pricing in particular, and to the understanding of how to pursue policy developments even when they can be controversial and portrayed as 'anti-car'.

Given his personal involvement in the decision-making process, the author had access to sources additional to those normally available through a desk exercise. These include insider discussions, attendance of meetings of campaign groups and public debates, conversation with the public during campaigning activities and campaign materials, such as leaflets and posters, that are not available in public archives. The author was personally engaged in the debate through a group that was part of the campaign supporting the plans. This affords a level of familiarity with, and closeness to, the field that would be difficult to achieve using retrospective approaches.

*Sustainable Housing and Urban Studies Unit, University of Salford, UK

The chapter begins by looking at the ways our relationship with transport is both a burden and – looking to the future – an opportunity. It considers the lens of car dependency and employs this as a way of understanding the potential contribution of GM's plans and of placing them within the context of sustainable transport policy in the United Kingdom and internationally. It then looks at the reactions to the proposals, describing the viewpoints expressed by those in favour of and opposed to the plans. Particular challenges in progressing the plans are then explored, first in terms of media and communications and, second, government, business and politics. It then looks at the situation following the vote, firstly by outlining policy developments in GM since 2008, and secondly through a discussion of implications and lessons for road pricing as a whole: what does the experience mean for transport policy in GM and other cities, and for our understanding of car dependency?

7.2 Transport as challenge and opportunity

Transport is at once the lifeblood of our towns and cities and a burden to them, lubricating our economies and keeping people on the move whilst bringing with it a wide-ranging set of well-established impacts. Such impacts include environmental damage, such as climate change [5], local air pollution [6] and biodiversity loss; social issues, including social exclusion [7], loss of community [8] and inactivity [9]; and economic impacts, including congestion [10] and healthcare costs [11].

The notion of car dependency provides a useful lens [12] and Hickman and Banister [5, pp36] argue that this is 'still likely to be the reality, both now and in the short-term future'. In describing car dependency, commentators point to a vicious circle in which, as motorised traffic grows, and urban form shapes itself around it – sometimes dubbed 'carchitecture' [11, pp41] – walking and cycling can become less attractive options [9], public transport becomes stuck in the same queues as the car, and 'the fragmented city becomes hostile and can even force reluctant users into their cars, thus exacerbating the problem' [13, pp26]. The dominance of cars enables the speeding-up of common tasks – although congestion constraints limit this – such that 'with the remaking of space around the car comes a restructuring of temporalities' and, as more journeys become difficult or impossible without private transport, the car can 'create the conditions of its own necessity' [14, pp492].

In car-dependent societies, lack of access to a car is a key defining factor in social exclusion [7], since it becomes increasingly necessary to own a car to access essential services such as education, health care and employment. In the most deprived wards in Manchester, 60% of households have no access to a car or van.[1] Across cities, communities enjoy varying levels of access to public transport networks, with implications for participation in society: 'the availability or

[1] The 2011 UK Census results indicate that 44.5% of Greater Manchester households have 'no cars or vans in households', and that figures for the Ardwick, Moss Side and Harpurhey wards are 63.3%, 60.2% and 58.3%, respectively. Available from the Office for National Statistics [15].

unavailability of transport... shapes people's life opportunities' [16, pp55]. High levels of private mobility can contribute to social exclusion through environmental degradation, health impacts and community severance [17]: those experiencing a lack of access to opportunities also suffer some of the worst impacts of the transport system – 'they are both "less travelled" and more "travelled-upon"' [18, pp5]. The pervasiveness of the car, it is argued, instils a cultural acceptance of its costs [9].

Whilst the notion of car dependency is compelling, however, there are aspects of car use not directly related to utility. In this sense 'the car is associated with both instrumental and affective emotions' [5, pp32], implying that policy to promote alternatives to car-dependence needs to consider not only getting from A to B, but also the experience that car-based mobility offers and the culture that surrounds it.

A range of measures are available to policy-makers. Santos *et al.* [19] divide these into physical policies, such as developing infrastructure for public transport, walking and cycling, and integrating with land use planning; soft policies, such as car sharing and car clubs, eco-driving training and awareness-raising; and knowledge policies, such as incentivising research and development into low-emissions and electric vehicles. Banister *et al.* [20] classify policy approaches by their entry points into society, whether through lifestyle, market, regulation or public infrastructure and public transport. In the literature, there is increasing recognition of the importance of lifestyles and behaviours [21,22], and the importance of understanding transport as a socio-technical system [14,23].

Within this mix, road pricing and congestion charging are options that seek to disincentivise driving, at least at certain times and in certain places, and, through potential revenue generation, provide improved alternatives to the car. It is fair to say that this idea has long been considered a policy option but 'has enjoyed only spasmodic success in terms of actual implementation, more often failing to construct any climate of acceptance, and instead generally experiencing public and political rejection' [24, pp1145].

As the contents of this book evidences, Manchester's attempt to implement road pricing sits within an international story of the development of this approach. Prominent examples of implemented schemes include Gothenburg, Singapore, London, Stockholm, Milan and a set of Norwegian cities (Chapters 6, 8, 14 and 18). Non-implemented cases include Copenhagen, Hong Kong, New York and Edinburgh [25–27] (Chapter 8). In Edinburgh, plans were developed but, like Manchester, were rejected in a public vote. In Copenhagen and New York, the congestion charge was discussed and rejected by politicians [28].

Hong Kong ran a 21-month pilot stage trail of an electronic pricing system between 1983 and 85 but public opposition meant that it was not continued [29]; Manville and King [30] reflected that voters did not believe government promises to rebate toll revenue to drivers, but see Chapter 8, Section 8.4.5, for more recent developments. In Edinburgh, car use appeared to be a clear determinant of voting against the plans and Gaunt, Rye and Allen identify a range of factors including complexity and limited public understanding, lack of faith that this would be an effective solution to congestion, and even that the public transport improvements were 'insufficient, irrelevant or ill-defined' [31, pp100].

This chapter, then, contributes to ongoing work in understanding why schemes involving congestion charging often fail the public acceptability test, rather than the body of work that looks at the efficacy and impact of congestion charging itself. In reflecting on Manchester's experiences, familiar themes of public trust, understanding and acceptance recur, and place it within this international debate, whilst particularities of political structure, spatial distribution and public transport provision emphasise the importance of situating the local within a global context.

7.3 Greater Manchester's Transport Innovation Fund (TIF) bid

In July 2007, the Association of Greater Manchester Authorities (AGMA) placed a bid to the UK Government's Transport Innovation Fund (referred to here as GMTIF). AGMA is the local government association for GM, with representation from the ten boroughs in the conurbation. The decision to go ahead with the bid followed a period of public engagement. Less formalised discussion started earlier than this formal consultation, following the Manchester Evening News (MEN) headline on 24 January 2007 'Congestion Charging: It's coming our way' (Figure 7.1) [32].

From a UK point of view, the plans followed the earlier introduction of congestion charging in London in 2003. This happened alongside significant improvements in an already well-regarded public transport system [33]. It was implemented by Mayor Ken Livingstone and represented a 'highly visible, very deliberate, very political effort to reduce car use and congestion in the capital' [34, pp473].

Manchester Evening News
North west newspaper of the year
Wednesday 24 January, 2007
CITY EDITION
manchestereveningnews.co.uk

Congestion charging: It's coming our way

Pay-to-drive in Greater Manchester within five years

Fifteen main routes into city will trigger fee of up to £8 each day

Figure 7.1 Front page of Manchester Evening News, 24 January 2007

The Transport Innovation Fund (TIF) was aimed at local authorities and required a focus on demand management and anti-congestion packages, reflecting government recognition of the London congestion charge combined, arguably, with reluctance to commit to a national scheme [35]. This followed a statement from government to work alongside local authorities, with the Transport Act 2000 providing the necessary powers to local authorities to introduce charging regimes subject to approval by the Secretary of State (i.e. the Minister of Transport) [36].

In AGMA's bid, £1.5 billion from TIF would be made up to approximately £3 billion by borrowing on revenue from a congestion charge over 30 years [37]. This charging element was the most prominent part of the package, and would be weekday peak-time only, in the direction of flow of peak-time traffic, chargeable on driving through either or both of two cordons. Depending on when and how many cordons were passed, the charge would be between £1 and £5 over a day (see schematic, Figure 7.2).

In so far as the congestion charging element of the plans could be considered a 'stick', other elements of the plan could be recognised as complementary 'carrots', with incentives for public transport and active travel, including extensions to the Metrolink tram network, an expanded cycle lane network, additional yellow school buses, new train carriages, improved interchange facilities and better integration of the bus network. Such investment would create opportunities for currently car-dependent commuters to take other options.

The plans included closer integration of services and smart ticketing and these, combined with an increase in the number and frequency of services, were intended

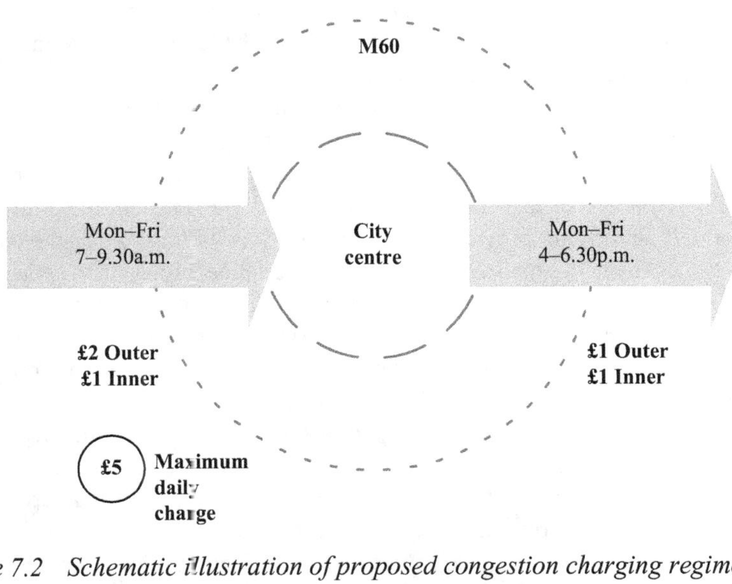

Figure 7.2 Schematic illustration of proposed congestion charging regime. Based on billboard advertisement 'Would you need to pay the congestion charge' shown in [38]. The M60 is the outer motorway ring road around Manchester

to make public transport a more attractive proposition. With the currently privatised, multi-operator system of buses, trams and trains, the area can appear to have a collection of services rather than a network, a situation that has been observed in England and Wales outside of London [39].

The plans also included personalised journey planning and promotional activities. These 'soft measures', aiming to bring about behavioural change through information about consequences and alternatives [19], were intended to act alongside the infrastructure improvements and help commuters explore options, helping to tackle some of the behavioural aspects of car dependency, such as lack of experience of public transport and confidence when cycling.

7.4 The plans as response to the challenge of car-dependent cities

The GMTIF plans can be understood to be a response to the challenges of car-dependency in cities and as an attempt to mitigate the environmental and social impact of increasing car use. However, whilst with hindsight it is possible to see these potential societal impacts, it has been noted that the discourse around the referendum was focused heavily on economic dimensions: 'given that one motivation for the proposed congestion charge scheme was protecting the environment through reducing car use, the campaign on the referendum had been overwhelmingly dominated, both online and off, by an economic discourse' [40, pp89]. Nonetheless, references were made to broader societal goals, including reducing carbon emissions, improving air quality and reducing social exclusion.

One way in which the plans responded to these challenges was the implication of modal shift. As mentioned, there was a clear 'carrot' and 'stick' approach. By making driving more expensive at certain times, and investing in more sustainable modes, the plans had the potential to bring about modal shift by, for example, 'produc[ing] a virtuous cycle in shifting motorists to public transport' [19, pp53]. It should be noted however that the 'stick' was primarily a financial one – there was no impediment on driving provided the individual was willing to pay the charge – whereas the 'carrots' related to quality of experience rather than price – for example bus coverage and frequency, ticket integration and cycle lane provision.

In this sense, the plans represented more than the provision of incentives and disincentives, since the intention was that they would give people more options to travel without a car. This can be understood in simple terms, in that new bus and tram routes would present the opportunity for some current car commuters to utilise public transport for some or all of their journeys – whether they would take this opportunity is of course a different matter, and some research has suggested that the provision of public transport itself does not affect vehicle kilometres travelled [41]. Similarly, new and improved cycle routes may have made cycling a more realistic or attractive option for some. At a more systemic level, indirect effects could be anticipated: a new school bus route may have freed a parent from driving a child to school, therefore reducing the need for trip chaining, making public transport a more viable option; reduced traffic levels during the rush hour could make walking and cycling a more attractive option.

Moreover, GMTIF represented something more than a trickle of transport improvements, important as they are individually. It would have been a programme of major investment over a 5-year period with substantial changes to the public transport offer in the conurbation followed by the introduction of the congestion charge. This would create opportunities for transport provision to be a high-profile news item, at least in the time leading up to the introduction of the charge; for the personalised journey planning service to help people consider their modal choices; for businesses to work with employees to help them think about their commute and potentially introduce flexible working policies to facilitate modal shift. Accepting, again, that they remain a theoretical possibility, they resonate with Hull's [42, pp96] observation that the cumulative effects of car dependency 'warrant a coherent and focused intervention' and perhaps provide 'window[s] of opportunity in which travel habits (e.g. driving) are disrupted and must be renegotiated' [43, pp485]. In this sense, the impact of GMTIF would have been more than the sum of its parts.

As a theoretical proposal, GMTIF contained within it many principles that transport researchers would associate with facilitating modal shift. This is not to claim that it was the perfect, complete package however. There were a number of omissions that could be seen as missed opportunities. For example, the plans contained nothing explicit that would encourage the use of cleaner vehicles. Whilst the publicity made claims around improvements in air quality and carbon emissions, the opportunity was not taken to reduce the congestion charge for electric or hybrid vehicles. There was also nothing explicit to help people on low incomes, apart from a short-term reduction on the congestion charge of 20% for the first year [44]. Additionally, the authorities could not guarantee that public transport fares would be cheaper.

In many ways, the complexity of the GMTIF scheme reflects an attempt to overcome some of the criticisms commonly made of road charging, as Vigar *et al.* agree [2]. By using a loan to enable public transport improvements before introducing charging, there was potential to overcome a common 'chicken and egg' dilemma: public transport improvements being required upfront, yet the money from charging needed to fund them becoming available only downstream. The planned charging regime could be seen as less 'blunt' than, for example, the London charge, with the charge being incurred only when crossing one of the two cordons, only at certain times, and only in the direction of peak-time traffic. This meant that the charge was in principal more precisely targeted at congestion itself, at the times and directions it occurred, and as a result afforded more options to drive and avoid paying the charge.

7.5 Reactions to the proposals

Once national government had accepted the bid in June 2008, subject to public approval, there was a period of debate and a formal consultation on the proposals.

Campaigning for and against had been continuing and there were calls for a referendum, particularly from opponents [45].

In July 2008, AGMA announced that there would be a referendum in December of that year. A representative sample survey of residents run by Ipsos Mori in August found before the referendum that just over half (53%) of residents were in favour of the congestion charge scheme [46]. However, the result announced on 12 December 2008 was strongly against: 78.8% against and 21.2% for, with an overall voter turnout of 53.2% [47]. The proposals were therefore not implemented. The discrepancy between the opinion polls and the final result may stem from a number of factors: the time between them, which would have been the period of most intense campaigning, during which voters had time to change their minds; the likelihood of those with only a mild interest in the issue responding to the poll but not being motivated to vote; and the difference between an 'in principle' commitment in an opinion poll and a more calculative vote that feels more real and anticipates the ways the charge would 'hit the pocket'.

Although in the initial stages, there were no plans for a referendum, 'pro' and 'anti' alliances formed, and these could be mapped onto the subsequent 'yes' and 'no' positions. An early presence in the 'pro' campaign was Clean Air Now (CAN), a coalition led by environmental campaigners, and later joined by the official 'yes' campaign and United City, a coalition of businesses. The 'anti' campaign was initiated by campaign group Manchester Against Road Tolls (MART), backed by the national Association of British Drivers, and later joined by a coalition of businesses, the Greater Manchester Momentum Group (GMMG) [48]. Shark and squirrel characters became mascots for the 'no' and 'yes' campaigns, respectively, with the shark attending consultation road shows and appearing in viral videos taking money from people as they tried to buy a beer or go on holiday; the squirrel responding with the motto 'Say no to £3 billion? You must be nuts!'. The arenas for campaigning included the local press, with some newspapers taking particular campaigning stances; local radio and television; various organised public meetings; the tables of the area's pubs, on beer mats; and billboards around the conurbation.

As a methodological note, it is important to remember that the viewpoints expressed on either 'side' represent campaign messages rather than being, necessarily, an accurate representation of public opinion. The views of the public are not necessarily the same as the claims being made by powerful interests and active campaigners. To this end, the claims have to be recognised to be as much tactical – i.e. the claims that the groups thought would help them achieve their goal – as representative. Hepburn [49] looked at the ways in which the referendum was discussed in online forums, for example. He noted the ways in which 'no' campaigners had been observed '"supporting each other online" and endorsing others' comments and 'harassing' any 'Yes' campaigners who came online' [49, pp60]. In fact, a 'No' activist associated with the manchestertolltax.com website acknowledged to him that 'they had organised their supporters from across the country' [49] to interact with this debate.

Whilst the two 'sides' did not represent homogenous constituencies of opinion, there were overarching themes that characterised the arguments being made.

Figure 7.3 Photo of Manchester Against Road Tolls campaign billboard. Photo: Anonymous

We cannot, of course, assume that businesses and organisations who opposed or supported the plans necessarily signed up to all the claims made.

Within the set of viewpoints and campaign angles, there are many stances that directly contradict. Where those against the plans argued that the congestion charge would benefit the rich, for example by 'keeping the roads free for the rich' (Figure 7.3) for whom the £5 proposed cost would be proportionally less of their income, those in favour argued that the plans contained socially progressive elements and that it would be the poorest who would benefit most from better public transport links and reduced air pollution.

An early advert claimed that the plans were tantamount to telling nurses, firemen [sic] and teachers to get 'On Yer Bike' (Figure 7.4), highlighting that the congestion charge would represent a higher proportion of the wages of lower paid workers, whilst those campaigning in favour organised a 'Tour de TIF' bike ride covering the whole of GM and highlighting the potential for 'cleaner, cheaper and more convenient public transport options will take cars off the road and make cycling safer and more enjoyable' [50]. On the one hand, then, the plans could result in more people feeling that they needed to cycle to work; on the other, this may help to make cycling a more attractive option and have health and environmental benefits.

In the discourse against the plans, concerns about the costs to businesses were raised. Kellogg's, for example, welcome the public transport improvements in principle but argued that 'they will not help us move cornflakes' and therefore, the MEN reported, that it would be 'too difficult to shift cereals from its Trafford Park plant' [51]. A coaster produced by the GMMG argued that 'even if you don't commute by car, additional costs to business will be passed on through shops, pubs and other outlets' (Figure 7.5).

Figure 7.4 *Photo of Manchester Against Road Tolls campaign billboard. Photo: Anonymous*

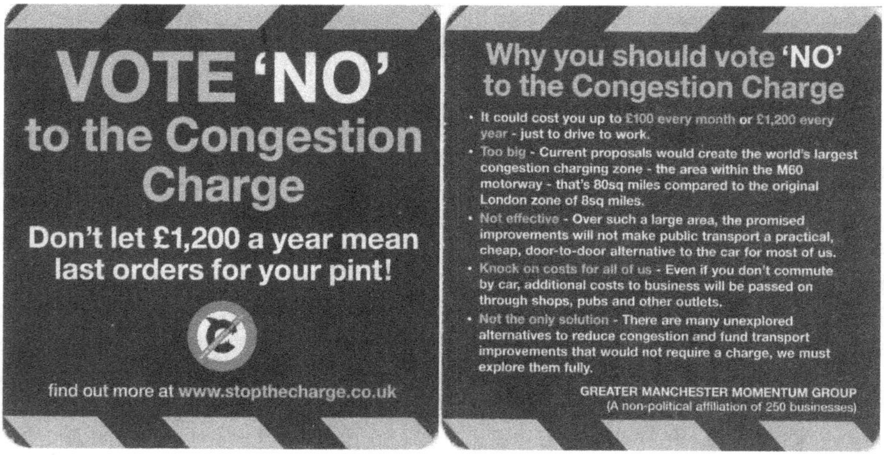

Figure 7.5 *Campaign coaster produced by Greater Manchester Momentum Group*

Figure 7.6 Campaign billboards produced by the YES Campaign. Design by Creative Concern, Photography by Graeme Cooper

Conversely, campaigners in favour highlighted the potential positive impact on businesses and the economy in terms of the costs of congestion and unreliability, as shown in the 'That's why I'm voting Yes' billboards (Figure 7.6): 'I can't afford to be late for any more meetings' and 'I want my deliveries to be on time'. It was also part of the core messaging of the 'yes' campaign to claim that 30,000 jobs in the area were at risk from congestion and that there is a general case for the economic health of the area: 'We can all have a laugh, but we are talking here about tens of thousands of people's jobs and the future of the region' [52]. 'Even drivers will benefit the argument goes', reported the MEN, 'because they will see their journey times reduced' [52].

Costs to individuals were also highlighted. A viral video showed a man standing at the bar in a pub with a beer and a shark coming up behind him and taking the drink with the message 'If the Congestion Charge comes in, it could cost you up to £100 a month .. Don't let the Congestion Charge be last orders for your pint'. Conversely, pro-campaigners highlighted that many people would not have to pay the charge, such as older people travelling off peak: 'And it is those who will not pay who feature in the Yes Campaign's big poster campaign – the off-peak traveller, the bus and train commuter – with the message "9 out of 10 people won't pay" (see Figure 7.7 for example of this slogan). The campaign attempts to convince people that the charge is the fairest way to pay for public transport improvements' [52].

Figure 7.7 Leaflet produced by YES Campaign. Design by Creative Concern, Photography by Graeme Cooper

Those opposed to the plans referred to the financial risk the city was taking, with half the GMTIF money being in the form of a loan, paid back through the congestion charge. Conversely, campaigners in favour of the plans highlighted the opportunity and potential of the investment: an example being the leaflet 'You wait years for some transport investment and then £3 billion comes along at once. Going to let it go by?' (Figure 7.8). Whilst opponents tended to frame this as a 'bribe', proponents tended to emphasise the opportunity presented by the money.

An objection raised by those opposed was the extent to which the congestion charge apparatus would entail tracking drivers: an early advert contained the words 'We're watching you: Tagged, Tracked and Taxed. The Toll Tax'. The 'yes' campaign did not focus on this issue, but did seek to reassure people that the mechanism would not be 'tracking' per se, but would instead register whether cars pass cordons during the hours of charging: there was an implication, however, that this was a reasonable price to pay for the potential benefits of the scheme.

A prominent theme for the proponents of the scheme was a reduction in the emissions from transport and the impact upon air quality and climate change. CAN, a coalition of groups and businesses, focussed their campaign on this, arguing that the plans, in controlling traffic levels and introducing newer more modern buses, would contribute to reducing air pollution. A campaign towards the end of the

Figure 7.8 Leaflet produced by Clean Air Now

period featured children in gas masks with the caption 'Is a congestion charge too much to pay for clean air?' [53].

These environmental issues did not feature in the campaigns of those opposing the plans. There were some examples of attempts to downplay the significance of air pollution for asthma and to claim that modern cars are cleaner, for example on the Manchester section of the National Alliance Against Tolls. 'They say that tolls are needed because it would result in a tremendous improvement in air quality and eliminate asthma and other respiratory diseases. But there is little evidence that external air pollution is a significant asthma factor though it may have an effect on those who already suffer from it. Cars are many times cleaner than they were . . . ' [54].

It is noteworthy that the arguments presented by those in favour of and opposed to the proposals are often two sides of the same coin, or the same issue perceived differently: whilst the charge might 'keep the roads free for the rich', it would allow for improvements that would give low-income commuters more options not to drive; whilst the plans seemed to be saying 'On Yer Bike' to poorer commuters, facilitating active travel can be seen as a positive way of boosting health and social inclusion; whilst the loan element was a risk, the funding was an opportunity; whilst businesses may incur charges, the proposals would help to reduce congestion, therefore potentially helping businesses make their deliveries on time.

Fundamentally, these opposing viewpoints can be seen to reflect different conceptualisations of justice. The close association between mobility and freedom translated into claims of social justice, albeit differently framed. The justice claims of the opponents can be understood to stem from an emphasis on mobility and a close association between this and private car use. The potential for the charge to affect poorer drivers the most was highlighted, with the implication that the roads would, in effect, be 'free for the rich' (Figure 7.3) and the poor would not be able to access work without resorting to the other modes of transport, such as the bicycle (Figure 7.4).

Conversely, the proponents tended to present non-car modes positively. CAN's 'Tour de TIF' cycle ride highlighted the opportunity to promote cycling as a healthy and sustainable mode. A leaflet produced by CAN suggested a 'yes' was an early Christmas present and highlighted the extent of the public transport investment. They argued that it would be the poorest in society, who do not have cars, who would benefit the most from the investment. They also tended to make reference to environmental issues such as air pollution and climate change, and argued that quality and reliability of journeys would improve.

The claims, then, represent alternative approaches to addressing social exclusion related to car dependency. The 'no' campaigners argued that making car travel more expensive exacerbates social exclusion, whereas the 'yes' proponents argued that, by investing in alternatives and reducing the growth of car use, policy can seek to be more inclusive.

7.6 Communicating complex proposals

7.6.1 Introduction

As discussed above and previously [4], the proposals were relatively complex. Although GMTIF was routinely presented as a congestion charge plan, the reality was that it was also a major public transport investment package that was contingent upon the introduction of a congestion charge. The project was 'characterised by a complicated message' [3] in a field of planning that is already a complex, and contested, field.

This discussion is premised on the assumption that the media have a significant effect on opinions and attitudes [55]. 'In our media-mediated world', argue Brand and Fischer [56, pp13], 'small chunks of information sell . . . information for the masses has to be made accessible and straightforward, which pushes towards a simplified picture of a complex scenario'.

Whilst the level of complexity reflected a level of sophistication that made the scheme more precisely targeted at congestion, and this should arguably have made it easier to justify and 'sell', the reality was that this complexity appeared to hinder rather than help its passage through political and public discourse.

Others have commented on the lack of penetration of some of the more complex elements of the plans, and highlighted a degree of simplification. Vigar *et al.*, in their survey of media in the run-up to the vote, agree that 'it is apparent that many dimensions of GMTIF were under-explained' [2]. Concerns over the clarity of communication have led some to question the extent to which people fully understood the plans and, in particular, the longer term implications [35]. 'Very few people', argued Roy Newton of the Joint Transportation Policy Team, 'got the message that a significant investment in public transport was the major component, and even fewer people knew that public transport investments would have to be fully implemented before the congestion charge would begin' [interviewed in 54]. Gaunt, Rye and Allen made a similar observation about Edinburgh's proposed two-cordon system: 'It was clear that it was too complicated to be understood, never mind supported, by a majority of the public' [31, pp100].

There are several reasons to suppose that the public received a rather simplified version of the plans through publicly available discussion [4]. First, there was a particular focus on the congestion charge element of the proposals with the process repeatedly referred to with variations on 'congestion charge referendum'. This is clearly unsurprising, and certainly with the benefit of hindsight. An amount of £3 billion of investment in transport is not in itself controversial, and therefore, the vote was always going to be primarily about the congestion charge element. Vigar *et al.* agree, also finding that the improvements were often presented as a 'sweetener', with the congestion charge as a 'bribe' [2]. This framing was common from early in the process, an example being 'The government is offering GM up to £3bn for public transport improvements in return for the peak hour only charge of up to £10[sic]' [57].

Second, the plans themselves were routinely simplified, in both campaign materials and media reports, particularly the financial aspects. The complexity of the charging system created opportunities for simplification. Some advertising on billboards, beer mats and viral videos repeated annual and daily figures, typically £5 per day and £1,200 per year: in each case this was the maximum charge and, as discussed above, was contingent on time and direction of travel, but individuals would have to dig into the detail to find this.

The official 'yes' campaign, which could have broadened the debate to draw attention to the wider benefits, seemed to be pushed to respond at the level of how much the individual would pay. It circulated leaflets in the run-up to the referendum and these included information on the public transport improvements, but the core messaging of these and the billboard adverts that accompanied them was '9 out of 10 people won't pay – will you?' (Figure 7.7) apparently countering the 'hitting your pocket' focus of the opponents by appealing to those who would not, given their travel patterns or personal situations, pay the charge.

Third, comparisons with the London system belied the comparative complexity of GMTIF. The GMMG coaster (Figure 7.5) states that it would 'create the world's largest congestion charging zone – the area within the M60 motorway – that's 80 sq miles compared to the original London zone of 8 sq miles' – yet the schemes were quite different, with London's based on a zone and Manchester's around two cordons.

Throughout the almost 2-year period of debate, there are examples of potentially misleading information, whether wilful or not. One example is the MEN front

Manchester Evening News

manchestereveningnews.co.uk

Congestion charging

You say NO

- Two out of three reject plans to bring in pay-as-you-drive
- But 59 per cent back move - if it brings cash for Metrolink
- Biggest-ever poll on road pricing plan carried out for MEN Media

See pages 4 and 5 for the complete results of our exclusive study

Figure 7.9 Front page of Manchester Evening News, 21 March 2007 (extraneous news items removed)

page in March 2007 ‘Congestion Charge, you say NO’ on which it was reported that ‘two out of three reject plans to bring in pay-as-you-drive’ and ‘but 59 per cent of you back move [to congestion charge scheme] – if it brings cash for Metrolink’ [57] (Figure 7.9). The plans did include funding for Metrolink, so the headline would more accurately have read ‘you say YES’. Whether intentional or not, such reporting would not have helped the public to clearly understand the proposals, and again seemed to consider the congestion charge in isolation from the proposed public transport improvements. Some media organisations openly ran campaigns against the proposals. This is not to argue that the media consistently misrepresented information or was always against GMTIF. Rather, it is to highlight the charged and potentially confusing way in which the details of the GMTIF were communicated to the public.

Also of interest in seeking to understand how the plans were communicated is the ways in which road users are framed. Given the focus of the media on the congestion charge element of the plans, ‘the driver’ seemed to take centre stage in media coverage, implying a simplified conceptualisation of mobility. The repeated headlines and claims like 'Drivers to Pay £1,200' a year were simplifications of the proposals and belied the potential for public transport investment to create feasible alternatives for many 'drivers'. One of the first headlines in the MEN, for example, was 'Is THIS really an option for drivers?' with a photo of a crowded tram, implying not only that being a 'driver' is an identity, rather than a performance of the practice of driving, but also that public transport was in some way not good enough for them. However, as King *et al.* [58] argue, although the charging is likely to 'harm most drivers, no one is only a driver' – a 'driver', or their family, could benefit from being able to use a new tram route, or benefit as a parent or resident from lower traffic levels. Constructing and perpetuating this identity of 'driver' can be seen to be problematic, considering that a core principle of sustainable transport is a shift from emphasising one mode (uni-modality) to embracing plurality (multi-modality) [11], such that the mode of transport is determined by the demands of a particular journey rather than habitual use.

Some proponents did question this. After the MEN launched a survey to let 'drivers' report problems with congestion, Manchester Friends of the Earth, the Community Network for Manchester and the Environment Network for Manchester signed a letter to them arguing that 'No one is simply a "motorist" – we all have a range of transport modes available to use, depending on the journey we are making' [59].

Vigar *et al.* found that in general 'the press resorts to a narrow repertoire of words and metaphors to define transport issues' [2, pp478] and identified a need to better understand how these are interpreted by the public. Brand [23] argues that the question that is asked, and by extension the way the question is framed, can delimit the perception-space of answers, such that 'limits to describe are limits to imagine' [23, pp67]. It could be argued, then, that a certain poverty of vocabulary was evident in the press coverage, and such poverty may have translated into a poverty of argument and vision.

7.7 Political leadership and governance

Ahmed makes the point that by this time the equipment for congestion charging was already available and proven and that 'the emphasis has shifted to public and political acceptability as the key constraints' [36, pp2]. Clearly part of this challenge rests with the decision-makers of a conurbation. Observations above of the GMTIF debate present a picture that is a long way from notions of the well-informed electorate and communicative rationality [60], and highlight the need to recognise power and knowledge disparities, strategic behaviour and deception [61]. They also highlight the ways in which governance structures can contribute to inertia in decision-making and the limitations of a referendum process in representing opinion on a complex issue.

It cannot of course be known what the outcome would have been under different circumstances, but it is clear that aspects of GM's governance undermined efforts to forge a decision to go ahead with these plans. GMTIF was born into a

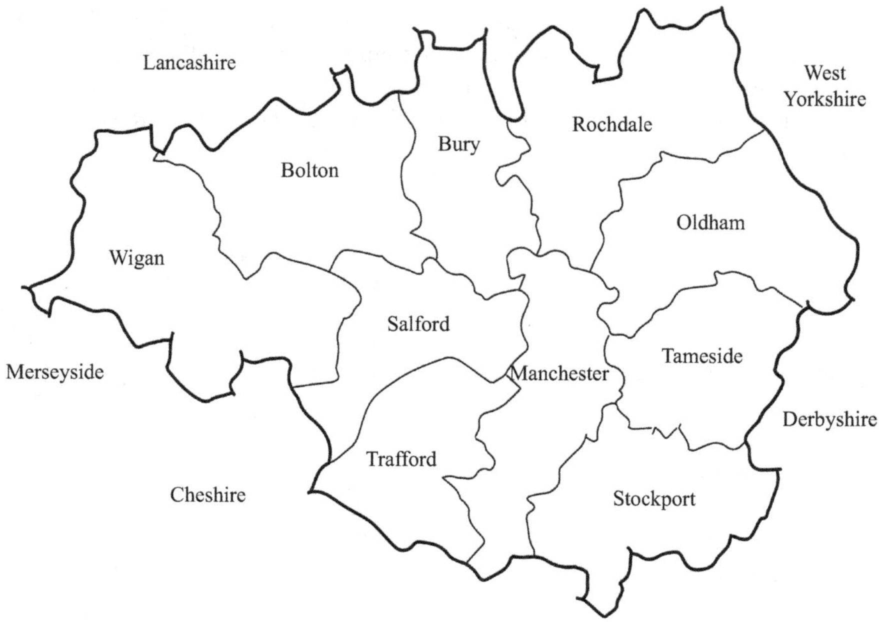

Figure 7.10 Map of Greater Manchester showing borough and city boundary and neighbouring Counties

fragmented governance setting. The ten local authorities in AGMA (see Figure 7.10) were a combination of Labour, Liberal Democrat and Conservative and AGMA sought agreement between these political parties in order to progress the plans. For this purpose, AGMA amended its normal consensus vote to a majority vote [62] and decided, with a majority of eight, to bid for TIF funding. Central government's decision to award the conurbation the funding was influenced by AGMA being able to demonstrate broad support for the proposals, evidenced by an early consultation exercise[2] [63]. Nevertheless, the months that followed were characterised by further political debates. At the time of the referendum, three councils were opposed: Stockport, Trafford and Bury [48]. Stockport Borough Council was vocal in its opposition [64] and in the period running up to the vote attempted to arrange bus advertising to oppose the plans [65].

The proposals lacked a high-profile, top-level figure to champion them, in contrast to London, where Ken Livingstone personally championed the congestion charging plans, and it was he, rather than national government, who took the political risk [35]. The Greater Manchester Passenger Transport Authority (GMPTA) and AGMA were expected to maintain a neutral stance, co-ordinating

[2] 57% of residents surveyed agreed with the principle of paying charges in return for the package of public transport investment. 68% agreed that a bid should be submitted to TIF, but that no decision should be made on the charging scheme until more details were available.

the consultation and disseminating neutral information about the plans. In such a charged political climate, they could not be seen to be taking sides. Councillor Roger Jones, arguably the public figure most closely associated with GMTIF in 2008, lost his seat in the 2008 local elections and subsequently stepped down as chair of GMPTA. During the election, much was made of his involvement with GMTIF. Having publicly supported the plans, this was seen by some in the media as something that should 'send a message to his colleagues and potential successors' [66] and, by MART, as confirmation that 'the congestion charge is a cast-iron vote loser' [66].

It is striking that this political 'mud' stuck. Even in June 2011, on re-election, the local press referred to him as 'Congestion Charge Mastermind' [67] and referred to him having previously been 'ousted', illustrating the longevity of the reputational risk of supporting a controversial scheme. This legacy affected the Manchester Labour Party as a whole: when the Greater Manchester Combined Authority was created in 2010, it was decided that all issues would be decided on a simple majority, except for road pricing, which would require unanimity [68]. A local councillor commented that 'this is a very good safeguard against Labour leaders and non-elected civil servants within some councils trying to force a congestion charge upon GM' [69]. More recently, when GM Labour mayoral candidate Andy Burnham was tweeted a question 'What is your stance re congestion charge for Manchester', his response was 'Completely off the agenda'.[3]

It is also important to shed light upon the interface between business and government. A group of prominent businesses in the area established the GMMG to campaign for a 'no' vote, and another group of supportive businesses established United City. Hepburn describes the characteristics of opponent and proponent businesses, noting that the former, on the whole, had 'economic interests in freight haulage and retail parks located on the edge of the urban area' and included multinational businesses, including Kellogg's and Unilever as well as major national real estate companies such as Peel Holdings, owners of the Trafford Centre, a shopping centre. The latter, he continued, were more likely to be located in the city centre of Manchester and 'whilst numerous were generally not in the same financial league as those businesses opposing the proposal' [49, pp51].

In 2008, Peel Holdings was allegedly carrying out voter research in the constituencies of politicians related to transport, such as Councillor Roger Jones and Transport Secretary (Minister of Transport) Ruth Kelly, asking voters if they supported congestion charging [71] and, reportedly in one case, if they were aware of the position of Councillor Jones on this issue [72]. Another local company, Kellogg's, was criticised for emailing its employees with a 'cut and paste' section of text with which to respond to the official consultation and reflect their employer's stance against the proposals [73].

Hepburn [49], in his analysis of online discussions, observed that 'the exercise of political power being wielded by dominant economic interests' had muted the voices

[3] Tweet available at [70]. Note that Andy Burnham was elected as mayor of Manchester in May 2017.

of, in his example, environmental campaigners, and notes that 'this undoubtedly helped to deprive the civic debate of an important environmental dimension'.

Whilst businesses have a legitimate voice on policy in a city region, this example highlights the sometimes opaque nature of their interventions and the need for a good understanding of the 'behind the scenes' influences on policy and decision-making. Whilst business often has a privileged position in policy-making, as 'government relies greatly upon business to carry out basic functions such as employing people and organising the economy' [74, pp99], its views are not necessarily unbiased or representative, and powerful interests with financial resources may 'skew the outcomes of policy debates' [74, pp99], either directly, as stakeholders, or indirectly by influencing the electorate. Given Hepburn's observation that national and multi-national concerns were disproportionately on the opposing side, one has also to question the extent to which this referendum can be seen as genuinely local democracy.

7.7.1 The referendum

The decision to hold a referendum, which followed campaigning primarily by opponents, appears to have been in part a result of the failure of the ten boroughs to reach agreement. Sir Richard Leese, Manchester City Council leader, had not been in favour of a referendum, arguing that it 'would not work' for such a complex issue and preferred the decision to be made through the 'system based on elected representatives who have to take difficult and important decisions' [75]. Some felt that the decision to hold the referendum was an 'admission of failure' [76] or buck passing. Similarly, in the Edinburgh case, it was argued that the referendum was a 'politically expedient way to distance the then ruling Labour group from making a decision on the congestion charging scheme' [77]. Whilst the GM authorities may be criticised for lacking decisiveness in pursuing the programme, some local politicians felt that the buck had already been passed from central government [76], which did not have the boldness to push congestion charging directly but had instead made anti-congestion measures essential components of bids to TIF.

The experience of the referendum illustrates the difficulty of translating a strategic long-term macro-level decision into individual decisions. It calls into question not only the suitability of a referendum process, but also the ability of the democratic system to bring about radical change, or, in lock-in [78] terminology, a discontinuity. The binary choice, and outcome, of the referendum arguably became a simplified representation of a complex issue and did not facilitate deliberation of the finer details, advantages and disadvantages, nor was voting able to communicate anything more than a 'signal' [79].

There was also an issue with familiarity with the subject matter. As Pierson [80] argues, many participants in politics engage in activities sporadically. He refers to studies that indicate that actors who operate in contexts of high complexity and opacity tend to filter information into existing 'mental maps'. This is pertinent to GMTIF: the majority of voters will not have engaged in transport planning prior to the referendum and will have had only a short amount of time to digest the plans and get to grips with the complex set of issues. They will have been bombarded by campaign messages and media content which, as has been seen, often did not do justice to this complexity. This would also apply to elected members outside of the transport and related committees. At the same time as the GMTIF was being debated, and concerns about the extent of the use of public finances and the debt were being raised, GM Waste Disposal Authority, serving nine of the ten AGMA boroughs, had signed a deal for waste services over 25 years with the value of £3.8 billion, including a mixture of grant funding and finance [81] – such high-value packages are not alien to local government.

7.8 Developments following the referendum

This section outlines developments in transport policy relevant to GMTIF following the rejection of the plans in 2009. Although some had emphasised that there was no 'plan B' for transport in the conurbation, the debate surrounding TIF had emphasised the importance of transport to the region's economic prosperity and AGMA asked the Clerk to the Authority and the Chief Executive of the PTE to 'examine alternative approaches designed to create investment capacity in transport and to maximise the contribution transport could make to the delivery of GM's employment and Gross Value Added growth potential' [82].

It was recognised that the decision to withdraw from TIF precluded GM from accessing associated funds from Department for Transport (DfT). An initial options document highlighted the potential for a workplace parking levy and supplementary business rates as a fundraising tool, but this was rejected by AGMA executive in 2010 [83]. AGMA chairman Lord Peter Smith related this decision to the economic downturn [83].

To the extent that the 2009 GMPTA[4] document 'Post TIF Investment and Funding Strategy' [82] represents local government thinking about ways forward, it suggests a business and economy focus, referring to transport as a 'growth constraint'. The exercise of applying to and planning TIF had furthered the evidence base for public transport investment and added to the bottom drawer of planned schemes. The document recommended the establishment of a Greater Manchester Transport Fund (GMTF) with a clear focus on economic growth: 'to focus scarce existing and potential new sources of funding on the growth and productivity agenda' [82, pp20].

A later report [84] detailed the components of this fund, which was established in May 2009. The GMTF, it stated, is a total investment package of £1,512 million and is described as a 'Major Transport Scheme Prioritisation and Funding Strategy' [85]. This comprised £448 million of Regional Funding Allocation; £165 million grant from the DfT relating to the South East Manchester Multi-modal Strategy (SEMMMS) road scheme; a Block LTP funding over a period of 9 years from 2010/11; up to £775 million from a combination of borrowing, and partly from

[4] GMPTA (now consolidated under Greater Manchester Combined Authority).

local/third party contributors. In Local Transport Plan 3 it is stated that Transport for Greater Manchester (TfGM) will repay the borrowings from a combination of Metrolink net revenues, annual ring-fenced levy contribution, and local revenue contributions [85]. This is noteworthy since it means that despite debt being one of the concerns that apparently influenced the referendum result, GM pursued a partly finance-based package, and public transport revenue, rather than congestion charging, was part of the prepayment strategy.

The GMTF had a list of capital projects attached, many of which are recognisable from the original GMTIF plans. Metrolink extensions are included but not the additional expansion to Trafford Park and the Trafford Centre. At the time of writing, work is beginning on this line [86], with funding through an earn-back model in which extra funding is secured if certain levels of economic growth are achieved [87]. Some, but not all, of the new interchanges and bus ways mentioned in GMTIF made it into GMTF. Additionally, four road building projects were amongst the priority schemes.[5]

In terms of capital spend then, the authorities were able to find funding for some major schemes, particularly the Metrolink extensions, but not to provide all that was proposed under GMTIF. Other elements, like personalised journey planning, cycle infrastructure, walking routes and yellow school buses do not feature in the plans. The funding amounted to around half of what would have been available under TIF and, as would have been the case with TIF, about half was a loan. Interestingly, the loan is being paid back through Metrolink revenue and Council Tax rather than revenue from a congestion charge. There is no 'stick'; in fact, extra road capacity being allocated in funding could be seen as a nod to the motoring lobby. It is also important, when considering the significance of GMTF as a response to GMTIF, that much of this money already existed in GM or the North West region, in the form of Council Tax or the Regional Funding Allocation. GM has also managed to attract other transport-related funding and to gradually roll out other improvements in, for example, cycle infrastructure, but this has been in response to standalone government grants, such as the Cycle City Ambition Grant of £20 million and the Clean Bus Technology Fund of £0.68 million [88], rather than a coordinated package of measures.

TfGM superseded Greater Manchester Passenger Transport Executive (GMPTE) in 2011. Its Consultation Draft of 'Greater Manchester Transport Strategy 2040: A Sustainable urban mobility plan for the future' [89] is a reasonable starting point from which to view the conurbation's ongoing plans for transport. The organisation clearly recognises the need for integrated transport planning and for modal shift away from private car use and recognises, to an extent, a role for demand management: 'We will focus on measures that encourage people to travel, or freight to be moved: at a different time; on a different part of the network; or to change to a different mode to make more efficient use of available capacity, particularly during peak periods' [89, pp16]. Whilst this is a policy goal to which a

[5] South East Manchester Multi-modal Strategy (SEMMMS) route, the Mottram Bypass and Glossop Spur, Ashton Northern Bypass Stage 2, and Wigan Inner Relief Road.

congestion charge could contribute, there is no mention of charging or of any real constraints on car use or parking. Whilst there is recognition that 'many of the negative impacts of transport, such as congestion, high emissions, noise and road traffic casualties, are a consequence of our over-reliance on the car . . . ' [89, pp20], its responses to this can be classified as 'carrot' rather than 'stick'.

This is not to say that there have not been policy steps in the direction of a more sustainable and multi-modal city, not least the Oxford Road bus priority scheme, which facilitates cross-city bus travel as well as featuring segregated dedicated cycle ways heralded as 'Dutch style' and, most relevantly for this discussion, closes a busy city centre route to general traffic between 6 a.m. and 9 p.m. seven days a week [90]. Whilst a single example that must be viewed in the context of the wider 2040 strategy, this does at least show some willingness to limit space available to the private car.

7.9 Lessons and implications for road pricing

This section develops the themes discussed above with a view to drawing lessons from GM's experience. This experience, and outcome of the referendum, highlights the extent of public scepticism towards road pricing and serves as a reminder of the powerful lobby that exists to challenge measures that are seen to penalise 'the motorist'. What then could GM do differently if revisiting plans that involve road pricing, what can other cities learn from its experience and what does GM's experience contribute to our understanding of car dependency?

These lessons concern the political and public acceptability of the plans rather than the plans themselves: the efficacy of the package was never put to the test. Whilst these lessons do not assume that other cities would hold a referendum – in fact, it is suggested here that they should not – it does assume that there will be a need to win over public and political support: the introduction of road pricing is a socio-political as well as a techno-economic process.

The GM referendum was the product of particular circumstances, a particular economic and political climate, and a unique set of plans. It formed a human story, with passions, fears, irrationalities and agendas much in evidence – in hindsight if not at the time.

Whilst we should not necessarily expect the same politics to play out in the same way in another setting at another time, it is important not to discount the national context, and to recognise that any local debate will be situated in ongoing national, and international, developments. Campaigns against restrictions on driving, be they anti-speed cameras, anti-parking restrictions or anti-road pricing, are in evidence nationally, and we have seen that these campaigns infiltrated the local discussions. We have seen that multi-national concerns were active in the opposition campaigning. Similarly, concerns about air quality, climate change and fossil fuel dependency are national and international concerns. One of the major players in the 'yes' campaign was Manchester Friends of the Earth, part of a network of Friends of the Earth groups nationally. Whilst locally rooted, GM's experience is also part of ongoing dialogues on a larger scale and provides an indication of the types of issues that are likely to arise.

Our first lesson then, and a first step in any proposal for road pricing, relates to the importance of understanding each city's own configuration and its place in a national context, and of identifying important actors, arguments and agendas that are likely to shape a debate on this approach. This also implies that car dependency cannot be understood as a purely local phenomenon (see also Chapter 4).

When reflecting with people involved in the debate, their first response is normally that there should not have been a referendum and that this was the death knell of the plans. The limitations of referenda in dealing with complex issues over a diverse population have been discussed above. It may have been possible to plough ahead with the plans had political consensus been achieved but it is not clear what the political fallout from such a move would have been and if difficulties that followed would have resulted in the plans being stopped or diluted.

The decision to hold a referendum can be seen as the result of a particular political situation in which agreement could not be reached and this in itself therefore relates to the city's particular characteristics. The enduring implication of the referendum, however, is the knowledge of a clear public vote against the proposals, and this makes it very difficult in the future for politicians to engage with this issue. Sustainable transport planning requires a diverse toolkit that includes carrots and sticks, and the political reaction to the referendum result has been to in effect rule out certain categories of stick: the simplistic nature of the vote limits the potential for complexity in transport planning not only at the time but over years to come. **A referendum is an extremely limited way to make such a complex decision, provides no guarantee of the most sustainable or fair outcome and limits the potential to revisit road pricing in the future.** This is not to discount democracy – as opponents of referenda are commonly accused of – but to point out the relative benefits of persevering through existing democratic structures.

It is worth noting, however, that not all referenda are equal. There was potential to hold a referendum after charging and improvements took place [36] to demonstrate the benefits before the vote, as was the case in Stockholm (Chapter 14) and Milan (Chapter 18). The difficulty for GM's situation was that the investment in public transport was dependent on finances from the congestion charge. Similarly, voting could have been restricted to those inside the charging rings, therefore affording greater weight to the opinions of residents than commuters from outside. This may have been an option, although the political climate would have made it very difficult to convince opponents that this represented the level of democracy being demanded. These possibilities notwithstanding, referenda remain an extremely limited way to accommodate and balance diverse views.

Timing then becomes an important issue. If GM was not influenced by national timescales to develop and approve the plans, could a slower pace have enabled a calmer, more deliberative decision-making process in which the ten local authorities could have arrived at a mutually acceptable plan, consulting the population in a more conventional way and not putting it to a binary vote? This would have avoided the disadvantages of a referendum, although it seems clear that the same issues would be fought, and other ways may have been found by opponents to derail the plans.

The difficulty in securing agreement can be seen in part to reflect the fact that GMTIF was a response to national priorities and was presented as such. If it were part of an ongoing strategy for GM, or if more time had been allowed to develop one, then there would have been more time to develop consensus and the congestion charging policy may not have been so unexpected. In reality, GMTIF was seen by many to be an opportunistic grab of national funding. **There is a lesson here for national policy to avoid being counter-productive by expecting towns and cities to develop relatively quick and competitive responses to national calls. There is a related lesson for conurbations to be proactive in developing transport policy strategically, allowing time and democratic space for the involvement of businesses, citizens' groups and political representatives: far from an appeasement process – which opponents may cast it as – this should be seen as a collaborative development of transport policy and an opportunity to get to the roots of car dependency.**

One of the things that is striking on reflection is how quickly the debate centred around the congestion charge and how quickly it became polarised, even before it was known that a referendum would make the final decision. It has been noted above that this was inevitable – since it was the congestion charge that was the controversial element and without it calls for a referendum would have been extremely unlikely. This has important and problematic implications.

First, it has been established that many votes were contingent on the quality of public transport and the media attention on the congestion charge limited the extent to which people knew about the public transport improvements. I have spoken to people since the vote who had no knowledge that there were substantial public transport improvements proposed. Even if the congestion charge was the most controversial part of the package, knowledge of the whole package was – or should have been – an important part of decision-making. It is therefore insufficient to simply argue that since the congestion charge was the most controversial element, or that it was somehow acceptable that this was the media's focus.

Second, debate air time and column inches focused on the congestion charge and there was little if any opportunity to discuss the detail of the public transport improvements. The other parts of the plans were denied both oxygen and critical attention. It is fair to say that much of the energy of the pro-GMTIF campaigners therefore went into defending the charge and countering claims rather than building and promoting an alternative vision for transport in the conurbation of which the congestion charge was one element. This contributed to certain perverse outcomes such as cycling activists saying they would vote against the plans because the cycling plans were not as they would have liked.

The lesson is to ensure that the congestion charge is seen in the context of a wider strategic transport goal for the area. This, as the GM experience

indicates, is not a simple task and implies the need for a proactive approach to communication. This should be based on full knowledge and anticipation of a backlash against the plans, but also avoid being on the defensive. The GM experience at least helps to identify the likely points of contention.

Important elements of such a communications strategy include working alongside businesses to develop the plan that meets their interests and reassures them about their concerns. This requires time, and therefore relates to the previous points about developing transport strategy over time. These businesses then become potential allies in selling the plans and it becomes harder for opponents to attack weaknesses. It would be naïve, of course, to expect all businesses to become supportive of a congestion charge, and there would need to be a close eye on anything that could be seen to be unreasonable behaviour on behalf of businesses. The GM experience indicates that environmental organisations are likely to be allies in this process and it would be beneficial to build early and ongoing relationships with them.

The formation of the proposals would allow for the development of a more informed and positive relationship with the media about transport issues, and congestion charging in particular. In this process, congestion charging becomes one part of a strategy for sustainable transport rather than an individual 'shock' on which the media fixates. This is not, however, to underestimate how controversial it is likely to be and how much proactive work is needed. This demands an ongoing coverage of transport in the media and a widening of the narrow vocabulary often associated with it [2].

It seems also that the GMTIF package lacked an attractive name. This may seem unimportant, but left it open to opponents to, apparently very effectively, impose their own media-friendly branding with terms like 'C Charge' and 'Toll tax', and these helped to reinforce the focus on the charging element. Alternatives like 'clean air charge' or 'accessibility charge' may have helped to move the debate to the positive impacts of such a charge, but were not forthcoming.

One of the unknowns when reflecting on the referendum relates to the role of the local authorities and transport authority: what if there had been stronger, clearer leadership from the authorities with outspoken support for the plans? Whilst GMPTA and AGMA had submitted the plans in the form of a bid to government and could therefore be assumed to be supportive of them, they were expected to take a neutral role in the debate, limited to providing information on the plans. On reflection, this stance did not help to dispel any sense that the plans were somehow not to be trusted and were being enacted 'on' the city, rather than part of its ongoing transport strategy. Close to the referendum vote, for example, a TV broadcast purporting to give a balanced account of the options was pulled off air following claims of potential bias [4]. **It is possible therefore that stronger leadership and political consensus on the plans may have helped instil a greater sense of public confidence in them.**

An issue that arose in the course of the debate was clarity over the purpose of the plans. Whilst the potential to reduce congestion was clear, and this was the primary message, claims were made about the potential to improve accessibility and air quality. However, it was pointed out that there were no specific elements of the plans that related to pollution levels, in particular there were no concessions to low emissions and electric vehicles. Apart from a temporary discount on bus fares following the introduction, there were also no clear concessions for low-income travellers. It could of course be argued that the modal shift resulting from the plans would have helped to reduce air pollution, but this was not explicit. As with other perceived weaknesses of the plans, these apparent inconsistencies are both a potential source of doubt for the voter and weaknesses that can be exploited by opponents. It is also likely to have been helpful to be more specific and committal regarding the public transport improvements, as Gaunt, Rye and Allen [31, pp100] noted in the Edinburgh context: 'whilst reduced congestion and improved alternatives to the car were abstract, the prospect of being charged was very much more tangible'. **The lesson here is to make sure the plans and claims about their impact are not inconsistent and that impact is clear rather than implied.**

The final point is more difficult and less tangible but important in terms of the societal context in which the debate is framed. It is necessary to build a vision of sustainable transport and to link this with notions of the good city, and what it is to be a resident of GM. It is in the interests of opponents to keep the debate at the level of whether Josephine public will pay 50p, £1 or £5, or how many times larger the 'zone' is than London's. These encourage a focus on the costs to the individual driver rather than the broader societal balance of costs and benefits, and also refer exclusively to the perceived unfairness of the 'new' regime rather than that of the current one, the extent of transport provision and who pays for it.

At the time of a referendum, it is this 'new' regime that is being considered, but its fairness can only be understood in the context of the fairness of existing transport provision. Could low-income communities with unreliable and patchy public transport and asthmatic children living on busy roads be framed as victims of the current system in the same way that 'drivers' are framed as victims of road pricing? Are people already paying an 'air pollution tax' that is imposed on them because there is a lack of restriction on driving? These vital questions are difficult to raise in a simplified debate with simplistic media coverage. This connects with the notion of developing transport policy strategically over time and emphasises the importance of education and awareness-raising through public engagement.

This is challenging since it pertains to societal change in how the transport system is seen and how fairness is conceptualised in relation to it, but it points to widening the debate on transport to make it clear that this is not simply about the 'driver' verses the rest of society but about how a diverse range of people access the services they need, breathe clean air and are protected from the impacts of climate change.

These points reinforce an understanding of car dependency that transcends the individual's mobility experiences and connects with the ability of the city to adapt and transition to more sustainable transport profiles. It hints at a condition of lock-in [78] in which cities experience systemic barriers to radical change and sheds light on some of the ways this is operationalised. Poor public transport becomes locked-in, as it erodes the public's confidence in the transport authority's ability to provide a quality service. Businesses that thrive in a car-friendly city flex their muscle when they feel proposals may challenge the status quo. Relatively simplistic media coverage reinforces the focus on the car and provides a narrow lens through which to view a complex issue.

With these points in mind, it is worth reflecting on the current situation in GM. A lot of the above reflects a holistic and strategic approach to transport planning that takes a longer term view and meaningfully engages stakeholders along the journey. If GMTIF were proposed now in 2017, would it have a different reception?

On the one hand, there has been substantial investment in public transport, particularly the Metrolink but also bus priority routes and smartcard ticketing is gradually being rolled out. Yet congestion continues to be a problem – TomTom have rated it the 8th most congested city in Europe for example [91]. There is also increasing awareness about air quality and health, as evidenced by the Client Earth legal case [92] in which the UK Government was taken to court over its record of tackling air pollution, and the Volkswagen emissions controversy in which it was found that cars had been programmed to perform better when under emissions testing than on the road [93].

On the other, perceptions and experiences of public transport remain problematic, especially in areas that have not enjoyed so many improvements. Ironically, the expansion of Metrolink could weaken the case for another GMTIF, since residents can see that a 'big bang' of public transport has happened to some extent, although there is plenty of scope for further extensions of the network. Economically, we remain in times of austerity with the potential of the United Kingdom leaving of the European Union providing further uncertainty. A major factor may be 'Devo Manc' in which the conurbation will get increased powers to manage its own affairs, including transport, and a mayor. Andy Burnham, who was elected the first mayor of the conurbation in May 2017, will still have to work in conjunction with the ten local authorities, but in principle has the power to be a strategic lead. This, combined with the Bus Services Bill progressing through Parliament [94], presents an opportunity for greater control and therefore coordination, integration and price capping of buses, trams and trains. The 2040 strategy discussed above also indicates a willingness to at least in principle accept the need to challenge the growth of car use, and the closing of part of Oxford Road to general traffic during the day could be seen as an initial step [90,95]. This would be more convincing, however, if there were evidence of a strategy for further such developments.

7.10 Conclusion

This chapter has explored the challenges faced by GM in decision-making around its plans for a package of measures that would include congestion charging and public transport investment. It has shed light on some of the ways in which party politics, local campaigning and the media contributed to shaping public opinion.

In looking at related developments in GM following the vote, it has shown that whilst some of the intentions of the proposals have been realised as part of subsequent policy initiatives there remains a reluctance to pursue the charging element. As someone who supported the GMTIF proposals, seeing their value for the conurbation in social, economic and environmental terms and recognising the central role of the congestion charge within them, I note that there is still a need for greater investment in public transport and active travel and to recognise that moving towards these more sustainable modes requires sticks as well as carrots. Whilst bold action on GM's car-centred transport networks would therefore be welcome, it must be recognised that the referendum result makes that more challenging politically than it would otherwise be – though it is of course not impossible. Whilst the use of a referendum might enable politicians to defer to the 'will of the people', the challenges of meaningfully addressing the congestion in our cities and moving towards sustainable mobility remain.

References

[1] Åkerman J, Gudmundsson H, Sørensen CH, *et al.* 'How to manage barriers to formation and implementation of policy packages in transport'. OPTIC. Optimal policies for transport in combination: 7th framework programme: Theme 7 transport (Chapter Deliverable 5); 2011.

[2] Vigar G, Shaw A, and Swann R. 'Selling sustainable mobility: the reporting of the Manchester Transport Innovation Fund bid in UK media'. *Transport Policy.* 2011 **18**(2):468–479.

[3] Swanson J. *Policy brief. Gaining public support for congestion charging: lessons from Europe for U.S. Metropolitan Areas.* Report to the German Marshall Fund of the United States 2009.

[4] Sherriff G. 'Voting on sustainable transport: communication and governance challenges in Greater Manchester's "congestion charge" referendum'. *Local Environment.* 2014 **20**(12):1507–1530.

[5] Hickman R, and Banister D. *Transport, climate change and the city.* Abingdon: Routledge; 2014.

[6] Royal College of Physicians. 'Every breath we take: The lifelong impact of air pollution. Report of a working party'. The Royal College of Physicians; London, 2016. https://www.rcplondon.ac.uk/projects/outputs/every-breath-we-take-lifelong-impact-air-pollution (Accessed 31/5/2017).

[7] Lucas K. *Running on empty: transport, social exclusion and environmental justice.* Bristol, Policy Press; 2004.

[8] Rosen P. 'Towards sustainable and democratic urban transport: constructivism, planning and policy'. *Technology Analysis & Strategic Management.* 2001 **13**(1):117–135.

[9] Sloman L. *Car Sick: solutions for our car-addicted culture.* White River Junction; Green Books, Cambridge; 2006.

[10] Hutton B. *Planning sustainable transport.* Abingdon: Routledge; 2013.

[11] Schiller PL, Bruun EC, and Kenworthy JR. *An introduction to sustainable transportation: policy, planning and implementation*. Routledge, Oxford; 2010.

[12] Newman P. 'The ten myths of automobile dependence Peter Newman & Jeff Kenworthy'. *Editorial Board*. 2000 **6**(1):15–25.

[13] Banister D. *Unsustainable transport: city transport in the new century*. Routledge, Oxford; 2005.

[14] Watson M. 'How theories of practice can inform transition to a decarbonised transport system'. *Journal of Transport Geography*. 2012 **24**:488–496.

[15] Office for National Statistics, www.neighbourhood.statistics.gov.uk. (Accessed 08/8/2017).

[16] Martens K. *Transport justice: designing fair transportation systems*. London: Routledge; 2016.

[17] Preston J, and Rajé F. 'Accessibility, mobility and transport-related social exclusion'. *Journal of Transport Geography*. 2007 **15**(3):151–160.

[18] Sustainable Development Commission. *Fairness in a car dependent society*. London: Sustainable Development Commission; 2011.

[19] Santos G, Behrendt H, and Teytelboym A. 'Part II: Policy instruments for sustainable road transport'. *Research in Transportation Economics*. January 2010 **28**(1):46–91.

[20] Banister D, Stead D, Steen P, *et al*. *European transport policy and sustainable mobility*. Abingdon: Routledge; 2000.

[21] Anable J, Brand C, Tran M, and Eyre N. 'Modelling transport energy demand: a socio-technical approach'. *Energy Policy*. February 2012 **41**:125–138.

[22] Schwanen T, Banister D, and Anable J. 'Rethinking habits and their role in behaviour change: the case of low-carbon mobility'. *Journal of Transport Geography*. September 2012 **24**:522–532.

[23] Brand R. *Synchronizing science and technology with human behaviour*. Earthscan, Oxford; 2005.

[24] Dudley G. 'Why do ideas succeed and fail over time? The role of narratives in policy windows and the case of the London congestion charge'. *Journal of European Public Policy*. 2013 **20**(8):1139–1156.

[25] Vonk Noordegraaf D, Annema JA, and van Wee B. 'Policy implementation lessons from six road pricing cases'. *Transportation Research Part A: Policy and Practice*. 2014 **59**:172–91.

[26] Sørensen CH, Isaksson K, Macmillen J, Åkerman J, and Kressler F. 'Strategies to manage barriers in policy formation and implementation of road pricing packages'. *Transportation Research Part A: Policy and Practice*. 2014 **60**:40–52.

[27] Borjesson M, and Kristoffersson I. 'The Gothenburg congestion charge. Effects, design and politics'. *Transportation Research Part A: Policy and Practice*. 2015 **75**:134–146.

[28] Schaller B. 'New York City's congestion pricing experience and implications for road pricing acceptance in the United States'. *Transport Policy*. 2010 **17**(4):266–273.

[29] Hau TD. 'Electronic road pricing: developments in Hong Kong 1983–89'. *Journal of Transport Economics and Policy*. 1990 **24**(2):203–214.

[30] Manville M, and King D. 'Credible commitment and congestion pricing'. *Transportation*. 2013 **40**(2):229–249.

[31] Gaunt M, Rye T, and Allen S. 'Public acceptability of road user charging: the case of Edinburgh and the 2005 referendum'. *Transport Reviews*. 2007 **27**(1):85–102.

[32] M.E.N. 'Congestion charging: it's coming our way'. *Manchester Evening News 24/01/2*007. January 2007.

[33] Givoni M. 'Re-assessing the Results of the London Congestion Charging Scheme'. *Urban Studies*. 2012 **49**(5):1089–1105.

[34] Shove E, and Walker G. 'Governing transitions in the sustainability of everyday life'. *Research Policy*. 2010 **39**(4):471–476.

[35] Dudley G, and Chatterjee K. 'The Dynamics of Regime Strength and Instability: Policy Challenges to the Dominance of the Private Car in the United Kingdom'. In: Geels F, Kemp R, Lyons G, editors. *Automobility in Transition? A Socio-technical Analysis of Sustainable Transport*. Routledge, Oxford; 2012:83–103.

[36] Ahmed B. 'Congestion charging scheme for the city of Manchester: lessons learnt'. *Journal of Civil Engineering*. 2012 **40**(1):1–9.

[37] Greater Manchester Future Transport. 'Greater Manchester's Future Transport: Why the Transport Innovation Fund? An analysis of the rationale behind the Greater Manchester TIF package. Information Pack Document 1'. 2008.

[38] Hetherington P. 'Manchester congestion charge plans about to go to a public vote'. *The Guardian*. 17th September 2008.

[39] Sorensen CH, and Gudmundsson H. 'The impact of governance modes on sustainable transport – the case of bus transport in Greater Manchester, UK'. *World Review of Intermodal Transportation Research*. 2010 **3**(1):8–25.

[40] Hepburn P. 'Local democracy in a digital age: lessons for local government from the Manchester Congestion charge referendum'. *Local Government Studies*. 2014 **40**(1):82–101.

[41] Duranton G, and Turner MA. 'The fundamental law of road congestion: evidence from US cities'. *American Economic Review*. 2011. NBER Working Paper 15376.

[42] Hull A. 'Policy integration: what will it take to achieve more sustainable transport solutions in cities?'. *Transport Policy*. 2008 **15**(2):94–103.

[43] Whitmarsh L. 'How useful is the multi-level perspective for transport and sustainability research?'. *Journal of Transport Geography*. 2012 **24**: 483–487.

[44] Ottewell D. 'Green light for C-charge poll – by just ONE vote'. *Manchester Evening News*. 1st November 2008.

[45] M.E.N. 'Bury blow for congestion charge plans'. *Manchester Evening News*. 13th December 2007.

[46] Ottewell D. 'Secret poll reveals 53pc support for road charge'. Manchester Evening News 19th August 2008.

[47] Ottewell D. 'C-charge: a resounding "NO"'. *Manchester Evening News*. 12th December 2008.

[48] Forster A. 'They've made their cases, now it's up to the public to decide'. *Local Transport Today*. December 2008:3.

[49] Hepburn PA. 'Is this local e-democracy? How the online sphere of influence shaped local politics. Empirical evidence from the Manchester Congestion Charge referendum'. *JeDEM – eJournal of eDemocracy and Open Government*. 2012 **4**(1):45–66.

[50] Clean Air Now. 'Tour of Greater Manchester: Tour de TIF' [Internet] http://www.manchesterfoe.org.uk/can/tourdetif.php. *Manchester; 2008 [cited 9th March 2017]*. Available from: http://www.manchesterfoe.org.uk/can/tourdetif.php.

[51] Salter A. 'C-charge: Kellogg's say "no" – Manchester Evening News'. *Manchester Evening News*. 22nd January 2008.

[52] Manchester Evening News. 'C-charge: the yes campaign – Manchester Evening News'. *Manchester Evening News*. 3rd November 2008.

[53] Clean Air Now. 'Is a congestion charge too much to pay for clean air? (Poster)'. Clean Air Now, Manchester; 2008.

[54] National Alliance Against Tolls. 'Manchester Toll Tax or "Congestion Charge"' [Internet]http://www.notolls.org.uk/manchester.htm. *2017 [cited 9th* March 2017]. p. 1–18. Available from: http://www.notolls.org.uk/manchester.htm.

[55] Ryley T, and Gjersoe N. 'Newspaper response to the Edinburgh congestion charging proposals'. *Transport Policy*. January 2006 **13**(1):66–73.

[56] Brand R, Fischer J. 'Overcoming the technophilia/technophobia split in environmental discourse'. *Environmental Politics* 22(2). October 2012:1–20.

[57] M.E.N. 'Congestion charging, you say NO'. *Manchester Evening News 21/03/2*007. March 2007.

[58] King D, Manville M, and Shoup D. 'The political calculus of congestion pricing'. *Transport Policy*. March 2007 **14**(2):111–123.

[59] Clean Air Now. 'CAN Responds to M.E.N. "Commuter Survey"' [Internet] http://www.manchesterfoe.org.uk/can/pressr.php?id=8. 2007. Available from: http://www.manchesterfoe.org.uk/can/pressr.php?id=8.

[60] Campbell H, and Marshall R. 'Utilitarianism's bad breath? A re-evaluation of the public interest justification for planning'. *Planning Theory*. January 2002 **1**(2):163–187.

[61] Hay PR. *A companion to environmental thought*. Edinburgh University Press Edinburgh; 2002.

[62] Ottewell D. 'C-charge: majority voting'. *Manchester Evening News*. 20th November 2007.

[63] Ottewell D. 'C-charge gets public approval'. *Manchester Evening News*. 20th July 2007.

[64] Stockport Express. 'Congestion charge drives hard bargain for Stockport'. *Stockport Express*. 11th June 2008.

[65] BBC. 'Anger over C-charge bus adverts'. *British Broadcasting Corporation*. 29th October 2008.

[66] Ottewell D. 'Transport boss loses seat'. *Manchester Evening News*. 5th February 2008.

[67] Kirby D, and Welsh P. 'Return ticket: congestion charge mastermind Roger Jones set to return to transport committee'. *Manchester Evening News*. 6th February 2011.

[68] Hayman A. 'Councils throw city-region into doubt'. *Regeneration and Renewal*. March 2010.

[69] The Bolton News. 'Safeguard to prevent C-charge is welcomed'. *The Bolton News*. 19th December 2010.

[70] Andy Burnham Tweet on the congestion charge for Manchester https://twitter.com/andyburnhammp/status/821751320797396992. 2016

[71] Binns S. 'Peel holdings researches C-charge advocates' wards'. *Crains Manchester*. 11th March 2008.

[72] M.E.N. 'Trafford centre "wants pay-to-drive leader out"'. *Manchester Evening News*. 11th March 2008.

[73] BBC. 'Kellogg's defends C-charge e-mail'. *British Broadcasting Corporation*. 4th October 2008.

[74] Dryzek JS. *The politics of the earth: environmental discourses*. Oxford University Press; 1997.

[75] Ottewell D. 'C-charge: You could still have say'. *Manchester Evening News*. 18th April 2010.

[76] Swanson J. *Gaining public support for congestion charging: notes from Europe on the implementation of bold transportation policies* report to the German Marshall Fund of the United States, 2009.

[77] Rye T, Gaunt M, and Ison S. 'Edinburgh's congestion charging plans: an analysis of reasons for non-implementation'. *Transportation Planning and Technology*. 2008 **31**(6):641–661.

[78] Unruh GC. 'Understanding carbon lock-in'. *Energy Policy*. 2000 **28**(12): 817–830.

[79] Warren M. *When, where and why do we need deliberation, voting, and other means of organizing democracy? A problem-based approach to democratic systems*. NY: Rochester; 2012 July.

[80] Pierson P. 'Increasing returns, path dependence, and the study of politics'. *American Political Science Review* 94(2). 2000:251–267.

[81] GNWDA. 'GMWDA our aim is zero waste'. *GMWDA Newsletter*. 2009.

[82] GMPTA. 'Post TIF investment and funding strategy'. 2009.

[83] Scheerhout BYJ. 'Pay-to-park plan axed'. *Manchester Evening News*. Manchester; 12th January 2016:19–21.

[84] Manchester City Council. 'Manchester City Council Economy, Employment and Skills Overview and Scrutiny Committee Item 8: 6th January 2010 – Greater Manchester Transport Fund'. 2010. p. 12–41.

[85] Greater Manchester Combined Authority, Transport for Greater Manchester. 'Greater Manchester Local Transport Plan 3'. 2011.

[86] Transport for Greater Manchester. 'Green light for new Trafford Park Metrolink line'. 14th October 2016:2016.

[87] Greater Manchester Combined Authority. 'Greater Manchester Agreement: Devolution to the GMCA & transition to a directly elected mayor'. 2015.

[88] Transport for Greater Manchester. 'Greater Manchester Local Transport Plan 3: Capital Programme 2015/16–2020/21'. Transport for Greater Manchester; 2011.

[89] TfGM. *Greater Manchester Transport Strategy* 2040. Transport for Greater Manchester, 2016.

[90] Transport for Greater Manchester. 'Oxford Road update: December 2016'. Transport for Greater Manchester, 2016:1–2.

[91] TomTom. 'TomTom Traffic Index 2016 – The Results Are In! Mexico City Takes Crown of "Most Traffic Congested City" in World from Istanbul 22nd March 2016' [Internet] http://corporate.tomtom.com/releasedetail.cfm?releaseid=961546. 2016 [cited 24th March 2017]. Available from: http://corporate.tomtom.com/releasedetail.cfm?releaseid=961546.

[92] ClientEarth. 'ClientEarth wins air pollution case in High Court 2nd November 2016' [Internet] https://www.clientearth.org/major-victory-health-uk-high-court-government-inaction-air-pollution/. 2016 [cited 24th March 2017]. Available from: https://www.clientearth.org/major-victory-health-uk-high-court-government-inaction-air-pollution/.

[93] Hotten R. 'Volkswagen: the scandal explained – BBC News' [Internet] http://www.bbc.co.uk/news/business-34324772. 2015 [cited 24th March 2017]. Available from: http://www.bbc.co.uk/news/business-34324772.

[94] House of Lords. 'Parliamentary Bill: Public Transport'. 2016 (March).

[95] Transport for Greater Manchester. 'Bus Priority: Oxford Road'. 2015.

Chapter 8

Case studies of communication and consultation strategies for road pricing schemes

Andrew Pickford and John Walker***

8.1 Introduction

The development and delivery of stakeholder consultation and public communication programmes for any form of travel demand management (TDM) schemes, including those that depend on pricing, cannot be independent of the context of the intended (or operating) scheme but must be integrated within in it. It could be argued that since TDM would invariably form part of a transport policy and its realisation as a 'package' of measures, then these broader policies or packages should form the basis of stakeholder consultation, rather than focusing on one part such as the TDM measures. The impacts of road pricing (RP) are often complex, impacting on entrenched habits of commuters, commercial operators and other businesses. Restricting vehicle access through road closures, imposing emission (or weight) limits or by charging a fee for vehicle entry to a well-defined zone needs to be legitimate, not as the outcome of a scientific, technical or law-making process but as a policy that mitigates the effects of widely accepted problems, either increasing congestion or declining air quality.

If we consider a broader definition of RP and instead expand the geographic area of interest from a town or city to a subregion or national jurisdiction, then we may consider 'all road' policies where charges are differentiated according to the category of road, a road's environmental context (i.e. traversing an area of outstanding national beauty or scientific interest) or the characteristics of a vehicle or status of a driver. Alternatively, in an era of declining revenue from vehicle taxation driven by reduction in vehicle miles travelled (VMT) and improved fuel consumption, then the potential to replace the proven (and largely accepted) 109-year-old policy of fuel duty[1] with a distance-based charge provides yet another rationale to introduce RP.

*MVA Hong Kong Ltd., Hong Kong
**Transportation Research Group, University of Southampton, UK

[1] The United Kingdom introduced fuel duty as one of a range of new motor vehicle-related taxes in 1909 with the Finance Act of 1908 to enable the national road network to be self-financing; in the United States, Oregon became the first state to tax motor fuels in 1911.

Whether we consider the application of restraint or pricing to a defined area, vehicle type or user group, such measures will incentivise stakeholders that depend on travel and transport to question their widely held beliefs and their entrenched patterns of behaviour. The process of consultation implies interaction with relevant stakeholders and, as various potential scenarios for a pricing regime are developed, it is likely that policy objectives and operating strategies need to be refined – not only to demonstrate that the consultation process is meaningful but to resolve inconsistencies as early as possible, and to develop and maintain political and public acceptance. Consultation of stakeholders is delivered through communications and the process of communications could be a simple act of broadcast of information, informed dialogue with high granular segments of affected parties or somewhere in between. Generalisation of the process would be as challenging as management of the process itself and so this chapter aims to present the (often) complex context in which consultation processes are delivered, the shape of a process and its resourcing.

Most importantly, the perspective of the stakeholder needs to be understood and in all RP schemes, there are many stakeholders, each developing their perspectives on the costs and benefits of measures that may impact their behaviour – and most of all, asking the question 'what's in it for me?'

8.2 Policy context, themes and consultation variables

8.2.1 Policy, legal and regulatory contexts

Pricing or restraint measures generally apply to small well-defined areas (e.g. congestion charging) or to all navigable routes within national boundaries. Whatever their size, territorial regions are complex in many dimensions, including:

- their jurisdictional structures;
- entrenched travel habits;
- patterns of demand;
- variations in land use patterns, growth limitations caused by natural restrictions (such as rivers or other physical boundaries);
- influencing power of coalitions of interests (e.g. road user associations, residents' cooperatives and political parties);
- national regulations on car ownership (such as a tax on first registration, annual taxes relating to carbon emissions or engine size) and vehicle usage (such as fuel tax, insurance and parking fees);
- rivalry between local economic regions and
- competition for government resources for other local projects.

Creating a single package to address congestion or worsening air quality or both – or more complex national policies such as fuel taxation replacement or network tolling – would be so challenging that, potentially, it should not be attempted.

By definition, each stakeholder group may have a different perspective on the impact (including costs and benefits) which implies that consultation should be equally targeted. Like any organisation, understanding ones' 'clients' through segmentation and targeting is critical. Therefore, answering the question 'what's in it for me?' can only be understood by initially defining who 'me' is.

Stakeholder consultation, including the means of communications, is likely to face one or more of the following:

- visible transport-related problems, such as declining air quality, worsening congestion, increasing delays and increasing journey time variability;
- demands for new capacity and sometimes a lack of public funds to do so (the trigger for the Oslo toll ring and other projects in Norway);
- locally relevant precedents, such as Singapore's Area Licensing Scheme (ALS) that enabled the evolution to Singapore's Electronic Road Pricing (ERP) policy (see Chapter 6), congestion charging in Stockholm that paved the way for a similar scheme in Gothenburg (Chapter 14) and many of Europe's electronic truck tolling programmes that have evolved from earlier paper-based vignette[2] schemes to a version that swaps a paper ticket for an On Board Unit (OBU) (everything else being 'invisible' to users), and the role of toll rings in Oslo and Trondheim as pathfinders for Norway's 50 urban and interurban tolling projects;
- the existence of proven enforcement processes, particularly those that relate to moving traffic violations where compliance checking is largely automated and where image-based evidence is acceptable to the courts;
- the availability (or lack) of alternative routes through traffic, workforce flexibility to change time of travel, or inflexibility as faced by fleet operators experiencing customer demands for just-in-time deliveries or other time-based performance targets; and high quality alternatives such as Bus Rapid Transit, light rail or mass transit services and
- the stage of the economic cycle where there may be some (or no) appetite for investments in new transport policies (e.g. apparent in the introduction of South Africa's urban tolling scheme).

In addition, the impact of the regulatory context on public perception and implementation strategies cannot be underestimated, one example being the Directive 2008/50/EC on air quality within the European Union (EU) sets limits on PM_{10} Particulate Matter requiring member states to comply, or otherwise risk being fined. The EU Transport Plan, published in 2014 [1], focused on the potential use of congestion charging and Low Emission Zones (LEZs) and, from a process of consultation, reported that 61% of respondents agreed that LEZs are an effective measure to improve air quality in urban areas (opposed by 30%). Congestion charging was regarded as an effective measure to improve urban accessibility by 50% (with 30% opposed).

[2] Defined by the European Commission as road infrastructure charging for heavy goods vehicles 'based on distance-related tolls and time-based user charges' levied on vehicle owners consistent with 'the principles of fair and efficient pricing [to reflect] the costs associated with road use'.

Regulatory precedents also reduce the 'risk premium' on the introduction of successive schemes. In the United Kingdom, London followed Durham's smaller congestion charging scheme[3]; Gothenburg followed Stockholm's; Singapore's ERP followed its own paper licence-based RP scheme which followed its own Area Licensing Scheme (ALS); the German truck tolling scheme 'LKW Maut'[4] followed a paper licence-based vignette scheme, amongst many others.

Whilst a technology developer may see a new opportunity for disruptive change in technology, a regulator or policy-maker may seek an evolutionary change in regulation or new regulation without precedents. Each is prone to implementation risks and it is argued here that consultation needs to include references to both, potentially to the point of being routinised. For example, in Norway, the combination of national enabling legislation (focused on tolling to fund infrastructure rather than pricing or LEZ) not only led to a robust precedent for urban and interurban tolling schemes but resulted in a suite of specifications on how to do it backed up by the implementation assistance of the Norwegian Public Roads Administration. Similarly, in Germany, the precedents for access restrictions based on a vehicle's emissions category led to an approach for implementation of an emission control area in Berlin, Cologne and Hanover and similar (although smaller) policies in 80 other German cities and towns up to now. In China, the development of 'a systematic evaluation toolkit' for TDM was the focus of a 4-year project that concluded in 2015 that aimed to inform similar approaches in at least five other Chinese cities.

However, a regulatory precedent may afford a baseline level of support and awareness, but it cannot guarantee that a 'successful' consultation process will enable a scheme to be implemented.

8.3 Stakeholder consultation and communication processes

8.3.1 Perceptions and attitudes

The attitude of stakeholders is informed by their perceptions. If a stakeholder perceives that the impact of restrictions on time or location will not be compensated as a direct, local and visible benefit then the attitude towards the restriction may be unsupportive. Similarly, if pricing is implemented, then users that cannot afford to pay the price, do not have the means to avoid it or perceive no apparent measurable benefits (to them), could regard the scheme as unfair (to them).

Amongst many other factors, stakeholders are likely to reflect on how a scheme will serve their self-interests (such as the time savings or benefits, enabled by investment funded by revenues collected). Therefore, the perceptions of stakeholders will be influenced by any information that is provided during the

[3] The Durham Congestion charging scheme was launched in October 2002; London followed in February 2003 largely underpinned by the same regulations.

[4] LKW is the abbreviated form of *Lastkrafwagon Maut*, may be translated literally 'heavy goods vehicle tolls'.

consultation and market communication process. This also means that the media and other opinion-leaders will influence stakeholder perception. Then, as the process of consultation unfolds, stakeholders develop their views and the focus on developing awareness is replaced by developing understanding and finally, as the ultimate aim of the scheme sponsors, a sufficient level of acceptance. At any time, if people's awareness is weak then their understanding may be weak also, and in turn so could their level of acceptance. Perception informs attitudes and any consultation process would benefit from measuring these attitudes, as explained further in Section 8.3.4.

8.3.2 Types of consultation, levels of engagement and setting expectations

Consultation should be a process of dialogue (prompted by structured questions) with selected stakeholders that may be prioritised as 'internal' (within the administration and related public organisations) and 'external' (other public authorities and the public, including business). Each phase of a consultation process should, amongst other attributes:

- have a specified start and end date;
- inform decisions on the objectives, scope and implementation for the scheme;
- have the effect of encouraging greater involvement from stakeholders; and
- help to refine the policy objectives and scheme operations strategy.

Consultation may include a range of formal and informal mechanisms, and if a scheme is implemented then consultation may need to be continuous to measure ongoing level of support or to inform potential changes to policy or operations. Informal consultation helps to shape policy through dialogue with a limited range of stakeholders (mostly internal) and formal consultation helps refine emergent policies and gain support – or may uncover sufficient resistance that could render all or part of a scheme infeasible.

8.3.3 The shape of a consultation programme, themes and phasing

Consultation and interaction with the public with communication campaigns is likely to be an on-going process, not just during the pre-implementation phase, but during piloting and subsequent full-scale implementation. Overall, the consultation process aims to shape policy through dialogue with a limited range of stakeholders (mostly internal); then refines policies in a structured manner based on informing and securing feedback from other political and public stakeholders and, during scheme operation, confirms that expectations continue to be met.

Ideally, these steps should be sequential – the internal/informal stage resulting in sufficient maturity and confidence to proceed with structured external/formal consultation, followed by ongoing 'expectation management' to maintain and consolidate support for a policy and its subsequent evolution (Figure 8.1).

Of course, the level of maturity of internal policy shaping that is necessary to proceed to external consultation depends on the context and the perception of

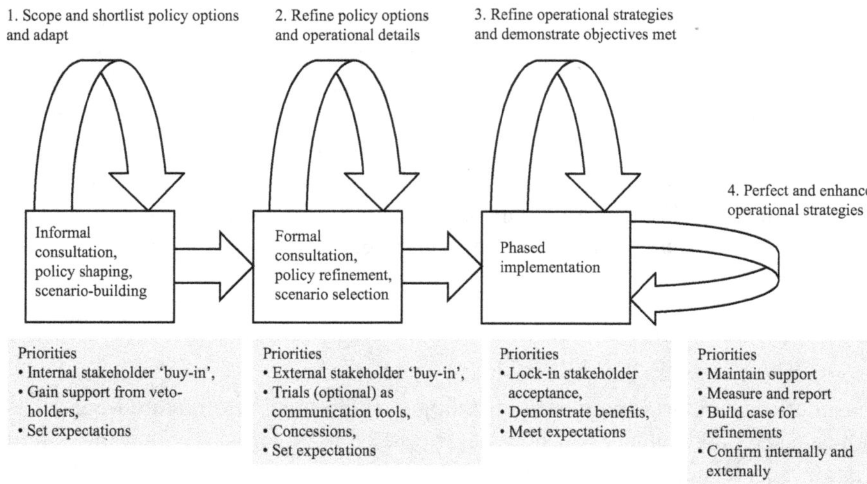

Figure 8.1 General consultation and communication framework: informal, formal and ongoing performance monitoring/reporting and ongoing policy refinement

benefits and costs that may be perceived by external stakeholders. Transport for London (TfL) stated 'one of the most useful precedents [for the introduction of congestion charging] was privatisation ... since the public would need to go through a steep learning curve and there was no second chance' [2]. This campaign is elaborated further in Chapter 13. This suggests that for urban congestion charging in London at least, the perceived public impact and the challenge of consulting on a package of measures aimed to mitigate congestion in favour of improved quality of public transport and other complementary measures could be complex.

With the above in mind, an external/formal consultation process should aim to

- develop a channel for dialogue with all relevant stakeholders, including road users and non-road users;
- raise awareness of the potential forms of the scheme and their relative impacts, based on one or more 'storylines' (outline concepts of operations) described from the perspective of each stakeholder; employ a public demonstration to test political support (e.g. Jakarta), or to raise awareness and to seek feedback as part of the implementation programme (e.g. Singapore);
- present the benefits that are meaningful to each stakeholder, potentially as individual communication campaigns;
- confirm the scale, scope and plans for implementation and the timing of benefits delivered (sometimes benefits may be planned to be delivered earlier than any perceived restrictions) potentially resulting in the delivery of benefits *before* access restrictions apply to build support (e.g. extra buses in London and Stockholm);

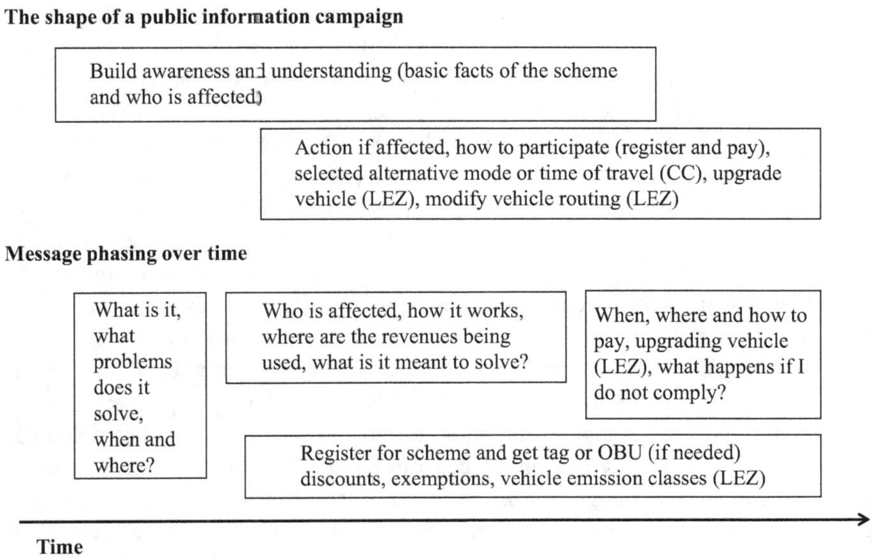

Figure 8.2 The phases of a public communication campaign leading up to 'go live' (operations commencement) date

- confirm the level of acceptance by stakeholders at each stage of the consultation process, particularly because of any changes to messages delivered by means of structured surveys (e.g. New York, Stockholm and London);
- confirm the potential scope of integration with other stakeholders' initiatives (e.g. public transport and infrastructure improvements) and other complementary measures and
- confirm the approach to governance of the scheme, use of revenues[5] collected (assuming the scheme is not revenue-neutral), how the scheme will be monitored and reported.

Perceptions drive attitudes and attitudes may be measured through interviews, surveys, panels and focus groups. It is likely that the media may be regarded as reflecting the many dimensions of public attitudes.

The phasing of a consultation process could follow the example in Figure 8.2, a generic form of that used by TfL in the period from 2002 to 2003, prior to 'go live' in February 2003 [3], extended to show where additions for a LEZ scheme could be included. In this example, the liable party (i.e. a driver or vehicle owner) is identified by means of either the vehicle's number plate (as in London), or a secure identification device such as tag or OBU.

As shown, the 'shape' of a communication campaign progresses from building awareness of the basic facts and who is affected, through to actions on how to

[5] Noting that the treatment of income from enforcement processes may not always be regarded as 'revenues' – the treatment will be specific to each jurisdiction, for example to operate an enforcement process so that it covers its own costs. The business model is outside the scope of this treatise.

participate without being subject to the risk of enforcement. Within these broad themes, the phases move from statement of policy objectives to ensure early alignment of stakeholder with them, through to explaining the solutions, benefits and practical statements about what stakeholder need to do to comply – and what will happen if they don't. As highlighted in Section 8.4, the schemes that were implemented included a public communication programme where vehicle restrictions were promoted as part of a comprehensive package of transport-related and other complementary measures. In New York, public support was poor for congestion charging when presented as 'the policy' but, when included in a comprehensive transport package that included substantial investment and related broad-based benefits, the package secured majority public support.

As explained above, consultation (and the continued measurement of stakeholder attitudes and other contextual metrics) should also apply during implementation *and* operation to help explain and refine policies. The corollary of this is that if any substantive changes to policy are introduced during scheme operation (e.g. emission limits, geographic boundaries, etc.), then a consultation process should be pursued again although this can be targeted to a narrower range of stakeholders, and can already assume a sufficient level of support for the scheme that is the subject of the change.

8.3.4 Communications, media management and measurement of stakeholder attitudes

Monitoring the media is not only to capture a general level of support or otherwise of a transport package, but also to provide a means of communication and stimulating public debate. For this reason, the minimum response from an administration to media reporting should be to correct any errors of fact, as well as to promote objective views on the scheme implementation.

Whilst perceptions cannot be directly measured, attitudes can. The complex relationship (and time-based lags) from communications through to measurement of sentiments, thoughts or views, collectively represented as 'attitudes', is highlighted in Figure 8.3. Attitude measurement based on interpretation of public

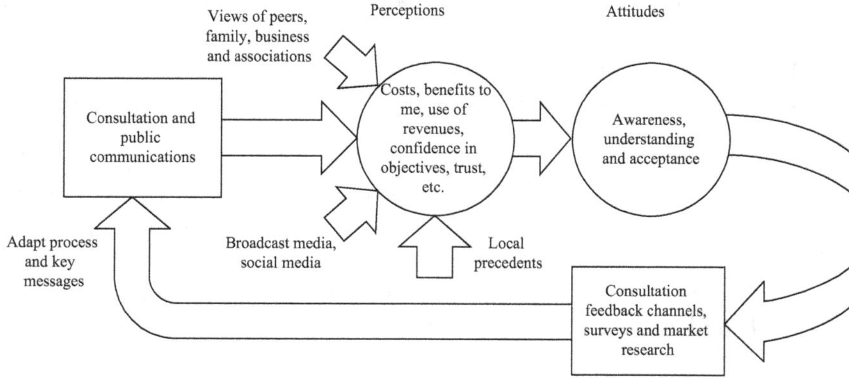

Figure 8.3 The role of formal stakeholder consultation, public communication and attitudinal measurement as part of the development of policies

messages transmitted via social media channels such as Twitter and WeChat may also be beneficial, particularly as insight could be provided in near real time which, as a minimum, helps address misunderstanding or errors quickly.

Communication methods generally depend on multiple channels such as leaflets, media events/interviews, Internet/social media (e.g. WeChat and Twitter), on-street trials, public hearings and road shows that satisfy stakeholder curiosity, many of which should permit the exchange of views which underpin any stakeholder consultation process. Communication helps present the underlying rationale (ideally based on a clear statement of objectives) and the benefits to each stakeholder group. To assist with this, the recommended starting point to help shape the consultation material is to understand the position of the stakeholder group itself, namely to be prepared to respond to the question 'what's in it for me?'

During the operation period itself (which implies that sufficient support was secured to proceed), ongoing measurement of stakeholder views and sentiment should also be aligned with the performance of the scheme against stated objectives. For example, expectations established with stakeholders that claim improvements in air quality can then be compared with the results themselves. It was the recognition that the highest priority 'problem' in Milan was traffic congestion rather than air quality that led to a disruptive change in policy from the original *Ecopass* emissions-related charge (2008–11) to the *Area C* congestion charge (2012 onwards) – see Chapter 19. The former policy was unsustainable, partially undermined by excessive levels of exemptions, but the latter scheme has since benefitted from majority support for the scheme and its rules thereby underpinning the sustainability of the scheme itself.

8.4 Case studies

Short case studies on the role of communications and consultation are provided below, including references to other chapters where applicable.

8.4.1 London

The Greater London Authority Act 1999 introduced new powers on road user charging and workplace parking in London that a future mayor could put into practice. London Mayoral candidate Ken Livingstone adopted road user charging in his manifesto, the potential approach for which had been described in the Road Charging Options for London report [2]. Following his election in 2000, the mayor was required to consult the London Assembly (the city government authority) before other stakeholders or the public on the strategy for charging in central London, informed by invited witnesses and organisations based on the expected costs and benefits. Public consultation was held in two phases; an initial public consultation 'Hearing London's views' in July 2000, a second stage focused on detailed proposals in November 2000. Finally, from late 2002 to early 2003, an extensive public information campaign was undertaken to inform the public on the aims, benefits and other details of the scheme [3]. This included a multimedia campaign, targeted marketing, leaflets sent to all households, road shows and motorway service stations. In parallel, research monitored levels of awareness (up to 97% awareness two weeks prior to 'go-live'). Messages were fine-tuned throughout and the press were fully informed, leading to 317 articles being published in October 2002 alone [3,4]. The use of 'paid for' channels maintained the integrity of TfL's messages and generally followed the flow presented in Figure 8.2.

Congestion charging began operation on 17 February 2003, as part of a package of complementary measures including investment in surface public transport. The charge was originally £5, but has been raised several times since, and applies between 7 a.m. and 6 p.m. on weekdays, with discounts for residents. The scheme is applicable to all vehicles within the zone on public roads, parked or being driven, whether or not they cross the boundary (Figure 8.4). Chapter 13 provides further details on scheme operation.

Although the Mayor was elected, public opinion was initially equivocal; but after it was introduced, opinion shifted in favour, and opposition decreased. Subsequently, the proposal for an extension of the charging zone produced a reduction in support, but it increased following a campaign promoting the benefits [5]. In 2007, the scheme area was increased in size, into what was known as the Western Extension Zone (WEZ) covering 205,000 residents generally based on the same charging policy as the original zone of about 130,000 residents [6]. The incoming mayor Boris Johnson was elected with a manifesto to consult on the WEZ. Despite the benefits of reduced traffic and resident's discounts within the WEZ, the consultation showed that 67% of the public and 88% of business would abolish it [7] suggesting that familiarity does not always breed acceptance [8]. The WEZ was removed in December 2010.

Figure 8.4 Entry to London's congestion charging zone

Each change to the published strategy for charging (embodied in the Mayor's Transport Strategy) requires internal than external consultation. The introduction and evolution of congestion charging in London, now in operation for 15 years, was underpinned by public information campaigns, formal and informal consultation, above-inflation increases in the charge and refinements to scheme operations to improve customer service. In parallel, a comprehensive broad-based monitoring programme for the first 5 years of its operation maintained awareness in scheme costs and benefits.

8.4.2 Edinburgh and Manchester

In 2005, the residents of the city of Edinburgh in Scotland voted in a referendum on the implementation of a two-cordon (inner and outer zone) congestion charging scheme – following on from several years of planning and modelling by the city authorities. The public voted against the proposed scheme by a ratio of 3:1 so the scheme was abandoned. The result was surprising because there was an extensive public information and consultation campaign, and surveys showed significant support from residents [9]. (Though, as Goodwin and Lyons [10] point out, survey responses depend on the precise questions asked and the people who reply.)

Gaunt *et al.* [11] did a 'post-mortem' study showing that car use determined voting: car owners strongly opposed the scheme; non-car owners (public transport users, who stood to benefit from the scheme) only weakly supported it. Also, limited understanding of the scheme increased the opposition:

- The maximum charge was £2 but 38% of respondents thought it could be higher;
- 20% of respondents' journeys would not have been charged but respondents thought they would be;
- 37% incorrectly thought that outbound traffic was charged.

Furthermore, people did not think that the scheme would achieve its objectives of reduced congestion and improved public transport, though 75% recognised that congestion was a problem. Gaunt *et al.* [11] suggested that a simpler, more easily communicated scheme than the dual cordon would have convinced residents, particularly public transport users, of the benefits. If people do not understand what is being proposed or why, they vote against a proposed scheme even though they would benefit from it.

The situation was similar in Manchester in 2008, a scheme design and consultation process for a dual-cordon charging scheme sponsored by the Association of Greater Manchester Authorities. Support for the scheme, or rather lack of it, was represented by a referendum vote of 4:1 against the proposed RP scheme which also meant turning down £1.5 billion of transport infrastructure investment, the equivalent of 50 years' government funding through 'normal' mechanisms. The budget for the consultation was £3 million, about 1% of the value of the package that funded a consultation brochure that was distributed to 1.2 million households, a mobile exhibition bus for each of the ten authorities, public meetings, telephone enquiry service, web enquiry service, updates sent to 30,000 business, a curriculum pack for schools, 65 static exhibitions and a website. A survey immediately following the first round of external consultation revealed a minority in favour and a majority split between opposed and undecided. Despite the extensive consultation and refining, the policy through selective discounts and price capping, it seems that Manchester did not learn any lessons from Edinburgh – most particularly the advice to 'keep the scheme simple', at least initially. It was also suggested that Manchester simply consulted and provided information about the proposed scheme, but was not able, or did not feel able, to campaign aggressively in favour of the scheme – a handicap not experienced by scheme opponents. Other conclusions are that the consultation and education process should be a long-term affair (it takes time for people to understand and accept the arguments), that referenda should be avoided if possible, and that if they are held, it should be only after people have had the chance to experience the benefits, as they did in Stockholm. (For more details of the proposed Manchester scheme see Chapter 7.)

Marsden [12] suggested that the proposed congestion charging plans might have been more popular if they had included a reduction in parking charges. It also seems that a bad time to hold a referendum is immediately before the scheme is due to start – public opinion is at its most negative; 'after initial support acceptability decreases the closer and more specific the proposal gets' [13,14].

8.4.3 New York

To tackle regional congestion costing USD13 billion, a congestion pricing scheme was proposed in 2007 by New York City (NYC) Mayor Michael Bloomberg as part of a wider sustainability plan (PlaNYC, [15]) to create a 'greener, greater New York'. It was advocated by transport planning, business and environmental groups, academics and elected officials. NYC residents supported the proposals by 67% to 27%, if the revenue was used to expand public transport. The plan was also supported by the State Governor, a coalition of 135 civic, businesses, labour, environmental and advocacy groups, the editorial boards of all four major newspapers, suburban elected officials, the NY City Council and the State Senate – though not crucially the NY State Assembly. There seemed to be little opposition, apart from the four NYC boroughs outside Manhattan, especially Queens and Brooklyn, which were more car-dependent than other neighbourhoods, with less good public transport (Schaller [16,17]).

Most attention was focused on the congestion charging proposals, which originally were: $8 on weekdays 6 a.m. to 6 p.m. for cars travelling in Manhattan south of 86th Street, though subsequently simplified to cover the smaller area south of 60th Street. Payment would be via the well-established E-ZPass electronic tolling system, or other payment channels. Bridge and tunnel tolls would offset the congestion charge so that no driver would pay more than $8 per day.

Although congestion pricing generated strong feelings, press and public reaction was surprisingly positive, probably because of the wider sustainability plan as revealed in polls [18]. When it was explained that it would mitigate increases elsewhere, congestion charging become steadily more popular. CapitalNewYork.com explained this as 'The more that you explain congestion pricing, the less scary it becomes' [19]. VMT in the charging area would be reduced by 7%, with 30% reduction in congestion in the zone and 20% in adjacent areas. Projected annual net revenues of USD491 million were to be devoted to public transport improvements. In addition, the city and the Metropolitan Transport Authority were allocated USD354 million from the Federal Department of Transportation to implement the scheme.

The NYC council voted 30 to 20 to adopt the plan; but congestion pricing needed authorisation under state law; this was blocked in the State Assembly; the plan failed to meet the Federal funding deadline, and was abandoned. Schaller [16] asks why this wide support failed to get the scheme approved: 'The short answer is that a relatively small group of auto users believed that congestion pricing was against their best interests... The intensive interests of one group were thus able to overcome widespread public support'.

Schaller also makes the following points. First, congestion pricing is supported only if it makes sense for society and individuals – leading to surprising results such as public transport users opposing if they think it is unfair to drivers. Second, drivers thought that public transport was not a viable alternative, that the travel time savings were not worth the USD8 charge, and that the societal benefits would not materialise. He draws the following lessons for acceptance of RP:

- build support for the proposals by extensive public outreach and education;
- involve the public in the plan;
- have strong leadership from the Local Authority, with a high-profile project champion;
- have a comprehensive approach including public transport and land use;
- acquire external funding if possible.
- In designing pricing proposals:
  - be aware of the disproportionate power of small groups of vocal opponents;
  - offer a free-of-charge alternative route.
  - convince drivers that they personally will benefit from the scheme, irrespective of the overall societal benefits.

In 2015, a new group, led by Samuel I. Schwartz, a former city traffic commissioner, proposed a rationalisation of tolls on the city's bridges combined with additional tolls for vehicles entering the city, which would be adjusted depending on congestion levels. The new scheme ('The Move NY Fair Plan' [20]) would generate $1.5 billion in annual revenue for transport investment. At the time of writing, it is not known whether this scheme will be implemented.

Details on other US schemes are provided in Chapters 5 and 18. The rationale for, and implementation of, interoperability in the United States is described in Chapter 5.

8.4.4 Singapore

Singapore implemented an Area Licensing Scheme (ALS) in 1975 that required vehicle owners to purchase a period-based paper licence to enter the Restricted

Zone, otherwise known as the Central Business District. ERP replaced the ALS scheme in 1998 and the Land Transport Authority (LTA) is currently underway with the implementation of the replacement 'ERP2' system that will enable more complex pricing measures and valued-added services. It is anticipated that ERP2 will be implemented progressively from 2020.

The consultation process for the first-generation ERP scheme to functionally replace the 23-year-old ALS was underpinned by political consensus that had already been developed. The legislation already supported charging for the use of road space, enforcement of violators, the appeals process and the application of charging to foreign vehicles.

After an aborted procurement in 1991, a new tender led to three consortia being shortlisted for on-road trials that were exploited as part of a 1-year structured outreach and public consultation programme and to help de-risk scheme operations. Throughout the implementation phases, LTA closely liaised with Members of Parliament, residents (whether they were car owners or not), road users, city councils, public transport operators, the vehicle registrar and infrastructure departments with the common theme that ERP would form part of '[the] management of the car population, and at the same time, optimise the use of [the] roads, through a judicious mix of ownership and usage management'. An education programme was also set up in schools to teach children about transport solutions and the benefits of RP.

The main marketing messages were focused on improved traffic management (rather than revenue generation), reliability (proven during the trial period), flexibility (replacement of a flat charge with a usage-related charge) and usability (no need to apply for daily licences). Brochures were sent to all motorists before system launch and the system was switched on early to enable motorists to familiarise themselves with the new system and the operation of the In-vehicle Units. Adverts were placed in the print media and on TV to raise awareness. This was also extended to all motorists entering from Malaysia. A hotline and customer service centres were maintained to further allow feedback to come from road users.

Following implementation, reports demonstrated that the original policy expectations were being met, including a reduction in traffic volumes on expressways and an increase in average speeds from 35 to 55 km/h, further enabled through periodic adjustments in tariffs. Public acceptance was also dependent on the usage charges mitigating some of the increases in vehicle ownership charges, as well as fairness (charging and enforcement accuracy was critical), privacy and demonstrable reductions in congestion. In addition, vehicle owners received periodic rebates on their annual vehicle taxes, funded by ERP revenues. In 2010, the LTA declared: 'While a congestion pricing scheme may be justified from a technical perspective and can be rationally argued for, utmost importance must be given to communicating the rationale of the scheme to road-users and the communities including the businesses. There is [...] never too much communications and publicity when it comes to road pricing schemes' [21].

The approach to consultation for the new satellite-based ERP2 scheme will follow similar principles and the challenge will be promoting the benefits of an increased level of pricing complexity, potentially enabling every road segment to be individually charged. In line with the evolutionary approach practiced on road user charging in Singapore, a step-by-step approach will also be employed throughout the consultation process.

For more details of the existing and proposed Singapore schemes, see Chapter 6.

8.4.5 Hong Kong SAR

Hong Kong SAR (pop. 7.2 million) published its first Comprehensive Transport Study in 1976, the contents of which were endorsed, after successful public consultations, in a Government White Paper in 1979, including a significant investment in transport infrastructure, the development of a new subway system and the use of ERP to restrain the usage of private vehicles in favour of commercial goods and public transport. Successive increases in motoring-related taxes led to the then Governor to describe such taxes as 'crude' and that usage-based charges should be imposed instead. Three potential ERP schemes based on cordon charging were outlined and presented to panels of road users to understand relative preferences but each was based on charging private vehicles rather than commercial or public service vehicles.

On-road trials were conducted over 21 months between 1983 and 1985 using Government vehicles and volunteers (in total 2,500 vehicles) to assess usability, to gain further feedback from the public and to confirm technical feasibility [22]. What became the world's first technical trials of ERP systems showed that they were highly efficient: it could be provided/operated at low cost, was technically feasible and was 99% reliable. At that time, the technology for charging was crude, utilising an Electronic Number Plate read by 130 in-road inductive-loop-based roadside systems, enforced by camera. Revenues from such a policy would have been regarded as a tax and not hypothecated into transport investments.

Consultation within the newly formed District Councils (now known as 'District Boards') showed only 2 of the 19 supported ERP as a TDM measure. Extended consultation, based on no changes to the proposed policy, resulted in little change of opinion. The results of the trial were not made public and one commentator stated that '[they] did not provide [any] practical and understandable illustrations of how much different types of travellers might benefit from the ERP scheme' [23]. Government attempts to respond to the public's concerns on lack of privacy was based on offering to keep data confidential (and to destroy it in 3 months) and the location of charges would not be included on bills. A policy focus on private vehicles only contributed to the perception by private vehicle owners that they were being targeted unfairly. At that time, Hong Kong had no environmental policy and the trials were perceived by a public whose confidence was at the lowest levels (because of recent increased motoring taxes, recession and lack of consultation on ERP) to be no more than a technology test. Plans for ERP were dropped shortly after.

The second Comprehensive Transport Study published in 1989 included a proposed variation of one of the three scheme options that had been presented earlier. The results of the consultation on this option were reflected in a 1990 White Paper, including the use of smart cards to resolve the privacy fears – the most difficult objection that had been raised in 1985. The study did not result in a further consultation and plans for ERP were not progressed any further. The consultation process lacked elements of dialogue with political and public stakeholders, and contemporary reports show that the Government's communication of the intentions of ERP did not appear to be matched by efforts made to understand public concerns nor resolve these in a timely manner.

In the 1990s, further internal consultation, followed by a stated preference survey in May 1997, revealed a shift in demand towards public transport and apparent support for a range of ERP policies. No public consultation was conducted and the results of a comparative technical trial into DSRC- and GNSS-based charging received only limited publicity. The final report claimed 'there will likely be public resistance to ERP arising from suspicion about the government's motive, the reluctance to accept new initiatives and road users' resentment about the imposition of new charges' yet, supported further public consultation. A combination of political indifference and economic recession reduced momentum in any further policy development for ERP, with Government emphasising that measures to reduce congestion for Hong Kong and Kowloon would not be envisaged until 2006 and 2011 at the earliest, respectively.

Following the announcement of three pilot LEZ, measures to reduce congestion were reported by Transport Department in December 2014, including a recommendation to 'start planning for a congestion charging pilot scheme by developing a conceptual plan for engagement with the public and relevant stakeholders' [24]. The results of the consultation into ERP conducted in March 2016 [25] resulted in 515 submissions and included confirmation that a feasibility study on a pilot scheme (the third since 1984) will be conducted, potentially followed by new legislation and funding for such a scheme. The Government commenced procurement for technical advisory services in July 2017 in anticipation of implementing a Proof on Concept of the scheme in Central and its surrounding area.

8.4.6 The Netherlands

The Netherlands (pop. 17 million and 9 million vehicles) formally recognised the impending problem of congestion in 1977 when RP first appeared on a national policy document. The debate continued to gather momentum from 1992 onwards, as highlighted in three attempts up to 2001 to introduce RP as a regional or national policy. The introduction of each new policy informed the development of the next, in a process described as 'policy-oriented learning' attributed to advances in the underlying science and technology for RP but also the need to seek consensus [26].

In October 2003, a national 'platform' (the Nouwen[6] Committee) was established as a professional discussion forum comprising members of all major interest groups and local governments that was used to define and consult on a

[6] Named after (and chaired by) the main opposer to *Rekeningrijden II*, Mr Paul Nouwen (ANWB).

Kilometerheffing (lit. 'kilometre charge'), to replace vehicle ownership taxes with usage charges based on distance travelled, time, location and environmental impact. Low cost operation was emphasised, referenced by the low cost (*c.* 3%–5% of revenues) of collecting existing motoring-related taxes. Fuel duty would not be impacted (contrary to US attempts at maintaining period taxes and introducing a fuel duty replacement mechanism). Earmarking revenues and the use of time and place-based charging fuelled most disagreements with the Committee. To inform the development of a national policy, potentially permitting local variations, the two-stage consultation process included a fact-finding mission, development of a reference architecture covering systems design and operations, surveys on international best practice and the development of policy 'storylines[7]'.

In 2007, the Cabinet confirmed its intention to implement RP, thereby triggering the development of a technology trials procurement programme, which would also serve to test operational strategies and, in this case, to obtain feedback of in-vehicle technology performance under realistic conditions. In parallel, a draft *Kilometerheffing* Act was prepared to enable a national policy for RP that would commence in 2012 for trucks and from 2013 to 2016 for private cars. To help develop a package of measures that would operate alongside the pricing element, some of which would help fund the scheme roll-out, a value-added services stream was initiated to encourage market players to define new services. Large scale trials included 25,000 users, yet a survey [27] conducted by *De Telegraaf* was summarised as a 'crushing no', revealing that the majority were opposed to national RP, including concerns that it would not solve congestion[8], the costs of driving would be too high and that the system would threaten user privacy. Two months later in March 2010, a survey amongst the four million members of the Dutch motoring organisation, *Algemene Nederlandse Wielrijdersbond* (ANWB, the Dutch Touring Club) showed a large majority accepted the principle of paying according to road use, and that someone who drives a lot should pay more than someone who seldom drives. The *De Telegraaf* and ANWB Internet surveys were heavily criticised since respondents were self-selected (thereby contributing to polarisation of views, an effect also highlighted in an Internet survey conducted for the proposed Cambridge congestion charging scheme).

The Government fell in February 2010 following political disputes unconnected with RP and, shortly after, the new Minister of Transport, Public Affairs and Water Management declared the project as 'controversial' and then gave instructions to abandon any plans to roll out the RP scheme.

In summary, the consultation and communication process in the Netherlands confirms that societal support is a key driver, and that planning needs to be realistic and its operation (from a user perspective) must be simple. Public communications

[7] A storyline describes a feasible implementation and operations scenario to stimulate discussion and comparison of options.

[8] This was also a concern in Edinburgh. But we know, from London and Stockholm, that it does indeed solve the congestion problem. Education of voters and politicians is needed on this point. Road traffic is not like the flow of water, seeking a new route if it is blocked. Drivers have more options than just an alternative route; they can choose not to travel, or take public transport, or travel at a different time, or telecommute.

must employ clear messages on why the scheme is necessary and feedback from stakeholder consultation must be accommodated within the scheme design. In 2015, one analysis [26] of policy development of RP in the Netherlands stated that 'Dutch road pricing policy process has been marked by "political conflicts" and road pricing policies involve "complex technological issues"'. Arguably, technological issues are now largely solved, requiring detection and measurement of vehicle usage. However, in all cases, systems for RP require localisation, particularly to ensure enforceability of scheme rules, elaborated further in Chapter 20.

In 2012, six of the ten political parties in The Netherlands included reference to some form of RP policy in their election manifestos. Although the Intelligent Transport Systems (ITS) Plan for the Netherlands (2013–2107) made no reference to RP, the need for focused efforts on improving mobility remains. Rotterdam has since implemented access regulations and 13 of the Netherland's largest cities have implemented LEZs.

8.4.7 Milan

The heavily industrialised *Commune di Milano* ('Milan') (pop. 5.2 million) suffers from some of the highest vehicle and motorcycle ownership rates in the world and faced the visible problems of worsening congestion and transport-related air quality. Milan had acquired the reputation as Europe's most polluted city: the environmental standard for PM_{10} was being exceeded on 150 days (limit was 35 days) per year. In 2006, a steering committee was set up to assist with the regional political consultation process to address the challenges of poor air quality as part of broader investments in transport and complementary measures.

Shortly after a 'conference of mayors' and regional representatives reached a consensus on the scope as a reaction to winter smog and intense lobbying campaigns[9], the Mayor took the unpopular step to ban 170,000 cars and motorcycles manufactured to pre-*Euro* requirements from being used in a restricted area known as the *Zona a Traffico Limitato* (ZTL). This had been informed by surveys that showed a roughly even split of support[10] and opposition to the scheme and a majority agreeing that something needed to be done to improve air quality. In general, there was a lack of information on the plans and the media became the main channel to keep the public informed. The initial 60 km^2 scheme was reduced to 8.2 km^2 (thereby reducing the complexity of political negotiations from the region to the city) and a residents' discount was added.

A 2-year *EcoPass* trial differentiated more finely between emission classes entering the ZTL, with fees set at €2, €5 and €10. Payments and renewals were handled via the Internet, newsagents, garages and banks. Policy aims were clearly stated: a reduction in particulate matter by 30% and to improve air quality for everyone. The initial response to the scheme was favourable although this was undermined by various operational failures, including 160,000 fines being issued in

[9] Led by pro-cycling, anti-pollution groups, outspoken doctors and Italy's Green party.
[10] For example, in a poll conducted by the '*Corriere Della Sera*' newspaper in 2006, a majority of citizens (53.9%) were opposed to the scheme.

the first few days of operation to road users that claimed to be unaware of the new regulations.

Public communication made no reference to the other benefits such as reduced accident rates or the increased speed of public transport within the ZTL but instead had focused on concentrating its public communication on reporting variables (i.e. road network performance and air quality). A planned public referendum was replaced by a survey of 500 citizens in November 2009 that confirmed continuing concern about environmental pollution. Despite this, the following 2-year period witnessed traffic congestion and harmful emissions returning close to former levels, not helped by an increasing share of exempted vehicles from 50% in 2007 to 90% in 2010.

After a change of mayor in 2011, a formal consultation on a range of transport measures commenced, resulting in a financial package of improvements worth about €30 billion, including public transport and an access fee. Buoyed by a referendum that confirmed a 95% acceptance, *EcoPass* was subsequently replaced by *Area C*. The reduction in exempted vehicles to 12% accompanied a reduction in vehicle entries of 34%, a 5% increase in average speed of public transport, 52% reduction in accidents and a 19% reduction in PM_{10} emissions. The public communications for *Area C* were based on flyers, letters to households, road signs and advertisements on radio, TV, website and public information meetings.

Arguably, the focus on stakeholder communication on a widely acknowledged visible problem (air quality and related adverse impact on health) and policy flexibility to migrate to congestion charging underpinned the success of the scheme. For more details, see Chapter 19.

8.4.8 Berlin

Berlin (pop. 3.5 million) is Europe's second most populous city (after London with 8.3 million). In 2004, the Senate Department for Urban Development and the Environment published and consulted on a draft Air Quality Plan before being adopted by the local parliament. The relatively long (2.5 years) planning and stakeholder consultation period of Berlin's Air Quality Plan ensured sufficient levels of acceptance, reflected the complexities of a larger metropolitan population, the more complex regional jurisdictional structure[11] and some delays in securing additional funding from the national government.

At a local level, in 2008, Berlin established a roundtable that included representatives from other government institutions, environmental NGOs and business associations which were consulted to confirm feasibility, potential implementation costs and socio-economic impacts. The main variables were the boundaries of the area, timing of introduction and emission requirements for compliant vehicles. Funding to subsidise the fitment of particulate traps on diesel vehicles was also sought. Consultation was based on marketing (e.g. extensive use of flyers, newspaper

[11] Berlin is a city-state that has its own Parliament. The state is divided into 12 boroughs (*Bezirke*) representing 96 localities (*Ortsteile*), each comprising several neighbourhoods (*Kiez*).

articles, advertisements and a comprehensive web-based Internet resource) coupled with seeking feedback (e.g. ad-hoc surveys, meetings with businesses and truck operator associations) to inform the development of Berlin's *umweltzone.*

The final Air Quality Plan introduced concessions such as the use of temporary exemptions[12] for commercial fleet operators that could demonstrate potential severe economic harm from the implementation of an *umweltzone.* Following the initial consultation and implementation, the Berlin scheme commenced operation in 2008, followed by a second phase in 2010 based on more stringent emission levels. The migration to the second phase was achieved by a continuous monitoring programme and communication with stakeholders (particularly commercial fleet operators) that informed a tightening of regulations to Euro 4 for all light and heavy diesel vehicles. The natural replacement rate of vehicles and resulting migration towards Euro 6 and beyond meant that tighter emission targets could be progressively introduced.

Overall, the use of enabling legislation, a framework for phased public communications and consultation, proven implementation timetables, increasing national public acceptance, have all contributed to implementation of *umweltzone* in Berlin and their adoption throughout Germany. Many cities adopted Berlin's roundtable approach, supplemented with marketing activities as described above. However, cities that have implemented an *umweltzone* appear to be regionally clustered potentially due to the different rates of the development of critical mass of political will in each region.

As an example of best practice in Germany, Berlin had developed (and currently maintains) a comprehensive web resource [28] showing measured emission levels at individual street level, baseline data files, a selection of future scenarios that refer to achievable targets, draft plans to further improve air quality for transport (including improved traffic management, speed limits, tax incentives, etc.) and reduction in harmful emissions from other non-transport sources.

8.4.9 Summary

To varying degrees, each of the schemes described in the case studies highlighted a visible 'problem': the need for road infrastructure, investment in public transport (or other complementary measures), increasing congestion or worsening air quality – or some combination of these, and willingness to change the scheme design in response to feedback and (mostly) structured internal dialogue followed by some form of interaction with the public. The structure and sequence of each phase (Section 8.3.3) starts with internal (usually politically consultation) and then progressively intensifies the interaction with the public to inform detailed scheme design, often punctuated by further 'internal' dialogue to confirm and refine political

[12] According to the Berlin state government: 'Euro 3 vehicles for which there is no suitable particulate filter for retrofitting and upgrading to a green sticker, and which therefore cannot be retrofitted, may be driven in the environmental zone with a yellow sticker provided a certificate to this effect is obtained from a technical inspection agency and clearly displayed in the vehicle'. The exemption expired on 31 December 2014 after which time these Euro 3 vehicles may not enter Berlin's *umweltzone.*

support. As a means of comparison, it may be helpful to refer to three distinct but non-mutually exclusive levels of interaction with stakeholders as described by the Organisation for Economic Co-operation and Development (OECD) [29]: 'notification', 'consultation' and 'participation', demonstrated by all schemes to varying degrees. This distinction of type of interaction is widely shared [30,31].

However, the scope and intensity of stakeholder interaction is unlikely to be a good predictor of the likelihood that a scheme will be implemented or not. For example, Edinburgh had a highly structured, multiphased, extensive consultation process but this revealed significant public dissatisfaction in the use of congestion charging as revealed in its referendum – even though congestion was widely accepted as being problematic.

Referenda appear to introduce an element of risk, as the experiences of Edinburgh, Manchester, Stockholm and Gothenburg showed (the approaches to public consultation and communications for Stockholm and Gothenburg are detailed in Chapter 14 and are not addressed further here). Flexibility in scheme design and use of public consultation to shape this may be a better determinant of the success (or otherwise) of a proposed scheme and its path to implementation, as shown in Milan.

Consultation on infrastructure projects in Asia, including ERP in Hong Kong, Singapore and early plans in Beijing [32] have tended to focus more on developing coherent government policies and the support from key officials and other key political stakeholders. Public opinion on scheme design is emphasised less and many projects lack any extensive dialogue. Instead, structured notification is the main form of public engagement. In the examples selected, there is not enough evidence to draw any conclusions on whether an individual 'champion' such as an elected Mayor is necessary (although arguably London and Milan benefitted from this but this was not sufficient in New York). There is some evidence that a city authority itself can perform this role, although as an implementing authority it would be limited in how much it can market Congestion Charging (CC) or LEZ – arguably, Berlin, Stockholm, Gothenburg and Singapore benefitted. Instead, it is likely that the development of a positive and supportive ecosystem is fundamentally important and the role of stakeholder communications underpins this.

8.5 Recommended approaches

The consultation process needs to be positive and proactive and must inform and educate people – not just present them with a few facts and leave it at that. It really needs a positive 'pro' campaign – and needs to allow plenty of time – sometimes years – to allow people to absorb and accept the information, noting that more information is better than less. They also need to be convinced that a CC or LEZ scheme, perhaps as part of a package of measures, will reduce congestion and harmful emissions rather than just divert traffic and emissions elsewhere. It is important that consultation is not carried out too late in the policy design process.

Execution needs to identify relevant stakeholders, define the multistage consultation process from 'internal' to 'external', decision on which forms of interaction suit the stage of the process (e.g. mass-market leaflets, public visits to on-road trial sites, community workshops, one-to-one dialogues with key opinion-leaders, etc.) and the development of policy 'storylines' to aid communication. The tools available include communications and surveys, the former to inform and to set expectations on objectives, and the latter to measure the development of public opinion.

The execution process needs to identify public acceptance challenges, revise policy themes/operational strategies and budgets whilst progressively revealing more and more detail and, most importantly, gather feedback to check levels of understanding and acceptance to be fed back into the policy design process. Despite the challenges of stakeholder consultation, it develops legitimacy and, in turn, legitimacy secures public confidence, acceptance and scheme sustainability.

It is also recommended that the potential benefits of public acceptance on scheme sustainability be studied and that scheme design is flexible enough to adapt to feedback gained from the stakeholder engagement process. Interactions with stakeholders provide a cost-effective means of gathering data on travel patterns and preferences – and provide the means to assess the acceptance of a policy or the relative acceptance of a range of policies. Informed feedback also allows operational strategies for potential CC or LEZ schemes to be refined before a scheme is implemented and to maintain sufficient public support throughout, including the operations period.

8.6 Conclusions

The case studies highlight the variety of social, political and environmental contexts within which behaviour change programmes were planned and implemented. From this, the type of consultation process will need to reflect the cultural background of the region in which the package or scheme is to be delivered. In common though, all the international case studies highlight how consultation (and related communications) is woven into the policy development process, split into 'internal' (to define and refine possible policies) and 'external' (to shortlist and refine a single policy) within an integrated transport package. In some cases, these well-founded processes failed to create a critical mass of support. Therefore, as the examples demonstrate, stakeholder engagement is a necessary but not sufficient prerequisite for political and public acceptance.

Also, the *process* of consultation and its *outcomes* are equally important and we can conclude that the minimum requirement for a successful consultation is to achieve high levels of public awareness and a process that is designed to facilitate political and public acceptance. The successful schemes, even those that faced public opposition to the use of CC or LEZ, depended on public communications that included comprehensive notification, promotion, intensive monitoring of changes in public attitudes, timely adjustments in scope and policies – and most of all, willingness to change.

Stakeholders need to be identified, interactions recorded and the information provided must be adapted to the information needs of each stakeholder group. Consultation should also extend into each phase of scheme delivery for reassurance and to help maintain a critical mass of support. During operations, monitoring is critical not only to measure public support but to help refine scheme objectives, to provide reassurance that stated objectives are being met and actions that will be taken if not. If changes are required, then a new (often smaller) consultation process should start again based on the existing scheme as a starting point. Also, as the examples highlight, a successful implementation in one city may become best practice for the implementation in many others. In the case of Milan (Chapter 19), a former scheme became the reference for a disruptive policy change and succeeded (and continues to succeed) due to an unwavering focus by successive administrations on pragmatically targeting the most pressing problems of congestion through a well-defined package of measures, which included congestion charging. On the other hand, Singapore demonstrates policy evolution that started in 1975 and continues to this day with its second-generation scheme (Chapter 6).

The examples were selected from different social, economic and political contexts demonstrating common themes employed in public communications, including comprehensive notification, promotion, intensive monitoring of changes in public attitudes and timely adjustment. Change cannot be forced, it must be nurtured: messages need to be refined, the constituent elements of the scheme policy may need to be changed and the response from stakeholders needs to be monitored. Political and public acceptance also depends on policies that mitigate visible problems in a way that satisfies the self-interest of informed users sufficient to permit scheme implementation as part of a package of measures and thereafter ensure readiness to allow scheme policies to be refined to meet, and continue to meet, expectations of all stakeholders.

Acknowledgement

We wish to acknowledge the World Resources Institute (WRI) for permission to use some of the text that was generated as part of an earlier study by the lead author. Further information at http://www.wri.org.

References

[1] Results of the public consultation, *The Urban Dimension of the EU Transport Policy*, European Commission DG Move, April 2013.

[2] Road Charging Options for London: A Technical Assessment, Government Office for London, November 1999.

[3] *Impacts Monitoring – First Annual Report*, Transport for London, 2003.
[4] *Impact Monitoring – Fifth Annual Report*, Transport for London, June 2007.
[5] Dix, M. *An update from London: the evolution of the scheme.* In EU Road User Charging Conference, London, 23 January 2007.
[6] Ward figures, 2001 Census. National Statistics Office. As reported in Draft Mayor's Transport Strategy Revision: Central London Congestion Charging – westward extension, Response from London Assembly (undated).
[7] Dix, M. *Key changes and current challenges for road user charging within the UK.* 6th Annual EU Road User Charging Conference, London, 26–27 January 2009.
[8] Walker, J., *The Acceptability of Road Pricing*, RAC Foundation, May 2011.
[9] Cain, A. and Jones, P. M. (2008). Does urban road pricing cause hardship to low-income car drivers? An affordability based approach. Transportation Research Record, 2067: 47–55.
[10] Goodwin, P. and Lyons, G. (2010). Public attitudes to transport: interpreting the evidence. Transportation Planning and Technology, 33(1): 3–17.
[11] Gaunt, M., Rye, T. and Allen, S. (2007). Public acceptability of road user charging: the case of Edinburgh and the 2005 referendum. Transport Reviews, 27(1): 85–102.
[12] Marsden, G. (2009). *Congestion pricing, do we need a new approach to parking policy? Lessons from Europe*. Presentation in Session 167 (Nexus Between Parking Pricing and Congestion Pricing) at TRB Annual Meeting, Washington, January 2009.
[13] Goodwin, P. (2006). *The Gestation Process for Road Pricing Schemes*, Local Transport Today, 444: 1 June 2006.
[14] Owen, R., Sweeting, A., Clegg, S., Musselwhite, C. and Lyons, G., Public acceptability of road pricing, *Final Report for UK Department for Transport*, UK Department for Transport, 2008.
[15] *A Greener, Greater New York, PlanNYC Full Report*, The City of New York, April 2007.
[16] Schaller, B. (2010). New York City's congestion pricing experience and implications for road pricing acceptance in the United States. Transport Policy, 17: 266–273.
[17] Schaller, B., *Battling Traffic: What New Yorkers Think About Road Pricing*, Center for Rethinking Development, Manhattan Institute for Policy Research, 2006. http://www.manhattan-institute.org/html/rdr_03.htm, accessed on 21 April 2017.
[18] University of Quinnipiac, *'State Voters Back NYC Traffic Fee 2-1, If Funds to Transit, Quinnipiac University Poll Finds Voters Back Millionaire's Tax 4-1'*, Press Release, 24 March 2008.
[19] Paybarah, A., *The More that You Explain Congestion Pricing, the Less Scary It Becomes*, CaptalNewYork.com, 15 February 2013.
[20] Move NY, *The Move NY Fair Plan,* February 2015.
[21] Kian-Keong, C. (Land Transport Agency), *The Singapore experience: the evolution of technologies, costs and benefits, and lessons learnt.* International Transport Forum, Paris, 4–5 February 2010.

[22] Transport Department (Hong Kong), *Feasibility Study on Electronic Road Pricing – Final Report*, Transport Department, April 2001.

[23] Gómez-Ibáñez, Jose A and Small, KA, *Road Pricing for Congestion Management: A Survey of International Practice,* Publ. Transportation Research Board, National Research Council, 1994.

[24] Transport Advisory Committee (Hong Kong), *Report on Study of Road Traffic Congestion in Hong Kong*, Transport Advisory Committee, December 2014.

[25] *Electronic Road Pricing Pilot Scheme in Central and its Adjacent Areas – Public Engagement Report,* Transport and Housing Bureau/Transport Department, January 2017.

[26] Ardic, O, Annema, JA, and van Wee, GP. (2015). Non-implementation of road pricing policy in the Netherlands: an application of the 'Advocacy Coalition Framework'. European Journal of Transport and Infrastructure Research, 15(2): 116–146. ISSN: 1567-7141.

[27] De Telegraaf, *Verpletterend NEE; Ruim 89% Tegen Kilometerheffing Angst Voor Fraude*, De Telegraaf, 26 January 2010.

[28] 03.11.3 *Clean Air Plan, 2011-2017 – Scenario Calculations for the Assessment of the Effectiveness of Selected Measures for Air Quality along Streets* (Edition 2012) at http://www.stadtentwicklung.berlin.de/umwelt/umweltatlas/eic311.htm, accessed on 21 April 2017.

[29] *Background Document on Public Consultation,* Public Governance and Territorial Development Directorate, OECD, undated.

[30] Fife Council, *Consultation Methods*, Section 4, at www.fifedirect.org.uk/consultationhandbook, accessed on 21 April 2017.

[31] Rietbergen-McCracken, J. and Narayan, D., *Participation and Social Assessment: Tools and Techniques*, Published by World Bank, Washington DC, USA, 1998. documents.worldbank.org/curated/en/673361468742834292/pdf/multi0page.pdf (accessed 18 October 2017).

[32] Transport Demand Management in Beijing, a programme implemented by Deutsche Gesellschaft für Internationale Zusammenarbeit (Germany – GIZ), funded by Federal Ministry for the Environment (Germany – BMU), https://mitigationpartnership.net/transport-demand-management-beijing, accessed on 21 April 2017.

[33] For Low Emission Zones in Germany, go to: http://www.umweltbundesamt.de/en/topics/air/particulate-matter-pm10/low-emission-zones-in-germany, accessed on 21 April 2017.

Further reading

[1] For Impact Assessments and Other Reports Relating to London's Congestion Charging Scheme go to https://tfl.gov.uk/corporate/publications-and-reports/congestion-charge, accessed on 21 April 2017.

[2] For Copies of Impact Assessments of Congestion Charging in Sweden, go to Trafikverket (Swedish Transport Administration) at http://www.trafikverket.se, including the public leaflet 'This Is the West Swedish Agreement' at www.trafikverket.se/contentassets/.../this_is_the_west_swedish_agreement.pdf, accessed on 21 April 2017.
[3] Göteborg Region Association of Local Authorities (GR): http://www.grkom.se/toppmenyn/omgrgoteborgsregionen/inenglish, accessed on 21 April 2017.

Chapter 9
Road pricing standardisation

Jan Kersten and Jasja Tijink***

9.1 Introduction and scope

Road pricing systems, their subsystems and components use national and international standards in various areas, like telecommunications, mechanical, payment systems, safety, construction and many others. They all serve specific needs for specific areas. The aim of this chapter is to introduce the framework of standards that have been developed explicitly and primarily for use in road pricing systems. These are standards for the field of Electronic Fee Collection (EFC).

The clauses below will start a tour at the history of standards in EFC, why standards in EFC are important and which initiatives contributed (and continue to contribute) to the establishment of the standards framework that now exists for EFC. Most important standards are explained in more detail following a structured approach that starts with the architecture standard for tolling systems. The further journey will range from base standards defining major characteristics of the applied communication technologies, via interoperable application profile standards up to respective conformance evaluation standards. Standards for both major tolling technologies applied in Europe – short-range communication [5.8 GHz Dedicated Short-Range Communication (DSRC)] and autonomous Global Navigation Satellite Systems (GNSS) – will be encompassed in this chapter.

The scope of the chapter is limited to those standards which are applicable in the context of EFC technologies falling under the European Interoperability Directive 2004/52/EC [1]. It should be noted that many of these standards are applicable in other regions of the world as well. However, communication technologies and related standards used in EFC primarily in Asia or in the Northern American market are outside the scope of this chapter – but see Chapter 12.

*Trafficteq, Germany
**Kapsch TrafficCom AG, Austria

9.2 History and status of standardisation

9.2.1 Why standards in road pricing?

Technical standards seek for harmonisation of interfaces between technical components – such as devices and subsystems – of technical systems. Standards ensure that products and services are safe, reliable and of good quality. For business, they are strategic tools that reduce costs by minimising waste and errors, and increasing productivity. They help companies to access new markets, level the playing field for developing countries and facilitate free and fair global trade [2].

Road pricing as currently in use applies technical standards in many regards. However, one can clearly identify two areas in which standardisation had a major influence on the evolution of road pricing schemes from pure manual and cash dominant concepts towards cashless systems which work without barriers. The objectives of making payments for the use of road infrastructure as convenient and fast as possible with minimal impact on travel times and traffic flow was a major driver in applying and developing international standards in road pricing. These two areas are:

- cashless payment means and processes, including the related interfaces and
- EFC by means of on-board units (OBUs).

Accepting commonly used cashless payment means such as credit cards, fuel cards, Radio Frequency Identification-based (toll) payment cards and OBUs is one step towards increasing the traffic throughput at toll infrastructure (e.g. toll plazas) while keeping the number of lanes and required plaza operators' staff and consequently investment and operations costs at acceptable levels. Road pricing business gained benefits from existing harmonisation and standards for payment cards and will still be benefiting from technologies like Near-Field Communication that make payment with cards even faster without jeopardising security. Moreover, operators of road pricing infrastructure use the same (harmonised and standardised) backbone for payment and clearing processing as other business do.

Even more contribution to the evolution to road pricing schemes since the 1990s is made by EFC, using a device mounted inside the vehicle – a so-called OBU – that stores vehicle, user and payment related data required for toll payments. Such OBUs exchange information either with a transceiver station installed alongside or above the road infrastructure or via mobile/cellular networks to the central system of a toll operating company. Depending on the data exchanged, the toll operator determines the toll fee due and charges the respective road user account.

The tremendous advantage of EFC is obvious. Main areas are

- no need for barriers at toll plazas;
- massively increased traffic throughput per lane;
- allows for multi-lane-free-flow tolling/open road tolling;
- less investment and operations costs.

International standards for EFC have been developed for

- business architecture (role model);
- security framework architecture;
- underlying communication technology (if no existing technology is reused);
- processes and services to be made available at application interfaces;
- harmonised EFC specific application data; and
- related conformance evaluation standards.

Note that such standards are not only applicable for the charging-related processes and interfaces but also cover supporting services like enforcement.

9.2.2 Early initiatives in standardisation in road pricing

The first steps towards EFC were made in the early 1990s in Italy. Italian road concessionaires suffered from massively increasing traffic at their existing toll plaza infrastructure which resulted in congestion and long waiting times. A supporting factor for a harmonised EFC technology approach can be found in the de facto monopoly situation in Italian road concessionaire business at that time. The base EFC communication layer specification was finally adopted and published as a National Italian standard [3]. This national standard was the early and first adopter of a frequency regulation made by the European Communications Committee (ECC) in 1992 in which a 20 MHz band in the 5.8 GHz industrial scientific and medical (ISM) radio band was allocated on a non-exclusive basis to applications in road traffic telematics – including tolling [4]. However, this National Italian EFC standard was later not taken as the basis for the international standards developed by International Organization for Standardization (ISO) and European Committee for Standardization (Comité Européen de Normalisation, CEN).

The allocation of the 5.8 GHz band for EFC road pricing was substantially supported by research work carried out in technology leading companies, like Bosch (Germany), Marconi (UK), CGA (France), Combitech (Sweden), Micodesign (Norway) and others. The same group of technology frontiers demonstrated first prototype products and system concepts in EC co-funded research initiatives and Intelligent Transport Systems (ITS) World congress events. Following the work done in the PROMETHEUS research program [5], it become obvious that applying standardised communication channels and technology for the purpose of data exchange between moving vehicles and road infrastructure is a precondition for market acceptance and penetration for all ITS applications, including tolling. Consequently, ISO and CEN jointly – applying the rules of the Vienna Agreement [6] for co-developing standards – created technical committees ISO TC 204 and CEN TC 278 basing their works on programs dedicated for creating standards in Road Traffic and Transport Telematics (RTTT)/ITS applications (refer to Section 9.2.3). As for the communication technology, for short-range applications, the DSRC technology had been proposed for standardisation in Europe by CEN TC 278, as a suitable technology to support road transport and traffic telematics applications [such as Automatic and EFC, Access Control, Traffic & Travel Information and Dynamic Route Guidance (DRG)].

Table 9.1 EFC standardisation supporting and complementing projects

Project	Years	Major tasks and achievements	Reference
VASCO	1996–98	• Validation of DSRC technology as proposed by CEN TC 178 WG9 • Taking into account the requirements from all potential users of DSRC in regards to the different envisaged applications • Carrying out simulations, conformance testing, laboratory and field trials • Validating for correctness, consistency, completeness and technical viability	[7]
A1	1998–99	• Extending the results of the project VASCO from the communication channel towards a system approach with payment application, video enforcement and coordinating functions • Development of EFC applications in an interoperable context • Validating the results in a field trial at three test sites (Germany, France, Sweden) • Demonstrating and dissemination of the project results	[8]
CARDME (I–IV) CARDME support	1993–2001	• Identification of obstacles in regards to achieving cross-border interoperability in EFC in Europe • Creating a specification for a common payment service opting for short implementation time and fewer changes in existing infrastructure • Taking recommendations and findings from other projects (e.g. CESARE) and early standardisation into account • Definition of a common payment service for different road geometries and natures and different road users (light and heavy vehicles) in regards of the system architecture, processes and technical specifications	[9]
CESARE (I–IV)	2001–09	• Multiphase project co-funded by EC and ASECAP	[10]
The MEDIA project	2002–08	• Practical solution for interoperable HGV fees in selected countries	[11]
RCI – Road Charging Interoperability	2005–08	• Developing of an architecture for the European Electronic Toll Service (EETS) • Identification of the (interoperable) interfaces between the systems in the realm of different roles • Applying the achievements in regards to technical standards for EFC of different technologies (DSRC and autonomous) developed by CEN and ISO • Applying the achievement of the conceptual and role model for EFC as developed by the CARDME and CESARE projects	[12]

The activities of CEN and ISO were supported and complemented by projects that have been carried out under the umbrella of EC research framework programs. A non-exhaustive list in Table 9.1 provides an overview on the main characteristics of the individual projects.

The exchange of ideas, concepts, specifications and results between the activities carried out in the respective standardisation groups and the research projects went in both directions. While standardisation work is a great tool to achieve and publish harmonised specifications for technical elements like interfaces and data, the standardisation bodies have no mandate to validate their results directly e.g. in laboratory tests, field trials or similar undertakings. Such tasks have been carried out in the early times of EFC standardisation by projects like VASCO and A1. On the other hand, standards need to be based on and enable business models, the related split of roles, responsibilities and tasks. Such business models have been developed by projects like CARDME and CESARE and consequently formed the basis for application and back-office interface standards for EFC.

Further support and contribution to EFC standards – in early times in which the EFC standards were still in a premature stage and contained specification gaps – was received from a group of manufacturers that issued a joint specification for DSRC, called Global Specification for Short Range Communication (GSS). GSS enabled the first real interoperability of EFC devices designed and produced by different suppliers.

In addition, field trials, primarily carried out in the 1990s (e.g. A555 in Germany, Tauernautobahn in Austria, DIRECTS project in UK), contributed with their results and findings directly or indirectly to standards in EFC.

9.2.3 Standardisation organisations currently active in road pricing standards

9.2.3.1 CEN TC 278 WG1

Within the European Standardisation Body CEN, the working group CEN TC 278 WG 1 has got the tasks to create and maintain international standards in the field of EFC and Access Control. The working focus of WG1 remains at the higher level while other standardisation bodies provide standards for lower layers in the protocol stack of the applied communication protocols and supporting technologies [e.g. DSRC, GNSS, cellular communication (CN)].

Most EFC standards are developed as joint work items with ISO TC 204 WG5, under CEN lead.

The current standardisation work within CEN TC278 WG1 includes defining application interfaces between equipment, information flows between EFC operators, architecture, security, conformance evaluation and test standards, personalisation of on-board equipment (OBE), performance indicators and enforcement/ monitoring support. A key objective of the working group is to support the European Commission in regard to the development of technical specifications that are required for the European Electronic Toll Service (EETS) [1]. See also Chapter 10 for information on REETS, the Regional European Electronic Toll Service.

EFC standardisation in CEN TC 278 WG 1 is divided roughly into the following sub-working groups and key areas:

- SG1: technology-independent standards (role model, architecture, information flows between the roles);
- SG2: standards for DSRC-based EFC (interoperable application profile, security enhancements, personalisation of OBE and related test standards);
- SG5: standards for autonomous EFC (exchange of charge reports and toll context data, provisions for location augmentation support and compliance checking and related test standards).

The EETS is being set up for interoperable EFC systems in Europe called for by Directive 2004/52/EC [13] on the interoperability of electronic road toll systems in the Community. This Directive lays down the conditions necessary to ensure the interoperability of electronic road toll systems in the European Union. CEN TC 278 WG 1 provides the standards that form the main building blocks for defining this pan-European interoperable service under mandates of the European Commission.

9.2.3.2 ISO TC 204 WG5

ISO TC 204 WG5 is a joint working group to CEN T 278 WG1. ISO and CEN share similar objectives in regards to standards for EFC. Consequently, both working groups arrange their work in one common group and run their meetings jointly with the same experts being involved in the work. Most of the work items are joint ISO and CEN work items. However, room is available for CEN-only activities (like specifications for EETS) and ISO specifics (like the use of smart cards in tolling applications).

9.2.3.3 ETSI TC ITS

ETSI TC ITS is working on global standards for co-operative ITS, also in coordination with CEN TC278 WG16/ISO TC204 WG18. Some selected standards are related to EFC:

- ETSI TS 102 792 [14] defines the Mitigation techniques to avoid interference between European CEN Dedicated Short Range Communication (CEN DSRC) equipment and ITS operating in the 5-GHz frequency range based on test and measurement in ETSI TR 102 960 [15]
- ETSI TS 102 916 [16] provides test specifications for the methods to ensure coexistence of Cooperative ITS G5 with RTTT DSRC. The published version covers only selected tests performed at the EC Joint Research Centre in Ispra (Italy). A new version is currently under work
- ETSI TS 102 486 [17] provides test specifications for the CEN DSRC data link layer and application layer, and are referenced by EN 15876 [18,19]
- ETSI ES 200 674-1 [20] defines both the technical characteristics and test methods for High Data Rate (HDR) DSRC.
- ETSI TS 102 708 [21] provides test specifications for the CEN HDR DSRC data link layer and application layer.

9.2.3.4 ETSI EMC and Radio Spectrum Matters (ERM) TG37

ETSI TC ERM is in charge of coordinating ETSI positions on an adequate level of electromagnetic compatibility and an effective and efficient use of radio spectrum; and producing ETSI deliverables dealing with radio spectrum parameters concerned with inter-system characteristics. Such ETSI deliverables may include harmonised standards intended to be used for regulatory purposes.

Task Group 37 specifically deals with radio-related matters in ITS. In particular, EN 300 674 provides one voluntary means for EFC equipment of conforming to the essential requirements of Directive 2014/53/EC [22] on the harmonisation of the laws of the Member States relating to the making available on the market of radio equipment and repealing Directive 1999/5/EC [23].

In addition, ETSI provides certain technical standards on the use of the frequency spectrum and testing that are vital for the EFC area.

9.3 Paving the path towards interoperability

9.3.1 DSRC-based EFC systems

The successful proliferation of DSRC-based EFC systems has been made possible by a consistent set of standards and interoperability specifications that ultimately lead to the production of the standard EN 15509 [24] (refer to Section 9.4.4). This standard defines, mainly by reference to other underlying standards, all requirements needed for OBUs and road side equipments (RSEs) from different vendors and/or operators to communicate seamlessly and with highest performance. It is an umbrella standard which represents the single reference for interoperability for DSRC-based systems and eliminates the need for regional dialects. First, it is the result of more than 10 years of standardisation and European research projects. It incorporates by reference the 4 'application agnostic' CEN DSRC communication standards originally developed by CEN TC278 WG9 (now dormant) for the DSRC physical layer, data link layer, application layer and profiles. These standards are used by many other applications as well such as Automatic Vehicle Identification and Digital Tachograph monitoring. Second, it uses EFC functions and data types of EN ISO 14906 [25] (see Section 9.4.5), the toolbox standard and major cornerstone of TC278 WG1 for DSRC-based EFC. Third, it standardises the security mechanisms for DSRC that have been defined and tested in major EC projects.

The benefit of the use of EN 15509 [24] for EFC-based systems is evident:

- Provides a solid base for second-sourcing of OBUs and avoids single-vendor lock-in.
- Provides technical interoperability between OBUs of Toll Service Providers (TSPs) and RSE of toll chargers (TCs).
- Is required for compliance with the Interoperability Directive 2004/52/EC [1] and the EETS Decision 2009/750/EC [13].

Table 9.2 Set of specifications enabling interoperability in short-range communication for road pricing

Level	Body	Specification	Reference
Frequency band	ERC	• Common frequency allocation for Europe of a 20 MHz band in the 5.8 GHz ISM band	[4]
Communication protocol stack	TC 278 WG9	• EN 12253 DSRC physical layer • EN 13372 DSRC profile • EN 12795 DSRC data link layer • EN 12834 DSRC application layer	[26] [27] [28] [29]
Application interface and application data definition	TC 278 WG1	• EN ISO 14906 containing specifications for – Service primitives – Security services – Application data, including ○ Vehicle data ○ User data ○ Payment data ○ Receipt data • For details refer to Section 9.4.5	[25]
Interoperable application profile (IAP)	TC 278 WG1	• EN 15509 specification of those elements and services to be applied from the lower layer tool box standards in terms of – Communication system parameters – Security services – Application data • For details refer to Section 9.4.4	[24]
Conformance evaluation and test specifications	ETSI TC ITS and CEN TC 278 WG1	• Several test standards for evaluating the conformance of OBU and RSE to the respective standards (see lines above), including EN 300 674 (multipart standard) for assessing the conformity to DSRC physical layer parameters and frequency regulation requirements	

Compliance with EN15509 is nowadays a must in Europe, and also well accepted outside Europe. Many national specifications in EFC refer to the EN 15509.

EN 15509 is based on a layered set of standards which provide the communication protocol for DSRC-based EFC in Europe (see Table 9.2). However, it should be noted that in Italy, a different communication protocol applies for EFC on DSRC.

9.3.2 Autonomous systems based on CN and GNSS

First concepts for road pricing using cellular/mobile communication Global System for Mobile Communications (GSM) and its successors) as the primary communication media for data exchange between vehicle-mounted on-board devices and central processing facilities started in the 1990s. Field trails in Hong Kong and Germany investigated potential system concepts and required technologies. In contrast to DSRC-based EFC, in autonomous systems, the use of the respective toll-liable infrastructure is determined by either the OBU or central processes using vehicle position data that are captured by the OBU using a GNSS system (e.g. Global Positioning System (GPS)), Galileo, GLObal NAvigation Satellite System (GLONASS) or others). For both – GNSS and CN – standards or specifications exist.

In order to achieve interoperability, EFC specific standardisation needs have been identified in the areas of application data (charging data, toll context data, etc.) and back-office interfaces. First initiatives in the early 2000s in CEN TC 278 WG1 and in a specification attempt called 'Mister' failed due to a too complex approach without having a harmonised and well accepted business and role model available. After such models and the related technical architecture for an interoperable EFC landscape had been developed in mainly in CARDME [9], CESARE, RCI [12] and CEN TC 278 WG1 and published in the standard EN ISO 17573 [30], a first standards suite EN ISO 17575 [31–33] for autonomous EFC was published and accepted in 2010 in a first version. In regard to some applications, data autonomous tolling makes use of data initially defined for DSRC tolling in EN ISO 14906 [25].

However, it is important to understand that standards for autonomous tolling do not directly support a multivendor concept for OBUs communicating with the central system of the TSP, nor support the direct exchange data via mobile communication between the OBU of the TSP and the central system operated by the TC (except for enforcement purposes). The reason for this situation is multifold. The OBU is under control of the TSP and highly integrated in data exchange, device management and security processes of the entire TSP system. TSPs usually have performance based contracts with the TCs. Such contracts may result in heavy penalties in case of underperforming of the TSP. Consequently, TSPs do not allow any external party interfacing directly with the OBU (as a major means for charge data generation). Furthermore, in autonomous tolling the Front-End concept of so-called thin-clients is a preferred approach. In this architecture, the OBU functionalities are mainly limited to vehicle location data collection and HMI (Human Machine Interface) functions, while the toll charging processing algorithms are operated centrally in the back-office part of the Front-End of the TSP. However, the exact split of the Front-End internal processes between the OBU and the central system is made by each TSP on an individual basis. There are no standardised or public interfaces available within the Front-Ends.

Thus, the second major interface that asks for standardisation in (autonomous) EFC systems is the back-office interface for data exchange between the TSPs and the TCs. TC 278 WG1 has published a standard EN ISO 12855 [34] and interoperable application profiles CEN/TS 16986 [35] for this interface. It is expected that this interface and these two specifications become a 'back-bone' for EETS-based interoperability in Europe.

9.4 Existing framework of road pricing standards

9.4.1 Overview

Standards for EFC as being developed and maintained by CEN TC 278 WG1 (jointly with ISO TS 204 WG5) can be structured in a matrix (Table 9.3) in which one dimension of the matrix follows the different communication technologies applied for EFC, whereas the second dimension emphasises the different level and objective of the respective standard.

Following this concept, the technology areas are

- DSRC-based EFC;
- autonomous EFC and
- technology agnostic/independent standards.

Furthermore, the different objective areas can be identified as

- framework standards;
- toolbox standards and
- (interoperable) profile standards.

In addition, there are a set of technical reports which have been developed by experts of WG1. Such technical reports typically address potential (future) areas of EFC and can be understood as early analysis and investigations of potential trends in EFC with the ambition to identify possible needs for setting up international standards. Furthermore, test standards complement the requirements standard. Test standards provide specifications for evaluation of conformity to individual requirements standards. Most of the test standards consist of two parts; a part 1 defining the test purposes and test suite structure and a part 2 containing the test specifications usually provided in Testing and Test Control Notation code.

Table 9.3 Structured overview of major EFC requirements standards

	EFC in general	DSRC-based EFC	Autonomous EFC
Interoperable application profile standards	CEN/TS 16986 [35]	EN 15509 [24]	CEN/TS 16331 [36] EN ISO 12813 [37] EN ISO 13141 [38]
Tool box standards	EN ISO 12855 [34]	EN ISO 14906 [25]	EN ISO 17575-1 [39] EN ISO 17575-3 [33]
Frameworks	EN ISO 17573 [30] CEN/TS 19299 [40]		

9.4.2 EFC architecture, role model and interoperable interfaces

9.4.2.1 Architecture

The role model for (interoperable) EFC systems was developed in the CESARE project and is currently standardised in EN ISO 17573 [30]. In this role model, three main roles have been identified: Toll Charging, Service Provision and Service Using (see Figure 9.1). In addition, the (optional) role, System Management is present in this model.

The toll charging role is the provider of the tolled infrastructure or transport service and, hence, the recipient of the road usage charges. The TC is the actor associated with the toll charging role.

Main responsibilities of the toll charging role are

- defining the toll charging principles (e.g. tariff system and rules);
- in DSRC Toll Domains: providing and operating the DSRC-RSE infrastructure (e.g. gantries);
- in GNSS Toll Domains: defining the geographical details of the charge objects inside the Toll Domains and
- operating processes and systems for exception handling including enforcement.

TSPs (i.e. the actor performing the Service Provision role) provide toll service contracts and issue OBE to the users (i.e. Service Usage role) of the tolled infrastructure or transport service. TSPs are responsible for operating the OBE that will be used for collecting data enabling the TC to send a claim to the TSP for the usage of the infrastructure or transport service.

Main responsibilities of the Service Providing role are

- entering into contractual relations with the TC(s) including respective performance parameters;
- offering toll service contracts to interested users;
- providing the OBU that allows the user to use the Toll Domain(s) included in the toll service contract(s);
- operating and maintaining the OBU including the OBU-related functions, communication channels and data;
- managing customer relationships with the users;

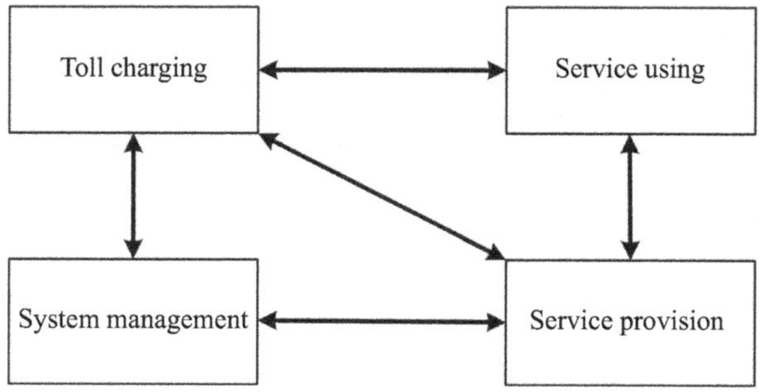

Figure 9.1 The EFC role model [after EN ISO 17573]

- collecting payments from the users for the tolls due;
- providing the collected funds to the respective TC(s) depending on the road usage of the users and
- managing and monitoring all own systems and interfaces including agreed Service Level Agreements (SLAs) with the TCs.

In an interoperable environment, each TSP delivers toll declarations to several TCs, and each TC receives toll declarations from more than one TSP. In DSRC systems, the TCs get the toll declarations from their own RSE and only supplementary charging data, if required, from the TSP's back-office.

The Service Using role is typically represented by a driver or the owner of a vehicle. The service user uses the toll liable road network with a toll liable vehicle.

- Main responsibilities of the Service Using role are: entering into the toll service contract with the TSP;
- providing toll relevant vehicle data that are required for correct toll determination;
- providing a payment means which the TSP can use for settlement of tolls due;
- using the OBU and interacting with it according to the rules set by the Service Provider and/or the TC(s);
- behaving according to the rules and obligations applicable for the respective Toll Domain;
- paying the tolls due based on the toll statement received from the TSP and
- cooperating with the TC and Service Provider in enforcement cases.

System Management in Figure 9.1 (also called Interoperability Management or in ISO 17573 Management of the toll charging environment) comprises all specifications and activities that in common define and maintain a set of rules that govern the overall toll charging environment.

Main responsibilities of the System Management role are

- defining certification requirements;
- introducing and maintaining common and unambiguous ID and numbering schemes and
- defining procedural rules amongst the actors and managing disputes related to these rules.

The above-identified roles Toll Charging, Service Provision, Service using and System Management interact with each other, typically via defined interfaces and using dedicated processes. The roles and their respective actors reside and operate in an (toll charging) environment. This environment is composed of objects like authorities, EFC equipment suppliers, financial systems, certification bodies, standardisation bodies, position systems, in-vehicle systems and sensors, other ITS-systems and services, telecommunication systems.

The high-level technical architecture of an EFC system showing the main interfaces, major data flows and international standards applied on these interfaces is provided in Figure 9.2. The two actors – TC and Service Provider – interact with each other by means of standardised interfaces. The architecture figure is applicable for both major technologies applied in EFC

- autonomous systems using localisation technologies in the OBU (GNSS), mobile communication (any type of suitable CN such as GPRS, 3G, LTE) with the TSP central equipment and DSRC technologies for compliance check communication and optional RSE-based location support and
- DSRC-based systems using short-range communication between the OBU and the RSE of the TC for OBU identification and localisation in combination with charge data and compliance check data exchange.

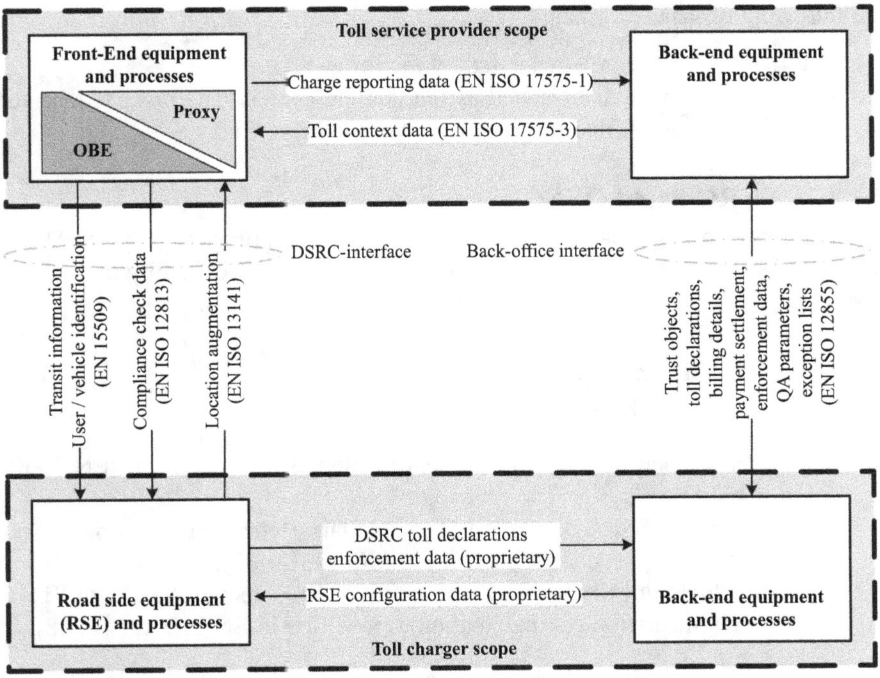

Figure 9.2 EFC high-level architecture, interfaces and major interface standards [after EN ISO 12855]

The description below explains the characteristics of major subsystems and components in a multi-technology EFC system (refer also to [41]). Please note that not all subsystems which compose a full road pricing system are shown in Figure 9.2 and explained in the text below. Missing subsystems are for instance (non-exhaustive list) customer service channels (e.g. point of sales, call centres), payment/invoicing systems, logistics functions.

9.4.2.2 Autonomous OBE (EFC Front-End)

The Front-End or autonomous OBE for EFC is composed of an OBE and a proxy, according to the general EFC Front-End architecture defined in EN ISO 17575 (all parts) [32,33,39]. The Front-End is the equipment through which the charging data are collected. Depending on the technical concept, the charging data are completely generated on the OBE or only raw data are collected by the OBE and the charging data are generated on the proxy. The autonomous OBE is in most cases for enforcement reasons and localisation support equipped with a DSRC interface.

The Front-End generates:

- information to be transmitted to the user, via the OBE's HMI;
- charging data to be further processed in the TSP back-office.

The Front-End interfaces to the TSP back-office (EN ISO 17575) and provides for the following exchange of data:

- charging data which are generated by the Front-End;
- EFC applications update as well as toll domain context data updates received from the TC back-office for deployment at the Front-End.

9.4.2.3 DSRC-based OBE

The OBE for a roadside-based EFC system has as its main function only as a DSRC interface for tolling and enforcement purposes. The main functionality in this EFC system approach, collecting charging data, is located at the RSE.

9.4.2.4 Road side equipment

The RSE has three different tasks:

- collect road usage data (charging identification, specified e.g. in EN 15509 [24]);
- perform vehicle and OBE (i.e. user) compliance check [CCC (compliance check communication)] (specified in EN ISO 12813 [37]);
- support vehicle localisation by transmitting RSE localisation data [LAC (location augmentation communication)] (specified in EN ISO 13141 [38]).

Compliance checking is done by data read out through the DSRC interface of the OBE, in combination with vehicle classification and automatic number plate recognition technologies.

9.4.2.5 Toll Charger back-office

The TC back-office calculates the payment claims either from the toll declarations received from the service provider back-office or from the charging data collected by its own RSE. The TC back-office manages and distributes also the EFC context data to the TSPs (EN ISO 12855 [34]).

It also receives information on black-listed or white-listed road users (the exception lists) of all connected TSPs and possibly manages and maintains the full exception list relevant for this particular Toll Domain. Exception lists will be distributed to the RSE of the TC or the enforcement authority for immediate identification and enforcement of vehicles and users that are on the exception lists.

9.4.2.6 Toll Service Provider back-office

The TSP with the support of its own back-office facilities is responsible for managing and deploying the OBU and keeping all parameters in the OBUs at an up-to-date state. For autonomous OBUs also, the respective toll context data in the Front-End need to be managed and maintained. In an autonomous OBU system, it collects the toll declaration of the Front-End and forwards this data to the corresponding TC back-office. The payment claims and billing details received from the TCs are used to calculate the invoices to the users.

The TSP Back-office implements the Customer Relationship Management, billing and invoicing processes of Service Users for EFC. Payment (collection of the fee) is handled between TSP and Service User. The TSP guarantees the payment to the TC. The back-office routes charge data generated by the Front-End summarised as toll declarations to the TC back-office.

The TSP back-office also manages exception lists of OBEs and its users which are no longer valid for collecting charging data (e.g. for users which have no valid contract or payment means registered with their contracting TSP; i.e. the user is on the so-called black list). This list is distributed to the TCs.

9.4.3 Application interface specification for autonomous systems

The scope of the EN ISO 17575 standards suite is to specify technical requirements for the exchange of data between a toll charging Front-End and a related back-end in autonomous systems. This interface is located within the TSP's realm. However, standardisation concepts for this (internal) interface relate back to the late 1990s and early 2000s in which first concepts for autonomous systems optionally include the use of intelligent (so-called 'thick client') OBUs that maintain all necessary toll context and account related data on board the OBU and are able to detect the use of toll-liable road infrastructure without the support of central back-office processes and charge the road user account. For various technical, operational and commercial reasons this concept was never adapted in real EFC system implementations. The main obstacle to a more extensive use of thick-client OBUs in tolling is the need to maintain and update the toll-domain-specific data and charging processes which are individual per toll domain. Such maintenance can be more efficiently and securely done in the central processes of the TSPs. In such a concept, only 'thin client' OBUs are necessary which collect position and preprocessed data, transfer such data to the central system and leave the process steps of charge object identification and charging to accounts to the back offices of the TSPs. Thin client concepts require more data communication from the OBU to the central system of the TSPs.

However, EN ISO 17575 is important in order to potentially achieve a common ground for technical requirements that can be 'translated' into expected capabilities of the entire Front-End of the TC. Furthermore, data definitions specified in EN ISO 17575 are imported from and reused by EN ISO 12855 (see Section 9.4.9) which is the back-bone standard for interoperability.

The EN ISO 17575 suite consists of the following parts:

- EN ISO 17575-1 [39]: Charging. This part defines the data elements for the transfer of usage data from the Front-End to the back-end.
- EN ISO 17575-2 [32]: Communication and connection to lower layers. This part defines basic communication services for data transfer over the OBE air-link or between Front-End and back-end.
- EN ISO 17575-3 [33]: Context data. This part defines the data to be used for a description of individual charging systems in terms of charged geographical objects as well as charging and reporting rules.

Functionalities that were specifically introduced to support roaming in different (also overlapping) Toll Domains have been initially specified in EN ISO 17575-4. However, in the systematic review of EN ISO 17575, these specifications have been moved into part 3. Consequently, EN ISO 17575-4 has been withdrawn from the set of EN ISO 17575 standards.

The location of the EN ISO 17575 interface in the technical architecture and the major data flow is outlined in Figure 9.3.

In EN ISO 17575, some common definitions have been introduced which allows the structuring of Toll Domains based on the type and nature of the tolled road infrastructure.

In *road section charging*, the toll is due if predefined sections of a road network are driven on. One can distinguish between sections with a length for each toll-liable section that is predefined by the TC (e.g. from map data or from a road network database) and a principle in which the OBU determined the toll-liable distance while actually using the road infrastructure.

Toll fees in area charging are due for road usages within a defined area. Three options are considered and supported by the standards: (1) a 'flat fee' is due for using the road infrastructure within the area (as in London); (2) the toll fees depend on the duration in time the toll-liable vehicle remains within the area; (3) the toll fees depend on the distance travelled on a defined road network within the area. This third option is similar to one of the road section charging principles.

Finally, in *cordon charging,* the toll fee is due for crossing a cordon, i.e. the border of a defined area. Neither the duration of stay nor the distance travelled within this area has an impact on the toll fee. A 'flat fee' is due for using the road

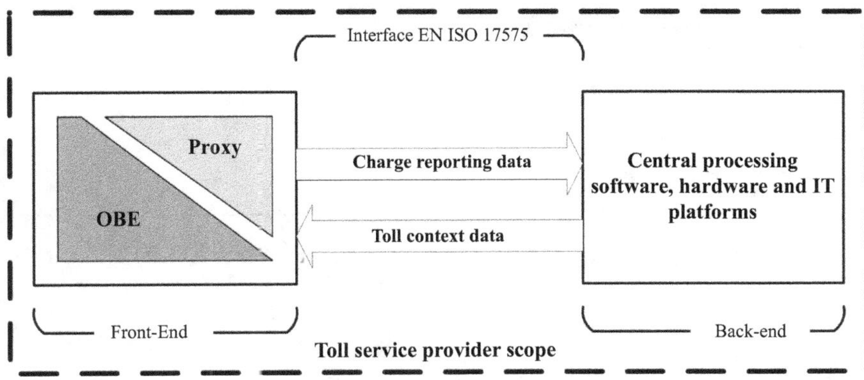

Figure 9.3 Location of the EN ISO 17575 interface in a technical architecture

infrastructure inside the cordon, though the fee may vary by time of day, as in Stockholm and Gothenburg (Chapter 14).

Depending on local needs, TCs may require more or less processed data to varying levels of detail. Privacy considerations, enforcement approach and legal nature of the charge will also influence the choices agreed between TC and TSP regarding the requested contents of charge reports (refer also to [39]). Consequently, EN ISO 17575-1 supports:

- reporting of lower level usage data (e.g. position itineraries);
- reporting of charge objects being used by the toll-liable vehicle;
- reporting of road usage sessions (e.g. individual trips) done by a toll-liable vehicle;
- reporting of aggregated values for mileage and/or a sum of toll fees over a certain period of time.

Furthermore, EN ISO 17575-1 includes specifications for responses from the back-end to the Front-End as a result of the charge reporting data, security provisions, on-board account management, version information and support for compliance checking processes.

EN ISO 17575-3 contains specifications for toll context data, which have been structured into the following areas:

- context overview data;
- tariff information;
- context layout and
- reporting rules.

In addition, specifications for message protocol structures, protocol and data versioning and security measures have been introduced in this international standard.

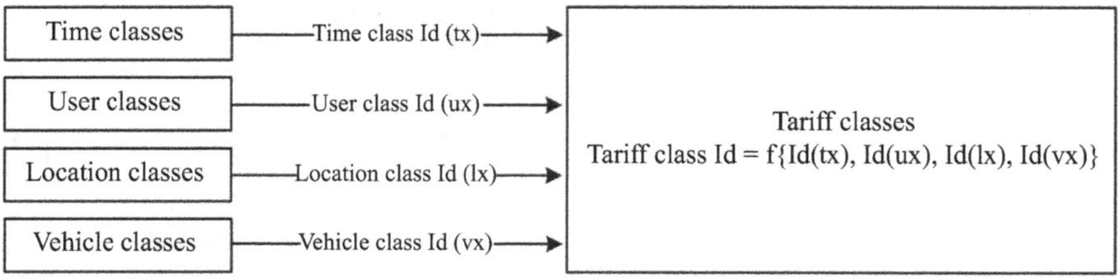

Figure 9.4 Tariff class parameters [after ISO 17575-3]

Context overview data provide basic and administrative information about the respective Toll Domain to the Front-End. Such information includes TC identification, operational status of the respective toll scheme, type of the Toll Domain (section, area, cordon), potential support for partitioning and overlapping of Toll Domains, time zone information and geographical extent of the Toll Domain.

Tariff table information, including respective toll fees per tariff class and required definitions of all tariff determining parameters, are embedded in the *tariff information* data set. EN ISO 17575-3 supports tariff classes that may consist of combinations of one or more of the following determinants: vehicle classes, time classes, user classes and location classes (refer to Figure 9.4). Moreover, the tariff table data definition allows for provisions for Value Added Tax (VAT), multi-currency use and fee rounding rules. Vehicle class definitions reuse data specifications introduced in EN ISO 14906 [25].

The geographical layout of the toll-liable road network including their borderlines and parameters of areas and cordons are specified in data group *context layout*. In regard to format and presentation of geolocation data, ISO 17575-3 [33] allows a range of different formats, ranging from simplified proprietary data specifications, via full three-dimensional geo-coordinates up to reuse of geo-information provided in digital maps and made available e.g. in Geographic Data Files format.

By means of *reporting rules,* the back-end can configure the Front-End in regards to the expected content, options and transfer rules for so-called charge reports (as specified in EN ISO 17575-1).

Test standards that are of relevance for the EN ISO 17575 suite are:

- EN ISO 16407-1: EFC – Evaluation of equipment for conformity to EN ISO 17575-1 – Part 1: Test suite structure and test purposes [31];
- EN ISO 16407-2: EFC – Evaluation of equipment for conformity to EN ISO 17575-1 – Part 2: Abstract test suite [42];
- EN ISO 16410-1: EFC – Evaluation of equipment for conformity to EN ISO 17575-3 – Part 1: Test suite structure and test purposes [43];
- EN ISO 16410-2: EFC – Evaluation of equipment for conformity to EN ISO 17575-3 – Part 2: Abstract test suite [44].

9.4.4 Interoperability application profile for DSRC-based systems

The IAP refers to a selection of functionality from the base DSRC standards and the Application Interface Definition for DSRC-based EFC systems. The goal of the profile is to enable interoperability on the OBU to RSE interface by narrowing down the options of the base standard and providing one consistent set of requirements. The IAP can also be seen as the umbrella standard for EFC over DSRC, or as the glue that binds together all base standards for EFC over DSRC to provide a consistent selection of features.

More specifically, the interoperability application profile provides, by reference and selection, definitions of:

- DSRC communication requirements, through selection of DSRC parameter profiles from the DSRC standards;
- DSRC application layer and EFC functions, through selection of the mandatory application layer function from the DSRC layer 7 standard and EFC functions from the EN ISO 14906 [25];
- data requirements, through the selection of mandatory data objects from EN ISO 14906 EFC-AID to be supported by the OBU;
- security requirements, security levels and pertaining security mechanisms and crypto algorithms;
- transaction requirements, by reference to the transaction model of the EN ISO 14906.

The approach for the EFC transaction itself enabled by the standard is commonly referred as 'pick what you like', i.e. the RSE can freely use functions to address data objects in the OBU within the scope of the profile. The OBU shall support all transactions that are possible within the scope of the profile.

The set of standards for the interoperability application profile comprises the base requirements standard as well as a multipart suit of test standards:

- EN 15509:2014 EFC – IAP for DSRC [24];
- EN 15876-1:2010+A1:2012 EFC – Evaluation of on-board and roadside equipment for conformity to EN 15509 – Part 1: Test suite structure and test purposes [18];
- EN 15876-2:2011 EFC – Evaluation of on-board and roadside equipment for conformity to EN 15509 – Part 2: Abstract test suite [19].

9.4.5 Application interface definition for EFC-DSRC systems

The application interface definition for EFC-DSRC systems (EFC-AID) is the standardised application protocol for the exchange of data between the OBU and the RSE for charging purposes (refer to DSRC-interface in Figure 9.2). In essence, it defines how the suites of DSRC standards can be used for EFC, i.e. how the application processes on OBU and RSE side communicate with each other in a standardised way by making use of the interface to the DSRC application layer.

The EFC application processes use the services provided by the DSRC application layer. These services represent an abstraction of the generic (i.e. non-EFC-specific) DSRC communication functionality.

The most important elements defined as part of the EFC-AID are

- the EFC transaction model, i.e. the definitions of the phases of a OBU to RSE transaction;
- the definition of specific EFC functions that can be invoked by the RSE, i.e. a pair of service primitives, a request and its associated response service primitive that are defined as a further detailed specification of a DSRC application layer function;
- the definition of a dictionary of information objects for EFC in the form of EFC attributes; EFC Attributes are composed of one or more data elements of specified data types in Abstract Syntax Notation One (ASN.1) and have a unique name and identifier.

The set of standards for the application interface definition comprises the base requirements standard as well as a test standard for the OBU side:

- EN ISO 14906:2011 EFC – Application interface definition for dedicated short-range communication and ISO 14906:2011/Cor 1:2013 and ISO 14906:2011/Amd 1:2015 [25];
- EN ISO/TS 14907-2:2016 EFC – Test procedures for user and fixed equipment – Part 2: Conformance test for the OBU application interface [45].

9.4.6 Standards for compliance check communication in autonomous EFC systems

Toll chargers have the need to check whether the tolled infrastructure is used in compliance with the applicable rules in the local toll regime by users in order to keep the rate of fraud within a defined limit. For this purpose, the use of (i.e. the presence of a certain vehicle on) the tolled infrastructure must be detected by trusted personnel or by some equipment capable of recording a proof of such usage (i.e. one or more photos), e.g. fixed, portable or mobile equipment. This presence must be compared with tolling data made available by the Service Provider that ensures that payment will or has been provided corresponding to the usage. There are different options to request and receive this data:

- Compliance Check Communication: querying the OBU about its tolling relevant status;
- Secure monitoring compliance checking (option 1): querying the OBU about its recorded itinerary;
- Secure monitoring compliance checking (option 2): automatically or manually read the vehicle's licence plate from the picture and querying the Service Provider about the vehicle's recorded itinerary.

An operational EFC System can use a combination of those options to keep misuse under control effectively.

Compliance check communication refers to a standardised application protocol for the exchange of data between the OBU and the RSE (refer to DSRC-interface in Figure 9.2), when the OBU is within the communication range of the OBU on the tolled road network. It provides the possibility for the RSE to read a set of data objects from the OBU, independently of the model, type or OBU architecture. Those data objects represent general aspects of the OBU's tolling functionality which may be of importance to TCs. This information allows the TC to judge if the user is complying with the applicable rules, such as having a functioning OBU activated, having declared the correct vehicle class information, etc. The underlying assumption is that the OBU's answer can be trusted, because it is assumed that it is secure enough to resists attacks in the unsecure environment (either because such attacks are absent, or because there are measures against such attacks, or a combination of both), and that the Service Provider can be trusted to operate correctly.

Compliance check communication can be performed via various short-range communication technologies, notably CEN DSRC according to EN 15509 is the first choice.

The set of standards for Compliance check communication comprises the base requirements standard as well as a multipart suit of test standards:

- EN ISO 12813:2015 EFC – Compliance check communication for autonomous systems [37];
- EN ISO/TS 13143-1:2011 EFC – Evaluation of on-board and roadside equipment for conformity to EN ISO/TS 12813 – Part 1: Test suite structure and test purposes [46];
- EN ISO/TS 13143-2:2011 EFC – Evaluation of on-board and roadside equipment for conformity to EN ISO/TS 12813 – Part 2: Abstract test suite [47].

9.4.7 *Standards for secure monitoring in autonomous EFC systems*

Secure monitoring within the context of compliance checking refers to a standardised application protocol for the exchange of data between the OBU and the RSE (refer to DSRC-interface in Figure 9.2), and the TC and Service Provider's back-end (Figures 9.2 and 9.5). It provides the TC with the possibility to verify its confidence in the reliable provision of tolling data by a service provider using an OBE operated by the service user, while respecting the privacy of the user.

In the case of communication between the OBU and RSE, i.e. when the OBU is within the communication range of the OBU on the tolled road network, the RSE is able to read out a data object which represents the latest itinerary (trace of the vehicle locations) driven by the vehicle and to check that the observed road usage of the vehicle has been correctly registered. The communication can be performed via various short-range communication technologies, notably CEN DSRC, accordingly EN 15509 is the first choice.

In the case of communication, between the TC and Service Provider's back-end, the TC is able to request from the TSP the data objects representing the itinerary corresponding to the observation of a vehicle.

In both cases, the TC is also able to claim incorrectness of the data to the Service Provider, who is responsible for the correctness of the tolling data.

Secure monitoring also is a concept that allows the TC to evaluate the trust it has in tolling data that are produced by an OBU that has no defined security level, and a service provider that has no defined level of trust. The concept has different options:

- the use of a trusted recorder in the OBU (secure monitoring option 'SM_CC-1' and secure monitoring option 'SM_CC-2') (refer also to [48]);
- the use of data declarations by the service provider to the TC that ensure that the data cannot be modified nor repudiated (secure monitoring option 'SM_CC-3a' and '-3b') (refer also to [48]).

The trusted recorder is a secure element that performs cryptographic operations which are trusted by the TC, due to some sort of certification. The OBU is required to feed the tolling data it registers to the trusted recorder in real time, so that records are marked with a unique ID and cryptographically signed to make any modification or deletion impossible without being detected. In the jargon, this is called 'real-time freezing'.

Declarations are digital commitments that use transmissions of the data itself or hashes of it to the addressee, so that the data cannot be modified afterwards. In the jargon, this is called 'freezing per declaration'.

In both cases, the TC has a means to spot-check the generation of tolling data and to prevent data being modified afterwards. Due to the fact that spot-checks can happen at any time, the system is obliged to freeze the data periodically. If the system is detected not to freeze, a suspicion is raised. If the system freezes its record periodically, faulty data cannot be corrected later in case of spot-check and incorrectness can be detected.

The set of standards for secure monitoring is:

- CEN/TS 16702-1:2014 EFC – Secure monitoring for autonomous toll systems – Part 1: Compliance checking [48]
- CEN/TS 16702-2:2015 EFC – Secure monitoring for autonomous toll systems – Part 2: Trusted recorder [49]

9.4.8 Location augmentation communication in autonomous EFC systems

Tolling systems that use some sort of positioning system in the OBU may need some assistance in localisation, or additional localisation information at places where satellite-based localisation accuracy or availability is insufficient. In some cases, a TC might directly want to provide an indication to the OBU about the identity of the relevant charge object.

Location augmentation communication refers to a standardised application protocol for the exchange of data between the OBU and the RSE (refer to DSRC-interface in Figure 9.2), when the OBU is within the communication range of the RSE on the tolled road network. It provides the possibility for the RSE to transmit localisation information to the OBE without any further need for identifying the OBE and/or processing OBE specific data; it is a simple 'push' functionality that may provide both charging context independent location information as well as context-dependent identification of charge objects.

Localisation augmentation communication can be performed via various short-range communication technologies, notably CEN DSRC so accordingly EN 15509 [24] is the first choice.

The set of standards for Location augmentation comprises the base requirements standard as well as a multipart suite of test standards:

- EN ISO 13141:2015 EFC – Localisation augmentation communication for autonomous systems [38];
- EN ISO/TS 13140-1:2011 EFC – Evaluation of on-board and roadside equipment for conformity to EN ISO/TS 13141 – Part 1: Test suite structure and test purposes [50];
- EN ISO/TS 13140-2:2012 EFC – Evaluation of on-board and roadside equipment for conformity to EN ISO/TS 13141 – Part 2: Abstract test suite [51].

9.4.9 Back-office data exchange between Toll Chargers and Service Providers

An interface that is of major importance in an EFC environment in which the roles of Toll Charging and Service Provisioning (see role model in Figure 9.1) are performed by two individual entities is the interface between these two roles. As EETS by definition applies such a split between a TC and (multiple) TSPs, this interface is also one of the backbone interfaces in the EETS frame and thus consequently references the respective EETS-regulations (e.g. the EETS Decision of the European Commission [13]). By nature, the interface facilitates the exchange of data between the two back offices, one being under possession of the TC, the second being under the scope of the TSP(s).

The back-office data exchange between TCs and Service Providers is specified in

- EN ISO 12855 (EFC – Information exchange between service provision and toll charging [34]).

The location and scope of this interface is outlined in Figure 9.2. Application data in scope of this interface are indicated in Figure 9.5.

Application data are structured in so-called Application Data Units (ADU). The application data specification in EN ISO 12855 provides a considerable level of flexibility and options. Furthermore, EN ISO 12855 is designed in a technology-agnostic manner, i.e. it can be used with any EFC communication technology applied in the respective Toll Domain of the TC (i.e. DSRC or autonomous) and irrespective of type and nature of the Charge Objects within the Toll Domain (road sections, areas, cordons).

In addition to the application data, there are supporting parameters defined in EN ISO 12855, e.g. for status information, acknowledge data, return values, etc.

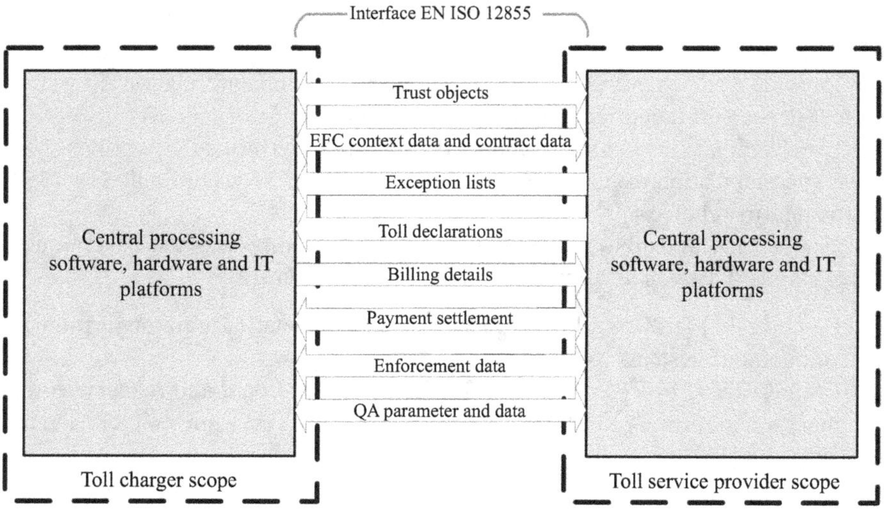

Figure 9.5 Back-office interface location and scope [after EN ISO 12855]

For the purpose of managing the exchange of ADUs between the central equipment of the TC and Service Provider, a message protocol framework in the form of Application Protocol Data Units (APDUs) is defined. APDUs provide for message exchange management functionalities and message authentication. For the applied communication channel and the communication protocol lower layers, EN ISO 12855 references a variety of existing and available technologies. EN ISO 12855 comes along with a full ASN.1 data specification.

9.4.10 Interoperable application profile for the back-office data exchange between Toll Chargers and Service Providers

One advantage of EN ISO 12855 [34] is the flexibility in terms of supporting different Toll Domains (refer also to Section 9.4.9). However, such flexibility is achieved by a large number of optional specifications. Most of the existing and upcoming EFC systems in Europe however, could already be supported with a smaller subset of the EN ISO 12855 specifications. In addition, EN ISO 12855 does not specify the details of the application data exchange process. Following the split of tasks and responsibilities between TCs and TSPs (see also in EN ISO 17573 [30]), the TSP which has the ambition (or the obligation) to cover many Toll Domains has the burden to implement a back office and respective interface which supports almost all options defined by EN ISO 12855. Implementation, testing and certification of such complex interfaces is costly and may jeopardise business cases of TSPs. Hence, CEN TC 278 WG1 has created an interoperable application profile standard for EN ISO 12855 which narrows the required functionalities and options to a minimum and in addition specifies details of the processes of data exchange between TC s and Service Providers.

The interoperable application profile for the back-office data exchange between the TCs and TSPs is published as

- CEN/TS 16986 EFC – Interoperable application profiles for information exchange between Service Provision and Toll Charging [35];
- a respective multi-part test standard suite is currently (as of 2017) in preparation at CEN TC 278 WG1 and is planned to be published in 2018 as CEN/TS 17154-1/-2.

A major achievement of CEN/TS 16986 is the introduction of a transaction concept for the exchange of ADUs, including the specification of relevant parameters, like triggers or starting transactions, time parameters, etc.

9.4.11 EFC security framework

The EFC security framework standard CEN/TS 19299 [40] is a framework for all organizational and technical entities of an EFC system; it provides a toolbox of EFC-specific security requirements and security measures that can be used by system owners to secure the operation of their system. Since it is a generic framework, it does not assume any particular security policy or perform a risk analysis of a given system.

A system owner shall define its own security objectives and policy statements in a security policy and perform a threat analysis with risk assessment to define the applicable security requirements. The EFC security measures can then be added to the catalogue of security measures for the toll charger's or service provider's system: this enables the system owner to include the EFC business processes in the scope of a security certification e.g. according to ISO 27001 [52] and to officially demonstrate compliance to the EFC security framework.

The security framework standard is available as Technical Specification CEN/TS 19299 EFC – Security framework [40].

9.4.12 Additional EFC standards

Work in EFC standardisation has gone beyond the major framework of standards as presented in the previous sections which is mainly determined by providing specifications for major interfaces in EFC systems and between roles and actor.

Additional areas which have been investigated and for which specifications or reports have been created are for example

- interfaces between DSRC OBU and external in-vehicle devices (i.e. ISO/TS 16785 [53]);
- performance parameters in EFC systems (i.e. CEN/TS 17444) [54];
- interoperable application profiles for autonomous EFC systems (i.e. CEN/TS 16331) [37];
- personalization and related interfaces for OBU (i.e. CEN/TS 21719 [55]);
- EFC in ITS-in vehicle stations (i.e. CEN/TR 16690 [56]).

9.5 Outlook for future work in standards for road pricing

ETSI ITS WG4 is currently working on TR 103 440 [57], investigating possible solutions to improve the radio characteristics of CEN DSRC and high data rate DSRC to be more robust against interferences from other radio applications. Those results will be of great importance for the future of DSRC-based EFC.

CEN TC 278 is currently working on

- The revision of many of the previously mentioned standards, in order to keep them up to date with the latest developments.
- A suite of standards for personalisation of the OBU, comprising a general requirements framework, and specifications for personalisation over various communication media.
- A technical report to identify emerging charging policies and technologies and gaps between those and the currently standardised technologies for EFC, and consequently recommend development or update of standards.
- A technical specification that defines the architecture of a toll system environment in which a TC can act to support traffic management with the use of a certain tariff scheme
- A technical specification that enables multi-modal transport services by using common payment media.

For an up-to-date overview of the published standards of CEN TC 278 refer to [58].

For an up-to-date overview of the published and draft standards of ETSI ERM refer to [59].

References

[1] European Commission, Directive 2004/52/EC on the interoperability of electronic road toll systems in the Community, Brussels, 2004.

[2] International Organization for Standardization, "ISO," [Online]. Available: http://www.iso.org/iso/home/standards.htm. [Accessed 04 01 2017].

[3] UNINFO, UNI10607 Road traffic and transport telematic. Automatic dynamic debiting systems and automatic access control systems using dedicated short-range communication at 5.8 GHz, 1996.

[4] European Radiocommunications Committee, Frequency Bands to be designated for the coordinated introduction of Road Transport Telematic Systems, 1992.

[5] The EUREKA secretariat, PROMETHEUS – Programme for a European traffic system with highest efficiency and unprecedented safety, 1987–1995.

[6] ISO and CEN, Agreement on technical co-operation between ISO and CEN (Vienna agreement), 1991.

[7] Cordis, "Cordis archive. Telematics applications for transport. VASCO TR1062," [Online]. Available: http://cordis.europa.eu/telematics/tap_transport/research/projectsum/vasco.html. [Accessed 08 01 2017].

[8] Cordis, "Cordis archive. Telematics applications for transport. A1 TR4001," [Online]. Available: http://cordis.europa.eu/telematics/tap_transport/research/projects/a1.html. [Accessed 08 01 2017].
[9] Cordis, "Cordis archive. Telematics applications for transport. Cardme TR4002 and Cardme support TR1109," [Online]. Available: http://cordis.europa.eu/telematics/tap_transport/research/projects/cardme.html. [Accessed 08 01 2017].
[10] The CESARE project, http://www.asecap.com/projects/9-en/74-cesare-iii.html
[11] Implementation of the media toll service, https://www.rapp.ch/wAssets-de/docs/trans/fachartikel-referate/2006/dokumente/its-congress-london/media_itslondon-doc.pdf
[12] RCI – Road Charging Interoperability, "Transport Research and Innovation Portal," 2008 [Online]. Available: http://www.transport-research.info/sites/default/files/project/documents/20101007_170546_77859_RCI%20-%20Final%20Brochure.pdf. [Accessed 08 January 2017].
[13] European Commission, Decision 2009/750/EC on the definition of the European Electronic Toll Service and its technical elements, Brussels, 2009.
[14] ETSI, ETSI TS 102 792, Intelligent Transport Systems (ITS). Mitigation techniques to avoid interference between European CEN Dedicated Short Range Communication (CEN DSRC) equipment and Intelligent Transport Systems (ITS) operating in the 5 GHz frequency range, 2015.
[15] ETSI, ETSI TS 102 960, Intelligent Transport Systems (ITS). Mitigation techniques to avoid interference between European CEN DSRC equipment and ITS operating in the 5 GHz frequency range. Evaluation of mitigation techniques, 2012.
[16] ETSI, ETSI TS 102 916, Intelligent Transport Systems. Test specifications for the methods to ensure coexistence of Cooperative ITS G5 with RTTT DSRC, 2012.
[17] ETSI, ETSI TS 102 489, Intelligent Transport Systems (ITS). Road Transport and Traffic Telematics (RTTT). Test specifications for Dedicated Range Communication (DSRC) transmission equipment, 2008.
[18] CEN, EN 15876-1 Electronic fee collection. Evaluation of on-board and roadside equipment for conformity to EN 15509. Part 1: Test suite structure and test purposes, 2010.
[19] CEN, EN 15876-2 Electronic fee collection. Evaluation of on-board and roadside equipment for conformity to EN 15509. Part 2: Abstract test suite, 2011.
[20] ETSI, ETSI ES 200 674-1; Intelligent Transport Systems (ITS); RTTT; Dedicated Short Range Communications (DSRC); Part 1: Technical characteristics and test methods for High Date Rate (HDR) data transmission equipment operating in the 5.8 GHz ISM band, 2013.
[21] ETSI, ETSI TS 102 708; Intelligent Transport Systems (ITS); RTTT; Test specifications for High Data Rate (HDR) transmission equipment operating in the 5.8 GHz ISM band, 2010.

[22] European Commission, Directive 2014/53/EC on the harmonisation of the laws relating on marking available on the market of radio equipment, Brussels, 2014.
[23] European Commission, Directive 1999/5/EC on Radio and Telecommunication Terminal Equipment, Brussels, 1999.
[24] CEN, EN 15509 Electronic fee collection. Interoperability application profile for DSRC, 2014.
[25] CEN and ISO, EN ISO 14906 Electronic fee collection. Application interface definition for dedicated short-range communication, 2011.
[26] CEN, EN 12253 Road Transport And Traffic Telematics. Dedicated Short-Range Communication (DSRC). Physical layer using microwave at 5.8 GHz, 2004.
[27] CEN, EN 13372 Road transport and traffic telematics. Dedicated short-range communication (DSRC). Profiles for RTTT applications, 2004.
[28] CEN, EN 12795 Road transport and traffic telematics. Dedicated Short Range Communication (DSRC). DSRC data link layer: medium access and logical link control, 2003.
[29] CEN, EN 12834 Road transport and traffic telematics. Dedicated Short Range Communication (DSRC). DSRC application layer, 2004.
[30] CEN and ISO, EN ISO 17573 Electronic Fee Collection. Systems architecture for vehicle-related tolling., 2010.
[31] CEN and ISO, EN ISO 16407-1 Electronic fee collection. Evaluation of equipment for conformity to EN ISO 17575-1. Part 1: Test suite structure and test purposes, 2011.
[32] CEN and ISO, EN ISO 17575-2 Electronic fee collection. Application Interface definition for autonomous systems. Part 2: Communication and connection to the lower layers, 2016.
[33] CEN and ISO, EN ISO 17575-3 Electronic fee collection. Application interface definition for autonomous systems. Part 3: Context data, 2016.
[34] CEN and ISO, EN ISO 12855 Electronic fee collection. Information exchange between service provision and toll charging, 2015.
[35] CEN, CEN/TS 16986 Electronic fee collection. Interoperable application profiles for information exchange between Service Provision and Toll Charging, 2016.
[36] CEN, prCEN/TS 16331 Electronic fee collection (EFC) – Interoperable application profiles for autonomous systems, 2017.
[37] CEN and ISO, EN ISO 12813 Electronic fee collection. Compliance check communication for autonomous systems, 2015.
[38] CEN and ISO, EN ISO 13141 Electronic fee collection. Localisation augmentation communication for autonomous systems, 2015.
[39] CEN and ISO, EN ISO 17575-1 Electronic fee collection. Application Interface definition for autonomous systems. Part 1: Charging, 2016.
[40] CEN, CEN/TS 19299 Electronic Fee Collection – Security framework, 2015.
[41] CEN, TR 19990 Electronic fee collection (EFC) – Guidelines for EFC applications based on in-vehicle ITS stations, 2014.

[42] CEN and ISO, EN ISO 16407-1 Electronic fee collection. Evaluation of equipment for conformity to EN ISO 17575-1. Part 2: Abstract test suite, 2011.
[43] CEN and ISO, EN ISO 16407-1 Electronic fee collection. Evaluation of equipment for conformity to EN ISO 17575-3. Part 1: Test suite structure and test purposes, 2011.
[44] CEN and ISO, EN ISO 16407-1 Electronic fee collection. Evaluation of equipment for conformity to EN ISO 17575-3. Part 2: Abstract test suite, 2011.
[45] CEN and ISO, EN ISO/TS 14907-2 Electronic fee collection. Test procedures for user and fixed equipment. Part 2: Conformance tests for the on-board unit application interface, 2016.
[46] CEN and ISO, EN ISO/TS 13141-1 Electronic fee collection. Evaluation of on-board and roadside equipment for conformity to EN ISO/TS 12813. Part 1: Test suite structure and test purposes, 2011.
[47] CEN and ISO, EN ISO/TS 13141-1 Electronic fee collection. Evaluation of on-board and roadside equipment for conformity to EN ISO/TS 12813. Part 2: Abstract test suite, 2011.
[48] CEN, CEN/TS 16702-1 Electronic fee collection. Secure monitoring for autonomous toll systems. Part 1: Compliance checking, 2014.
[49] CEN, CEN/TS 16702-2 Electronic fee collection. Secure monitoring for autonomous toll systems. Part 2: Trusted recorder, 2015.
[50] CEN and ISO, EN ISO/TS 13140-1 Electronic fee collection. Evaluation of on-board and roadside equipment for conformity to EN ISO/TS 13141. Part 1: Test suite structure and test purposes, 2011.
[51] CEN and ISO, EN ISO/TS 13140-1 Electronic fee collection. Evaluation of on-board and roadside equipment for conformity to EN ISO/TS 13141. Part 2: Abstract test suite, 2012.
[52] ISO/IEC, ISO 27001, Information technology. Security techniques. Information security management systems. Requirement, 2013.
[53] CEN, CEN/TS 16785 Electronic fee collection. Interface definition between DSRC OBE and external in-vehicle devices, 2014.
[54] CEN, CEN/TS 17444-1/-2 Electronic fee collection, Charging performance, Part 1: Metrics, Part 2: Examination and monitoring, 2012/2013.
[55] CEN, CEN/TS 21719 Electronic fee collection, Personalisation of on-board equipment, 2017.
[56] CEN and ISO, FprCEN ISO/TS 21917-1/-2 Electronic fee collection, Guidelines for EFC-applications based on in-vehicle ITS stations, 2014.
[57] ETSI, TR 103 440, Intelligent Transport Systems. Investigation to improve TTT DSRC, 2016.
[58] CEN TC278 WG1, http://www.itsstandards.eu/index.php/efc.
[59] ETSI, https://portal.etsi.org.

Chapter 10

The European Electronic Toll Service – EETS – and the REETS project

Mike Hayward and Hubert Resch***

10.1 Introduction – the EETS vision for interoperable Electronic Toll Collection (ETC) services

Across Europe, new electronic toll systems are constantly being introduced and existing systems being expanded. They address the need to provide new funding for road improvement and operation or to reduce the costs of operating existing toll collection facilities and increase user convenience. The demand for interoperability across Europe has risen as new systems have been installed and existing systems expanded. Thus, there was an imperative to take concerted action from a European perspective.

The vision of a driver being able to travel across Europe using a single electronic onboard unit and pay all road tolls using a single account has existed since the first electronic toll collection (ETC) systems were introduced in Europe in the early 1990s. This vision was described in European law initially within the 2004 Directive on interoperability of ETC systems [1], and then in more detail within the European Commission's 2009 Decision [2] on the definition of the European Electronic Toll Service (EETS).

The vision of EETS is one where users can maintain a single relationship with one service provider (SP) without the need to interact directly with every Toll Charger (TC) in every member state (Figure 10.1).

However, the vision has still yet to be completely achieved. So what progress has been made and what have been the obstacles that have had to be overcome.

This chapter provides a summary of how interoperability of ETC has progressed within Europe in recent years, first, describing the background from previous projects and standardisation work and how interoperable systems have developed.

The chapter then looks in detail at the recent REETS (Regional EETS) project which was co-funded by the European Commission (EC) in order to accelerate progress on the deployment of EETS. The chapter describes the key achievements

*Transport Logic Limited, United Kingdom
**ASFINAG, Austria

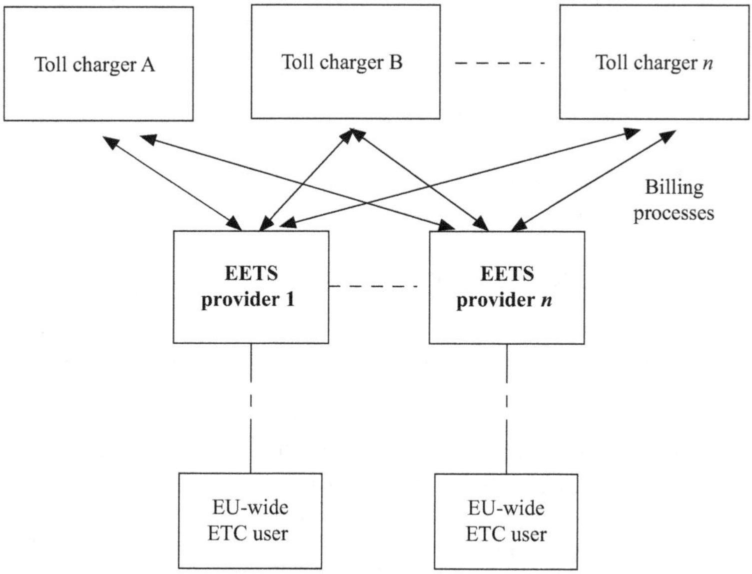

Figure 10.1 EETS interoperability architecture

and main areas of activity of the REETS project and how results of the project are being used.

The chapter concludes by presenting the outlook for the future of interoperability in Europe, what actions are planned for the future and the prognosis for the market of interoperable ETC in Europe. Technology trends are considered that may facilitate the more widespread introduction of interoperable ETC services as well as the disruptors that may affect the future direction of ETC interoperability.

10.2 The need – business and political drivers for interoperability of electronic toll collection systems in Europe

Achieving the interoperability of ETC systems in Europe has been an aspiration for over two decades. Since the first ETC systems were introduced by road operators in Europe, it has been recognised that implementation and operating costs could be reduced if systems covered more than a single toll domain.

In Europe, both passenger and freight vehicles frequently travel across administrative boundaries. This creates the need for services that allow users to pay tolls on roads in more than one jurisdiction or roads operated by different companies. Companies involved in toll road operation often operate roads in more than one member state or region. This means their users may expect to be able to use the same toll payment account when travelling across Europe.

A report published by the European Commission in 2014: 'Report on the State of the EU (European Union) Road Haulage Market' [3] identified that 'different

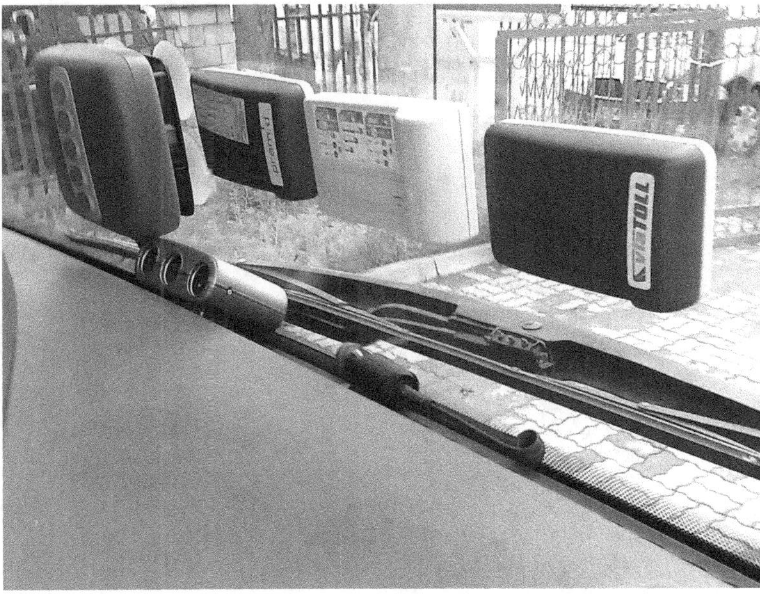

Figure 10.2 Multiple On Board Units (OBUs) required in European vehicles

charging regimes and technology have created a barrier to international transport'. The report identified that multiple in-vehicle units (Figure 10.2) and contracts are still required in order for users to operate in different countries, and this was not in line with the European Directive and Decision which requires that road users should be able to pay the tolls related to any road-charging scheme through a single contract with an EETS provider. The report cites the need to process multiple payments in different EU member states as being burdensome for back office staff in haulage companies. The report concluded that further work is required on achieving interoperability of toll systems across the EU.

Whilst the European Directive and Decision demanded the introduction of interoperable services for ETC, the systems implemented by different member states have included slightly different requirements. So no SP has been able to introduce a service available on an EU-wide level. So addressing the issue of interoperability and the implementation of EETS has become an even more urgent issue. Representatives of the European road haulage industry have requested via the European Parliament that this issue is progressed in order to improve the efficiency and competitiveness of goods transport within the EU and continent.

10.3 Background – the ETC landscape and the beginnings of ETC interoperability in Europe

The landscape of ETC systems which existed in Europe prior to the REETS project is illustrated in Figure 10.3 below.

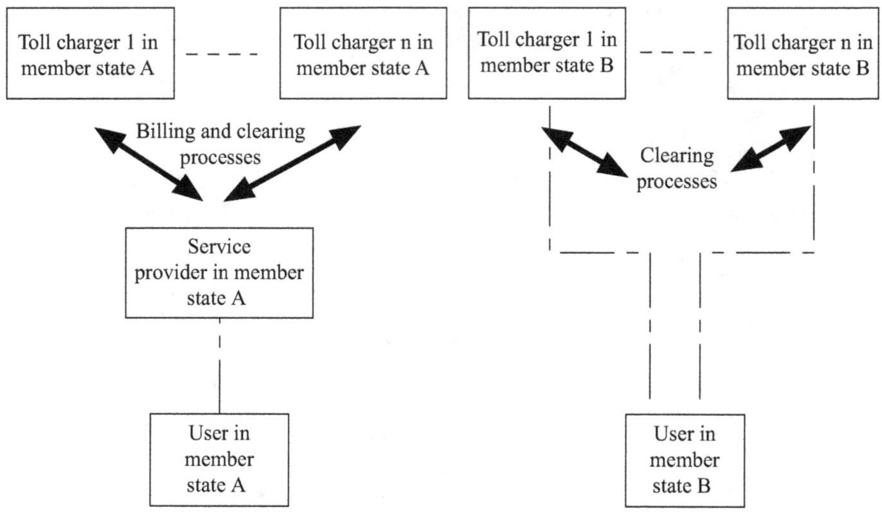

Figure 10.3 ETC landscape prior to REETS project

The ETC landscape has evolved through the collision of two tolling requirements: first, the need for speeding up toll collection at motorway toll plazas and, second, the need to collect distance-based charges for trucks.

10.3.1 ETC for motorway tolls in Europe

The first requirement was the introduction of ETC by operators of tolled motorways and toll bridges and tunnels. In these cases, the operators were motivated to introduce ETC to reduce the quantity of small cash transactions and delays at toll plazas and to improve overall operational efficiency. The ETC systems can mostly be considered to be electronic 'add-ons' to the existing systems used for collecting tolls. The first ETC systems in Europe were installed in Norway (1989), Italy (1990) and Portugal (1991), followed by Greece (2002), Spain (2003) and France (2005).

Initially, systems were not interoperable. Prior to the REETS project, the ETC landscape in Europe consisted of some onboard units ('tags') issued by SPs and some issued by road operators. In both cases, these onboard units were accepted only within a single country or on a single operator's roads within a country or even on a single road. Those SPs that issued onboard units that could be used across a whole country mostly did not provide the possibility for toll payments using the same onboard unit in other EU member states. Drivers required additional tags for each country or road that they travelled. This situation was in conflict with the goals described in the European Directive and Decision for users to be able to pay tolls across the EU using a single onboard unit, one account and one invoice.

However, there were some exceptions. Interoperability across member states was emerging in a limited way where significant amounts of regular users travelled between two or more member states, for example, on specific routes between Portugal and Spain, France and Spain, France and Italy and within Scandinavia.

10.3.2 ETC for national heavy vehicle charging

In parallel to ETC for motorway tolling, a second requirement emerged. This was for national freeflow systems for charging mostly heavy goods vehicles a distance-based charge for travel on all roads of a specific class within a whole country. These systems began to be implemented as freeflow systems not relying on toll plazas.

These national charging systems have been introduced in European countries where motorways are not operated by private concessions and hence are procured by national governments rather than individual motorway operators. The main reason behind national governments in Europe introducing these schemes has been the need to collect revenue from *all* heavy vehicles using a country's road network due to the wear and tear caused by increased heavy vehicle traffic originating in other countries. At the time of writing, systems have been introduced in Switzerland, Germany, Austria, Czech Republic, Slovakia, Poland, Belgium, Hungary and Belarus, with further systems planned by Slovenia and Bulgaria. Thus far, different electronic onboard units have been issued to road users by each government that has introduced such a system. These systems also use slightly different methods of registering the distance or roads covered during a vehicle's journey and applying the relevant charge.

However, services which provide interoperability between the various national heavy vehicle freeflow tolling schemes have, to date, only been introduced on a limited basis. Specifically, before the REETS project, only two examples existed. In the first, it is possible to pay the Swiss heavy vehicle tax using data declared from the Austrian onboard unit. In the second example, it is also possible to pay the Austrian GOMaut using the German onboard unit. However, in neither case was it possible to pay both tolls using the same SP. In the case of using the German onboard unit in Austria, it was possible to have a single service and account – known as Toll2Go – but the user still has two separate contracts with each toll charger.

So this was the landscape which provided the background for the introduction of EETS and also for the REETS project. EETS could not be created using a 'clean sheet', as a confusing array of tolling systems and SPs existed in several European member states. Each national freeflow ETC system presented individual legacy requirements, and in some other countries, the long-established presence of national SPs would impact on the commercial viability for EETS providers entering the market.

10.4 The route to EETS via REETS – the service provider concept and the emergence of regional interoperability services

10.4.1 The service provider concept

Paradigmatically, interoperability can be obtained either by co-operation of toll chargers or by an SP concept. The first approach relies on the concept of co-operation and hubs between toll chargers, whereas the SP concept goes beyond, by opening parts of the value chain to the private sector and competition. By way of example, current developments of interoperability between toll chargers in the United States are illustrated in Figure 10.4.

In Europe, the SP concept was introduced within the CESARE [**C**ommon **E**FC (electronic fee collection) **S**ystem for **a** **R**oad Tolling **E**uropean System] III project (Figure 10.5), which then became the basis for the EETS concept and further definition work carried out in the CESARE IV project [4]. This marked a change of approach from work on the definition of interoperability completed within earlier projects such as CESARE I and CESARE II which was based on the concept of direct co-operation between toll chargers [5].

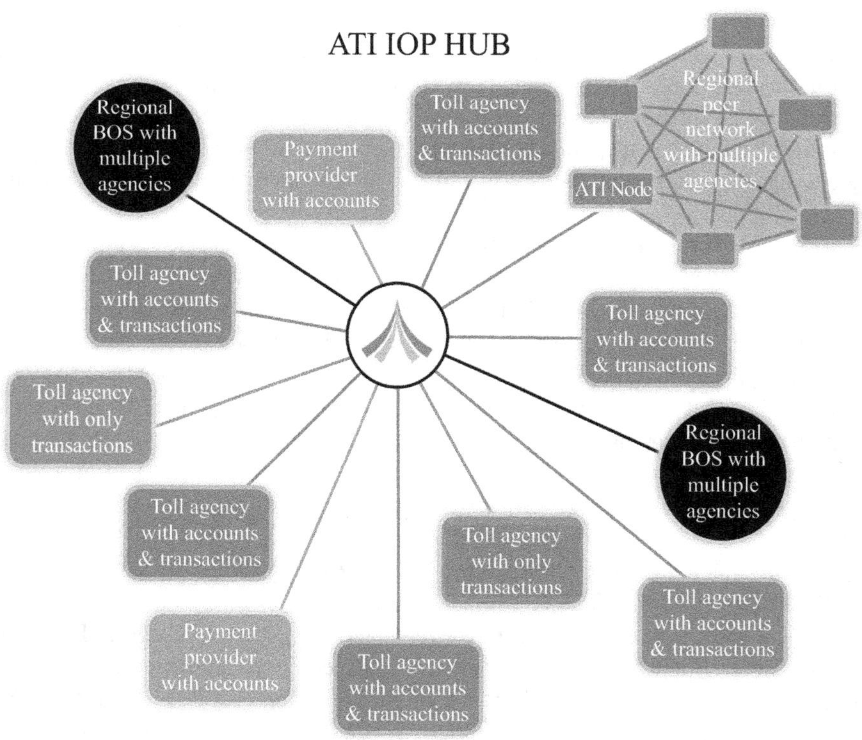

Figure 10.4 Possible US model provided by ATI (Alliance for Tolling Interoperability) hub

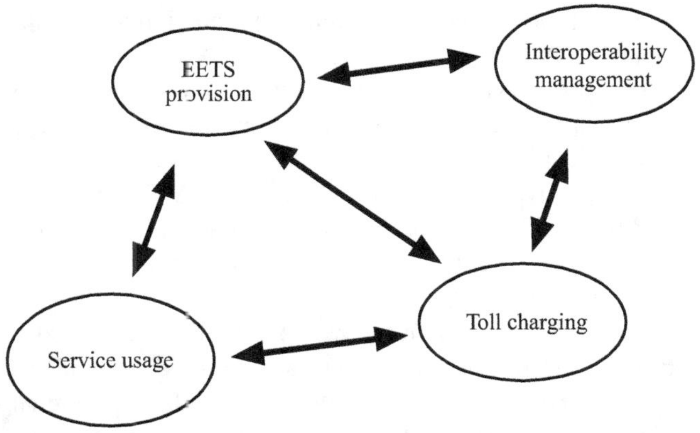

Figure 10.5 CESARE EETS service provision model

In the EC Decision on EETS, a 'toll charger' is defined as public or private organisation which levies tolls for the circulation of vehicles in an EETS domain, whereas the SP or EETS provider grants access to the service to the user.

In the SP model, onboard units are issued to road users not by the toll charger or road operator but by a third-party SP. This third-party provides an account for the user to pay tolls on roads operated not only by a single road operator but on all roads across a complete country or region. Interoperability of ETC onboard units issued within countries such as France and Ireland resulted in the creation of the 'SP' or 'Issuer' model for ETC. This SP concept was adopted by the European Commission when formulating its concept for the EETS.

The EETS service provision concept developed by the CESARE project participants was fundamentally based on the principle that road users would not have a contract directly with each road operator but with a tolling SP that provides users with a toll payment account and handles toll payments to a number of toll chargers.

10.4.2 The service provision model in France

France adopted the service provision model for its TIS (Télépéage InterSociétés)-Liber-t service for light vehicles – in operation since 2000 – and the TIS-PL (Télépéage Inter Sociétés Poids Lourds) service for heavy vehicles – in operation since 2005. Both systems allow a user to use a single DSRC (dedicated short-range communications) onboard unit throughout France and pay all tolls through one account with the user's chosen SP. Users receive only one invoice for travel on all tolled motorways and tunnels across France regardless of operator.

The Liber-t service is, at the time of writing, offered by 11 different SPs, has over 5 million subscribers and accounts for 42% of the total light-vehicle transactions at French toll plazas. The TIS-PL service is offered by five SPs, has over 640,000 subscribers and accounts for 85% of the total heavy vehicle toll transactions at French toll plazas. France has 14 separate motorway concession companies plus a further four tunnel-operating companies, operating over 9,000 km of motorways in total.

Onboard unit testing is carried out at a central facility operated by the French toll charger's association ASFA (Association des Sociétés Francaises d'Autoroutes). ASFA's members comprise the private concession companies which are toll chargers for motorways and tunnels in France. ASFA also writes and maintains technical standards for ETC interoperability within France. So once an onboard unit has been tested at ASFA's test facility, it is approved for use on the roads of all ASFA's member companies.

The French ministry of economic development also developed a nationwide freeflow electronic charging system known as 'Ecotaxe' (see also Chapter 16) intended to apply to heavy vehicles when driving on selected national class roads where tolls do not apply. The system was separate and in addition to, but also interoperable with, the TIS-PL ETC system. This system was due to be introduced in 2014 but was subsequently cancelled and is not in operation. However, the system was fully developed and implemented and used an onboard unit type combining the DSRC charging system used in TIS-PL system with a GNSS (Global Navigation Satellite Services) receiver for use on roads where the TIS-PL system was not operational. These onboard units were to be offered by SPs known as 'Société Habilité Télépéage (SHT)s', which would also act as SPs for and providing interoperability with the existing TIS-PL service on the concession-operated motorway network. Although the project was abandoned due to political reasons, the companies that participated in the project as SHTs have gone on to develop EETS services and register as EETS providers.

10.4.3 The service provision model in Ireland

Ireland's motorways have been constructed through public–private partnerships (PPPs), where the government utilises private sources of funding to reduce public sector borrowing requirements. Sections of motorway are operated under concession by private contractors, which use the toll revenue to secure financing for construction and maintenance of the road. As part of the PPP agreements, an obligation was placed on the road operators to issue ETC onboard units which would be accepted by other road operators. The National Road Administration also established an Information Exchange Agent with the role of clearing all transaction information occurring between operators. This resulted in a nationally interoperable system of electronic toll payment being deployed by 2007.

Also in 2007, the National Road Administration forced the removal of a toll plaza on the M50 Dublin Ring in order to alleviate severe queues. A freeflow system of ETC was subsequently introduced at this location.

Since the introduction of the M50 freeflow system in 2008, the numbers of drivers using tag accounts has increased dramatically. As part of establishing the M50 freeflow system, independent SPs were introduced into the market. Users of onboard units issued by these independent SPs could also roam on all other tolled motorways in Ireland. Interoperable tags are currently issued in Ireland both by independent SPs (parkmagic [6] and Easytrip [7]), as well as toll road operators eFlow for M50 and Direct Route (Limerick Tunnel and M8).

10.4.4 ETC interoperability and service provision in Spain and Portugal

In Spain, the toll network is 3,300 km long and is managed by 30 different toll companies. Tolls are based on the number of axles as well as vehicle category. Toll payment at plazas is by cash, payment card or ETC onboard unit. The nationwide ETC system is called VIA-T, and it is in operation in all tolled motorways. The 'VIA-T' system was introduced by Motorway Concession companies in 2003 to reduce queues at toll plazas and improve user convenience and acceptance of tolls.

VIA-T onboard units are issued to users in Spain by approximately 70 SPs, which collect tolls on behalf of all 30 concession companies. VIA-T SPs comprise two main groups:

- Financial issuers (more than 60 banks)
- Other SPs e.g. fleet card providers (RESSA, SERVISA, CEPSA CARD, SOLRED, INGENICO, DKV & PAGATELIA)

In the case of the financial issuers, three 'clearing houses' provide the interface with individual 'acquirer' banks who act on behalf of each toll charger. This mechanism is carried over from the banking system of Spain and was not established specifically for ETC. Its use was facilitated through the retail businesses of investment banks involved in the financing of Spanish toll motorway concessions.

In the case of fleet card providers, bilateral agreements are needed between each toll charger and each fleet card provider since they do not participate in the clearing system used by the financial issuers.

A memorandum of understanding (MoU) exists between toll chargers in order to define standardised requirements for SPs.

At the time of writing, there are 21 separate tolled motorway and bridge concessions in Portugal. A single interoperable ETC SP – ViaVerde – exists which is operated on behalf of all of the concession companies. Portugal was the first country in the world to introduce ETC throughout its motorway network. The Portuguese ViaVerde service has one of the highest rates of ETC usage with over 3.5 million subscribers. The ViaVerde service is now accepted at over 300 toll plazas throughout Portugal, as well as at 108 petrol stations, 113 car parks and 11 MacDonalds McDrive outlets [8]. Portuguese motorway operators have recently pioneered the widespread introduction of freeflow ETC for light vehicles on motorways.

Interoperability between Spain and Portugal has been achieved in the following ways:

- Spain to Portugal: two of the Spanish tag issuers have been approved by viaVerde Portugal who collect all tolls on behalf of all concessions in Portugal.
- Portugal to Spain: Tags issued by viaVerde Portugal are accepted on the roads of two concession companies and viaVerde Portugal purchase transits on behalf of their customers with Spanish tag issuers.

10.4.5 ETC interoperability and service provision in Italy

Italy was the first country to introduce ETC on a motorway in 1990. This system is operated by one SP, known as Telepass, on behalf of all toll chargers in Italy and provides an interoperable electronic payment system across all tolled motorways in Italy. Italy has 25 concession motorway operators, and a single interoperability agreement exists between them ensuring Telepass is accepted across the whole national motorway network of approximately 6,000 km. Telepass has over 9 million subscribers and accounts for some 60% of total toll payments [9].

Telepass uses a slightly different standard for its DSRC onboard unit to roadside infrastructure communications, an ETSI standard instead of the CEN TC 278 derived standard (ISO EN 15509), and onboard units therefore have to comply with ETSI ES 200674-1 and its related technical reports.

The European Decision 2009/750/EC on the definition of the EETS and its technical elements also requires that onboard units for a European-wide service must support both the ETSI and CEN TC 278 standards. Telepass has recently introduced an interoperable onboard unit which enables its users to pay tolls in France and Spain as well as Italy through their local SP.

10.4.6 Interoperability of ETC services in Scandinavia – EasyGo

EasyGo is a joint venture between toll chargers in Scandinavia and Austria. The partners include the Norwegian and Swedish Road Administrations as well as publically owned companies ASFINAG (Austria), Storebæltbridge (Denmark) and ØresundBridge (Denmark).

Its aim is to provide an interoperable ETC service to users throughout the EasyGo region, which currently includes all toll chargers in Norway, Sweden and Denmark as well as ASFiNAG in Austria. In total, more than 50 toll chargers are participants in EasyGo as well as ferry companies using DSRC onboard units for payment.

All of the toll chargers participating in EasyGo are linked via a joint venture agreement signed by all of the partners. Onboard units are issued by SPs and can be used throughout the EasyGo region. SPs are also subsidiary companies of the EasyGo toll chargers. Each SP enters into a toll SP agreement with the EasyGo Joint Venture. These agreements define common technical and commercial requirements for interoperability. Clearance of charging transactions between toll chargers and SPs is achieved via a bespoke 'hub' maintained by the toll charger for the Øresund and Storebælt bridge/tunnel links (Sund & Baælt) on behalf of all the EasyGo participants. The EasyGo service portal can be found at the following web page http://easygo.com/en.

10.4.7 Overall European interoperability architecture

So prior to the REETS project, no common architecture had been implemented which could provide interoperable ETC services across all European member states. In some cases, interoperability of onboard units existed only at the roadside, requiring users to register in multiple toll charger systems and maintain separate accounts. In other cases, interoperability existed within a single country because all onboard units were issued by a single SP. In further cases, interoperability was possible across multiple chargers in a country or region – and using onboard units issued by multiple SPs – but requiring transaction clearance via a national or regional hub.

No unified scalable system existed allowing any SP to transact with any toll charger in any European member state.

10.5 EETS is *more* than interoperability – advantages and disadvantages

As mentioned above, the SP concept does not just ensure interoperability but also opens parts of the toll charging value chain to the private sector and competition (Figure 10.6).

Compared to only ensuring interoperability between toll systems, which reduces complexity between the different players, the introduction of EETS is more challenging. EETS introduces a means of opening a market for service provision and is therefore more complex to achieve. However, EETS in operation has many advantages, which are as follows:

- For the toll charger, EETS simplifies the handling of international customers: Customer service and support for international customers will be mainly provided by EETS providers rather than the toll charger (note: an EETS provider is a SP registered to provide toll payment services in all European toll domains).
- Market, competition and cost efficiency: Parts of the services in the value chain will be open to competition.
- A better service: Competition between SPs usually provides for a better customer service.
- Easing implementation of new toll systems: One main challenge when introducing new electronic toll systems is to reach a certain penetration of the market with onboard units from the start. Therefore, introduction is considerably easier if a significant number of (international) customers is already equipped.

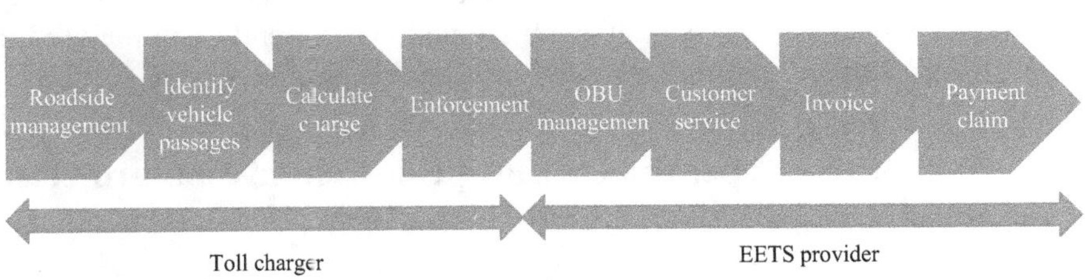

Figure 10.6 EETS value chain (DSRC)

- Easing migration of existing toll systems: As above, EETS customers do not need to be equipped again.
- Innovation by fostering competition of technologies: If a toll charger implements a new system, which may remain unchanged for around 10 years. Different onboard units from different suppliers integrating the required technologies foster innovation and competition of technologies.

10.6 Legal background/framework

In 2004, the European Commission published Directive 2004/52/EC on the Interoperability of Electronic Road Toll Systems in the community. Its objective was to ensure EU-wide interoperability of services. The directive allowed for the use of either DSRC or GNSS/CN (Global Navigation Satellite Services combined with Cellular Network) technologies.

In 2009, the European Commission introduced Decision 2009/750/EC against a background of member states introducing national electronic toll systems for heavy vehicles using different technical solutions. The decision, however, sets out a clear approach involving the separation of the roles of toll charger and SP in future systems. Toll chargers were to have no direct relationship with road users in the payment of tolls, whilst users would have a choice of SPs whom they could contract with to pay all their tolls. This model was largely based on the successful French TIS-PL model [10] which already allowed interoperability of toll payments for users on all French motorway concessions.

However, in the intervening years between publication of the directive and the decision, national systems of ETC for heavy vehicles had been introduced which were not interoperable. The decision included a provision that national tolling systems, either manual or electronic, could coexist with EETS services when they were introduced. However, in most cases, the national tolling systems for heavy vehicles had been established without the separation of toll charger and SP roles required by the EETS Decision. Onboard units and toll payment accounts were provided to users by the national operator of the ETC system, with no possibility for users to obtain an EETS service from these operators. Some limited service interoperability was later made possible through bilateral agreements between national operators e.g. the Toll2Go service between Germany and Austria, although users still required a contract with both toll chargers.

In 2016 Belgium became the first country to approve the use of an OBU issued by an EETS Provider – Axxes – although the unit can currently only be used in France, Spain, Portugal and Belgium with introduction also planned in Austria.

10.7 European co-operation – European Commission projects, standards, activities, committees, etc.

To support the development of interoperable services under the directive, the European Commission funded a series of working groups which published reports mainly in 2005.

These reports produced recommendations on topics including

- DSRC and GNSS technologies
- Vehicle classification
- Enforcement
- Equipment certification
- Payment and contractual aspects
- Enforcement and
- Security.

The following reports were produced by the EETS Expert Groups:

Expert Group 1: Microwave technologies (DSRC)
Expert Group 2: Vehicle classification
Expert Group 3: Enforcement
Expert Group 4: Equipment certification
Expert Group 5: Satellite technologies
Expert Group 6: Integration of onboard units into vehicles
Expert Group 7: The role of financial institutions – payment and contractual aspects
Expert Group 8: Final review of draft UNI (Italian) DSRC specifications
Expert Group 9: Specification of the EFC application based on satellite technologies
Expert Group 10: Recommendations on enforcement specifications and technologies
Expert Group 11: Definition of the EFC application based on DSRC
Expert Group 12: EETS security aspects

These reports also provided the basis for further work which led to the publication of the European Commission Decision (Figure 10.7).

As previously mentioned, work on road charging interoperability was also completed through the CESARE series of projects co-funded by the EC. These projects involved the participation of Europe's motorway concession operators led by ASECAP (Association Européen des Concessionnaires d'Autoroutes et d'Ouvrages à Péage), joined in the CESARE III project by the so-called 'Stockholm group' of national public toll chargers.

Standards work in Europe initially focused on the interoperability of DSRC systems. However, plans for the introduction of a system of tolling heavy vehicles in Germany based on GNSS/CN technology meant that technical standards for EETS had to be broadened to include both DSRC and GNSS technologies.

In Europe, work on standards for EFC systems was taken forward by CEN (Comité Européen de Normalisation) Technical Committee TC278 working group 1.

The committee has so far produced the standards listed in Table 10.1.

Currently, only CEN 15509 is referred to by the EC Decision defining the requirements for the EETS. However, other standards such as ISO 12855 are becoming more widely adopted by toll chargers and toll SPs enabling a clearer path to the realisation of EETS.

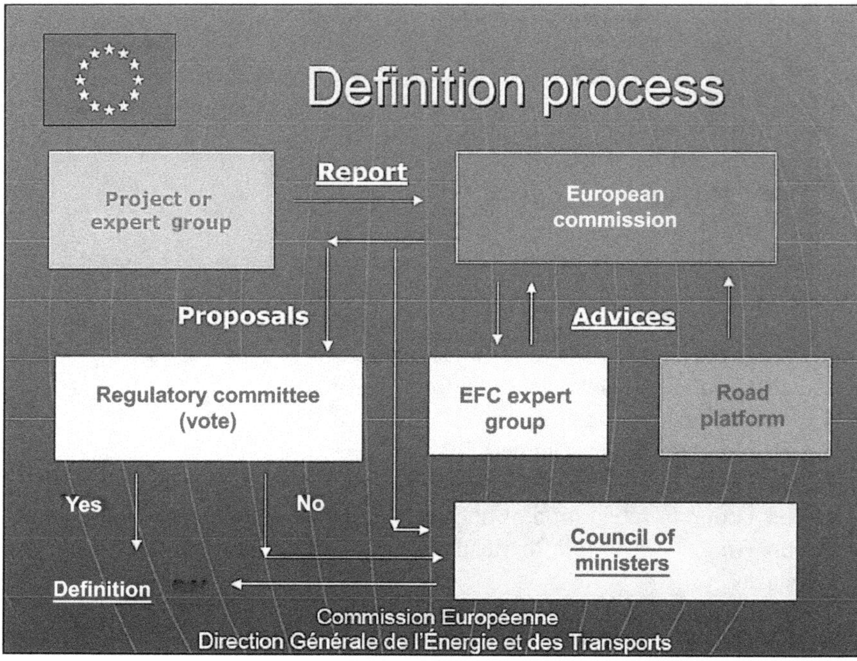

Figure 10.7 European Commission Decision – definition process (European Commission DG TREN 2006)

10.8 Getting EETS off the ground: the REETS project

Ten years after the European Directive had been published, EETS as a service had still not been implemented and offered to users anywhere, even despite the development and adoption of common technical standards.

The lack of emergence of EETS since the publication of the decision made it a priority for the European Commission to take steps to encourage deployment. The commission's ITS Action Plan identified the need for a 'stepwise approach to full deployment of EETS'. In practice, this meant a more pragmatic approach to its introduction by allowing the development of a regional version of EETS.

In 2014, the European Commission announced the co-funding of a regional project 'REETS'. The commission's aim was to accelerate the deployment of EETS services at an EU-wide level, through the incubation of services initially at a regional level. It launched the REETS project to bring together a 'critical mass' of toll chargers and potential SPs in order to solve the remaining problems. The project partners were bound by common interests in removing market uncertainties and barriers, and allowing interoperable services to be defined and introduced on a commercially viable basis.

REETS was initially supported by seven member states (Austria, Denmark, France, Germany, Italy, Poland and Spain) plus Switzerland (Figure 10.8).

Table 10.1 CEN technical standards for EFC [11]

	DSRC-based EFC	EFC technology independent	Autonomous EFC
Frameworks	ISO14907-1 Test procedures for user and fixed equipment	ISO 17573 Electronic Fee Collection (EFC) Architecture ISO 17574 Security Profiles ISO 19299 Security Framework	
Toolboxes	ISO 14906 Application Interface Definition (AID) for DSRC EFC [ASN.1] ISO 14907-2 Test procedures for DSRC OEU ISO 25110 AID for IC-cards ISO16785 Int. def. between DSRC-OBE and external in-vehicle devices [ASN.1]	ISO 12855 Information exchange between TC and TSP [ASN.1] ISO 17444-1 Charging performance indicators – Metrics ISO17444-2 Charging performance indicators – Examination framework	ISO 17575 AID for Autonomous EFC [part1][part2][part3] [ASN.1] Tests for AID Autonomous EFC part [1.1] [1.2/TTCN] [2.1] [2.2/TTCN] [3.1] [3.2/TTCN] CEN16702 Secure Monitoring [part1] [part2] [ASN.1]
Profiles	CEN15509 Interoperability Application Protocol (IAP) for DSRC EFC CEN15876 Tests for IAP for DSRC EFC [part1] [part2] [TTCN]	CEN/TS 16986 IAP for Information Exchange between TC and TSP [WSDL]	CEN16331 IAP for Autonomous Tolling ISO 12813 Compliance Check Communication (CCC) [ASN.1] ISO 13143 Tests for CCC [part1] [part2] [TTCN] ISO13141 Location Augmentation Communication (LAC) [ASN.1] ISO 13140 Tests for LAC [part1] [part2] [TTCN]

The project was led and developed jointly by the European association of motorway concession associations – ASECAP, and the Austrian Motorway operator – ASFINAG.

The project partners therefore represented a diverse set of road operators and types of tolling systems as is illustrated in Table 10.2.

In order to ensure the involvement of all the required stakeholders in the implementation of EETS, the project also included AETIS (**A**ssociation of **E**lectronic **T**oll and **I**nteroperable **S**ervices) [12] as the association of potential EETS providers. AETIS comprises 12 member organisations that have either registered as EETS providers or intend to do so. AETIS's member companies include

- Issuers of pan-European fuel payment/fleet services cards:
  - Total, LogPay, Ressa, EuroShell, OMV, UTA, WAG, DKV and Trafineo

Figure 10.8 Overview REETS-TEN participating toll domains (shown in dark grey)

- Toll payment SPs:
  - Axxes, Eurotoll and Telepass.

At the time of writing Axxes, Telepass, Eurotoll, Total, W.A.G., T-Systems and BroBizz have already registered as EETS providers.

Prior to the REETS project, one of the reasons for the lack of progress of EETS was the existence of key uncertainties surrounding the business model for SPs. For example, one of the requirements of the European Commission Decision was that organisations should obtain approval to offer EETS in all European toll domains within 2 years of registering as an EETS provider. This requirement was included in order to ensure that countries on the periphery of Europe were covered by EETS. However, in practice, the demand for EETS services in some toll domains is quite low, and a positive business case to enter these toll domains might not exist for SPs.

One of the key strategic thrusts of the REETS project was therefore to remove these uncertainties for potential EETS providers by bringing all stakeholders

Table 10.2 Toll chargers participating in REETS

EU member state/country	Road operators represented in REETS	Type of charge	System
Austria	ASFINAG	Heavy vehicle distance-based charge	Freeflow DSRC
Denmark	Sund & Bælt	Bridge/tunnel toll	DSRC at toll plaza
France	ASFA (association representing French toll chargers)	Motorway tolls	DSRC at toll plazas
France	Minstère de l'Environnement, de l'Energie et de la Mer	Environmental tax	GNSS
Germany	Bundesministerium fur Verkehr und digitale Infrastuktur	Heavy vehicle distance-based charge	GNSS
Italy	AISCAT (association representing Italian toll chargers)	Motorway tolls	DSRC
Poland	Generalna Dyrekcja Dróg Krajowych I Autostrad	Heavy vehicle distance-based charge	Freeflow DSRC
Spain	ASETA (association representing Spanish toll chargers)	Motorway tolls	DSRC
Switzerland	Federal Customs Administration	Heavy vehicle distance-based tax	DSRC + tachograph

together and entering a dialogue. SPs identified the main uncertainties being connected with the type of guarantees which would be accepted to cover toll liabilities.

The situation regarding technical standards also creates uncertainty for potential EETS providers. The EETS Directive and subsequent European Commission decision allowed toll chargers to base their systems around either GNSS/CN (telematics) or DSRC (tag and beacon) technologies. EETS providers therefore must issue onboard units that can operate using both types of system. However, in practice, there are different configurations and architectures possible even within these basic technology choices. This means that if an EETS provider issues an onboard unit for EETS, it may be forced to upgrade it to comply with new requirements from individual toll chargers in the future.

The communication from the commission of 30 August 2012 about the implementation of the EETS, as defined in Directive 2004/52/EC and Decision 2009/750/EC, addresses the progress achieved and lists outstanding problems. In order to facilitate the introduction of EETS and to promote the necessary co-operation the communication proposed a number of actions to be taken by the commission and the member states.

The communication provides for a stepwise approach towards EETS: As a first step towards full European interoperability, member states with significant volume of traffic on the trans-European network should encourage the cross-border interoperability of their electronic road toll systems. These early deployment project(s), on a regional basis, will be promoted such that they can be extended to cover all the electronically tolled road infrastructures in the EU at a later stage and can provide concrete experience in solving practical EETS issues.

Furthermore, the communication mentions the creation of an open comprehensive information-sharing resource platform as another provision for facilitating EETS.

10.9 REETS project organisation

10.9.1 Overall rationale and approach

The rationale of the project was that working in a concerted way to deploy EETS concepts over a limited region would accelerate the wider deployment of EETS. The project would therefore be cognisant of commercial constraints but provide a first step towards the ambition of pan-European interoperability of electronic toll payment services. It aimed to reduce barriers for the deployment of the EETS, in particular, by reducing business uncertainty for EETS providers. The expected outcome was the creation of an environment in which bilateral negotiations between toll chargers and EETS providers could proceed with greater certainty.

The fundamental approach of the project was that by bringing together toll chargers and SP partners from eight countries in the same project, it would be possible to develop a common understanding of the different roles required to implement EETS and the different organisational perspectives.

10.9.2 Project scope and objectives

The project objectives can be summarised as follows:

- Reduce barriers for EETS deployment by reducing business uncertainty for EETS providers in order to prevent potential market failure of European ETC services
- Create a basis for easing the bilateral negotiations between toll chargers and EETS providers
- Develop a common understanding of the service components provided by the different roles
- Demonstrate in practice how EETS can be deployed and operated at a regional level based on interfacing the different environments using the results from the activities
- Provide an information resource to support rolling out EETS across Europe (known as 'return-on-experience' activity within the project)
- Finding acceptable risk management rules.

The scope of the project included

- coordination and simplification of the interface components for establishing and operating EETS to the extent possible, taking into account the limits set by different collection processes and technologies in use, and the legal and contractual frameworks for toll charging in the different countries;

- carrying out a cost–benefit analysis before testing/implementation as a basis to decide on investments in testing and implementation activities;
- practical tests, pilot operations and contractual frameworks enabling transition to normal operation of the EETS in REETS toll domains;
- provision of return on experience to other stakeholders, as a basis for rolling out EETS beyond the region directly involved in the action.

The members of the consortium in charge of the action were EU member states, represented by the competent ministries and/or other competent authorities, toll chargers and their organisations, as well as organisations who have declared an interest in assuming the role of EETS providers ('potential EETS providers').

10.9.3 Project organisation

The project was organised in two phases: first, an analysis phase, second, an implementation monitoring phase.

In the analysis phase, the project carried out collaborative work on the topics of:

- Establishing a contractual framework and risk management guidelines for EETS stakeholders
- Clarifying certification processes across different member states
- Defining key performance indicators (KPIs) for service agreements
- Harmonising back office interface requirements
- Defining responsibilities and processes for interoperability management.

In the implementation monitoring phase of the REETS project, the management team monitored the deployment of EETS pilot services within the participating member states. The project's scope did not include coordination of EETS service implementation as this depended on the business plans of EETS providers and their commercial discussions with toll chargers. However, the aim was that continuing the precompetitive collaborative environment amongst the project partners would facilitate clarification of their commercial processes and business plans. The project focused on monitoring progress towards deployment of EETS at a regional level and the dissemination of results and information from the study phase as the basis for rolling out EETS across Europe.

Throughout both phases of the project, a 'horizontal' task entitled 'return on experience' was concerned with communicating the results of the project. This task organised several outreach events and activities to ensure engagement during the project with a wider range of stakeholders such as toll systems suppliers.

10.9.4 Analysis phase activities

The analysis phase covered five activities, which are as follows:

Activity 1 – Contractual framework and risk management
Activity 2 – Certification process
Activity 3 – KPIs

Activity 4 – Back office interfaces
Activity 5 – Interoperability management

In the analysis phase, the activities were led by different organisations and all organisations participated in one or more activities to ensure appropriate representation of toll chargers and SPs in each.

The analysis of contractual, procedural and technical issues was performed based on the known issues identified by previous work carried out in projects such as CESARE [13]. The output was a series of reports which provide recommendations for common solutions that if adopted would facilitate the easier introduction of EETS. The reports from the analysis phase are available on the REETS information platform created in the next phase of the project. The information platform also provides practical information about the participating countries' toll domains [14].

Activity 6 'return on experience' prepared information about the results of the analysis phase for inclusion in deployment phase activities of the project. This ensured continuity between the analysis and pilot monitoring phases of the project.

10.9.5 Information platform

A key deliverable of the pilot monitoring phase of the project was the information platform created within two activities, which are as follows:

Activity 7a – Information platform – In this activity, information was prepared for all the participating countries. First, a knowledge base was prepared and subsequently support was provided in the form of a helpdesk for users of the information platform

Activity 7b – Testing/implementation – In this activity, a framework was provided for the activities performed by individual toll chargers and EETS providers in order to implement REETS. It was performed by selected organisations representing the toll chargers. In this activity, prospective EETS providers acted individually (and not represented by their association) together with the toll chargers.

10.10 Implementation of EETS – compliant services and REETS pilot

10.10.1 Conditions for the pilot and roles of the actors

A key challenge for the REETS project was to establish conditions to begin exploration of the options for testing and implementation of pilot services within the commercial constraints and priorities of the partners.

From the very beginning, it was clear that a pilot should not be limited just to a practical demonstration but should be based on sustainable solutions which can be transferred into normal operation. No organisation, neither toll charger nor SP, would have made such large investments otherwise. Thus, it was clear that investments needed to be refinanced by revenues which brought the key element of the service onto the table – the question of remuneration. In the end, pilot operations were carried out only in those toll domains that committed at an early stage to pay a (fair) remuneration for the service.

Another key condition for the success of the pilot was that participants did not need registration as an EETS provider and that remuneration should be possible for the service.

Following the EETS model, it was crucial that individual toll chargers and EETS providers take responsibility for the implementation. This excludes an involvement – as contractual partner – of the toll motorways concessionaires' national associations as well as AETIS in testing/implementation. Furthermore, the REETS pilot was open to any potential EETS provider, independent of their participation in the previous activities of the project. This allowed for an open competition between the EETS providers, and AETIS – as an association – did not directly participate in the pilot activities.

Whilst any EETS provider can contact any toll charger in Europe to seek a contract for providing EETS compliant services, it was important nevertheless that the project make all results available to stakeholders. Within the framework of the REETS project, pilot activities were not monitored outside the planned geographical region.

10.10.2 Roadmap actions project coordination

The role of project management in this phase was a kind of catalyst: it monitored, coordinated and facilitated as far as possible. However, testing and implementation activities associated with the pilot services were managed and performed individually by each toll charger and EETS provider and not co-ordinated centrally within the consortium.

The REETS project coordination therefore initially

- prepared the analysis and discussion for the exploring of the options for testing/implementation with all REETS Trans European Network (TEN) toll chargers;
- discussed with all REETS TEN toll chargers options for testing/implementation in their toll domain according to market activities, including documentation of such activities;
- prepared a concept for the involvement of and communication with AETIS, its members and other service/EETS providers and
- started the involvement of individual service/EETS providers in the project with regard to implementation.

10.10.3 Coordination with REETS TEN toll chargers

As part of the preparations for monitoring the pilot activities, the project coordination team had prepared an example work plan covering the activities for a toll domain. This was the basis for discussions about the possible options and timelines for pilot activities. The work plan had to be adapted for each toll domain, based on its specific situation. Also, the start date for the activities depended on the situation in the respective toll domain. Starting implementation/testing at the same time in all toll domains would be inefficient or impossible, due to limited capacities, in particular of interested EETS providers.

Based on that from mid-March until May 2014, project coordination had meetings with representatives of all REETS toll domains. The focus of these meetings was to:

- collect information about activities and/or applications of SPs in the toll domain;
- discuss the options for testing/implementation and possible schedules according to market activities and the conditions in the toll domains and
- align in order to provide a concise picture of the current status and expected actions and/or developments in the toll domain.

10.10.4 Coordination with service providers

The coordination by the REETS project management with the SPs was a two-way communication with the objectives of

- Collecting information about the intentions of the individual SPs regarding the implementation of the REETS TEN pilot. This was useful for planning the activities and coordinating the next steps.
- Informing the SPs about the planned steps.
- Providing transparent information to all interested SPs whether member of AETIS or not and thereby ensuring equal treatment of interested companies.

In June 2014, project coordination sent information to AETIS in order to ensure a common understanding of the role for AETIS and the individual SPs during the Implementation monitoring phase. It was emphasised that the monitoring phase is focused on testing and pilot implementation activities performed by individual toll chargers and SPs. The role of the REETS project was focused on monitoring and coordination to support the pilot activities. The approach for direct communication between project coordination and individual SPs regarding the pilot activities – as opposed to a communication through AETIS – was outlined.

Furthermore, it was clarified that this approach for direct communication with individual SPs is limited to the pilot activities. This has no effect for the work and role of AETIS in other upcoming project activities (like Activity 5 – interoperability management) and in the consortium.

In July 2014, project coordination contacted the individual SPs to inform them about the planned REETS pilot activities and the role of the REETS project regarding monitoring and coordination of the testing and implementation activities. The second part of the communication is a request for information from the SPs. The main questions to the SPs were as follows:

- Are you interested in participating in a REETS pilot which shall be seamlessly transformed to normal operations after a certain period?

- If yes, in which of the REETS toll domains do you intend to operate first (which areas have priority for you)?
- If yes, how far have you got in the preparations? (i.e. when could you start a dialogue with the respective TCs?)
- According to your current plans, when could you be ready for first live tests?
- If not decided yet, when do you expect to make a decision?

Although due to the market approach, some of this information could have been seen as confidential by individual SPs, from the very beginning SPs declared their priorities which were summarised in a 'REETS Activity 8 Priority Matrix' (Figure 10.9). This enabled toll chargers and SPs to make a better planning of their activities and resources. SPs not declaring priorities for implementation were not considered further on in the activity (however with the possibility to step in any time if a decision for investment was made).

For toll chargers, the priority matrix gave the possibility to create regions within which SPs could introduce EETS – so-called 'certification clusters' – rather than across the entire EU as is required by the EC Decision – and allow the identification of indicative time slots for service introduction. Benefits are cost and time savings, transparency of the process and fair conditions in time to market.

10.10.5 Transparency of the process between the partners

The results of the contributions and meetings and the consequent statements were provided to all interested partners in the pilot in the form of unpublished half yearly progress reports to ensure continuing open exchange between partners during the project.

One essential tool, the Priority Matrix (Figure 10.9), was published on the REETS Information Platform. The matrix shows

(i) the countries where independent EETS/service providers are already in operation and (ii) where these providers see their priorities for the next service implementations (1st and 2nd priority).

The matrix is web based and dynamic; it is updated on a regular basis and the current version can be viewed at [15].

10.10.6 Status at the end of the REETS pilot

The REETS project concluded in December 2015 with the following stated achievements and recommendations (reproduced from the 4th Progress Report of Activity 8 31/12/2015):

- The REETS project had triggered a huge change process.
- (At the end of 2015) AGES and AXXÈS achieved registration as EETS providers; a further application was pending in France; others were also planned.
- The REETS community continued to grow after the end of the project. The Czech Republic, Belgium and Norway decided to join as observers, contacts with other toll domains are ongoing.
- The enlarged REETS community will continue moving towards EETS.

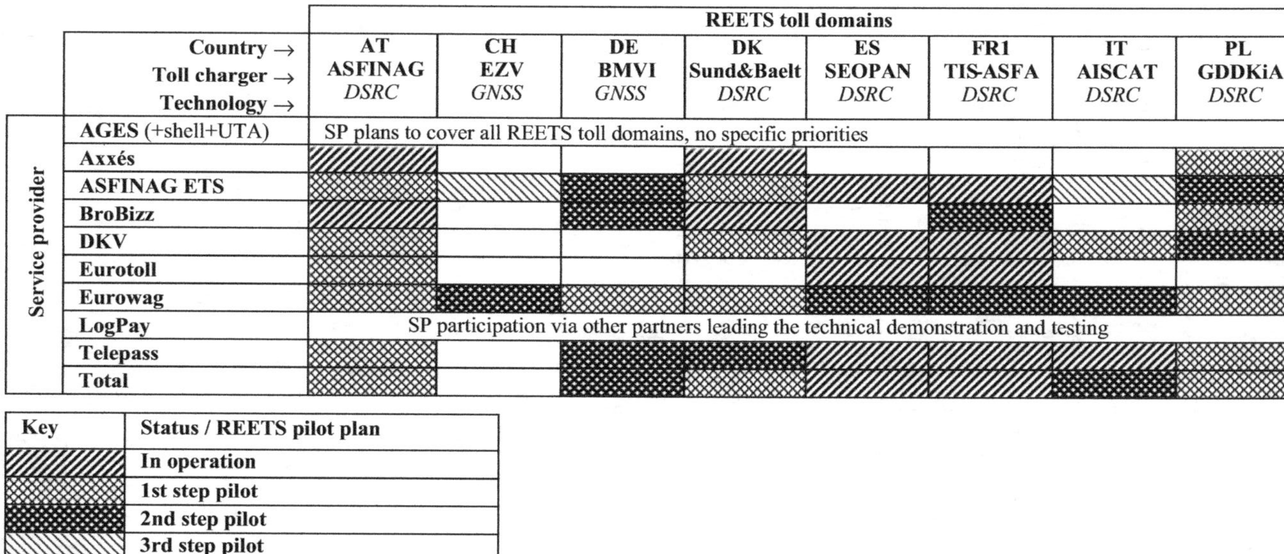

Figure 10.9 REETS pilot priority matrix

- Interim solutions towards EETS should be accepted and fostered rather than waiting for a perfect solution (e.g. more than just one EETS OBU will be needed initially).
- Evolution and amendment of the EU regulatory framework is needed to avoid market failure.
- Under given circumstances, REETS has achieved its objectives.
  - Analytical Activities 1–5 have been completed and provide a solid basis to build on.
  - EETS Information Platform is in operation [16].
  - Cross-border implementations started and will continue.

As a first step to the deployment of EETS, the information platform (Activity 7) has been set up and is running, and implementation activities have begun in the REETS member states. Therefore, the first step of the deployment of the EETS is complete.

10.10.7 Continuation under an EETS facilitation platform

Formally, the REETS TEN project finished at the end of 2015, though pilot activities which had already started, continued into 2016 and, step by step, are transforming into regular operations. Further preparations for pilot activities are expected to start, e.g. Switzerland and Czech Republic, and other toll chargers are interested and examining whether to join.

As the 'market approach' of the EETS will remain the way forward, the continued support and guidance for the EETS change process by the operation of the EETS Information Platform was secured beyond 2015. Stopping the activities would slow down the implementation of EETS, losing momentum and, therefore, these activities are continuing to provide a 'facilitation platform' for EETS which is described in the following sections.

10.11 EETS facilitation platform

The basis of the platform has been defined in a MoU concluded by around 40 parties (toll chargers, EETS/service providers, member states). Geographically, in addition to the REETS countries, Portugal, Belgium and Slovenia have joined. Since the end of the REETS project, and with the blessing of the EC, ASECAP have continued to work by extending and evolving the original REETS Information Platform and created the EETS facilitation platform. The EETS facilitation platform has therefore been established within the community of ETC interoperability practitioners in Europe, as the primary resource to provide information and guidance to any toll charger, SP or other interested party involved in the implementation of EETS.

Under the MoU, the structure illustrated in Figure 10.10 has been established. Tasks of the steering group and working group are as follows:

- Exchange of information regarding progress
- Facilitation of EETS pilot activities

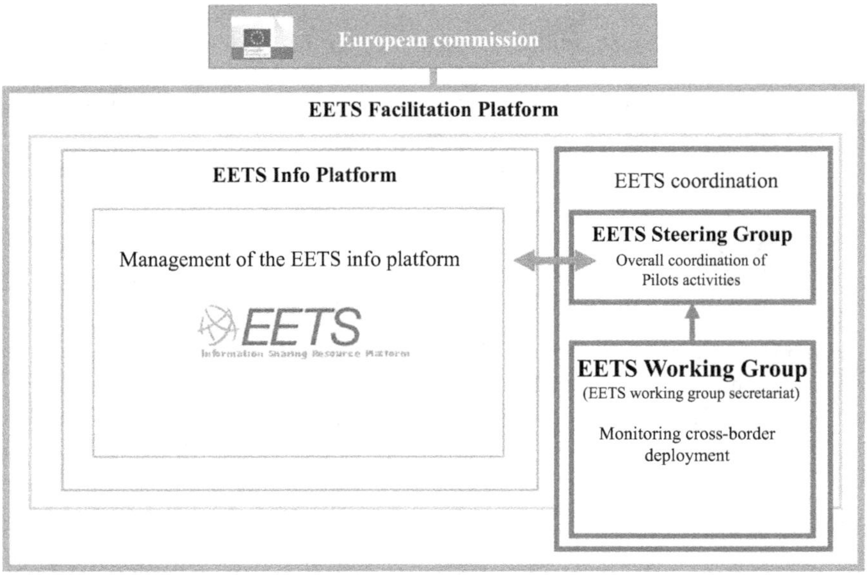

Figure 10.10 EETS facilitation platform

- Facilitation of the progress by coordination and communication
- Discussion of challenges/barriers for implementation
- Organise at least quarterly meetings of one of the groups
- Dissemination through the information platform
- Half yearly progress reports.

Objectives of the EETS steering group are as follows:

- Providing strategic guidance
- Facilitation of the EETS implementation and pilot operations
- Issuing of positions, statements and recommendations
- Continuation of communication and information exchange
- Platform for discussing the further evolution of the EETS.

Objectives of the EETS working group are as follows:

- Coordination of operational cross-border deployment activities, information exchange and
- Reporting to other stakeholders and to the EETS Steering Group.

This activity is coordinated by a secretariat held by a member of the consortium under the supervision of the EETS Steering Group.

The platform and its activities is open to

- any EU member state and other state where EETS according to EU legislation is applicable.

- any nationwide operating toll charger or national toll charger association with ETC systems in current or imminent operation in the above-mentioned states.
- any registered EETS provider or qualified SP having its seat in an EU state or an associated state.
- ASECAP and AETIS.

The platform will certainly continue to play a kind of 'catalyst' role in the transition to market and might develop to a future interoperability management (IM) platform which is certainly necessary even when markets are established. Future elements may be added over time.

10.12 Conclusions

The proliferation of existing and different solutions for ETC in Europe has delayed the introduction of interoperable services covering the whole of Europe. This is despite a directive issued by the European Parliament in 2004 requiring all ETC in Europe systems to be interoperable.

Demand for interoperability of ETC systems exists within Europe and is particularly driven by the need of heavy-vehicle-operating companies to reduce costs. However, until now, this demand has not been sufficiently high to justify the investment by SPs in launching EETS across Europe. Furthermore, the technical complexities of different toll system implementations and commercial uncertainties surrounding negotiations with toll chargers have increased the reluctance of the market to respond to the need for EETS.

The REETS project concluded at the end of 2015, and it has resulted in significant progress towards the introduction of interoperable services at least on a regional level. The European Commission now intends to lock-in this progress by reviewing the legislative framework surrounding EETS.

The project achieved its specific objectives:

- Activities in the analysis phase were successfully completed.
- The information platform was constructed and continues to operate.
- Some new cross-border services have commenced in the region covered by the REETS project partners.

The project has also triggered a huge change process within the market for ETC services in Europe and the situation in the market is now evolving.

- The SP model is operating across different member states.
- Three companies – Axxès, Telepass and Total – were actively launching EETS services at the end of the project with others likely to follow.
- Services are evolving and expanding their geographic coverage.
- The REETS project community has developed a legacy of a facilitation platform for EETS.

The community of toll chargers which formed the REETS project consortium has now been expanded through the addition of the Czech Republic, Portugal, Finland

and Belgium. SPs Via Verde and T-Systems have also joined the REETS community. Since the start of operation in April 2016, toll payers in the Belgian heavy vehicle charging system 'ViaPASS' are now able to use OBUs from several different SPs as well as the default national provider. This provides compatibility with EETS in the future and provides users with a choice of SP as use of an OBU in Belgium is now mandatory for heavy vehicles.

A key conclusion from the REETS project is that interoperable services should start using available technical solutions rather than waiting for the perfect solution. Specifically, this means that users should be offered interoperable services covering toll domains where existing onboard units work rather than waiting for onboard units that will function everywhere. This implies that more than one onboard unit might be necessary for fully Europe-wide EETS services for an interim period until the market matures.

Fair and effective competition is also essential to ensure affordability of services in the future. Regional providers will be of significant importance in driving down costs.

Another observation from the REETS project is the tension between commercial imperatives and the requirement for pan-European coverage. The conclusion was that the set-up costs of a pan-European service are too high for a totally market-driven approach, and there is no economic case for companies to introduce services in member states where demand is low.

The REETS partners also urged EU legislators to look for ideas and solutions that could be borrowed from other sectors, e.g. deregulation in airline industry or telecommunications.

10.13 Necessary changes in legislation

The REETS project made specific recommendations to open up regulatory requirements for EETS in order to encourage SPs to enter the market for EETS. In particular, it recommended relaxing the requirement for any SP registering as a provider of EETS to achieve approval in all toll domains in the EU within two years of registration. This provision of the decision currently in force is seen as far too onerous for SPs to achieve and introduces the risk that registration could be revoked if a service is not offered in all EU member states within two years of registration.

Evolution and amendment of the regulatory framework is necessary to allow further progress. Key issues are

- The current requirement for SPs to achieve full pan-European coverage of all toll domains within 24 months of registration
- Agreement on remuneration of SPs in all toll domains
- A single EETS OBU (DSRC+GNSS) as a criterion for registration
- Incomplete adoption of standards/common architecture models leading to different technical solutions especially across GNSS-based toll domains

- Does market demand exist for EETS for private cars and is there a need for European regulation?

10.14 Outstanding high-level issues

10.14.1 Continuing development of interoperability management

The European interoperability sky illustrated below (Figure 10.11) shows the different toll domains (just a few selected examples to show the paradigmatic approach) where implementation of EETS has to be done by co-operation between toll chargers and SPs. Some clouds are overlapping due to closer co-operation by the involved stakeholders. In the centre is the IM cloud integrating the two areas of IM according to the definition of Chapter 1.2 (Definition of Interoperability Management) of Deliverable D 5.1 (Report on Interoperability Management). All deliverables from the REETS project are available on the REETS website including recommendations on the topic of IM [17].

The work carried out on IM in the REETS project identified some high-level issues where action should be carried out. The EETS stakeholders considered that these issues still need to be addressed due to their specific importance and because in some cases there is a lack of supporting tools or conflicting interpretations of requirements are possible. These high-level issues are considered further below.

10.14.2 Consistency of registration procedures

The REETS partners identified a lack of consistency in the procedures for EETS provider registration and that further work should be done to harmonise these at European level. Once registered, an EETS provider will have access to the entire European market, and so registration processes should be applied to the same standard in every member state.

10.14.3 Back office interfaces

Another observation was that a lack of a common approach to the implementation of back office interfaces by toll chargers means that SPs have to ensure compatibility with multiple and different implementations. A common approach to implementation of options allowed within the relevant standards EN ISO 12855 and interoperability application profiles would be helpful and ensure that services can be delivered in a consistent way.

10.14.4 Notified bodies

The roles of different organisations and of standards within the certification process should be clearer. Due to uncertainty in the requirements, the certification process that notified bodies should use to assess compliance with standards for EETS is unclear. The REETS project recommended that the European Commission provides funding for additional work in the area of developing certification standards.

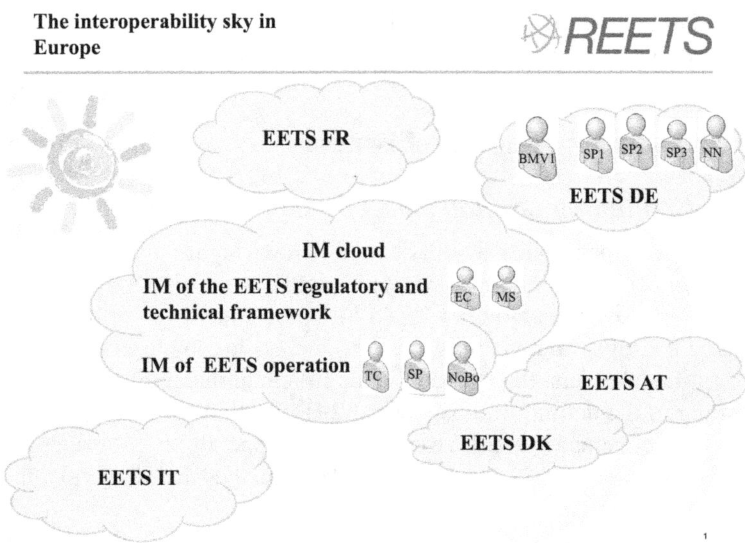

Figure 10.11 The interoperability 'sky' (IM is interoperability management)

10.14.5 Toll context data format

The format and content of toll context data should be harmonised. The REETS project recommended that future European legislation on EETS should provide a framework from which toll chargers can define toll context data for their toll domain on the basis of their political requirements. This would ensure that such requirements can be implemented within the available technical solutions and that enforcement can be carried out. For example, the framework should include the defined basis of the toll, use of a recognised vehicle classification scheme, definition of exemptions, vehicle parameters used in toll calculation, etc.

10.14.6 Cross-border enforcement

Common processes for cross-border enforcement should be defined; these are currently regulated only within bilateral agreements (the topic of cross-border enforcement is considered further below).

10.14.7 Conciliation procedures

Conciliation bodies should be aligned on a European level so that EETS providers can expect consistent decisions across different member states and toll domains. At the moment, it is not possible for EETS providers to consult conciliation bodies when establishing their services as these bodies are not established in every member state and where established are not yet active; this creates a degree of uncertainty and risk which has to be allowed for in business models. The REETS project identified the example of Ireland where a clear conciliation procedure exists in relation to disputes between EETS providers and toll chargers. Other member states were recommended to refer to this example and adopt appropriate aspects of its features. The topics of remuneration, certification process and financial risk management solutions should be included in the description of the conciliation procedures applicable to each toll scheme.

10.15 Other trends – light vehicle charging, smartphone apps, etc.

The European Directive on ETC interoperability requires the use of specific technologies. The European Commission Decision further clarified that EETS providers must implement both GNSS and DSRC interfaces in their onboard equipment. This was largely to ensure that legacy systems already installed by toll chargers would become part of any future interoperable European-wide service. The idea was that EETS providers would offer users an onboard unit that would function equally well in a toll domain which relies on DSRC-based roadside infrastructure such as toll plazas on Autoroutes, Autopistas and Autostradas, as well as providing GNSS-derived toll declarations in toll domains where this is required or allowed.

10.15.1 Smartphones

However, in the meantime, technology development has not stood still. At the time of publication of the directive in 2004, smartphones were not available in the market. The directive did foresee the use of mobile phone technology for charging and allows for 'mobile communications technology using the Groupe Spécial Mobile (GSM)-GPRS (General Packet Switched Radio) standard'. However, it makes no mention of 3G and other later technologies.

Currently, the market for smartphone apps for ETC in Europe is mainly limited to user convenience functions. There are many independently developed apps that provide users with information about the location of customer service points and toll stations.

Also, toll chargers and toll SPs in many EU countries offer smartphone applications which allow users to monitor their toll account balance and make top-up payments. For example, Kapsch provides such a smartphone app for the viaToll system in Poland [18].

However, there are also new charging concepts which are emerging based around the capabilities of smartphones.

One such concept is the Kapsch video enhancement app which is aimed at reducing the cost of manually verifying toll transactions that are based on vehicle numberplate recognition (Chapter 11) rather than an electronic transaction with an onboard unit. In some freeflow tolling systems in Europe, an electronic onboard unit is not a mandatory requirement. Users that do not have a tolling account linked to an electronic onboard unit are allowed to pay tolls based on information generated by automatic numberplate recognition cameras. For example, such systems exist in the United Kingdom, Norway and Portugal. The concept of the video enhancement app is that the user's location derived from the smartphone is reported to the back office in order to correlate this automatically with received video transactions and automate the process of verifying the video transaction. The need for manual verification of video transactions is therefore eliminated for users of the app. This reduces the operational cost for the operator and increases the reliability of toll invoices for the user.

In Hungary, a smartphone app is available to purchase route tickets [19]. It is debatable whether this actually qualifies as ETC since the route ticket is based on information manually entered in advance of the journey. The smartphone app doesn't declare the actual distance travelled using the location-based functions of the smartphone.

So for now at least, it looks as though the smartphone will not replace the onboard unit but will provide user convenience functions alongside the user's tolling account and onboard unit. See also Chapter 16, Section 16.6.6 where it is argued that GNSS-enabled smartphones are not in general a substitute for GNSS-based OBUs when it comes to mandatory fee collection.

10.15.2 RFID and light vehicles

Another technology not covered by the European Directive is the passive RFID device. However, its non-inclusion is not because it is an innovation but rather because it was not favoured by the European technical community as a solution for ETC. European technical standards for ETC require features which are not easily implemented using a passive RFID device. However, it has recently re-emerged as an alternative technology being considered by some toll chargers particularly for light vehicles. This is due to the much lower cost of the onboard unit compared to active alternatives and some of the more complex features thought to be necessary early on in the standardisation process e.g. completely anonymous accounts – have in practice not been implemented. The lack of a battery in the device also means that in-service issues are reduced compared to active devices. However, whilst individual toll chargers may consider the use of passive RFID devices, it is unlikely that they will be used in the context of EETS at least in the short term.

10.16 Summary, conclusions and further work

The REETS project has undoubtedly been a success through bringing together a core group of toll chargers and potential EETS providers to resolve some of the commercial and technical questions which were leading to uncertainty. The project reached a broad consensus between toll chargers on how to proceed on the main topics where potential EETS providers needed clarification in order to make service introduction more viable from a commercial perspective.

The success of the REETS project is evident from the fact that at the time of writing there are seven organisations that have registered as EETS providers. They are

- W.A.G. payment solutions a.s. (Czech Republic)
- BroBizz A/S (Denmark)

- Axxes SAS (France)
- Eurotoll SAS (France)
- Total Marketing Services SA (France)
- T-Systems EETS GmbH (Germany)
- Telepass S.p.a. (Italy).

These organisations are now actively developing services initially covering selected member states but which will in the fullness of time offer interoperability of toll payment services across the EU.

The European Commission recently carried out a consultation on the subject of the potential revision to Directive 2004/52/EC and Decision 2009/750/EC and at the time of writing is considering the results of this consultation.

In the meantime, several national ETC systems around Europe have been reconfigured to allow EETS operation and SPs continue to extend their interoperability offerings to additional member states.

References

[1] EC Directive. 2004. http://eur-lex.europa.eu/LexUriServ/LexUriServ.do?uri=OJ:L:2004:166:0124:0143:EN:PDF.

[2] EC Decision. 2009. http://eur-lex.europa.eu/legal-content/en/TXT/?uri=CELEX:32009D0750.

[3] European Commission/Aecom. "State of the EU Road Haulage Market Task B Report". 5 February 2014.

[4] http://www.asecap.com/staff-contacts/9-en/73-cesare.html.

[5] D1.2 Revised CESARE Model (October 2006). http://www.asecap.com/images/documents/D1.2-RevisedCesareModel-9October2006-Final.pdf.

[6] parkmagic. www.tolltag.ie [Accessed 08 August 2017].

[7] Easytrip. www.easytrip.ie [Accessed 08 August 2017].

[8] ViaVerde Figures, 2017 figures. www.viaverde.pt/servicos/mcdrive [Accessed 18 October 2017].

[9] AISCAT Figures, REETS meeting Rome. 29 April 2014. www.reets.eu.

[10] http://www.atec-itsfrance.net/userfiles/file/B_B01-TOULMINET_MALBRUNOT_SCHWAB%20_TEC227_GB.pdf.

[11] http://www.itsstandards.eu/index.php/efc [Accessed 08 August 2017].

[12] http://www.aetis-europe.eu.

[13] http://www.asecap.com/index.php?option=com_content&view=article&id=73:cesare&catid=9:en&Itemid=154&lang=en.

[14] http://www.reets.eu/index.php?option=com_content&view=article&id=12:information-platform&catid=8:activities&Itemid=101.

[15] http://www.eetsinfoplatform.eu/index.php?option=com_best_practice&view=bestpracticess&Itemid=145 [Accessed 08 August 2017].

[16] http://www.eetsinfoplatform.eu.

[17] http://www.reets.eu/index.php?option=com_phocadownload&view=category&id=8:activity-06-return-on-experience&Itemid=109#.

[18] http://www.viatoll.pl/en/mobileapp?utm_source=LV_EN&utm_medium=banner&utm_campaign=aplikacja_mobilna.
[19] https://www.hu-go.hu/articles/article/purchasing-a-route-ticket-with-a-smart-phone-1.

Further reading

[1] REETS Project Deliverables, 2015: http://www.reets.eu/index.php?option=com_phocadownload&view=categories&Itemid=109.
[2] European Commission Study on economic and social impact of the implementation of Directive 2004/52/CE on interoperability of electronic fee collection in Europe Final Report, 2006. (Ernst & Young/Ecorys Netherlands) http://ec.europa.eu/transport/themes/its/studies/doc/eets_socio_economic_impact.pdf.
[3] Road Charging Interoperability (RCI), 2008: field operational tests http://www.transport-research.info/project/road-charging-interoperability-pilot-project.
[4] Feasibility of a European Network of Certification Centres (ENCC), 2007: https://ec.europa.eu/transport/sites/transport/files/themes/its/studies/doc/european_network_of_certification_centres.pdf.

Chapter 11

Standardisation and implementation of ANPR – a practical guide

*Peter Vermaat**

11.1 Introduction to ANPR

11.1.1 Short history of ANPR

Virtually, in every jurisdiction in the world, vehicles which use public roads are required to carry a standardised identification plate, sometimes called a number plate, which uniquely identifies the vehicle to law enforcement officials and others who need to identify the vehicle. This identification plate is also used in some jurisdictions as a proof that the vehicle has been licenced, and hence is sometimes called a licence plate. The licence is sometimes the plate itself,[1] but more commonly, it comes in the form of a sticker on the plate. In this chapter, we will refer to the unique identification plate as a registration plate.

Historically, registration plates came in a wide variety of shapes and sizes, and made with a range of materials. In recent decades, there has been a tendency towards some standardisation in registration plates, particularly in Europe and the Western hemisphere, and now they most commonly fall into two size classes – see Figure 11.1:

- 'Euro' style, a plate about 520 mm wide and 120 mm high, with 6–9 characters typically about 80 by 50 mm and a stroke thickness of 10–12 mm. These registration plates are used throughout Europe and ex-colonies, and in many other countries.
- 'US' style, a smaller plate of about 300 mm wide and 150 mm high, with characters of about 55 by 30 mm and a stroke thickness of about 6 mm. These plates normally have an area above and below the characters which can be used for state/province identification, logos and even advertising. These plates are commonly used in North America and countries which fall under the US sphere of influence.

*Transport Research Laboratory (TRL Ltd.), University of Cape Town, South Africa

[1] For example, in the USA, most plates used are to be renewed every year as proof of vehicle licencing, with the expiry date stamped into the plate. The cost of issuing new plates every year means that this practice has largely stopped, with the licence replaced by stickers on the plate.

Figure 11.1 Euro (left, from the UK) and US (right) style plates

Registration plates are primarily designed to be read by humans, and their original design was based on requirements to make this as easy as possible. With the constant improvement in computer processing power however, it has become increasingly feasible to use image processing to read the registration plates.

Automatic Number Plate Recognition (ANPR) is the automatic reading of vehicle registration plates. Apart from ANPR (the common name in the United Kingdom), it is also called Number Plate Recognition, Automatic License Plate recognition, License Plate Recognition, as well as various other names and acronyms.

The first practical ANPR system was invented by the UK Home Office Police Scientific Development Branch and first trialled on the M1 motorway in 1984 [1]. In the early 1990s, several companies began marketing commercial ANPR systems, primarily to the security services for surveillance purposes.

Initially, the high cost of ANPR systems restricted their use to these well-funded security services, but inevitably the costs came down over time, and ANPR became feasible for a wider range of uses, including tolling, access control and eventually parking applications, where they are now common.

The use of ANPR in road pricing and tolling was foreseen in the 1990s as commercial companies sought to expand the market systems. An early use of ANPR in tolling was on the 407 ETR, Express Toll Route, in Ontario, Canada, which was opened in 1997. Since then, the use of ANPR in road pricing has expanded significantly.

11.1.2 How ANPR works – a short technical description

ANPR almost exclusively works by processing images from a video camera, and attempting to identify the registration plate and read the characters in the plate.

The processing chain will typically implement the following steps:

- Triggering
- Segmentation
- Classification
- Post-processing

These are described in a little more detail in the following sections.

11.1.2.1 Triggering

This process decides whether there is a potential registration plate in the image to process. Triggering may be a hardware process where an external vehicle sensor detects the presence of a vehicle and instructs the ANPR processor to find and read the plate, or software where a scanning algorithm processes each captured video frame for a plate-like object. In both cases, the trigger is used to inform subsequent processing that there is likely to be a plate in the image, and possibly (in the case of a software trigger) more or less where that plate is in the image.

11.1.2.2 Segmentation

This is the process of splitting up the image into features which can be recognised. In an ANPR system, the features to be recognised are the individual characters in the plate.

A range of algorithms may be employed to achieve the segmentation, including geometric normalisation where the plate image is converted to a horizontal image, character extent finding where the edges of each character are determined and size normalisation.

It is usual for the segmentation process to binarize the image by applying a threshold to each pixel to determine if it is foreground (a character stroke) or background. The binarization may be either explicit as a separate process applied to the whole plate image, or implicit where each pixel is evaluated as being part of the foreground or background as part of the character finding process.

11.1.2.3 Classification

On completion of the segmentation process, the registration plate should have been reduced to a set of (normalised) character images. These must now be converted to actual recognised characters, a process called classification. A range of algorithms are available, including template matching, neural networks, rule-based analysis and others, each with different advantages and disadvantages. The classification stage may also require a feature extraction process where each character image is analysed for specific features relevant to the classification process.

11.1.2.4 Post-processing

For each captured image, the ANPR process should have produced a string of characters, often with a confidence level for the classification result for each character. As a video camera may capture multiple images from a single passing vehicle, it is normal to get multiple results from the ANPR system for one vehicle. These multiple results must now be post-processed into a single result to be logged.

Where a system is externally triggered, the post-processing simply needs to use the multiple results and produce a single result of the highest possible confidence. For internally triggered system, the post-processing is trickier as it is not always clear when a vehicle has completed its passage through the field of view.

The post-processing must also resolve a range of ambiguities found in registration plates. For example, in the United Kingdom the same character is used for the letter ‘O’ and the number ‘0’, and it can only be resolved by analysing the spacing between groups of characters and applying the semantic rules[2] of UK plates.

It is often a requirement that the ANPR system must identify the nationality of the registration plate. While many vehicles carry a national identifier, these are not always visible in the image used for ANPR (see the next section on IR cameras for an explanation). In this case, the semantics of the plate format (which often varies between countries), or even analysis of the font used in the registration plate, can be used.

11.1.2.5 Logging of results

The post-processed result from the ANPR system needs to be logged to allow it to be used in some way. This can either be to a results database, a results file or files, or be to a remote internet location or cloud service.

Most ANPR systems will use internal algorithms to record some level of 'confidence' with each ANPR result. This can be based on the quality of the captured image, conformance of the plate to format rules and the self-measurement of the classification stage of the recognition process. The confidence figure may record a percentage likelihood that the result is correct (though this is extremely difficult to verify), or a simple confidence scheme where each result is given a score.

11.1.3 ANPR cameras

ANPR cameras have to operate within specific parameters to maximise the performance of the ANPR results.

As ANPR is normally deployed outdoors, they need to cope with the wide range of lighting conditions found there, from bright sunlight to streetlights, or even total darkness. This inevitably requires artificial lighting to supplement any natural light available.

As number plates are often moving, sometimes at speeds of 120 kph or greater, ANPR cameras need a high shutter speed, typically less than 5 ms in low speed urban environments, and 1 ms or faster at motorway speeds to prevent motion blur. This limits the range of adjustment available to cope with varying lighting.

How various camera types can be used is discussed below.

11.1.3.1 CCTV cameras

Early ANPR systems used standard PAL (Phase Alternating Line, the European standard for analogue colour television) or NTSC (National Television System Committee, the North American standard for analogue colour television) CCTV (Closed Circuit Television) cameras, sometimes supplemented by artificial lighting. A typical example of this is the ANPR system used as part of the first version of the 'ring of steel' (a security cordon) around the City of London, which used pole-mounted CCTV cameras on every entry road into the city.

[2] In most countries, registration plates are issued with specified formats, usually referred to as semantic rules. A common example may be that all plates must consist of three letters, followed by three numbers. Applying these rules can improve the accuracy of ANPR results, or can be used to help determine nationality.

While CCTV cameras can work well as part of an ANPR system, they do have significant limitations.

- They require supplementary artificial light, and this must normally be visible light, for 24 h operation
- Without artificial supplementary light, the images captured can suffer from motion blur, making them unusable
- High levels of artificial light cannot be used to read front registration plates due to the danger of blinding the driver
- They cannot be used at long-range (e.g. on motorways) in other than bright light conditions
- They can lead to cluttered images, which may compromise the performance of the ANPR algorithms.
- Registration plates in partial shadow can be almost impossible to read
- It is difficult to implement covert ANPR installations where powerful sources of artificial light are used to improve visibility.

Despite these limitations, in the correct conditions, typically slow moving traffic in well-lit areas, they can provide high quality images suitable for ANPR. They have a particular advantage that, because they work in visible light, they are less susceptible to the more esoteric evasion tactics which can be used against pulsed infrared (IR) cameras (see the next section). Furthermore, as pulsed IR cameras rely on the retroreflective properties of most modern registration plates, they cannot read non-retroreflective plates, which CCTV cameras can do.

11.1.3.2 Pulsed IR cameras

Early registration plates were mostly made of materials which tried to maximise visibility through using a high contrast between characters and the plate background. Many plates consisted of a black background with white or other light-coloured characters. In the 1970s, many countries started standardising on plates which used a white or other light-coloured retro-reflective background with black or dark characters. These plates had two distinct benefits: first, they were more visible at night in the headlamps of other vehicles making them more readable and second, a relatively large reflector on the front and rear of the vehicle improved vehicle visibility at night, improving road safety.

Retro-reflective material has the property that light falling on the surface is reflected directly back to the source of the light, rather than being dispersed as it would be on most surfaces. Good quality retro-reflective material can reflect 20–30 times more incident light power back in the direction of the light source than a dispersive surface.

The limitations of visible light CCTV cameras used for ANPR led to attempts to use the retro-reflectivity of plates to improve low-light readability. By co-locating the camera and light source, the power of artificial light required drops by a factor of 20 or more. Also by making use of IR light sources, the camera could be made more covert, and could also read front registration plates at night.

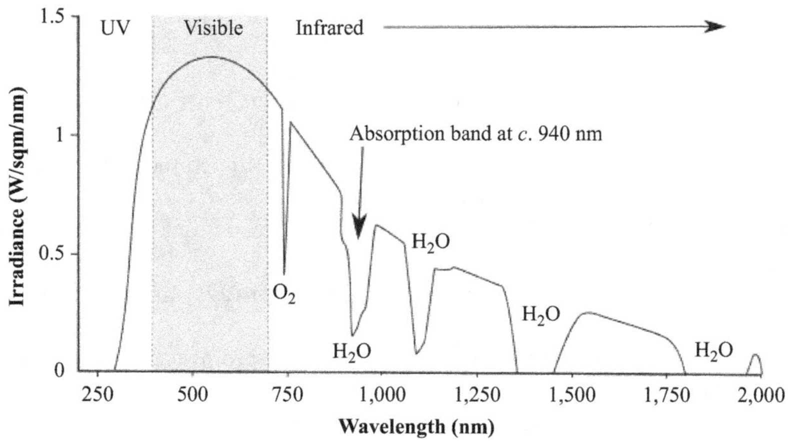

Figure 11.2 Solar spectrum

While using the retro-reflectivity of plates made possible lower power artificial lighting, cameras still need to cope with variable light levels varying from bright sunlight to night time with artificial light. This is difficult to achieve as the algorithms used to adjust camera sensitivity through shutter speed, aperture and electronic gain control tend to use a centre-weighted light-level detection. While this is appropriate for most normal video, and indeed is often fine for ANPR in high ambient light conditions, retro-reflective items in the field of view tend to be artificially bright, and can often end up over-exposed, particularly when the ambient light levels are low. The development of relatively high-power, low cost IR LEDs (Light Emitting Diodes) for the consumer remote control market led to them being adopted in commercially viable ANPR cameras for retro-reflective plates which overcome these exposure problems.

Figure 11.2 shows the spectrum of solar radiation, with the irradiance at the earth's surface being maximised in the visible light spectrum. It can be seen that at sea level there are several areas where the effect of atmospheric absorption significantly reduces the power levels from sunlight. There is a fairly deep null in the irradiance between about 900 and 950 nm due to the absorption by water vapour in the atmosphere. Rather conveniently, the IR LEDs used for remote controls mostly work at about 940 nm, with a half power spectrum of less than 50 nm.

By making use of an array of these IR LEDs, and inserting a matching narrow band (typically only 25–50 nm bandwidth) optical filter into the camera lens, it is possible to achieve a light level which is similar in intensity to, or indeed even higher than, sunlight at the wavelength of operation. The light level is now so consistent that the automatic exposure systems required by ambient light cameras are no longer required.

Figure 11.3 shows the difference between images from an ambient-light CCTV-style camera and from an IR ANPR camera. Apart from making for more covert, smaller and lighter cameras which work in all lighting conditions, the

Figure 11.3 Overview and infrared images

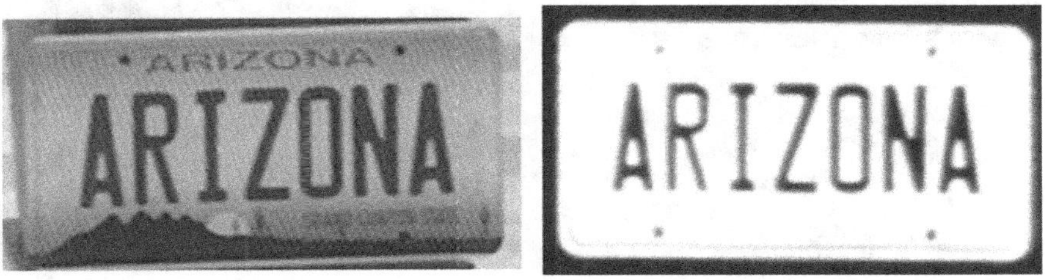

Figure 11.4 Removal of background using infrared

Figure 11.5 Effect of pulsing infrared illumination at different intensities

images produced by an IR ANPR camera are significantly easier to process by the ANPR software due to the much reduced image clutter.

As long as the entire background of the plate is made of a retro-reflective material, these cameras can also reduce clutter in the image. This is clearly shown (Figure 11.4) in the case of an Arizona plate where the entire background is retro-reflective, effectively removing it from the image.

While mostly beneficial, making use of retro-reflectivity does introduce its own problems. The retro-reflectivity of different plates can vary due to different materials being used, and aging and dirt on the plates, so cameras tend to vary the light-level to cope with this. This is normally achieved by pulsing the IR LEDs at (typically) three different powers. The effect of this is shown in Figure 11.5 which captured three successive images from a video of a plate from Virginia. The brightest image is severely over-exposed; however, a plate with a poorer

Figure 11.6 Artefacts introduced when using retro-reflective illumination

Figure 11.7 Arkansas (left) and Belgium (right) plates shown under normal (top) and infrared (bottom) illumination

retro-reflectivity (due to poorer material, dirt, aging, etc.) would be better exposed at this illumination level where the lower illumination levels may produce an under-exposed image.

Another issue that arises with these cameras is that they are more sensitive to damage and dirt on the plate as the retro-reflectivity of the background is easily compromised. A good example of where this can be an issue is with mounting bolts – in some countries (e.g. in the United Kingdom) the mounting bolts are sometimes placed within the character area of the plate, and the bolt head covered with a coloured plug to hide it. In normal light, the bolt head is all but invisible; however as the covers are not normally retro-reflective, they appear as dark in the plate image which can compromise the accuracy of the plate recognition software. This can be clearly seen in Figure 11.6, where the mounting bolts are placed between the J and the 3 on the left, and the W and P on the right, which makes the segmentation process far more difficult to achieve.

Another issue is that as the camera is working outside the visible light spectrum, features which are clearly visible under normal light may become less clear in IR. Two examples are shown in Figure 11.7, one from Arkansas on the left and one from an older Belgian plate on the right – in both cases paint is used to form the characters, and in both cases the paint used was fairly transparent in the IR spectrum. The upper images were captured using visible light, while the lower ones were captured under 940 nm IR illumination.

The Belgian plate image is all but unusable as an ANPR image.

It is also possible for those wishing to evade detection to make use of the properties of retro-reflectivity to compromise ANPR accuracy while not changing the visible appearance of the plate. This will be discussed in the section on ANPR shortcomings.

11.1.3.3 Digital cameras

Early ANPR systems, whether CCTV-based or pulsed IR, followed analogue PAL or NTSC standards, which restricted the size of the captured image. A digitised PAL frame has an effective resolution of at most 768 by 576 pixels; however, as a PAL frame (like NTSC) consists of two interlaced fields, the usable resolution is half this at some 384 by 288 pixels as only one field can be used. Making use of progressive scan (non-interlaced) cameras allows the use of the complete frame, but these were expensive and frame digitisers scarce, and the emergence of cheap, high resolution digital cameras means that progressive scan camera have never make a big impact.

The limited resolution of PAL and NTSC meant that the field of view of an early ANPR camera was quite small, less than a motorway lane width in most European countries and significantly less in the USA and other countries using small plates where the smaller features in the plates required a higher resolution to give a usable image.

Higher resolution digital cameras provide a proportionally greater field of view, allowing fewer cameras to be used, and many ANPR systems are now using digital cameras with horizontal resolutions of 1,000 pixels and greater.

For single lane applications, analogue cameras are still widely used, and have some advantages in terms of dynamic range and low light performance. However, digital ANPR cameras are common and can be expected to become increasingly dominant.

11.1.4 Effect of plate design on ANPR

Vehicle registration plates were primarily designed with human readers in mind, and hence ANPR has inherent shortcomings, limiting the ultimate performance of these systems. Some of the difficulties in reading registration plates have been pointed out in previous sections.

11.1.4.1 Plate formats and fonts

All countries specify particular formats of letters and numbers in their plates, though some countries also allow users to choose their own combinations of letters and numbers – so-called vanity plates – often at an extra charge.

The fonts used in registration plates are usually, though not always, specified by the registration authorities. The suitability of these fonts for machine reading varies between countries. Germany, for example, uses a font specifically designed to make it difficult to change one letter into another (e.g. by applying a piece

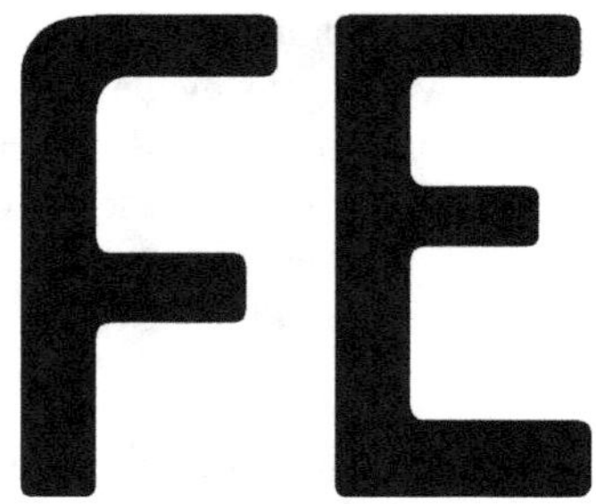

Figure 11.8 German plate font

Figure 11.9 UK plate ambiguity

of black tape to the plate). Look for example, at the F and E used in German plates (Figure 11.8).

Here, we can see that simply adding the lower stroke to an F to change it to an E would be obvious to the trained observer. The same characteristics which make it difficult to falsify German plates also make them easier to read by ANPR as the differences between characters are maximised.

The situation is different in other countries. In the United Kingdom for example, there was no legally binding font before 2001; characters simply had to conform to a particular size and shape. Since 2001, the standardised Charles Wright font has been used on UK plates. While an improvement, the difference between characters is not as marked as in Germany, making ANPR more problematic. A further issue in UK plates is that the same character is used for the letter 'O' and the number '0', as well as the letter 'I' (though this is only used in Northern Ireland) and the number '1', and the letters and numbers can only be distinguished by an understanding of the UK plate formats. This difficulty is illustrated by the plate in Figure 11.9.

This plate would, in the current format, be OO (letters OO denote an Oxfordshire registration) 01 (numbers 01 denoting date of first registration between March and September 2001) OOO (letters OOO being a serial).

Countries which only use specific formats of plate layouts do allow for a degree of error checking of ANPR readings as the result must fit a legal format. This is particularly the case where all plates are of the same length, for example the Netherlands, where all plates have six characters. A common problem where plates are not all of the same length is with partial reads. For example, UK plates between 1983 and 2001 (and still used on many vehicles) had a format A123BCD, where the first letter denoted the year of first registration. If an ANPR system were only to read the last six characters (123BCD) for some reason, the remaining characters still form a legal registration plate from before 1962.

This possibility of using plate format checking as a means of validating (to some extent) an ANPR result does not exist where users are able to choose their own plates, a common practice in North America.

11.1.4.2 Reading international plates

An ANPR system which only needs to read plates from a single country can make use of the limited plate formats required to be read, and the uniqueness of plates in a particular country, to identify the vehicle with a high confidence. If however, plates from multiple countries need to be read, this poses an additional burden on the ANPR system. First, the ANPR will need to be able to cope with far more plate formats, increasing the chance of an incorrect result. Furthermore, as some countries use identical formats (the six character plates using formats ABC123 and 123ABC are particularly common), the uniqueness of the results can no longer be guaranteed, unless the country of registration can also be determined. Reading the country code on the registration plate is far more difficult than reading the plate characters themselves. This problem also exists in the USA where, because plates are issued at the state level, it is quite normal for the same characters to appear in more than one state, making it important that the state of registration is also recognised.

11.1.5 Expected performance of ANPR systems

It is clear that the more accurate the results of an ANPR system are, the more useful it will be. However, determining this performance is not simple.

It would be simple to measure the performance of an ANPR system if it were only presented with clean, uniformly lit plate images of a predetermined size and geometry. Under these conditions, any ANPR system would be expected to read virtually all plates correctly.

Unfortunately, the reality of ANPR in real-world situations means that this ideal is seldom, if ever, achieved. Some of the issues compromising the performance of ANPR have been hinted at in the discussions above. We will examine the issues which can compromise ANPR performance in some more detail in this section.

11.1.5.1 Plate condition

Registration plates deteriorate over time, as a result of damage and ageing. The exact nature of the deterioration depends on the method of construction, but typically results in a reduction in the retro-reflectivity of the plate background, and/or a character becoming transparent as a result of the paint used wearing off.

Where characters are applied to the outside of the plate,[3] these can become damaged or fall off.

Probably, the most significant issue affecting the reading of plates is dirt. Even a thin layer of dirt on a plate can have a major effect on the reflectivity of the plate. This typically affects rear plates more than front plates due to the turbulence behind vehicles raising dirt from the road and depositing it on the plate surface. This is one of the reasons that ANPR systems are usually installed to read front plates, although in countries where front plates are not compulsory (e.g. many states in the USA), rear plates must be used.

11.1.5.2 Camera geometry

The ideal geometry for reading registration plates is for the camera to point directly at the plate. In most situations, this is not possible, so cameras are installed overhead or at the side of the road.

Overhead cameras result in a reduction in the perceived height of the plate, making it more difficult to read. Where plates are recessed, it is also possible that the plates could be partially obscured if the camera is mounted at too steep an angle. On the other hand, if the camera points too far down the road, it is possible for plates to be obscured if a vehicle follow a leading vehicle too closely. Figure 11.10 shows both situations.

For cameras mounted at the side of the road at more or less plate height as typically happens in car parks, a similar compromise between plate foreshortening and obscuration must be reached.

In cameras mounted at the side of the road on a pole mount, as typically happens in city installations, the combination of foreshortening in both downwards and sideways direction leads to an additional issue of plate distortion where the plate image is 'italicised', as shown in Figure 11.11.

Figure 11.10 ANPR camera geometry

Figure 11.11 Geometric distortion introduced by camera geometry

[3] Most plates are constructed by applying characters to a retro-reflective background, which is then covered with a protective film. Various other methods of construction are used, including gluing or riveting individual numbers to a reflective background.

When severe enough, this distortion can have a negative effect on ANPR performance.

11.1.5.3 Moving vehicles

If ANPR cameras use a low shutter speed, images can suffer from speed blur, leading to compromised performance. Modern pulsed IR cameras, which tend to use shutter speeds of 1/1,000 s or faster, do not suffer from this.

Fast moving vehicles also give the ANPR system fewer captured frames from which to recognise the plate as they move through the field of view of the camera very quickly. For example, at motorway speeds, a vehicle may pass through the field of view of an ANPR camera in less than 150 ms, resulting in as few as seven captured frames with the full plate in the field of view. If a three-level pulsed IR camera is used, it is possible that only two captured frames will have a good quality image of the plate. In slower traffic, far more frames will be captured, allowing multiple recognition attempts to be made, potentially resulting in higher performance.

11.1.5.4 Malicious manipulation

Those wishing to avoid detection by ANPR may resort to manipulation of the registration plate. Examples include:

- Plate cloning – where a registration plate on a vehicle is replaced with a different plate. (See also Chapter 13.) The most difficult to detect is where the plate is replaced with one from a vehicle of the same make, model and colour. This is particularly easy in jurisdictions where manufacture and distribution of plates is poorly controlled. Increasingly, countries are attempting to prevent this by strictly controlling the supply of plates. In the Netherlands for example, a single supplier is contracted to supply all plates. Where a user needs a replacement plate, this is visibly marked and the original plate is no longer valid.
- Changing of characters, where individual characters are manipulated. This can be very simple, for example, changing an E to an F by adding the bottom stroke with black tape. A more subtle manipulation could be where mounting bolts are strategically placed to make one character look like another – for example, placing a bolt between two 1's to make an H. See Figure 11.6.
- Re-spacing of characters on the plate. In some plate designs, characters can only be correctly recognised by taking into account the spacing between the characters. For example, in the United Kingdom a still legal pre-1962 plate could have the characters 10AB, depending on where large space is, this could be either read as 1-OAB or 10-AB.
- Manipulation of the retro-reflective background. Some drivers have realised that many ANPR systems depend on the retro-reflective background of most registration plates, and try ways of compromising this. Examples of this include covering the plate with see-though plastic films or spray-on coatings, or in more sophisticated ways like cutting away parts of the retro-reflective

background and replacing it with the same colour but non-reflective material. This is done in an attempt to fool the segmentation of the plate.

Most of these manipulation techniques are illegal in most jurisdictions, though enforcement varies from country to country.

11.2 The use of ANPR in road pricing

ANPR can be used as a primary road pricing technology where the ANPR system output is used to detect vehicles liable for a toll, or more commonly as a compliance and enforcement tool.

The use of ANPR in road pricing must take into account the fundamental limitations of ANPR performance, particularly when used as the primary charging mechanism. Therefore, it must be assumed that:

- A small percentage, typically 1%–10%, of ANPR results will be incorrect because of the limitations described in earlier sections.
- A small percentage of vehicles will produce no ANPR result, either as a result of failure of the recognition software, or because the vehicle has no readable plate.
- It should be assumed that at least some vehicles will have false/illegal plates, producing an incorrect result.

In all cases where ANPR is used, it is vital that a clear, human-readable image is stored along with the ANPR result for manual verification and evidence. Where IR cameras are used for ANPR, a colour camera is normally co-located with the IR camera for evidential images, and all ANPR results are stored together with a colour image of the vehicle enabling identification of the vehicle type and colour and, depending on the jurisdiction, showing an image of the driver's face as well. The storing of an image of the driver's face may come into conflict with privacy legislation in some jurisdictions; on the other hand, some jurisdictions may require it to prove who was actually driving the vehicle. Where both requirements are present, legislation may be required to successfully use ANPR for charging and enforcement.

11.2.1 Primary charging using ANPR

ANPR-based road pricing is in many ways similar to tag-and-beacon-based road pricing, in that it identifies the presence of a vehicle at a time and place which may incur a toll. This makes it suitable for area, either whole area or cordoned area, and segment-based pricing on limited road segments (bridges tunnels, etc.). Neither technology is particularly suitable for all-road tolling, or other large tolled networks.

Because of the limitations of ANPR described in the previous sections, ANPR-based charging systems must be carefully designed to mitigate the effects of these limitations. In tag-and-beacon systems, the reading of a tag is normally sufficient to incur a charge without any further verification as the probability of an incorrect read is extremely low – typically less than 1 in 10^{-6} or better. The system needs to be designed to cope with lost and stolen tags more than incorrect reading results. ANPR-based systems, on the other hand, need to include verification mechanisms to ensure that incorrect ANPR results do not lead to the creation of significant numbers of incorrect penalty fines.

Despite the limitations, a number of road pricing systems use ANPR as the primary means of detecting vehicles which incur a road pricing charge or toll – see for example, Chapters 13 and 14. Normally, the plate is registered with the charging authorities and linked to a user account, and the detection of the plate adds a charge to the user's account.

A process flow chart for a typical ANPR-based road pricing system is shown in Figure 11.12.

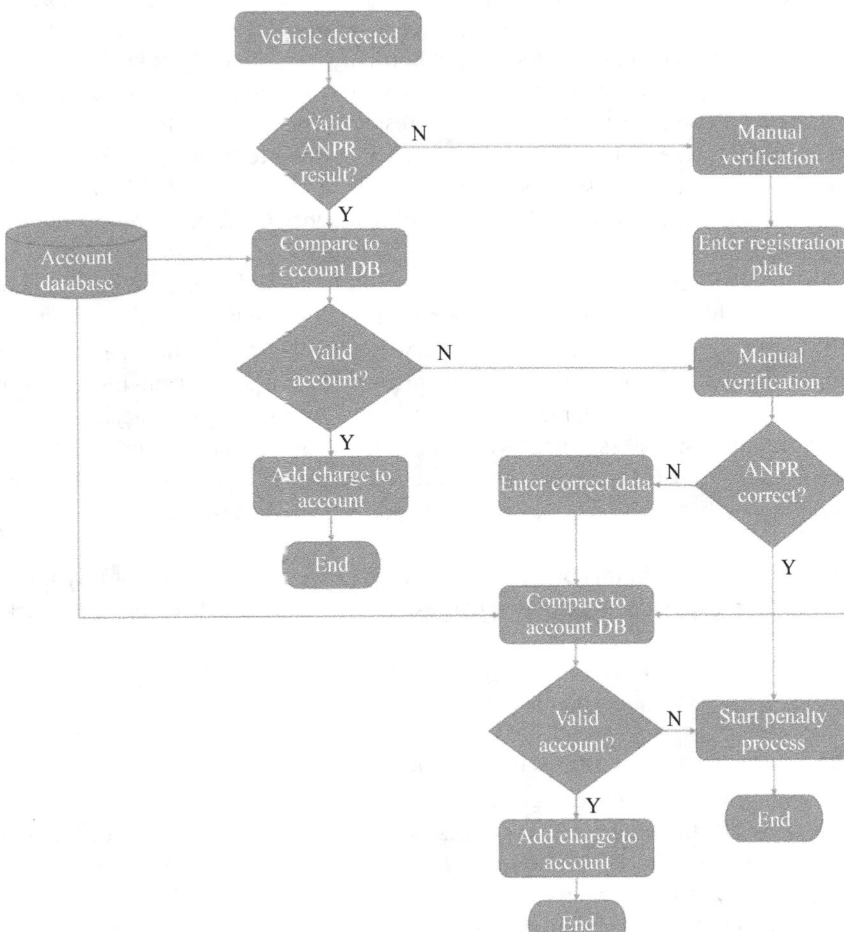

Figure 11.12 Primary charging process using ANPR

Normally, users will have an account where their registration plate is registered with the tolling operator and stored in an account database.

When a vehicle generates a charging event (e.g. driving through a cordon or being detected driving on a chargeable area or section of road), the ANPR system attempts to read the registration plate and generates an output. If the ANPR output is 'valid' by some criteria, for example, the result is of sufficiently high confidence and the plate is of the expected format, it is compared to the account database. If the plate is registered with the database, an account record is created and the account is charged for the event.

There is a finite probability that an incorrect ANPR result will be matched with a valid account, either through a 'cloned' plate being fitted to a vehicle, or because the ANPR has produced an incorrect result. Cloned plates can only be dealt with by allowing users access to an appeal system. The probability of an incorrect ANPR result being matched to a valid account through ANPR error is dependent on the accuracy of the ANPR system, and the percentage of valid plates which are registered with the tolling authority. As it is not normally economically feasible to check all accounts matched, users must have access to an easy-to-use appeals mechanism.

If the ANPR output is not 'valid', the result must be passed to a manual verification system where the ANPR result is checked before being sent for account checking. This is to process those occasions where either the ANPR system has produced a result which is considered to be of low confidence, or no valid plate has been recognised.

If an ANPR result does not produce a match to a valid user account, it is not normally sensible to directly produce a penalty charge notice based only on the ANPR result as the limitations of ANPR will result in a significant number of incorrect charges, leading to public dissatisfaction with the system and high costs in dealing with appeals. Hence, most systems will default to a manual checking phase for all account mismatches. The ANPR results from the mismatches are checked to ensure the ANPR result is correct. If not, the result is corrected and again compared against the database. If the ANPR result is correct, the penalty charge process is initiated.

Many variations on this process exist, but the core principles of not relying on ANPR results to produce penalty charges without manual checking are nearly always followed.

Two examples of where ANPR is used as a primary charging system are described below. See also Chapters 13 and 14.

11.2.1.1 Stockholm congestion charge

The Stockholm congestion charge was originally introduced as a 7-month trial in 2006 to evaluate the concept of a congestion charge as a means of reducing congestion in the inner city. A subsequent referendum saw a majority of voters support the charges. It was permanently instated in August 2007.

The Stockholm congestion charge is a cordon-based congestion charge operating on the main islands of central Stockholm. A time-differentiated charge is incurred by crossing the cordon at any of the charging points. Figure 11.13 shows

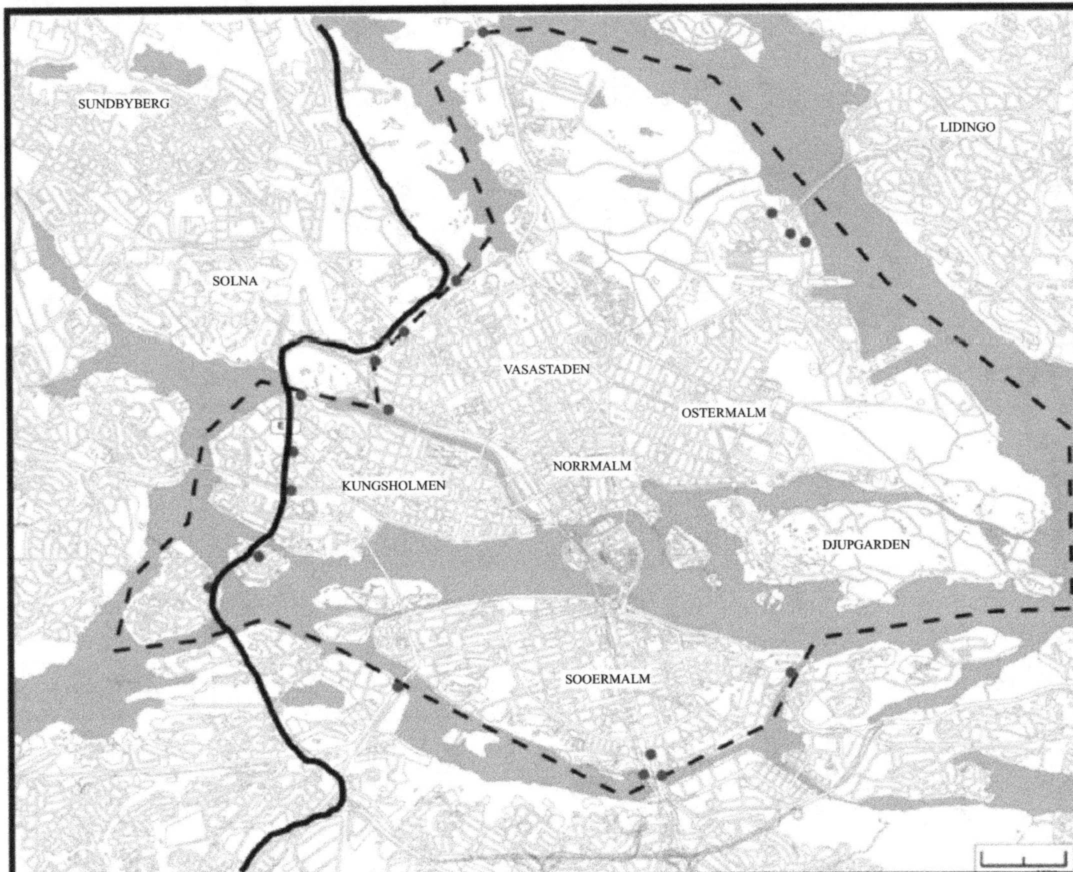

Figure 11.13 *Stockholm congestion charge. © 2009, Jonas Eliasson et al. Adapted, with permission, from original public domain colour image at https://commons.wikimedia.org/wiki/File:StockholmCordonMap.png*

the congestion charge zone defined by the dashed line, with the dots showing the 18 charging points. As the cordon defining the charge zone coincides with river crossings, implementation is simplified. Note, however, that the north–south Essinge bypass shown in black is not charged.

Identification of vehicles crossing the cordon is done exclusively using ANPR. The trial system used a combination of tag-and-beacon and ANPR technology, but results from the ANPR system were sufficiently good to drop the tag-and-beacon technology in the current installation. While exact figures for ANPR accuracy are not available, it is noted that in 2006, of the 115,100 daily tax decisions, only 100 were investigated by the Swedish Tax Agency and five were appealed, indicating that system accuracy is very high.

The ANPR identifies the registration plates of vehicles crossing the cordon, and a charge is calculated according to published charge rates. The vehicle owner is sent a monthly invoice of charges incurred.

The scheme has been very successful, with running costs significantly lower than originally anticipated. In 2006, future running costs were estimated at about SEK 220 million, though these have subsequently reduced, and in 2014, running costs for both the Stockholm and Gothenburg were estimated at SEK 250 million. The calls to the call centre, which is the biggest single item in operational costs, were 1/20 of that which was originally expected.

A cost–benefit analysis in 2009 estimated a net annual social surplus of SEK 650 million (about €65 million).

A detailed description of the Stockholm Congestion Charge can be found in [2]. See also Chapter 14.

11.2.1.2 Ontario Highway 407 Express Toll Route (ETR)

The Ontario Highway 407 ETR is a tolled motorway in the Canadian province of Ontario. It comprises a publicly owned segment as well as a privately leased segment which is leased to and operated by the 407 ETR Concession Company Limited and is officially known as the 407 Express Toll Route (407 ETR).

The road has no toll booths, being exclusively an electronically tolled route using a combination of tag-and-beacon and ANPR as the tolling mechanism. Users can choose whether they want to use a tag (also called a transponder), or ANPR.

Vehicles without a tag are recognised by ANPR cameras mounted on gantries over the road, though these incur a small surcharge, reflecting the additional costs of processing ANPR results. The system has a direct interface to various provincial and US state motor vehicle registries, allowing the toll operator to retrieve vehicle details and bill users.

As one of the first fully automatic free-flow tolling systems, this proved the viability of ANPR as a primary charging mechanism. Consequently, a number of other schemes have adopted ANPR as one of the primary tolling mechanisms (though often in combination with microwave tags) – for example, the Dartford (London) crossing and others.

11.2.1.3 ANPR used for compliance and enforcement

ANPR is most commonly used in road pricing as part of a compliance and enforcement system. The purpose of the ANPR in these systems is to detect those vehicles which are knowingly or accidentally avoiding paying a valid toll. How the ANPR helps depends on the type of charging system. We will consider three types, namely tag-and-beacon, Global Navigation Satellite System (GNSS)-based and self-declaration.

In tag-and-beacon systems, each vehicle is expected to carry a remotely readable 'tag' which can be interrogated by roadside or mobile equipment. The tags are normally microwave, and can be active (with a battery) or passive (powered from the radio waves received from the roadside beacon). Some older systems use IR tags.

Beacons are normally mounted on gantries over the road, or on roadside poles. ANPR cameras are normally co-located with the beacon. When a vehicle passes the beacon, the ANPR system reads the vehicle registration plate at the same time that the beacon interrogates the tag. If the vehicle has a valid tag (one for which an account exists, and it is not on a blacklist), the ANPR result is discarded, either immediately or at some later time. If the ANPR result identifies a registration plate belonging to a vehicle with a valid account, and the beacon has not read a tag, then it is possible that the tag is not working or not correctly mounted in the vehicle and the appropriate action should be taken. If the ANPR result identifies a registration plate belonging to a vehicle with a valid account, and the beacon has read a tag which is not matched to that plate, then it is possible that the user has stolen the tag, or has changed vehicles and not notified the authorities or some similar reason. Again, the appropriate action should be taken.

GNSS-based road pricing can be used for charging for the use of large networks of roads where gantries for tag-and-beacon systems will become prohibitively expensive; see for example, Chapters 16–18. GNSS-based charging can be very flexible, enabling road charges to be levied by distance driven, and varied by time, road type and vehicle type. These systems require vehicles to be equipped with an On-Board Unit (OBU) which uses GNSS to record all driven miles and calculate a charge based on the distance driven.

Most GNSS-based system use tag-and-beacon technology for compliance checking. Typically, a combination of fixed gantries and mobile compliance units will interrogate OBUs via the tag-and-beacon microwave link and match the detected vehicle with the records derived from the OBU. ANPR is used in these systems to detect vehicles which do not have an OBU.

Self-declaration systems are those where the user declares that he has or is about to incur a charge and volunteers to pay it. These types of systems are only used where a single event is charged, for example, an area-based congestion charge, or a tolled structure like a bridge or tunnel. In these cases, ANPR is used to detect all vehicles incurring the toll or charge. The ANPR result is compared to a database of declared toll liabilities, and those which have not declared their usage become liable for a penalty charge. The best known example of a self-declaration system is the central London congestion charge (Chapters 2 and 13), although this also has other payment options.

It is not unusual for self-declaration to be an option for occasional users of road pricing systems which primarily use other technologies. For example, the San Francisco Golden Gate Bridge has a tag-and-beacon-based bridge toll for regular users, but occasional users can elect to declare and pay for their usage online as an option to registering and paying for a tag. Similarly, in the GNSS-based German lorry road user charge system, occasional users who do not have an OBU can elect to declare and pay for their journey in advance.

11.2.1.4 Central London congestion charge

The London congestion charge zone (CCZ) is an area-based system where all vehicles which are within the charging zone at specified times are liable to pay a one-off charge. The charge is levied during working hours from Monday to Friday. There is no charge for evening and weekends. A map of the charging zone is shown in Figure 11.14 (the original in colour is available on the TfL website at http://content.tfl.gov.uk/congestion-charge-zone-map.gif).

The CCZ was introduced in 2003 with a standard £5 charge, though this has increased over the years and at the time of writing, the standard charge is £11.50 if paid on or before the day of travel. Late payment results in an increased charge and eventually a penalty charge of up to £195 after 28 days. Ultra-low emission vehicles (all-electric cars, some plug-in hybrids and any car or van that emits 75 g/km or less of CO_2 and meets the Euro 5 emission standards for air quality) are exempt from the charge. Residents living within or very close to the zone are eligible for a 90% discount if using Congestion Charge (CC) Autopay (see below).

The CCZ is primarily a self-declaration system, enforced by ANPR. If a vehicle is detected entering, leaving or being present in the charging zone, it is liable to a charge. A network of ANPR cameras is deployed at all entry points to the zone, as well as within the zone and on mobile enforcement vehicles.

A system of automated payment, called CC Autopay, is currently also in operation. This allows regular users to open an account with the operators and have payment automatically billed when their vehicle is detected. The introduction of CC Autopay has changed the system from a pure self-declaration system to a hybrid system of self-declaration and automatic vehicle detection based on ANPR.

As this is an area-based system, unlike the cordon-based system used in Stockholm, authorities potentially have multiple opportunities to detect vehicle within the zone. Indeed, it has been estimated that a vehicle using the CCZ on average passes six ANPR cameras during a typical journey. This means that the accuracy requirements of the ANPR system are not as stringent as with a cordon-based system where there is only one opportunity to detect the vehicle and read its registration plate.

11.2.1.5 German lorry road user charge system

Germany's LKW-Maut (Lastkraftwagen-Maut, literally 'Heavy Goods Vehicle toll') is a toll for goods vehicles based on the distance driven in kilometres. The toll was introduced in 2005 as a means of ensuring that the large numbers of foreign

Figure 11.14 London congestion charge zone. © 2016. Available at https://commons.wikimedia.org/wiki/File:London_congestion_charge_zone.png

lorries using Germany's roads contribute to the maintenance and expansion of the motorway road network.

Initially levied only on vehicles over 12 t, all vehicles over 7.5 t are now liable. The toll is only levied on federal roads.

LKW-Maut is a GNSS-based system where vehicles are charged for all kilometres driven, varied by the number of axles and emissions category of the vehicle. In this way, the vehicles which are the most damaging to the roads and/or the most polluting are required to pay a higher toll. Occasional users can pre-purchase cover for an individual journey at roadside toll stations, situated primarily at border crossings.

Apart from occasional users, all other users are required to have an OBU which uses Global Positioning System (GPS) to measure driven distance and GSM (Global System for Mobile communication, a digital mobile telephony system that is widely used in Europe and other parts of the world) communications to transmit the tolling information to the system operators. The emissions category of the vehicle is programmed into the OBU, and the driver can select the number of axles in use. Compliance checking and enforcement is by a combination of tag-and-beacon (using IR) communications and ANPR. Compliance checking equipment is mounted on over 300 enforcement gantries, as well as on a number of mobile enforcement units.

The ANPR is primarily used to detect vehicles without an OBU and to cross-check that occasional users have paid for the road they are using.

The LKW-Maut system was the first major GNSS-based road pricing system in the world and, despite numerous teething problems, has proved successful both in raising revenue and in reducing emissions.

11.3 Implementation of ANPR

11.3.1 General points

The implementation of an ANPR system is strongly dependent on the type of road pricing system into which it is being integrated.

The integration of ANPR into a road pricing system is largely an Information Technology (IT) problem, requiring well-defined interfaces between the ANPR and the rest of the road pricing system.

Where ANPR is the primary charging mechanism, performance and reliability are of paramount importance as any weaknesses in the ANPR system can have a direct bearing on both the performance of the road pricing infrastructure and the operating costs incurred. Matching of ANPR results to user accounts is crucial as mismatches can lead to incorrect charging. It is normal to concentrate on minimising the incorrect charging of users who have not incurred a charge, even when this may increase the probability of violators being missed.

Achieving the best possible performance from ANPR systems requires expertise on camera selection and installation. Assuming the camera and ANPR system are of a high standard, probably the biggest single contributor to ANPR performance is the camera installation. Here, an experienced installation professional will prove invaluable as quite subtle changes in camera angles, height, lighting, etc. can have a significant effect on performance.

Figure 11.15 *Entrance to London CCZ (Adapted under 'Creative Commons Attribution-ShareAlike 3.0 Unported' licence from original by user Mariordo on Wikimedia Commons)*

The following case study will illustrate some of the points, but it must be emphasised that each installation will pose unique problems to be solved.

11.3.2 Case study: London congestion charge zone

The CCZ has been described in Section 11.2.1.4; in this section, we will discuss the ANPR implementation issues.

The CCZ is an area-based charging system which means that a charge is incurred for being in the CCZ. This makes implementation somewhat easier than a cordon-based system as there may be multiple opportunities to detect the vehicle. For a cordon-based system, the vehicle must be detected crossing the cordon, providing only a single opportunity to detect it.

In the initial installation in 2003, a network of CCTV-based ANPR cameras were installed around the perimeter of the charging zone. Each entry road into the area was covered, though not always at the exact crossing point – the locations were chosen to maximise ANPR performance where possible. Figure 11.15 shows one entry point to the CCZ, where the chargeable area starts at the C painted on the road, but with the ANPR cameras arrowed in the background, well within the CCZ. In addition, further cameras were installed throughout the CCZ to detect vehicles which were missed at the entry point, or which were already in the zone. It has been estimated that on average each vehicle travelling within the CCZ is detected six times.

A crucial design decision made at this early stage was to locate the ANPR computers in a secure central location. Normally, it would be more efficient to locate the ANPR processors as close to the cameras as possible and transmit the ANPR results to a central database after processing using conventional data communication

Figure 11.16 CCZ updated ANPR cameras (public domain)

links; however, this could make the system vulnerable to legal challenge as it would need to be proved that the signal has not been tampered with. This risk minimisation was considered essential as this implementation was the first major ANPR-based congestion charge in the United Kingdom, and there was little legal precedent to guide the designers about potential challenges to the legality of the system. This decision required that the analogue video signals from each camera had to be connected back to the central ANPR processing centre on a dedicated fibre link.

Using a process flow similar to that described in Section 11.2.1 and Figure 11.12, the ANPR results were evaluated. As the majority of users were expected to prepay, ANPR results could be discarded early in the process for all prepaid users, leaving only those who had not yet paid to be subsequently processed.

The initial design decisions were vindicated as the system has operated without successful legal challenge from its inception.

In 2009, the system was updated with new ANPR cameras, shown in Figure 11.16. Extensive experience in the use of ANPR in various other schemes, and by the UK police, resulted in a decision to move to a more conventional ANPR solution where the processing is done at the roadside, and only results transmitted to the central back office, resulting in much lower operating costs.

11.4 Standards relevant to the implementation of ANPR

At the time of writing, no international standards covering the use and performance of ANPR systems exist, though national standards do. In this section, we will use the United Kingdom as an example of how standards can be used.

Standards cover two areas in ANPR, namely interface standards (to allow ANPR systems to interface to other systems) and performance standards which ensure that ANPR systems meet a minimum performance.

In the United Kingdom, standards for ANPR are created and maintained by police forces, as this is one of the main areas of ANPR usage. While not identical to road pricing, the principles are similar and can be used to illustrate how standards may be used in the specification of ANPR systems.

11.4.1 Interface standards

The National Police Chiefs' Council (NPCC) maintains the standards used by police forces in England and Wales. As policing in Scotland is devolved, they are not bound by the standards. General guidance to the use of ANPR in the police is given in [3], which also has links to the three most relevant interface standards, namely a data standard [4], an infrastructure standard [5] and a data access standard [6]. All the standards are freely available from the NPCC website, and the reader is invited to examine these for further details.

The data standard defines the data formats which must be used to interface to the UK's National ANPR Infrastructure. This standard defines the minimum data which must be included with each ANPR record and how the data (including metadata) should be encoded.

The Infrastructure Standard defines the ANPR infrastructure used by UK police forces and how they interface to each other. The standard defines different types of ANPR devices (static, mobile, CCTV-integrated systems, etc.), system capability and resilience, system connectivity, different databases used by the ANPR infrastructure, data retention and access control issues and performance evaluation. While many of the details are very police specific, the standard provides a useful insight into infrastructure issues affecting ANPR systems. It has been estimated that there are some 8,500 ANPR cameras in the United Kingdom, though this cannot be verified. According to a Freedom of Information request in 2014 [7], there were some 1,517 ANPR cameras controlled by Transport for London at that time; 646 of which were used for the enforcement and administration of the Congestion Charge, and the rest for enforcement of the Low Emission Zone and speed enforcement.

Finally, the Data Access Standard defines how ANPR records can be accessed and managed. This standard is of limited value outside police ANPR systems, though some issues regarding data privacy may be of interest.

11.4.2 Performance and deployment standards

The measurement of ANPR performance is notoriously difficult and error prone, affected as it is by camera type, installation, traffic conditions, weather, plate construction and lighting. Most ANPR provider will make claims about the performance of their systems, but these are normally heavily caveated by assumptions about the deployment and operating conditions. In the real world, it is not really possible to guarantee any level of performance under all conditions.

Nevertheless, it is important to understand the limitations and potential performance of any installed ANPR system. Police forces in the United Kingdom have many years of experience in the deployment and operation of ANPR systems, and have created a guidance document on the performance assessment and optimisation of ANPR systems [8]. This document gives practical guidance on the selection, installation and performance assessment of ANPR systems, as well as guidance to ongoing maintenance and checking.

11.5 Summary and conclusions

This section has shown that ANPR is a viable and valuable technology both as a primary means of charging, and as a tool to aid enforcement and compliance. It is also clear however that ANPR has, and will always have, some limitations in usability and performance, and these need to be taken into account when incorporating ANPR into a road charging system implementation.

While performance limitations may to some extent be mitigated by advancements in technology and processing algorithms, other limitations around the nature of registration plates need legal measures to mitigate, for example, to limit the possibility of plate cloning.

References

[1] Bridle, James. *How Britain Exported Next-Generation Surveillance. medium. com.* [Online] 18 Dec 2013. [Cited: 19 04 2016.] https://medium.com/matter/how-britain-exported-next-generation-surveillance-d15b5801b79e#.iharoktjy.

[2] Eliasson, Jonas. *The Stockholm congestion charges: an overview.* Stockholm: Centre for Transport Studies, KTH Royal Institute of Technology, CTS Working Paper 2014:7, 2014.

[3] National Police Chiefs' Council. *Automatic Number Plate Recognition.* [Online] 2016. [Cited: 08 09 2016.] http://www.npcc.police.uk/FreedomofInformation/ANPR.aspx.

[4] UK Home Office. *National ANPR Standards for Policing: Part 1 – Data Standards.* [Online] Jul 2016. [Cited: 08 Sep 2016.] http://www.npcc.police.uk/documents/20160722%20NASP%20Part%201%20Version%203.pdf.

[5] UK Home Office. *National ANPR Standards for Policing: Part 2 – Infrastructure Standards.* [Online] Jul 2016. [Cited: 08 Sep 2016.] http://www.npcc.police.uk/documents/20160722%20NASP%20Part%202%20Version%203.pdf.

[6] UK Home Office. *National ANPR Standards for Policing: Part 3 – Data Access and Management Standards.* [Online] Oct 2014. [Cited: 08 09 2016.] http://www.npcc.police.uk/Publication/ANPR/20141015%20NASP%20Part%203%20Approved%20version%202.pdf.

[7] Transport for London. TfL ANPR Cameras and Data. *whatdotheyknow.com.* [Online] 22 Dec 2014. [Cited: 8 Jan 2017.] https://www.whatdotheyknow.com/request/tfl_anpr_cameras_and_data.

[8] UK Home Office. *Guidance on ANPR Performance Assessment and Optimisation.* [Online] Mar 2014. [Cited: 08 Sep 2016.] https://www.gov.uk/government/publications/guidance-on-anpr-performance-assessment-and-optimisation.

Chapter 12

Engineering interoperability in the US: video tolling and multiprotocol tags and readers

*James J. (JJ) Eden**, **

I am hopeful that as you read this you'll get an idea of how and why the toll industry has grown in the United States. Perhaps, you will gain some insight into the future of surface transportation—and not just in the United States.

12.1 Introduction

Highways have always been an essential part of our history—from the horse and trails of yesteryear to the superhighways of today. During the 1930s, the Pennsylvania Turnpike was designed to improve automobile transportation across the mountains of Pennsylvania, using seven tunnels built for the abandoned South Pennsylvania Railroad in the 1880s. The road opened on October 1, 1940 between Irwin and Carlisle as the first long-distance, limited-access highway in the United States, leading to the construction of other limited-access toll roads and the Interstate Highway System. The Dwight D. Eisenhower National System of Interstate and Defense Highways is a network of controlled-access highways that forms a part of the National Highway System of the United States. The system is named for President Dwight D. Eisenhower who championed its formation. Construction was authorized by the Federal Aid Highway Act of 1956 (Figure 12.1).

We enjoy the mobility highways provide and rely heavily on them to carry goods and products. More than 90% of freight in the United States is carried by trucks—on local, state and federal highways. "We are nourishing a monster of great potential destructiveness.... the motor car... can spoil our civilization. But... the particular vehicle that stands outside the door (is) one of our most treasured possessions or dearest ambitions, an immense convenience, an expander of the dimensions of life, an instrument of emancipation, a symbol of the modern age. To refuse to accept its challenge would be defeatism.... we must meet it without confusion of purpose, without timidity over means, and above all without delay."[1]

*AECOM, US
**Alliance for Toll Interoperability, USA

[1] 1963 Sir David Crowther, foreword to UK Government's Buchanan Report, "Traffic in Towns."

Figure 12.1 Ground breaking on the Pa Turnpike. Images taken, with permission, reprinted from Reference [1]

However, we also have a love–hate relationship with highways. We love the speed, freedom and utility highways provide when they are operating smoothly, but we hate getting stuck in traffic when they are congested.

12.2 Interoperability when cash was king

As the only medium available for collection of fares at the time was cash, all toll authorities started with cash collections. Cash collection required the construction and maintenance of toll plazas, a central office to manage the operation and 24 h, seven-day-a-week staffing at the toll lanes.

With cash, everything is interoperable. Everyone carried cash in the "good old days." It didn't matter if you paid a dollar for a toll in New York or California, cash made everything interoperable. By today's standards, cash was very slow. On average, a toll collector is able to process 300 transactions an hour in mixed mode (cars and trucks). With cars only, you may be able to process 450 vehicles an hour.

During the lazy days of the 1950s and 1960s, I remember going for a Sunday drive with my father, stopping at a toll booth, talking to the toll collector about the weather and traffic and then going to Howard Johnson's for an ice cream before heading back home (Figures 12.2 and 12.3). It was a sign of the times, just a nice peaceful Sunday drive with the family on the Pennsylvania Turnpike. Toll collection was different as traffic and lane throughput wasn't a problem. The only "technology" agencies used were electro-mechanical devices for counting the traffic and tracking the revenue. Interoperability (IOP) is a problem that was brought on by the dramatic increase of traffic that mandated the advancement of technology to solve a congestion problem. In the early days of what is today considered modern toll collection, the majority of our revenue was collected by the interoperable medium of cash. However, cash was not a perfect solution. With cash came other issues such as the need to count not only paper money but the thousands of multidenomination coins that were collected daily. In addition, cash created other problems; first, you had to wait for customers to take out cash from their wallet and hand it to the toll collector. The toll collector then counted the cash, made change, issued a receipt and gave directions. All of this dramatically impacted the throughput

Figure 12.2 The 1950s toll booth

Figure 12.3 Pennsylvania Turnpike—the Tuscarora tunnel and the Blue Interchange. Images taken, with permission, reprinted from Reference [1]

of the now overcapacity toll plaza, adding to the cost of equipment and supplies as well as frustrating the customers. Initially, new toll lanes were built but as toll plazas reached their physical capacity, other solutions were needed.

As traffic increased, the business model of toll collection changed. Toll Agencies not only had to be concerned about collecting revenue and counting traffic, but also had to move traffic through collection points at a faster yet safe speed that was acceptable to the general public. Ironically, many toll plazas were landlocked by the commercial and economic expansion that their existence created. Building additional lanes to handle the growing traffic volumes was in most cases impossible. Industry experts had to come up with a solution to meet our new business reality.

Figure 12.4 Coin machine

The methods deployed varied from agency to agency but traffic-strapped agencies began to include modern coin machines (Figure 12.4) alongside the reliable toll collector in an attempt to improve traffic flow. The new automated coin or coin and ticket issuing machines promised to add traffic capacity while reducing staff cost. Now regular commuters would have change ready and be able to throw it into the coin basket as they drove through the lane. These coin machines represented the state of the art of automated tolls in the 1970s and early 1980s as their use expanded greatly. Coin machines didn't really improve processing speed as customers would pull up without exact change or they may have trouble hitting the basket with their fare. Coin machines also were not very popular with customers as they had to still roll down the window to pay their toll. These machines also required constant cleaning and maintenance. Toll operators were collecting, transporting and counting more and more coins daily. I remember touring the large coin counting facility on the Illinois Tollway where they counted 20 t of coins daily! Adding to the challenges of large volume, customers caused some of the reliability issues by throwing anything that was in a vehicle into the coin collection basket. Enterprising thieves were coming up with creative ways to cheat the systems or clog the coin machines. Coins trapped in the baskets could later be retrieved (stolen). Thus, began the deployment of CCTV cameras to watch over unattended plazas and lanes.

As tolls changed from a single coin to multiple ones, some agencies turned to other innovations such as tokens (Figure 12.5) and the paper version of tokens called script. This helped with throughput, though it added cost and complexity.

(a) (b)

Figure 12.5 Toll road authority metal tokens: (a) Harris County; (b) Garden State Parkway

First agencies had to "coin" the tokens and print script; then they needed to sell and distribute them. While solving the problem of making change and printing receipts in the lane, this added to the overall cost of collection and introduced another avenue for fraud. This marketing proved to be a precursor to the future of transponders.

Although cash remained king, most high volume toll agencies eventually used tokens and script. These new mediums were not interoperable as individual agencies each developed their own designs and programs. This not only led to tokens that were not interoperable but in some cases the same size or style token was used by two adjacent agencies. Interestingly, the new tokens could have different values in their respective agencies. So, a token issued in NY for $.25 might have been worth $.50 in NJ—offering the customer an unintended discount and the agency no way to collect the revenue. In fact, incompatible tokens were the first interoperability issue in US Tolling. While cash remained king, cash handling and security remained a great challenge: cash needed to be delivered to a bank where it was counted and deposited to the agency's account at a growing cost. Every time cash is handled, the potential of loss due to negligence or fraud was introduced.

Traffic continued to grow through the 1970s. By the 1980s, toll plaza congestion in some areas was approaching a crisis. Traffic was backed up during the morning and evening rush for hours. If there were any kind of roadway incident, a commute through a toll collection point could easily add 3 or 4 h to your journey. The news media were relentless in pushing for a solution. Agencies, particularly in the congested New York/New Jersey metropolitan area, could not build their way out of the traffic congestion and were scrambling for a solution. Bridges and toll plazas were landlocked and engineering solutions could not be found. To top it off, in 1983, there was a horrific toll plaza accident on the Connecticut Turnpike where traffic had historically backed up daily. These incidents had the politicians, public and the media pushing for an answer. One gubernatorial candidate promised to dismantle tolling in Connecticut if he won. In 1985, he made good on the promise and tolls were removed.

12.3 Electronic toll collection (ETC)—the early days

A new technology was evolving that many thought could be an answer. It was called Automatic Vehicle Identification or AVI. In 1989, the North Texas Tollway Authority in Dallas, Texas, was the first agency to go live with AVI in the United States and it was successful at moving traffic and collecting tolls. AVI was capable of almost double the speed of a manual lane, processing approximately 800 vehicles per hour. This technology was proving very effective with moving traffic. Another benefit soon recognized was that ETC greatly improved the safety around the toll plaza because vehicles were no longer starting and stopping to pay a toll. Around this time, customers were becoming more mobile as they were increasing their travel out of the region and through other states, crossing different toll agencies with different toll systems and business rules. This not only was confusing for the customer but created a delay in processing their payment in the lane and in some cases negatively impacted the revenue of a toll agency.

In the Northeast, it was becoming obvious that for toll agencies to survive in the future, they not only had to move traffic efficiently but also had to provide interoperability between the agencies that their customers traversed. In New York, it is not uncommon for a customer to drive through the lanes of multiple different toll agencies in the same day. Several states in the mid-Atlantic area decided to combine to investigate this new technology. So, at an IBTTA (International Bridge, Tunnel and Turnpike Association [2]) meeting in New York, a group of three states and seven toll agencies was assembled to tackle this problem. The group consisted of agencies from New York, New Jersey and Pennsylvania—which constituted two-thirds of the US toll industry. I was the representative from Pennsylvania.

12.4 The start: formation of E-ZPass

In 1990, the year in which the IBTTA also created an Electronic Toll and Traffic Management Task Force, these seven toll facilities joined to form an alliance that would later be known as the E-ZPass Interagency Group (IAG) [3]. These seven agencies represented almost 40% of all US toll transactions and nearly 70% of all the US $3 billion-year toll revenue. This effort included representatives from the New Jersey Highway Authority (operator of the Garden State Parkway), the New Jersey Turnpike Authority, the New York State Thruway Authority, the Pennsylvania Turnpike Commission, the Port Authority of New York and New Jersey, the South Jersey Transportation Authority (operator of the Atlantic City Expressway) and the MTA Bridges and Tunnels (Metropolitan Transportation Authority). The goal was to implement a regionally compatible, noninterfering electronic toll collection (ETC) system that would not only satisfy the different toll collection and traffic management needs of the seven participating agencies, but also provide regional mobility and convenience to their customers.

Over the next few years, a group of seven key individuals, one from each agency, met regularly with the agency executive directors to come up with a technological solution and a business rule concept that would work across multiple states and agencies.

The IAG wanted a governance model that would accommodate all agencies' participation regardless of size, yet allow the individual input needed to develop a multistate ETC system. An executive committee was formed that comprised of the executive directors from all seven agencies. They created a policy committee that would be responsible for developing the documentation, operating agreements and governance requirements in order to deploy and operate a multistate system.

The Policy committee started by addressing what many thought was the hardest part of the problem, the technology. Agency attorneys were always close as litigation was also a standard of the time. It had become so problematic that IBTTA had to conduct separate agency-only meetings to avoid conflicts, potential Intellectual Property violations and the constant turf issues within the supplier community.

The Policy and newly formed technical committees spent months looking at what Dallas had accomplished and what other potential technologies were on the market. Everyone realized that coming up with a solution that would work for all seven agencies with the potential of growing into a regional or even a national market would complicate the task. The realization came early on that there were limited common elements between the seven partnered agencies. Any solution would have to satisfy the individual needs of each agency with respect to technology functional requirements.

For instance, the classification systems that our seven agencies used were different. Pennsylvania used a nine-class weigh-in-motion schedule based on a ticket/closed loop system where the vehicle entered the toll road and was issued a ticket with the location, date, time, lane number and the class of vehicle which would be used to determine the toll rate depending on the distance the vehicle traveled. The New Jersey Turnpike and the New York Thruway also had ticket systems but with different types of axle-based vehicle payment classifications. The South Jersey Transportation Authority, Garden State Parkway and the MTA Bridges and Tunnels had Open/Barrier Axle-Based classification systems where the vehicles were required to stop at locations at varying distances along the toll road to pay incrementally, which ultimately accounted for the entire trip cost depending on distance and vehicle classification. Combining classifications to develop some commonality was almost an impossible task. Over the next few months, the working groups were able to come up with a concept that would allow tolling and classification across the region (Figure 12.6).

This concept is still in effect today. Realizing there would be a lot of large tasks and risks around any technology procurement the policy group set up functional committees with industry experts from within each agency. This group would manage all legal issues, public relations and business rules needed to procure a joint system. This structure allowed the group to have a streamlined, effective and functional structure that allowed us to rapidly evaluate and deploy

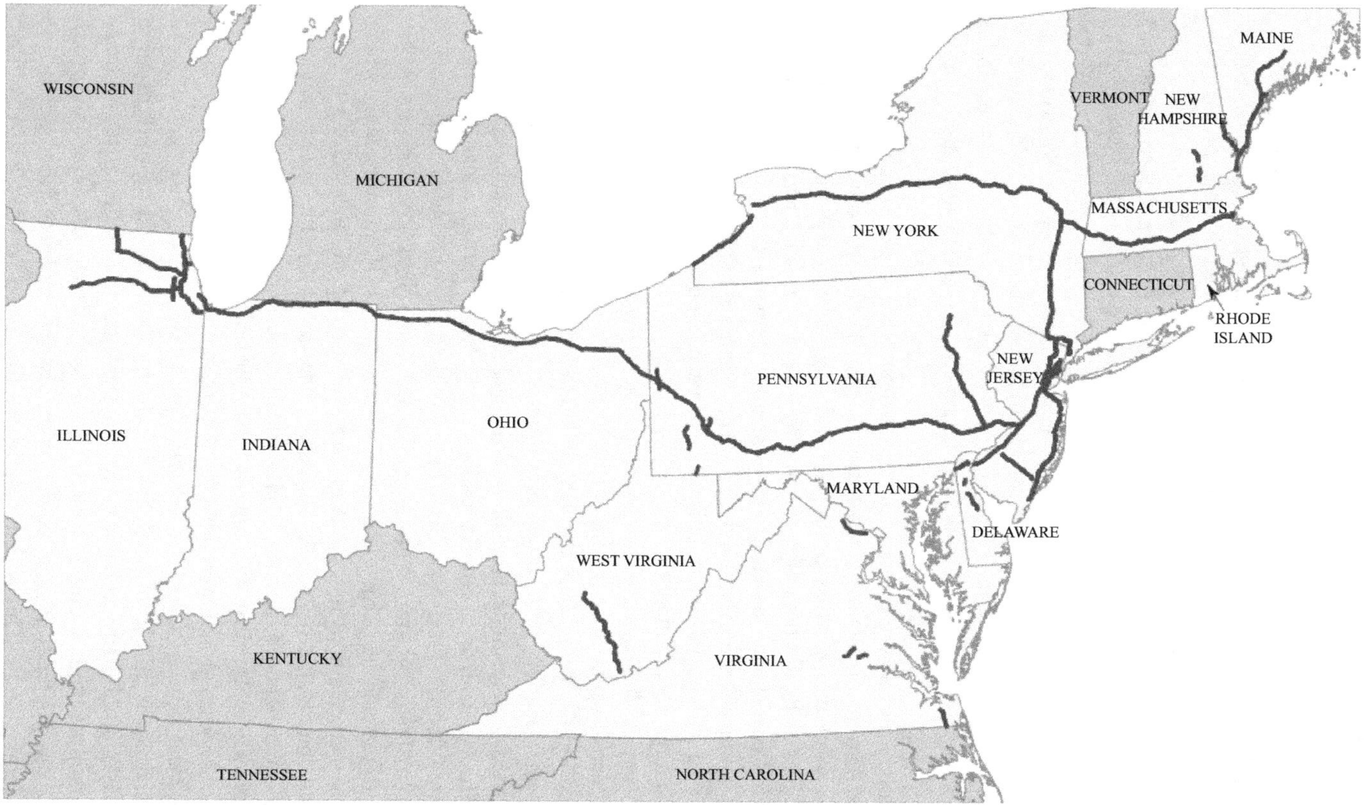

Figure 12.6 Map of US toll roads that in 2010 accept E-ZPass. Areas in light gray have at least one facility that accepts E-ZPass; those in dark gray do not. (Washington, DC does not have any toll roads.) North Carolina also accepts E-ZPass on its toll roads. Image and descriptive language taken, with permission, reprinted from Reference [4]

various solutions. This approach also allowed us to address multiple complex issues in deployment, procurement and rapidly changing technology and to utilize expertise from across the toll industry. Initially most our time was spent addressing technical issues and requirements; i.e. what would an acceptable read rate: 99.999% or 98.997%. Scenarios were run that assumed that everything under 100% was a complaint that had to be addressed and constituted lost revenue. Every agency individually assessed the financial impact to their agency.

Probably, the biggest most impactful discussion made was the selection of *read-write* technology over *read-only*. The Pennsylvania Turnpike, New Jersey Turnpike and New York Thruway all needed read-write technologies in order to mimic the existing ticket systems (data written to the transponders would be entrance location to the toll road, date and time, lane number and then it was read at the location/exit where the vehicle left the toll road so the correct toll due could be calculated and charged to the patron); the other agencies could utilize the less expensive and less complicated read-only technology. Unfortunately, a choice had to be made as the two technologies would not be compatible and allow for inter-operability. Ultimately, in order to accommodate most of the agencies, read-write technology was chosen.

One of the more humorous and lively discussions in the formative years focused on the colors of the transponders. Since few of the team had ultimate confidence in the technology, a color/class system for the transponders was developed. For instance, an orange transponder would indicate nonrevenue users whereas a white transponder would be a regular car transponder.

Branding was another discussion that took more time than allocated. It was challenging to come up with the name that would indicate ease and speed without indicating that you could speed through the lane. Several choices were vetted including "speed pass." Eventually, the brand E-ZPass was born. Other discussions with the marketing committee addressed growing public concerns over safety/security. Several news stations were reporting that a transponder and reader would be exposing customers to excessive microwave levels. And of course, there was always discussion over privacy and an agency's potential to monitor a customer's whereabouts.

These issues were addressed head-on as the IAG held several public forums and focus groups to discuss the issues and determine how best to publicize and launch this new product. Our public relations staff realized that it was difficult to explain the technology to the public. You have to remember this was pre-iPhone and the technology was pretty advanced.

During this time, due to public sensitivity over privacy, AVI was renamed Electronic toll collection or ETC.

Developing procurement across multiple agencies was more than challenging. Considerations had to be made for each agency's procurement regulations and laws while also considering interstate regulations and federal regulations on interstate commerce. At this point, specifications were being developed for both traffic lane equipment and a common back-office which put the IAG into the banking arena. Fortunately, there was a great procurement group that had experience in federal multiagency procurements. This talented group was able to look at all variables from all of the agencies and come up with a boiler plate that satisfied not only the operations groups but also procurement and legal staffs of seven toll agencies. The seven toll agencies actually were in three different Federal Court jurisdictions. So, any lawsuits that came about over this procurement would most likely be long, expensive and complicated.

The technical committee decided that extensive hardware testing would be necessary in order to ensure that this new advanced technology would actually work, especially in the volumes that were needed in the New York metropolitan area. The technical committee came up with a comprehensive test plan. This plan included all the possible scenarios that could be deployed to yield a reliable and effective ETC system. Hundreds of test transactions were run with multiple vehicle types. Tests were run at high speeds, low speeds, vehicles close together, and far apart, with fire hoses blasting at the windshield mounted transponders, hot temperatures, cold temperatures—anything anyone could think of. Some vendors thought that the test scenarios were unrealistic. In the end, the two vendors that offered systems for testing, Mark IV (now owned by Kapsch) and Amtech (now owned by TransCore), failed the first test. However, in a second testing, both companies passed. After analysis of the test results by both the technical and the policy committee, we found that the test results were too close to call to form the basis to award a winner.

At that point, the executive committee decided to look at other criteria such as value-added features. As an example, Mark IV had teamed with AT&T to test a smartcard solution in the launch of the Transportation Corridor Agencies' toll system in southern California. A similar pilot was taking place in Atlanta. The executive committee decided that the advanced features were not much of a benefit to tolling and moved on to price.

Evaluating total cost at this stage of deployment was not as easy as it initially seemed. No one had a firm idea of how many readers or transponders would be sold so some assumptions had to be made. The executive committee decided to assume tag readers would be installed in all available lanes. At that point, all member agencies were repolled and asked to estimate how many transponders they thought they would sell when fully deployed. Interestingly, the total number of transponders that all seven agencies anticipated selling was only 1 million. The selection was then based on the resulting pricing. The member agencies had one vote each. Given that we estimated one million transponders in the fully deployed system across seven agencies and a reader in every lane, we determined Mark IV to be the apparent winner as the first-generation toll transponder provider for E-ZPass.

Although the hardware procurement was difficult, the back-office team found that developing financial and business rules across multiple toll authorities, each with drastically different operating environments, business requirements and trust indentures, was going to be even harder. Looking across the seven agencies, they had different toll structures, several with ticket systems and some with barrier systems. There were different trust indentures (an agreement in the bond contract made between a bond issuer and a trustee that represents the bondholder's interests) that threatened to block any compromise coming together. Everyone had different procurement rules, different management structures and was managed by different political entities.

The subcommittees were tasked to start addressing these problems one at a time from legal to marketing to procurement. After much discussion, the team came up with business rules to address such things as tag sales and discounts so customers would not be confused about the program that was rolling out. The difference among trust indentures was the biggest hurdle. The challenge given to our legal teams was to find an answer to bring us together. After several months, they were able to come up with a solution in each of the jurisdictions that would make it work.

Deployment came in phases; the first to implement was the New York Thruway who actually deployed TransCore's technology as an interim solution in a couple of their toll plazas prior to the selection. The Thruway's initial deployment from cash collection brought several practical issues to light: including signage, safety and traffic management. The Thruway was sued by their union over safety during the deployment; however, the New York Thruway eventually won. The Thruway also experimented with drum signs, the first form of variable messaging signs. Although the messages proved effective, the mechanics proved difficult to maintain in the varied and sometimes harsh weather conditions. After several months, the project was scrapped.

After the E-ZPass Group selected Mark IV as the regional tag and reader, the New York Thruway proceeded to replace the TransCore equipment with the new IAG selected technology—a process that was now more complicated as it involved transponder recalls and reissue along with database conversions. The conversion went very smoothly, and to this day remains the largest single electronic conversion in history.

Another hurdle was whether the system would have one transponder in a vehicle backed by multiple accounts (validity would depend on an account with the toll agency being used) or have a way to exchange funds among agencies so the customer would only need one account. For simplicity's sake, the first option was taken as an interim process, while the "one tag, one account" Reciprocity System was developed and implemented.

As with most innovation, E-ZPass started out of necessity. Traffic delays at toll plazas were unacceptable and traditional solutions simply did not work. It is hard to believe looking back that the public not only accepted E-ZPass, but it *has actually become a recognized brand* for ETC. Effectively, it is the *Coke* of toll collection. Now E-ZPass has expanded to 26 agencies and has sold more than 45 million transponders. This would not have been possible without the leadership and dedication of the seven-agency's executive management.

The start of E-ZPass not only was a turning point in the development of modern toll collection but also marked the first regional interoperability solution. It is a tribute to interstate and interdepartmental coordination that this accomplishment was a world first that benefits millions of drivers daily.

12.4.1 The building blocks of interoperability

The basic requirements of interoperable toll operations were codified by the establishment and proliferation of E-ZPass. It is important to note that the following building blocks are more of a function of policy and business rules, and less so of technology:

- account identification,
- acceptance,
- transaction exchange and posting,
- reconciliation, and
- settlement.

The IAG created some business rules in the initial multistate interoperability effort. Probably, the most common are: the "home" agency is responsible for maintaining the toll account in question, and the "away" agency operates the toll facility hoping to collect its money from the home agency's accountholder.

12.4.1.1 Account recognition

Account recognition revolves around the technical ability to identify a vehicle passing a toll point (whether a traditional toll plaza or a toll zone on an open roadway). Subsequently, "account recognition" also came to include the use of license plates as a type of redundant system for vehicle identification. (See also Chapter 11.) Today, this area of technology has advanced radically and to the point that images of license plates are the primary source of identification of infrequent-use accounts or fleet accounts with constant changes in fleet composition.

12.4.1.2 Transaction and valid account acceptance

Transaction and valid account acceptance is the responsibility of the away agency (the toll facility used). Each home agency maintains a Tag Status File and a Customer License Plate File and distributes them to all other members. By presentation of these lists of transponders and license plates, the home agency guarantees payment to the away agency.

Away toll agencies in turn must determine if an identified transponder or license plate is "valid," i.e., is associated with an account that a home agency will recognize for payment.

The key to successful interagency account acceptance in the United States has not been in the details of the technology, but rather the early adoption of a business rule to require *pre-paid toll account balances.* In other words, American toll accounts are like debit cards—the money is already in the bank and use of the toll account just draws down on that money. The prepaid account balance is what allowed independent toll agencies in the 1990s to agree to honor each other's accounts—each home agency could promise to pay a toll because they already had the customer's money, and each away agency could allow a valid vehicle to pass because they knew they could count on the home agency to pay. This formed the basis for a Reciprocity Agreement signed by all members of the E-ZPass Group.

Indeed, this has proven to be a major hurdle in other countries where post-payment toll accounts are the accepted standard. Toll operators in these countries have struggled with interoperability and the willingness to guarantee payments to other toll operators.

American toll agencies today are considering various exceptions to the requirement for prepayment of tolls, so as to provide their customers new options which would increase toll program attractiveness.

12.4.1.3 Transaction exchange and posting

Transaction exchange and posting is then managed according to an IAG Interface File Specification that was originally developed almost 20 years ago, but is still in use today. Away agencies route toll transactions to their respective home agency. Home agencies then post the away transaction to the appropriate account to be handled in the same manner as home agency transactions.

When the IAG developed the Interface File Specification, the IT industry's ability to process and transmit millions of individual transactions was very limited.

Today, in new systems this function can be supported in near-real-time with independent exchange of the transactions themselves, which does eliminate some IT overhead functions to build and break down flat files. However, the main point is not whether one transmission protocol or another is the current state of the practice, but rather that there are many ways to agree to exchange valid toll transactions.

The other technology issue that has evolved with the growth of E-ZPass and other interoperable toll agencies is that of the hub-and-spoke routing of data, rather than the peer-to-peer approach as was developed originally. When the handful of original E-ZPass members agreed to exchange data, each agreed to send files overnight. Today, with 16 home agency Customer Service Centers this data transfer burden is becoming a challenge to accomplish daily. In a system based on exchange of individual transactions rather than large flat files, it would be simpler for away agencies to route all interoperable transactions through the hub and on to the correct home agency. As new interoperable groups or technology updates are considered, this hub-and-spoke approach is one of the key desired design features.

12.4.1.4 Reconciliation

Reconciliation refers to the process in which the home agency reviews and accepts the away agency transactions as valid for payment, makes corrections in cases where information is incorrect or transactions cannot post and acknowledges files to close the dialog.

12.4.1.5 Settlement

Settlement, finally, is the actual exchange of money from agency to agency.

Settlement frequency arrangements are made separately between interoperable parties and can vary depending on circumstances, volume of activity and business needs.

As capacity to move, store and process data all continue to increase, the technical details of these five steps will change, but the fundamental requirements of conducting the business of toll interoperability should remain intact.

12.5 A small greenfield project that changes everything: North Carolina

As traffic and congestion continued to grow, an added element that aided in the advancement of toll roads also began to emerge. Lack of funding for infrastructure projects was beginning to hit a crisis level. It was now becoming more difficult or impossible for Departments of Transportation to fund needed roadway projects. Bond financing of toll projects along with public–private partnerships begin to become more popular. These new roads had to be innovative and modern in order to stand the test of the public and the political entities that had to approve and use them.

Greenfield projects struggled to implement technologies and business rules that they could sell to their political overseers and customers while maintaining a solid financial bottom line. Litigation and lack of innovation within the toll industry threatened to impede the progress of modernization.

In 2004, there was a small 18 mi greenfield project in North Carolina that was destined to be constructed, the Triangle Expressway (Figure 12.7). This roadway would tie a key innovation center, Research Triangle Park, to a major population center. The North Carolina Department of Transportation [5] did not have the funding for the $1B project, despite several land donations from people and businesses along the right of way. Even if the legislature had been willing to set aside money every year, it would have taken an excess of 20 years before the project could be built with public funds. There was public and political support for the road but the people that would use the road wanted something that would save them time and have the technology that was going to take them into the future. A generational shift was also starting to take place in the area. The new millennial generation wasn't as concerned about privacy as previous generations were. This new generation of customers would give up potential privacy to gain time on their busy schedules.

An even bigger generational shift involved the use of cash. Fewer people were now carrying cash than even 3 years prior. The area around this road includes

Figure 12.7 The Triangle Expressway in Raleigh, North Carolina

several large universities; its residents include the largest per capita PhD population in the country. So, if any road was ready for an innovative approach, it was the Triangle Expressway in Raleigh, North Carolina.

There were also some challenges to building this project. The Triangle Expressway would be one of the first greenfield all-electronic toll collection (AET) systems in the country. The schedule was tight with a promise of 5 years from inception to completion. Adding to the complexity was the geography of the project; Raleigh sits between E-ZPass to the North and the SunPass toll system to the south and its customers were very mobile. The technologies utilized on E-ZPass and SunPass (Florida) were not interoperable but they were connected by a major northeast thoroughfare, Interstate 95.

Something innovative and out-of-the-box had to be done. But how would E-ZPass and SunPass become interoperable? What acceptable transponder and reader technologies could they deploy? How could they expect to have the two giants with operational systems accept any change or expense to accommodate a small startup?

The staff and consultants in North Carolina began to look at traditional, available tolling technologies. They knew that any implementation of AET would eliminate any cash collection in the lanes, greatly reduce the amount of right-of-way needed to construct toll collection points and reduce overall project costs.

To make this project work, North Carolina realized they would need a lot of help. Due to their location within the toll collection market they needed the cooperation of not only the states on their borders but all of the states up and down the East Coast. They decided to form an agency-based trade organization. Similar to the founding of E-ZPass, the conversation started at an IBTTA meeting in Boston, MA. Representatives from Florida, Texas and E-ZPass met to discuss the advantages of forming such a group. The initial meeting generated great interest as the participants realized that more interagency cooperation was needed as toll technologies advanced.

This was especially true in light of the growing public acceptance of AET systems. Three months later, there was a follow-up meeting in Dallas, TX, that included over 20 agencies. At that meeting, structure was added and it became formally known as the Alliance for Toll Interoperability or ATI. It was decided that it would be organized as a not-for-profit government entity and the membership would be limited to only North America toll agencies. ATI would be wholly owned by its agency members. The Board of Directors would be comprised of senior members of ATI's agency members.

The following mission statement and goals and objectives were established:

Mission Statement: ATI is a government organization dedicated to promoting and implementing interoperability between toll operators for the benefit of customers and agencies.

Goals and objectives:

1. Establish interstate customer video tolling interoperability (Figure 12.8).
2. Pursue Department of Motor Vehicles (DMV) plate exchange
3. Determine state legislation of ETC/DMV sharing information [6]

Figure 12.8 Video tolling equipment. Image courtesy of JAI

4. Investigate Radio-frequency Identification [7] toll interoperability
5. Consider future technology in the context of tolling interoperability—application of which would be within each agency's operating rules
6. Cost-effectiveness and efficiencies/controls

Several months later, over 45 agencies met in Orlando, FL, to discuss the next steps the Alliance for Toll Interoperability needed to take to become an effective group. It was decided that a strategic goal would be to develop and jointly write an RFP (request for proposal) to procure a North American HUB, a clearinghouse to exchange and settle toll transactions between HUB members' toll agencies. This HUB would be designed to initially identify customers' vehicles by their license plates (see also Chapter 11). If a customer from Pennsylvania drove through Florida, that vehicle's license plate number would be read in Florida, sent to the ATI HUB and matched to a license plate file provided by Pennsylvania. The ATI HUB would then send that transaction to Pennsylvania which would issue a payment to Florida through the ATI HUB. This initial system would be designed to provide customer interoperability despite having incompatible tag reader technology. At this point, the agencies also believed that anyone moving forward with AET would have video systems installed for enforcement reasons. Thus, using the ATI HUB would not necessitate any agency equipment purchases.

In 2011, after multiple agencies contributed to the development of the RFP for a North American HUB, a bid was issued. The bid required that the winning company develop and deploy for testing a transaction HUB that was capable of growing and supporting all of North America. Requirements also included that there would be no development costs or fees paid to the contractor and the HUB would be capable of connecting to any agency without the agency modifying any systems or business rules. There would be no guarantee that the HUB would ever be successful or that any agency would join. ATI received 11 bids from companies around the world and after much deliberation by a team of agency representatives, a partnership between EGIS (A French engineering and transport systems company) and SANEF ITS (the ITS division of the French motorway operator—now EMOVIS) was selected. After the HUB selection and intense negotiation between the winning vendors and ATI, a contract was signed. The business model for the contractor was to charge a transaction fee on each matched transaction. The projected volume of matches would provide the needed profit to sustain the model.

Even though ATI decided not to pursue interoperable technologies around the tag reader, the ATI Board highly encouraged agencies to go down that path. The ATI leadership felt that the technology was available to have multiprotocol tags and/or readers. Prior to this time, the vendors were telling toll agencies that multiple tags in a windshield would interfere with each other and degrade accuracy. Agencies were encouraged to test the validity of these statements by running their own tests with or without the cooperation of the vendor community. This would enable each agency to make its own judgments regarding if and how these systems would improve the operation of their agency. Any findings or results of any discussions or tests would be shared among all members. An ATI-hosted private bulletin board and file sharing system would be established for this purpose.

At that meeting, it was also decided that agencies were not giving enough direction to the vendors as to what they needed from technology in the future. The group set a plan in motion to move forward with having several equipment demonstrations that would not only allow vendors to show their wares but also allow agencies to share their input on what should be manufactured. The first demonstration to be held was a "video shoot-out" to be sponsored by ATI and held in North Carolina with the goal of producing better and more accurate toll video equipment. During the Florida ATI meeting, several discussions focused on the proprietary nature of the tag and reader. The vendors were invited in to give their side of the story which was at times conflicting. The explanations even conflicted with what was said from one agency to another. After these discussions, participants realized the importance of agencies talking one-on-one in order to assure that they were getting the best information, products and services going forward. Much discussion was held around the potential of producing a multiprotocol reader or tag that would allow at least E-ZPass and SunPass technologies to become interoperable. We were strongly warned by every vendor that there would most likely be litigation if any vendor or agency went down the path. Agencies openly discussed the necessity of interoperable systems particularly after all-electronic tolling was rolled out.

Discussions ensued around not only local and interstate enforcement but also, being able to identify users without a transponder in their vehicle. To do this, they would need a strong video system for enforcement and collection not only in North Carolina but from the toll operators in the states around them.

At this point in time, video systems were only around 80% accurate in Optical Character Recognition or OCR. This rate was unacceptable and would result not only in revenue loss but also in loss of confidence in the state's first toll road.

Vendors were also using nonstandard proprietary file types to store and collect images, making it difficult to switch vendors and integrate data with other non-compatible vendors. A proprietary license plate file was difficult to share with other agencies. When a dispute arose with a customer from another state over their vehicle's presence, a raw image would be required as proof and the customer's "home" agency may not be able to read the image file to verify the claim.

The team decided to take an innovative approach and advertised a series of what they termed a "video shoot-out." The shoot-outs were side-by-side demonstrations of existing video technology developed to capture and effectively read multistate license plates. Invited were all the traditional toll integrators along with some vendors that were providing video and OCR products for the military and border control. Although some of the military equipment was costlier it offered much better accuracy and reliability. These demonstration meetings attracted attendance from agencies around the world.

As more agencies were considering a switch to ETC, they needed a technology not only to help with enforcement of nonpaying customers but also to provide a bridge between incompatible transponders in neighboring states. During the shoot-outs, the camera vendors suggested that accuracy rates could be improved substantially if we worked more closely with the DMVs on the design of their license plates, in particular specialty plates. North Carolina had worked closely with their DMV in developing tolling processes, so it was natural to invite the DMVs to the shoot-out. This resulted in the realization that some of the problems associated with license plate recognition (LPR) were due to the plate's design.

The license plate readability issue is true not only in toll collection but also for law-enforcement. This series of specialized industry meetings helped to define what toll agencies needed while applying pressure on suppliers to deliver a higher quality but affordable product.

Within a year of the shoot-outs, LPR equipment being supplied to the toll industry was on a par with what had been seen in the shoot-outs from the military suppliers at prices the agencies could afford. Today, we're seeing accuracy rates above the 99% range. This development has allowed for the growth of video enforcement and video tolling, adding to our ability to be interoperable as an industry.

Video was a necessary component of ETC systems in particular as agencies went to AET. It was necessary for enforcement. If someone didn't have a transponder, how would that vehicle be identified and the toll collected? Video provided the necessary backup to transponder technologies. However, video, by itself didn't offer an answer to interoperability. Even with increased accuracy, video was not as accurate as a transponder. And there was no back up to video.

12.6 Multiprotocol tags and readers

Concurrent with the video shoot-outs, North Carolina conducted several tests of tag and reader technology. Multiple transponders of different protocols were placed on one windshield and read at varying speeds. Readers of different protocols were mounted side-by-side on the same gantry to see if the system would degrade. After thousands of vehicles, it was determined that no degradation took place so that any deployment of a multiprotocol tag reader would not be detrimental to the capabilities of the system to accurately collect tolls. At the same time, North Carolina's legal staff worked diligently with the state Attorney General to determine the validity of the intellectual property claims that were being made by various vendors. After many joint meetings with legal teams comprising agency and the Attorney General's office coupled with the state procurement office and the technical resources of both the North Carolina Turnpike and its consultants, it was decided that North Carolina would issue an RFP for a multiprotocol system. After this RFP was issued, there were many calls and letters that warned about litigation in violation of intellectual property rights. There were also many warnings about the degradation of the system due to the multiple frequencies being read. In addition, outside technical resources from some vendors warned that there wasn't enough time to read two protocols in a highway speed toll lane. The questions continued even into the prebid where one of the vendors asked, "What are you going to do about the lawsuit?"

The state's Deputy Attorney General answered that we have done extensive research on the validity of intellectual property claims for the items being procured and were confident that we would win any lawsuit brought to the courts of North Carolina. In the end, there were no protests or lawsuits filed. EPC Global is a standards industry to support Electronic Product Code identification using radio frequency. EPC Global in the supply chain industry utilizes ISO-18000-6-based standards (see Section 12.6.1) for radio frequency near-field communications between tags and reader which was used to provide reliable methods to effectively communicate with toll transponders and utilize toll transponder methodologies and products that leverage commercial applications that are strictly tailored to the toll industry. To further provide a programing standard that is interoperable in North America, a group of toll operators formed a 6c-working group that defined a toll programing standard based on the EPC global standard.

The procurement was successful with North Carolina awarding the contract for both multiprotocol tags and readers. This technology allows North Carolina to issue two transponders: one an inexpensive sticker tag that was ideal for customers that didn't intend to leave the local Raleigh area. The other transponder, though at higher cost to the consumer, is functional in both E-ZPass regions and in Florida. Now that North Carolina had the technology available so that its customers could travel easily between states on the East Coast with technology purchased in North Carolina, the challenge remained to negotiate agreements with the organizations and states in which these tags would be used.

ATI was instrumental in setting up negotiations with Florida and E-ZPass to negotiate testing of technologies to assure that these new multiprotocol systems were equal to or better than those already installed in neighboring states. Agreements were also needed for how and when transactions and money are exchanged. How were operational costs going to be covered and by whom? How would disputes be handled? Could agreements be reached that would satisfy not only Florida and all the members of E-ZPass but also other agencies in the corridor that would eventually be part of the interoperable group? Operational and financial impacts of issuing two different transponders at the same agency needed to be addressed. Inventory costs and the financial cost of maintaining that inventory had to be borne by North Carolina unless other agencies were also going to issue the multiprotocol tag. If other agencies were issuing, then agency codes would be needed on both transponders.

After several months of negotiation, agreements were reached and today the North Carolina multiprotocol transponder works in all E-ZPass states, Florida, Georgia, and South Carolina and has the capability of working in Texas and Oklahoma. However, the road to customer interoperability still needed the business rules and operating agreements between agencies to implement the technology. North Carolina also has an advanced video enforcement system that allows for the use of license plates for identification of vehicles without transponders and other state vehicles. This would not have been possible if not for the dedication of the leadership in Florida, E-ZPass and North Carolina.

At this point, E-ZPass provided interoperability from Maine to Virginia to Indiana. A successful North Carolina solution would eventually extend this interoperability from Maine down to Florida.

The need and value of interoperability was now a widespread belief. Pockets of interoperability existed, many concepts existed. But deployment was happening in baby steps.

IBTTA had heard from the leadership in the US Congress that there was interest in pursuing nationwide interoperability. IBTTA began formation of an interoperability committee at a meeting in Vienna (in 2007). The committee structure was formalized in 2009 and the initial open meeting was held in Boston in 2010. Based on that input, the initial steering committee, a group of 10 industry leaders, met at IBTTA HQ in Washington, DC. At this meeting, a strategic process began to identify the basic goals for the IOP committee and some objectives and structure.

12.6.1 A new tag and protocol—ISO-18000-6C

AET and interoperability was starting to take over toll collection in the United States. There wasn't a toll road being built or renovated that wasn't going to include ETC as its technology solution. ATI was still concentrating on the account while the multiprotocol tag and reader started to gain some ground. Most of us working on the lane solutions thought the technology selection was now in place with the two well-known protocols, Kapsch's E-ZPass and TransCore's SeGO® used in the Southern United States including SunPass in Florida.

Fast forward 15 years, and along came a new protocol, as outlined above. This new protocol was ISO-18000-6C, based on standards used in transportation and supply chain industries. This solution appeared to be as accurate and reliable as both the E-ZPass and the TransCore solutions, while being significantly less expensive. Further, the needs changed in the industry from building a transponder with advanced traffic information and other technologies that could be used besides tolling with earlier toll transponders to strictly a transponder that provides vehicle detection. The standard E-ZPass tag was still selling for around $26 where the TransCore sticker tag was selling for around $10. The newly introduced 6C sticker tag was being sold to agencies for around a dollar and continued to decrease in cost. This tag is also offered other form factors to be used with vehicles that have limitations due to electro-magnetic windshields.

Several agencies—Florida, Texas and California—began to test and eventually deploy the new 6C technology, further diversifying the already challenging technology choices that would have to be made in order to accommodate interoperability. For instance, with Georgia deploying 6C, the I-95 corridor now included three acceptable and deployed technologies instead of the two previous ones. Tag manufacturers started talking about a new complication in the tag battle. They were questioning the security of the new 6C tag. There was also new litigation around the intellectual property of all of the transponders. Further, these transponders were used by multiple industries and used open standards requiring the toll industry to further consider security standards to ensure the integrity of the toll transponder. A small group got together to discuss encryption and how to manage any new security that they would add to the tag. They decided that if they needed security, they could license it to the industry at no cost and still address this issue. 6C almost immediately advanced to a technology on par with that being utilized within the IAG and among TransCore customers.

IOP between tolling agencies was now a national issue. It was difficult to explain how the general public could use a cell phone by any manufacturer on anyone's network without an issue, yet an E-ZPass transponder issued in New York was not valid in Florida.

12.6.2 National interoperability

Around this time, the US Congress decided to get involved. ATI received a phone call and a follow-up letter from the chairman of the head of the US House Transportation Committee saying that it was going to mandate the establishment of a national ETC system and incorporate penalties for noninteroperable agencies in the new transportation bill. ATI immediately convened a meeting of its members along with key staff from IBTTA, the I-95 corridor coalition and ITS America to see how we could respond to the request from Congress.

At that point, Congress was looking at giving the agencies 1 year to comply with the request. The timeframe was not achievable even if a solution was known and in hand. It would take longer than that to adapt current equipment and business rules to meet the deadline. After several meetings in Congress, compromise was reached on the timing that would be allowed for agencies to become interoperable. Congress set the new date to be October 2016. This effort took a lot of coordination between all of the agencies, trade organizations and suppliers within the toll industry. However, that date has come and gone and the industry is waiting to see what Congress will do next to enforce the mandate. IBTTA and the agencies are continuing to work hard to provide a date that interoperability will be reached.

IBTTA formed a committee in response to the congressional mandate. This group decided to tackle the tasks of creating a standard for the tags and readers within the United States. They formed a steering committee to set direction and organize the effort. It was expanded to ensure coverage from all portions of the United States and Canada. Then, a technical subcommittee was created to develop a process to identify a preferred transponder/reader technology to function as a "national interoperability protocol" (NIOP) from among the seven different technologies in use in the United States. A back-office subcommittee was created to come up with a uniform set of business rules that would work across all agencies. These groups looked extensively at the current business rules and technical requirements of all agencies and identified where the overlaps and conflicts were. This was a monumental task as agencies had different baseline systems. For instance, as mentioned earlier, the variety of vehicle classification. Some agencies relied on weigh-in-motion, some counted axles while others relied on vehicle profiles. IBTTA information can be found in Ref. [8].

The original conflict that plagued the founders of E-ZPass once again became a problem to the planners. The issue was the technological differences between the agencies that needed a read-write transponder versus agencies that could use a read-only transponder. It was important that any national interoperable transponder be able to satisfy the needs of the open system agencies that use read-only technology and the closed ticket system agencies that needed read/write technology. A further complication was the need to develop a trip-based transaction from multiple open system locations to address High Occupancy Toll lanes. The business rules around the distribution of transponders were different. The penalties involved in not paying a toll varied widely, even the timeline in ways that agencies procured services and systems were different. It was determined that whoever provided the transponder and reader technology would have to agree to publishing their protocol so that multiple manufacturers could produce it, thus avoiding the creation of a monopoly.

The IBTTA technical committee developed a requirements-based document that identified an acceptable range of performance (a minimum standard) for the operation of the roadside technology read/write, which quickly eliminated it as an issue. It took one subcommittee meeting in Dallas. The real sticking point and what took so long was the institutional issue of what constituted an "open" protocol. Because many of the seven technologies in use were proprietary, there was significant push-back from the private sector. That's when the steering committee, almost unanimously, decided on using the FRAND approach (Fair, Reasonable and Non-Discriminatory). That decision allowed for the continuation of testing with equipment from Kapsch, TransCore and 3M.

After several months of hard work, the task force investigating customer service center interoperability issues had the skeleton of a national standard. As agencies looked at adapting, the standards became more refined.

The Committee looking at tags and readers came up with a test scenario that would look at the technical ability of each manufacturer's transponder and reader

national interoperable standard. This group also looked at the potential financial and operational impacts any one of these systems would cause to each specific agency.

The test plan involved independent verification of technical specifications and interoperability. An RFP was released. After evaluation, a vendor was selected to do the testing. This was going to cost the agencies $2.5 million. A fundraising effort was initiated and several agencies pledged to support the effort. As plans for testing were progressing and funding was being lined up, a congressional committee convened a hearing on the status of interoperability. A small group of invited agency professionals testified before the congressional committee and said that testing would be completed and protocols chosen by October 2016. During this testimony, it was also noted that ATI was technically ready to process transactions and had several agencies signed up.

During the hearing, status of the testing was detailed to the congressional committee. Also, the lack of funding for testing was discussed. A congressman requested that the US Department of Transportation try to find $2.5 million to fund this testing as a grant. Over the next 9 months IBTTA working with an independent consultant applied for and eventually was awarded the grant of $1.5 million. This money was combined with agency contributions and the technical committee of IBTTA proceeded to test three technologies that included the Kapsch E-ZPass, 6C and TransCore's SeGO. It is expected that, when testing is completed sometime in mid-2017, all three protocols will pass. At that time, it is expected that the committee will evaluate the potential cost and operational impacts of conversion to each as a primary national protocol. Agencies will still be able to use any other protocols as secondary or local transponders. The national protocol will permit national interoperability while the local tag can be used as a local or regional option. This would permit an agency to utilize a less expensive or less featured tag while maintaining national interoperability.

12.6.3 Governance of interoperability

IBTTA has also put extensive work into the governance of interoperability. Once technologies are chosen and business rules are in place, there will still need to be a governing body that administers interoperability. This governing body will also be responsible for dispute resolution and adaptation to future technologies or business rule changes. For now, IBTTA has determined it will be in the best interest of all agencies to keep this governance within the IBTTA board. When there are enough members actively participating in the new interoperable system the new governing model will be activated.

In recent years, we have seen strides made on the agency and on the private side. Kapsch made the E-ZPass technology open-access, adding to the open protocol 6C, both now being in the public domain. License plate payments are now being accepted by US Companies ATI, American Traffic Solutions and Highway Toll Administration. AET continues to expand with "brownfield" conversions at the Golden Gate Bridge; the MTAB&T (NY), Henry Hudson (NY), the

Massachusetts Turnpike and e-470 (Colorado). New greenfield projects also opened with State Highway 130 in the Triangle Parkway. The other big changes happening are the absorption of independent agencies into the Departments of Transportation. North Carolina Turnpike and Massachusetts Turnpike have recently been added to their states' Department of Transportation's responsibility and several other turnpikes are rumored to be about to follow.

12.7 The future of tolling in North America

There is no doubt that sometime in the future toll agencies will be fully interoperable. Because this transition has taken so long to implement, there are now other technologies and businesses that can not only take advantage of tolling's interoperability but provide an opportunity in tolling to expand their business model. There are companies on the transponder side that are now making products that can be read by every protocol. The same companies offer full back-office services so they in effect become the collection arm of any agency. There are also several companies that will provide a cell phone product that will utilize the agency's ability to capture a vehicle's license plate and turn that into a collection mechanism that they will manage. Other cell phone innovations take this a step further and use the GPS embedded in cell phones to Geo map toll sites thus allowing full toll collection without even the need of a gantry.[2] These businesses also could offer full-service collection for any toll agency.

An interesting twist in this model relates to when AET eliminated the need for toll plazas. GPS cell phone toll collection could now eliminate the need for even a toll gantry. This could open up a whole new business sector for toll collection where a Department of Transportation (DOT) or a private company could establish a toll road with little investment other than contracting with one of these businesses. These models have not yet been proven or even tested in the United States so time will tell which way they head. But see for example Chapters 6, 10 and 16–18 for relevant developments in Singapore, Europe and the Western US.

Currently, the United States has several focused startup companies who are concentrating on the commercial business and license plate based customers who infrequently use the road. PrePass and Best Pass are just examples of two companies who provide fleet efficiency solutions to commercial companies for management of their expenses including tolls nationwide. Other solutions such as PayTollo and BancPass provide cash-based solutions to infrequent customers who chose not to get an ETC account. These solutions will continue to provide niche solutions that can be adopted as more toll operators begin their transition to all-electronic tolling and move cash collections from the roadside environment to the mobile market and integrate toll payments into existing retail environments.

In addition to tolling, there is an entire market sector waiting for an interoperable payment system. ETC could very well in the future be an in-car payment

[2] But see also Chapters 6 and 16 for the problems of using of cellphones for tolling.

system. This system would allow a driver and their passengers to pay for not only a toll but also fast food, parking, carwashes and even oil changes as this technology would be open to any place which you were able to take your vehicle. It is interesting to look at the timing of all these technologies and business models. These new toll innovations coupled with new vehicle technologies like the rapidly developing autonomous vehicle could in the near future develop into the largest payment systems in the world.

If you look at the demographics of our customer and in particular our new millennials, they prefer a cashless technology-connected world where their payments are made automatically and without effort through the technology in their pockets.

In reality, toll agencies today that don't evolve into a connected interoperable cashless system that is well-run and coordinated could very well cease to exist as we know them within the next 10 years. We have to remember why toll agencies were created and have grown. We give the customer what they want and what they need. Today's millennial generation want mobility and convenience. They don't see a difference between toll agencies in the same state, and they don't understand why their cell phone works everywhere but their toll tag only works where they bought it. This generation of future customers wants mobility and will not rely on car ownership as its sole use of mobility. In the day of on-demand, shared mobility centers and the use of shared transportation options will drive how the next generations use the transportation network. Instead of asking how long will it take to get there by car at 7 or 8 AM, the question will be what transportation modes are available to get me to my destination by 8 AM? Millennials want convenience and are willing to pay for that privilege. They realize that yesterday's gas tax does not support the alternative fueling types of vehicles today, and that there may need to be something different such as a mileage-based user fee (see Chapter 18). All this can be easily incorporated into the tolling model of tomorrow if we in the industry are willing to work together. If we don't, some new innovative technology or business model can very well make us extinct.

12.8 Lessons for other countries

The lack of national standards in the United States has been the primary obstacle not only in slowing interoperability but also in the advancement of mobile payment and ITS systems. Effectively, we had 50 individual countries protecting their investments in the technology and unique business rules that they deployed years prior. Tolling in the United States grew up with every individual deployment believing that the concepts of their agency, state or region were better than anyone else's. In the early years, most agencies did not consider the need for cooperation between their agency and agencies or states beyond their confined regions. With the continued vision of those working on interoperability, this issue will be solved prior to the implementation of autonomous vehicles. This will bring mobile payments not only to tolling but also to commercial applications such as parking and fast food, not just in the United States but worldwide.

References

[1] Pennsylvania Turnpike Historic Photos. Turnpike History. N.p., n.d. Web. 08 August 2017. https://www.paturnpike.com/yourTurnpike/ptc_history.aspx

[2] Home. IBTTA, International Bridge, Tunnel and Turnpike Association. N.p., 01 October 1970. http://www.ibtta.org/ [Accessed 08 August 2017]

[3] Home. E-ZPass Group. N.p., n.d. http://e-zpassiag.com/ [Accessed 08 August 2017]

[4] E-ZPass. Wikipedia. Wikimedia Foundation, 28 September 2017. https://en.wikipedia.org/wiki/E-ZPass#/media/File:E-ZPass_system_map.svg. [Accessed 08 August 2017]

[5] Webmaster. North Carolina Turnpike Authority. NC DOT, North Carolina Department of Transportation. N.p., n.d. https://www.ncdot.gov/turnpike/ [Accessed 08 August 2017]

[6] Violation Enforcement. ATI. Srpout Media Lab, n.d. http://www.tollinterop.org/violation-enforcement/ [Accessed 08 August 2017]

[7] Radio-frequency Identification. Wikipedia. Wikimedia Foundation, 29 September 2017. https://en.wikipedia.org/wiki/Radio-frequency_identification [Accessed 08 August 2017]

[8] IBTTA Interoperability Committee (IOP). IBTTA, International Bridge, Tunnel and Turnpike Association. N.p., 01 October 1970. http://www.ibtta.org [Accessed 08 August 2017]

Chapter 13

London Congestion Charging – a personal account

Nick Patchett and Jeremy Evans***

13.1 Introduction to the project

Some background to the London Congestion Charging (CC) Scheme will be found in Chapter 2.

13.1.1 Our roles and the team

In 2000, Nick became the systems team leader for the London Congestion Charging project, supporting the newly formed Transport for London (TfL)[1] which now had responsibility for all transport including main roads and public transport in London. Nick was working at Deloitte and we had been appointed as advisers to the project. As a systems and telecommunications specialist with a procurement background and experience with cameras and image processing, the role was a perfect fit. Jeremy had 8 years' experience in the Congestion Charging Directorate in TfL as part of the Senior Management Team responsible for the design, procurement, implementation and operation of the London Congestion Charging Scheme and the London Low Emission Zone (LLEZ). The wider team involved legal specialists, highway and traffic engineers, enforcement specialists, transport planners and modellers, project managers, finance and policy makers and experts in signage and road markings. We had a project to deliver, with a set 'go-live' date in early 2003.[2] It was probably one of the most exciting and challenging projects to be involved in at the time.

13.1.2 Making it real

To date, congestion charging had been planned at a theoretical level and was based primarily on traffic and economic modelling principles. What we now had to do

*Pillar Strategy
**Harpur Evans Consultancy Ltd, formerly Head of Traffic Management and Enforcement Infrastructure, Congestion Charging Directorate, Transport for London, UK

[1] TfL was formed from in 2000 through a merger of London Transport, parts of the Government Office for London, parts of the Department of the Environment Transport and the Regions (DETR), the Traffic Director for London and parts of the Highways Agency

[2] Go live date for Congestion Charging was set by the Mayor

was make it practical, procurable and operationally feasible. As well as technology and procurement skills, we had to involve operations and process specialists; we had to build a commercial team; we had to design customer facing channels, websites, retail outlets and call centres; we had to examine what data we could get from DVLA[3] and agree security and privacy arrangements with the newly appointed Information Commissioner.[4] We also had to prepare a detailed programme timetable and estimate what it would cost to implement and operate, even though there were no clear benchmarks or precedents available; we had to keep TfL's leadership informed through a reporting cycle and board papers, and also satisfy the Greater London Authority's (GLA) scrutiny process [1]. Many of these additional 'practical' aspects also had a systems and technology angle to them, as they needed to be process driven by computers; so I (Nick) came to be involved in nearly every aspect of the project from how camera evidence was made secure, how the back office worked, helping to brief the Mayor about what data we wanted from DVLA, so he could ask the Secretary of State for Transport.[5]

13.1.3 Our leadership

The overall team was led by Derek Turner, Malcolm Murray-Clark and Michèle Dix. Michèle had been an author of the Road Charging Options for London (ROCOL), [2] report, Malcolm had been responsible for parking and transport for Westminster and Derek had recently been Traffic Director for London and introduced the Red Route network. All three knew all about London's traffic and congestion. In practice, the whole team had just come together to deliver this project to fulfil the election manifesto commitment [3] made by the new Mayor, Ken Livingstone, making use of new regulatory powers conferred by the GLA Act 1999 [4]. As a project, it had a quite different level of momentum and intensity from other public sector projects. This was definitely working at pace and there was a strong team spirit focused on delivery.

13.2 Why congestion charging

13.2.1 So why did London need congestion charging?

The centre of London essentially had 'all day' congestion; it wasn't just in the morning and evening peaks. It only took an incident, accident or inclement weather conditions to make the congestion even worse. The average speed of vehicles in London was less than 11 mph. This meant that it was difficult to predict journey times. It also meant that exhaust fumes from almost stationary vehicles were creating pollution with sulphur dioxide, nitrogen oxides (NO_x) and particulate matter. At this stage, emissions were considered as part of the charging policy, but

[3] The UK Driver and Vehicle Licensing Agency (DVLA), based in Swansea
[4] Information Commissioners Office established by the UK Data Protection Act 1998
[5] The Secretary of State at the time was Stephen Byers

congestion reduction was the main objective. The LLEZ came a few years later. The congestion meant that:

- There were many person-hours lost in traffic
- Delivery and trade vehicles could not fulfil their travel schedules reliably
- Buses could not run to timetable
- Congestion was costing London's economy, money and jobs and the vehicle emissions were damaging people's health.

13.2.2 What was the history?

Let's go back to the original proposals, which are summarised below; see also Chapter 2.

13.2.2.1 The history of the idea

Some key dates in the history of the London Congestion Charging Scheme are shown in Table 13.1.

So following on from his election manifesto, the Mayor published a consultation document, 'Hearing London's Views' [5], in Jul 2000. This was sent to about 300 stakeholders, including London Boroughs and business groups. In responding to this consultation, the Mayor proposed to reduce the charge for Heavy Goods Vehicles from £15 to £5, the same as for cars. This simplified the scheme as vehicle classes did not need to be identified. Also, a 100% discount was proposed for Blue Badge[6] holders, and 90% discount for residents of the charging area. Then the Mayor published his Draft Transport Strategy [6] in Jan 2001, which included the high-level plan for congestion charging as well as many other public transport measures, updating the 1996 Transport Strategy for London [7]. The Mayor had to

Table 13.1 Key dates in the history of the London Congestion Charging Scheme

- The Smeed Report assessed the feasibility of road pricing in a British City in 1964 (see Chapter 2)
- In 1965, the newly formed Greater London Council (GLC) drew up plans for a cordon charging scheme
- In 1973, a report assessed the potential benefits of such a scheme for traffic and the environment which led to the newly elected GLC developing a programme of public transport investment and introduction of traffic management
- In 1995, the London Congestion Research Programme concluded the city's economy would benefit from a congestion charging scheme
- The Road Traffic Reduction Act of 1997 required Local Authorities to study traffic and reduce congestion where it was damaging the economy and environment
- The GLA Act 1999 included the powers to implement a congestion charging scheme
- The ROCOL Report was published in 2000

[6] UK Blue Badge scheme lets disabled drivers park closer to their destination, in spaces not available to other drivers.

consult on the new strategy which meant the first half of 2001 involved getting stakeholder responses before publishing a final version (The Mayor's Transport Strategy [8]) in Jul 2001 which was approved and confirmed the 'go-live' date as early 2003. This document had to set out another level of detail about the proposed charging scheme, particularly how it would work and its likely impacts, costs and revenues.. It also had to trigger the preparation of a further level of legal detail known as 'Scheme Order' which described how the scheme would work from a process point of view in layman's terms and indicated the roads on which it would operate. Following consultation and confirmation of the Scheme Order in Feb 2002 [9], legal challenges were commenced calling for a Judicial Review by the City of Westminster and a group of residents in Kennington, south of the river Thames. Following a hearing, the two challenges were dismissed by Mr Justice Maurice Kay in Jul 2002.

13.2.3 Where does the scheme operate?

The area and roads on which the scheme operates are shown in Figure 13.1. This covers an area of approximately 8 sq mi (21 sq km). The outer boundary would not be charged, but driving inside the zone (the 'central zone') would incur a charge of £5. The initially proposed charging policy operated Mon–Fri 7.00 a.m.–7.00 p.m. This was subsequently modified to 7.00 a.m.–6.30 p.m. and then 7.00 a.m.–6.00 p.m. following consultation.

There were also various discounts, exemptions and fleet options: Residents living inside the zone were able to take advantage of a 90% discount. Drivers with 'blue badges' (disabled drivers) were exempt. Commercial vehicle fleet operators were able to open a fleet account.

13.2.4 What problem were we solving?

Table 13.2 indicates the problems we were solving by implementing a congestion charging zone.

Though we could model how congestion charging would operate, we could not be sure precisely how many drivers would respond to the charge by switching to public transport, and the exact level of traffic reduction. But we knew that some intervention was necessary. This meant the project was criticised by the GLA for not being able to 'exactly predict' the traffic impacts and therefore revenue. In practice, it was probably not feasible to predict the outcome.

13.2.5 What was the public's reaction?

In 2001, the Mayor consulted on the draft Scheme Order, and received a positive response allowing him to confirm it. But not all supported the scheme; particularly vociferous were:

- The motoring press
- Driving-related bodies

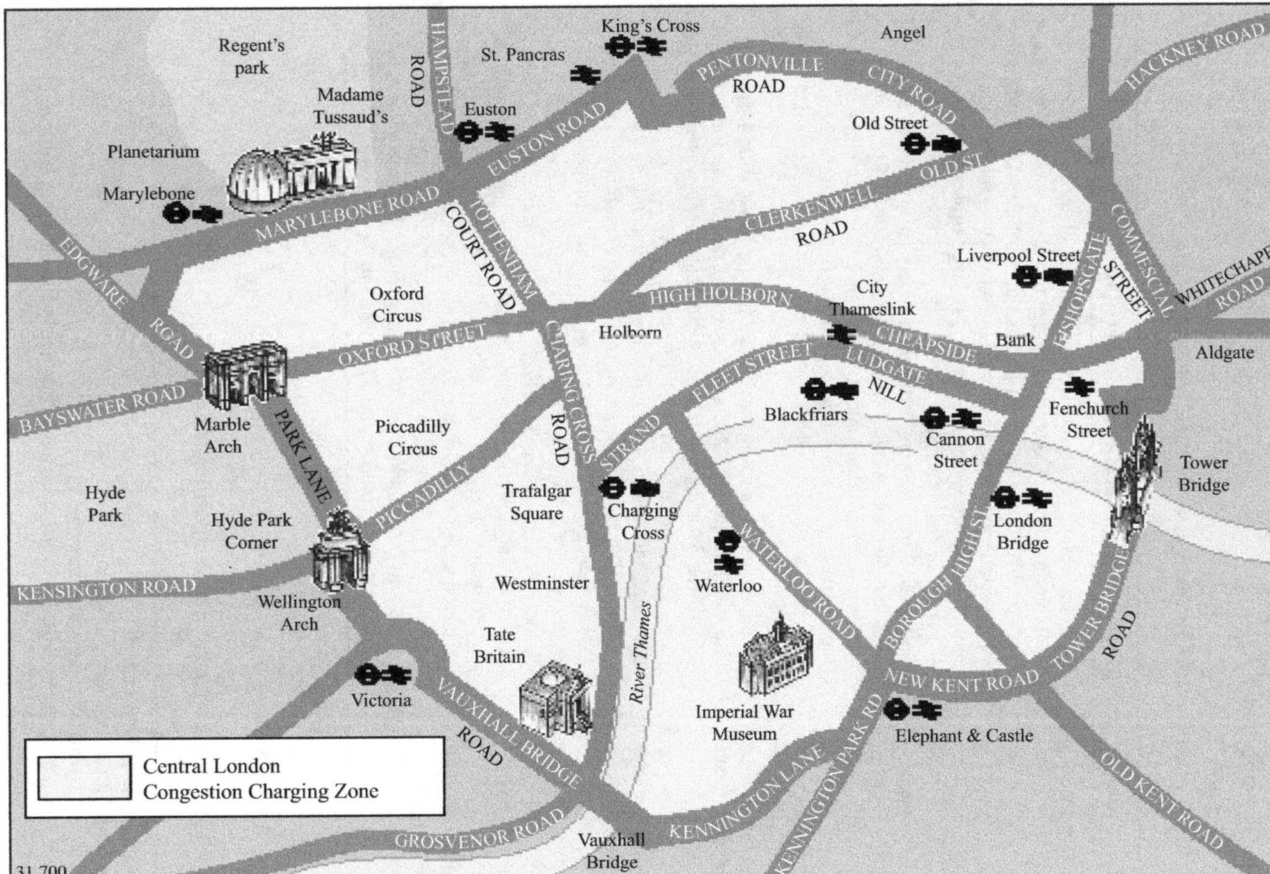

Figure 13.1 Overview of the central London congestion zone
[Source: TfL]

- Certain trade and professional representative bodies
- Westminster City Council
- A residents' group in Kennington
- The Evening Standard.

Table 13.2 The problems we were solving by a congestion charging zone

What and where	Measurements of traffic and congestion in 2002
Congestion inside the zone	The uncongested network travel rate for the charging zone is 1.9 min/km. But in 2002, the average travel rate in the zone was actually 4.2 min/km, which means that traffic inside the zone was experiencing delays or congestion of 2.3 min/km. This compares to an average delay across large urban areas in England of 0.4 min/km
Speed in the zone	Another way of assessing traffic conditions is in terms of speed distributions, or, in other words, the proportion of time spent driving within various speed bands. In 2002, during charging hours over half of the time spent by vehicles within the charging zone was spent either stationary or at less than 10 km/h
Congestion on the Inner Ring Road	The uncongested network travel rate for the Inner Ring Road is 1.8 min/km. In 2002, the average travel rate on the Inner Ring Road was 3.7 min/km, which means that traffic was experiencing delays or congestion of 1.9 min/km
Speed distributions on the Inner Ring Road	Data from 2002 shows a similar pattern as the charging zone, with about half of the time spent by vehicles on the Inner Ring Road being spent either stationary or at less than 10 km/h
Congestion on main roads in inner London	The uncongested network travel rate for main roads in inner London is 1.5 min/km. In 2002, the average travel rate on these roads was actually 2.8 min/km, which means that traffic was experiencing delays or congestion of 1.3 min/km
Congestion on main roads in outer London	The uncongested network travel rate for main roads in outer London is 1.2 min/km. In 2002, the average travel rate on these roads was 1.8 min/km, which means that traffic was experiencing delays or congestion of 0.6 min/km

Source: The First Monitoring Report, Published 2004, TfL.

Table 13.3 lists the key lessons learned from the scheme.

Table 13.3 Lessons learnt: why congestion charging?

1. There needs to be a well definable congestion problem to be resolved
2. Strong political leadership is needed
3. Supported by main stream acceptance of the scheme
4. There will always be opposition as it is inevitable that some stakeholders will prefer to use their cars
5. There needs to be a focused, well managed design and implementation team in place
6. Any profits derived from the scheme should be spent on projects to improve transport, particularly public transport

1. Customer channels	2. Vehicle detection
• Contact channels • Payment channels • Payment processing • Account management • Marketing and communications	• Gather vehicle passage records • Interpret number-plates • Create vehicle trip data • Manage detection data quality • Store evidence
3. Back office	**4. Enforcement**
• Reconcile vehicle transactions and payments/accounts • Manage exemptions and discounts • Process customer accounts and registrations • Manage customer data quality • Verification of vehicle detection data • Refer non-payers to enforcement • Billing of post-pay accounts	• Validate camera evidence • Gather keeper records • Issue penalty notices • Receive penalty payments • Manage enforcement data quality • Register debts • Manage input to appeals • Identify persistent non-payers • Initiate bailiff action • On-street enforcement • International debt collection

Figure 13.2 Main business processes
[Source: Pillar Strategy]

13.3 How does it work?

The congestion charging system is a little more complex than just having cameras on streets, with vehicles identified using Automatic Number-Plate Recognition; customer channels, 'back office' and enforcement are significant operations. At a technical level, there are four main business processes as illustrated in Figure 13.2.

From a technical point of view, the system worked at go-live as indicated in Figure 13.3.

In summary, the scheme works as follows:

- Customers pay the charge for the use of their vehicle on a specific date using the various payment channels
- The vehicles are detected by the cameras one or more times as they drive within, enter or exit the zone – note this is an area scheme, not a cordon scheme, so charges apply to any travel inside the zone, irrespective of whether the scheme boundary is crossed – the detection data also feeds real time traffic information as an additional benefit. The camera data is reviewed to check its integrity and remove misread errors
- The payments made are matched against the vehicles detected
- Images of vehicles that have paid are deleted
- Vehicles which have no payment made or which are not exempt are added to a suspects list
- Where multiple images of a suspect vehicle on a particular day are detected, the best one is kept and others are deleted
- Details for the vehicle keepers are retrieved from DVLA for the suspects

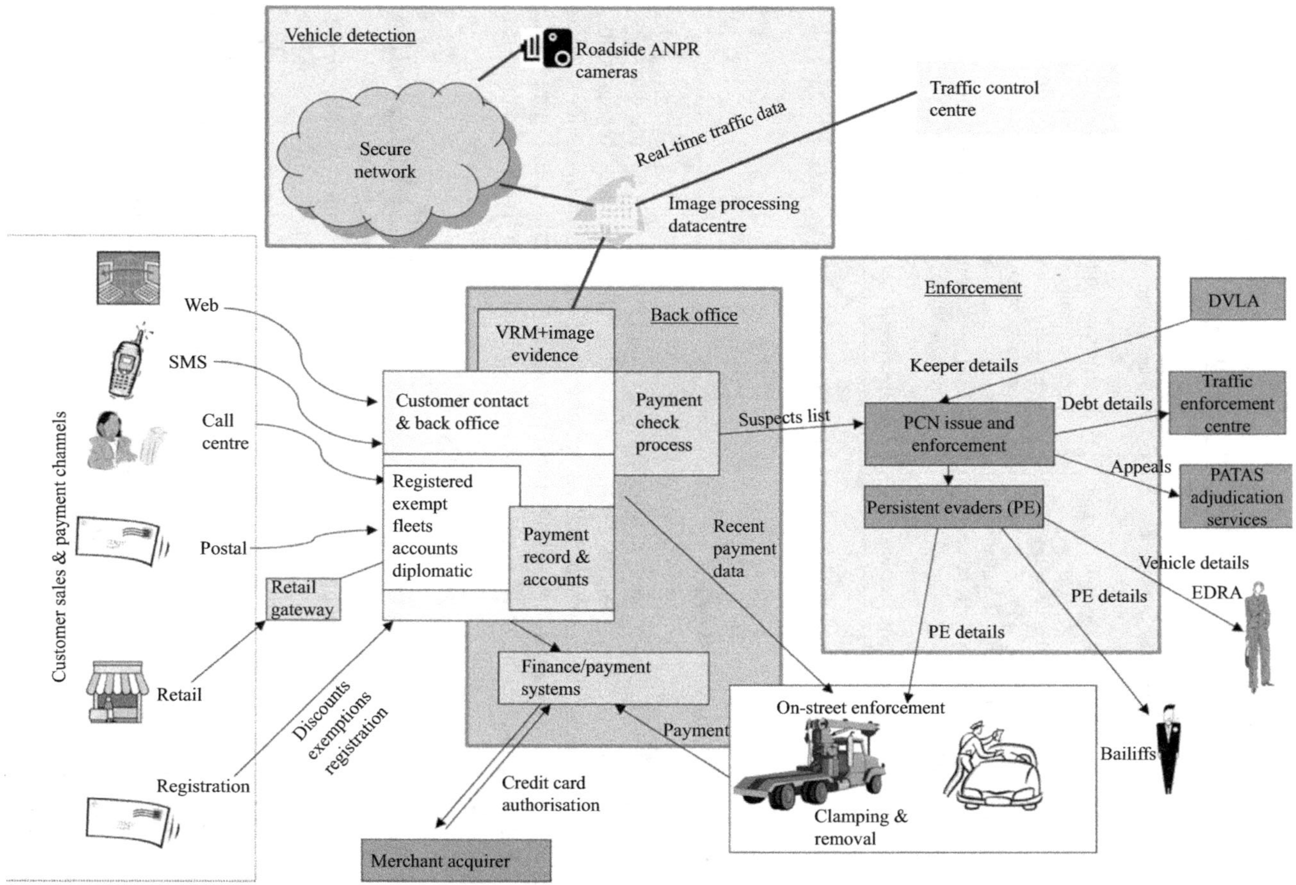

Figure 13.3 How it works – technical schematic
[Source: Pillar Strategy]

- The draft Penalty Charge Notices (PCNs) are prepared and visually checked against the DVLA data
- PCNs are issued to the vehicle's keeper; where there is no UK-based keeper, a European Debt Recovery Agent may be engaged subject to cost-effectiveness
- If PCNs are not paid within 28 days, the debt may be registered at the traffic enforcement centre (equivalent of UK county court)
- There is a process whereby a customer can challenge a PCN through making a representation and if the representation is rejected, an appeal can be made involving an independent adjudicator.

Since go-live in 2003, the main changes have been:

- The retail channel which originally played a significant role is not used
- A new CC Autopay account for regular users has been implemented.

Below we explore some key elements of this operation.

13.3.1 Customer channels and payment of a charge

The key objective is to use economic factors to deter drivers from using their cars to reduce congestion. This means that the charge has to be at a level that persuades a proportion of drivers to use other transport modes, or just not travel; the charge also needs to more than cover operating costs. This generates a difficult pricing decision that takes into account:

1. Whether the scheme is intended to generate revenue – many road pricing schemes have that as a key objective – the provisions of the GLA Act 1999 required that any surplus revenue must be applied to improve transport in London
2. Whether it is just intended to cover its costs
3. What proportion of the driver population is sensitive to the charge level – many drivers may make their trip irrespective of the price.

In London, the initial £5 charge was set primarily to persuade drivers to change mode on the assumption that it would also cover costs and generate a small surplus. The operation of customer channels drives much of the costs of operation – some examples of typical transaction costs are presented in Table 13.4. It is clear from this table that payment via a call centre will consume a high proportion of a £5 charge in transaction costs before allowing for the capital scheme costs and other operational aspects. Payment over the web is much more cost-effective. The customer channels available needed to be specially cost engineered to make them effective. It was important to incentivise customers to adopt the lowest cost channels.

The payment channel usage between go-live and early 2005 is presented in Figure 13.4.

This shift in channel usage also changed the operational costs. The most significant changes were the reduction in retail usage and increase in web-based payments.

Table 13.4 Typical marginal transaction operating costs (note these are not actual London scheme costs)

Event	Relevant assets	Illustrative calculation/ assumption	Indicative value
Contact centre call	• Call centre agent time • Capital and operating costs of call centre equipment & premises	3.5 min/call agent cost of £25k pa–£0.90 Equipment + premises @ £0.90 per call	~£1.80/call
Web-payment by credit card	• Capital cost of web portal • Merchant acquirer fee	2% of payment made Average payment of £5	£0.10/payment
Retailer	• Retailer commission	Commission of ~15% Payment of £5	£0.75

Source: Pillar Strategy.

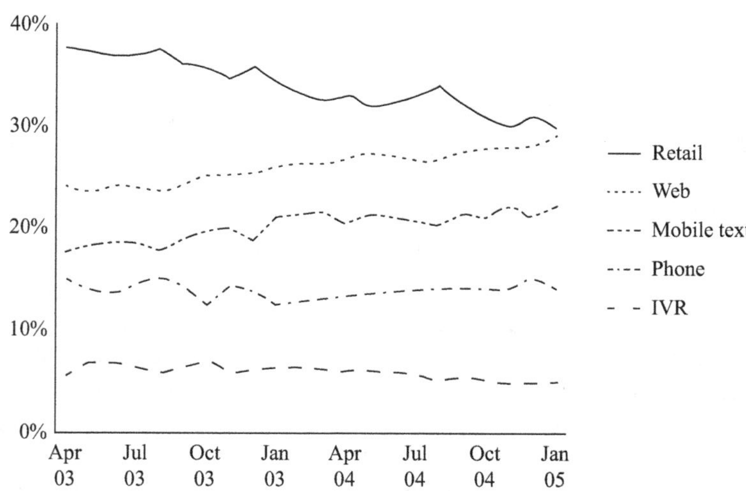

Figure 13.4 Payment channel usage in 2003–05
[Source: TfL]

The introduction of the CC Autopay (payment based on an automatic account set up by the driver) account in 2011 made a substantial difference to channel usage. By 2013, over 70% of payments were made through either fleet or personal accounts as shown in Figure 13.5.

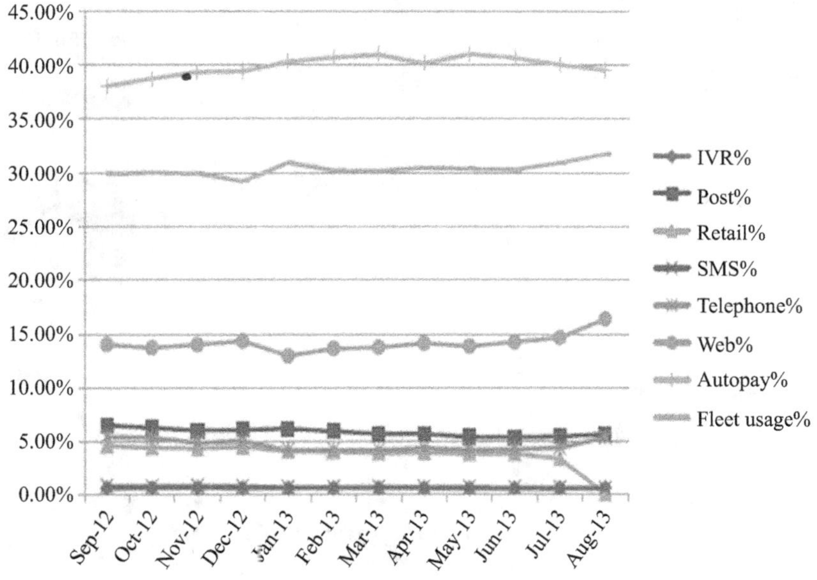

Figure 13.5 Payment channel usage in 2012–13
[Source: TfL]

The retail channel was used for less than 5% of payments and SMS payments were negligible. This shift meant the scheme's operational and contracting model had an opportunity to change.

13.3.2 Vehicle detection and camera images

When the scheme was designed, it relied on the new GLA Act 1999 for enforcement powers. The vehicle detection process therefore needed to follow established methods of gathering evidence that a vehicle was 'in a particular location at a particular time of day' to prove it was liable for the charge. The civil enforcement framework using camera detection of contraventions had been pioneered in the London Boroughs of Croydon and Newham, using powers conferred by the London Local Authorities Act 1996 [10]. Most other systems had to comply with the requirement for criminal enforcement:

- Civil enforcement – requires proof on the balance of probabilities
- Criminal enforcement – requires proof beyond reasonable doubt.

The principles had been established for the recent decriminalisation of parking [11].[7] This had two benefits: it removed the burden of parking enforcement from the police and meant that parking offenders did not get a criminal record. The liability for any contravention rested with the keeper of the vehicle rather than the driver.

[7] The Road Traffic Act 1991 provided for the decriminalisation of parking offences and allowed local authorities to apply to take over the enforcement of parking within Controlled Parking Zones.

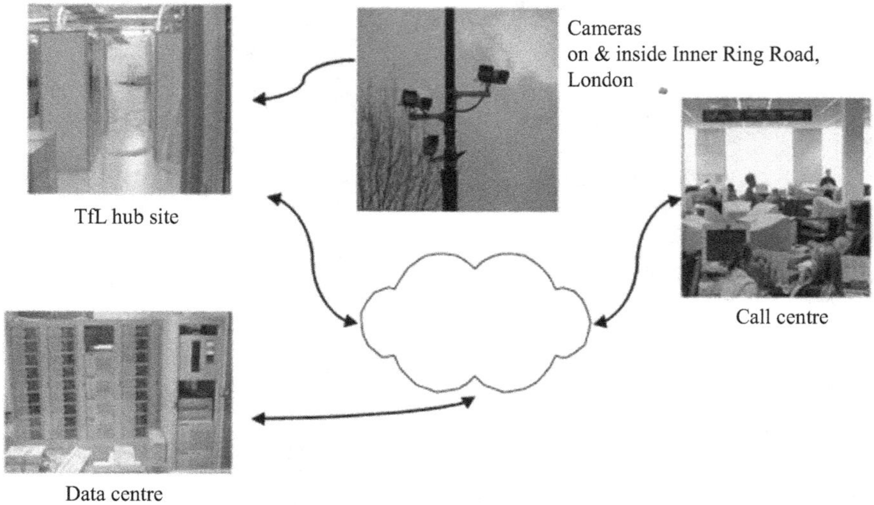

Figure 13.6 Camera infrastructure in 2003
[Source: Harpur Evans Consulting]

This meant the camera evidence (Figure 13.6) had to satisfy similar needs for integrity. The new development for congestion charging was to use fixed automatic cameras.

There are some key principles for designing how evidence is treated and processed that need to be followed to avoid openness to challenge:

- Evidence must be robust, disclosable and auditable
- It must also meet privacy requirements – the customer must have confidence that the evidence gathered will not be used for improper purposes that breach privacy
- On fairness – there must be an opportunity for the customer to dispute the evidence; the system is not always right – human checks and balances are needed.

The number-plate had to be clearly readable; the location and direction of travel needed to be established; the time and date needed to be clear; the image had to be secured so it could not be tampered with. Figure 13.7 provides an example set of camera images.

The original scheme had approximately 200 camera sites based on or near the boundary and within the zone.

13.3.3 When things go wrong

To provide an example of where checks and balances are needed, here is a set of images of a vehicle gathered in the first week of London Congestion Charging which went live on 17 Feb 2003. This vehicle had not paid the charge (Figure 13.8).

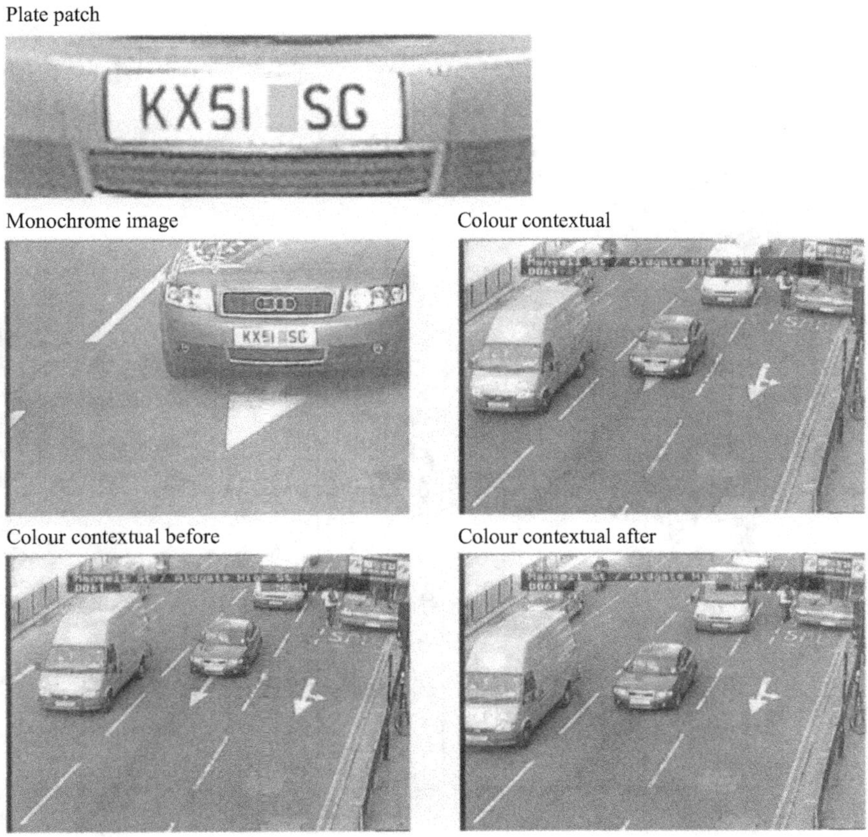

Figure 13.7 Example set of images for London – as at go-live in 2003
[Source: TfL]

This is actually a 'cloned vehicle' – in other words, a genuine number-plate has been copied and put on a different vehicle, usually of the same make and model, thereby attempting to evade the charge. Figure 13.9 shows an image of the genuine vehicle with that plate.

This was actually my (Nick Patchett's) own car, hence I can speak about it openly. Comparing the two images clearly shows the differences in registration plate:

- The clone has
  - Leading country logo
  - No motor trader details
- The genuine vehicle has
  - No country logo
  - Motor trader details.

ER view print

Mono-vehicle image	Colour contextual image	Colour contextual before	Colour contextual after

Mono-number-plate image

Interpreted VRM	S644XAY	**Confidence level**	95
Session ID	10008525	**Frame counter**	00000000205077
Date of capture	20/02/2003	**Time of capture**	14:32
Camera ID	1918	**Camera location**	Farringdon Rd - Charterhouse St
PCN		**Status**	PCN Pending

Session details

Session ID	Session start date time	Session end date time
10008525	20/02/2003 12:15:18	20/02/2003 14:45:00

Figure 13.8 Evidential record from 20 Feb 2003 – the clone
[Source: Nick Patchett]

Figure 13.9 The genuine vehicle
[Source: Nick Patchett]

The image evidence allowed

- The vehicle make, model and registration to be validated
- The clone of the number-plate to also be detected.

The manual check process revealed the difference.

It was estimated at the time that between 1 in 4,000 and 1 in 300 vehicles in the United Kingdom were 'cloned'. This led to a significant tightening in how vehicle registration was managed in the United Kingdom, which was previously relatively uncontrolled; for example

- New regulations were issued on number-plate manufacture and issue so that all retailers able to issue plates had to be registered and approved establishments
- A new vehicle owner registration form (known as the V5C) was introduced to require re-registration of all vehicles
- A new Ministry of Transport (MoT) Test inspection procedure was introduced so that only approved registered garages could issue MoT certificates.

There are significantly fewer clones on the road in the United Kingdom as a result.

The main conclusions that we draw from this example are that (Table 13.5):

- For the scheme to be fair, there must be checks and balances: moving away from images for enforcement is very challenging if fairness is to be preserved and with the current level of accurate vehicle registration.
- Duplicate registrations: the standard process has to deal with the risk of duplicate registrations either through clones or through multiple registration organisations or other countries using the same registration format
- Robustness of evidence: the images need to be supported by related timing and location detail which can be used to refute the risk that the evidence had been manufactured
- Image detail: the level of detail captured has to allow comparison of features such as location on the street and context and the features on the vehicle or plate, as in the cloned example above
- Enforcement process: the process needs to allow a customer response to reduce risk of incorrect enforcement and needs to allow the customer to see the evidence

- Privacy: the captured data needs to be kept private with only evidence of a contravention being retained
- Human involvement in enforcement: automated processes cannot always distinguish image differences or anomalies that arise – human involvement in checking is therefore necessary
- Auditability: the gathering and storage of the evidence and protection from potential tampering has to be auditable.

Table 13.5 Lessons learnt: how it works and operation

- The system is likely to have a significant cost to implement and operate driven primarily by customer channel processing costs. Most scheme designs start with optimistically low implementation and operation cost estimates
- The quality of vehicle registration data is key to successful operation
- The data processing can be highly automated but must have checks and balances to allow manual intervention to resolve anomalies and outcomes that were not intended

Provided the operation is designed in the correct manner, it can be scaled to operate in a larger area or serve multiple schemes

- To maximise the efficiency and cost-effectiveness of schemes so that more net revenue can be invested in public transport, a shared services approach is preferable rather than every new scheme having to implement its own back office infrastructure and customer payment channels
- Uncertainty of outcome – the impact of road pricing schemes cannot always be predicted; it is important to allow for flexibility in outcome and not implement an inflexible model. Adaptability is key
- The extent to which drivers are cost sensitive is surprising and depends mostly on journey purpose and affordability; each driver segment can be quite different
- Data quality is key to effective operation; poor data quality can make the operation infeasible and bring the system into disrepute
- Cost of operation is surprisingly high – there have been some ambitious targets quoted (like 5% of revenue collected) which we don't believe are feasible
- The supply chain and contracting model needs to also allow for flexibility and change because you cannot predict all the changes you will need to make

13.4 Complementary measures

Alongside the implementation of the congestion charging scheme, there were a wide set of complementary measures including:

- Provision of additional capacity on the bus network
- Implementation of controlled parking zones (CPZs) to prevent drivers leaving their cars just outside the zone or swamping local parking near stations

- Improvements to the road network, particularly on the Inner Ring Road, to reduce congestion and also complete much of the upgrade programme prior to go-live
- Ensuring that TfL's programme of major capital works at or near the zone boundary was complete prior to go-live date
- Rescheduling TfL's carriageway repair and maintenance programme to ensure that major routes would not be subject to planned road works for a reasonable period following the introduction of congestion charging
- Softer measures to promote usage of alternative modes of transport to the private car
- Improvements to signage
- The implementation of the London Traffic Control Centre
- Addition of journey time information gathering along the key radial routes.

Figure 13.10 illustrates the 'zone of influence' where the scheme was expected to have an impact and therefore where CPZs were considered and measures to reduce the risk of 'rat-running' were implemented.

The complementary measures implemented and the budget applied is summarised in Table 13.6.

Table 13.6 Summary of complementary measures

Type of measure	Number of schemes	Cost (£m)
Traffic management schemes	51	7.1
Home zones/20 mph zones	49	13.3
CPZs (inner area)	34	6.9
CPZs (outer area)	53	3.3
Capital works on TfL roads	27	22.2
Traffic signal schemes	53	2.5
Soft measures	22	2.9
Total	**261**	**58.2**

Table 13.7 summarises the lessons learned from these complementary/softer measures, which included the following examples:

- Lambeth North Personal Security Project
- Hackney Community Transport Scheduling Software for Minicab journeys
- Tower Gateway DLR Platform Lengthening
- Bus Spider Maps
- Green Travel Plans and Personalised Journeys Planning Project for NHS facilities inside the zone
- City Car Clubs.

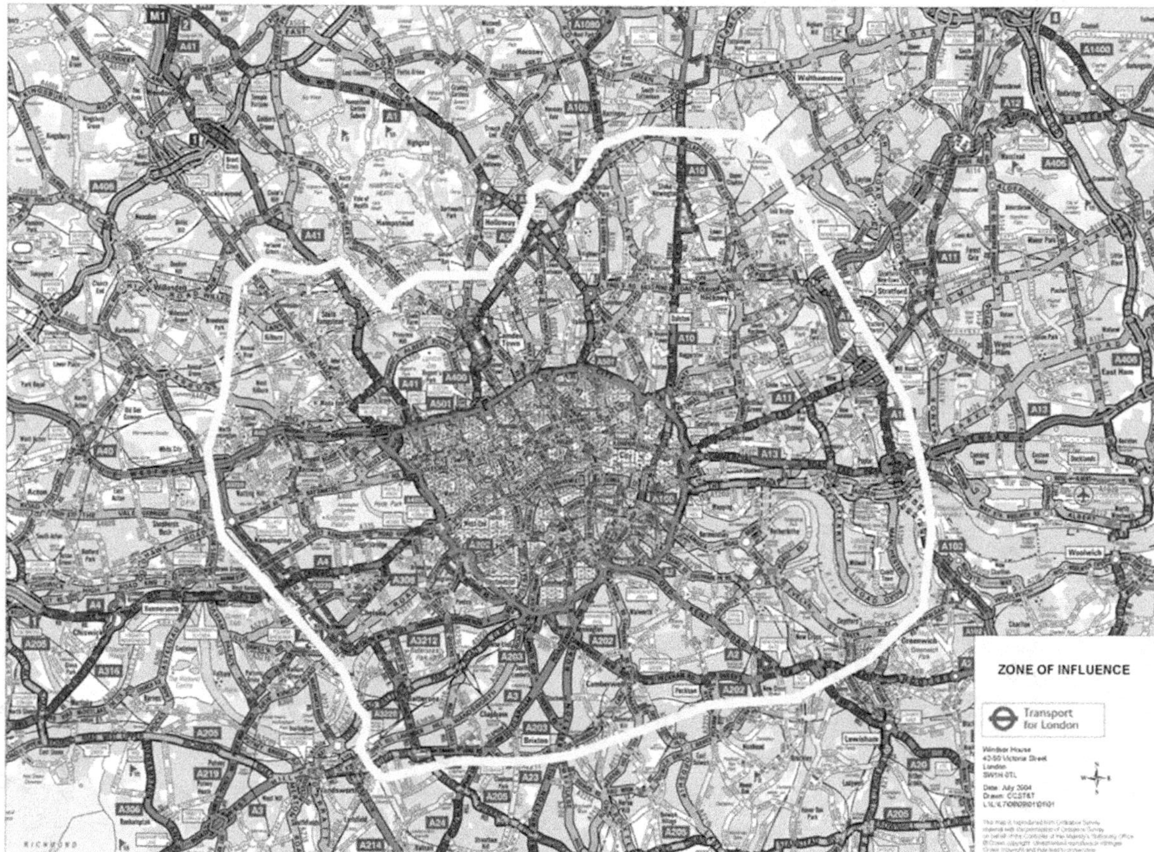

Figure 13.10   Congestion charging zone of influence
[Source: TfL]

Table 13.7 *Summary of lessons learnt: complementary measures*

- Need for public transport improvements to be in place prior to commencement of congestion charging scheme
- A comprehensive communications strategy needs to be devised to achieve meaningful dialogue with residents and businesses within the CC 'Zone of Influence'
- Need to ensure those residents and businesses located just outside the zone boundary derive some benefits from scheme – examples include establishment of Controlled Parking Zones, installation of traffic management measures to discourage traffic rerouting to avoid paying the charge
- Importance of Complementary Measures further from the zone boundary became more significant as it became apparent that drivers were making their decisions on rerouting a greater distance away from the CC zone than was originally forecast

13.5 Implementation

The implementation of congestion charging was probably one of the fastest delivery programmes in the UK public sector. To achieve the timescale, we had to do multiple activities in parallel. As the secondary legislation was being developed, the specification and procurement plan was also being implemented; while consultation was concluding, we were designing the scheme's operation; while preparing to tender contracts, we were testing the performance of different camera types and configurations to check out the state of the art.

From initial advertising of the operations contract in Jan 2001 until awarding a contract option in Dec 2001 was just 11 months during which we had carried out a full two-stage negotiated procurement procedure. That was 'lightning speed' for many similar scale projects. The procurement was regarded as a model example by the UK Office of Government Commerce and was regularly quoted as a case study [12].

In Dec 2001, we were not able to execute the contract as Scheme Order was not yet confirmed, so instead, we moved forward in 2–4 week stages that required individual approval, preparing the detailed system design, carrying out camera site surveys and generally preparing for implementation without committing to significant contracts or spend. Once Scheme Order was confirmed in Feb 2002, we could then execute the contract with our main service provider Capita as well as our camera and network providers. The preparation between Dec 2001 and Feb 2002 meant we could meet the key milestones for implementing, for example, the Enquiry Service, Registration for Residents, 'go-live' for the website, 'go-live' for the retail channels, 'go-live' for the call centre and then 'go-live' for the scheme operation. This series of 'go-live' events needed careful reviews of readiness, testing of the systems and processes and resolution of problems. We had to use strong project management techniques to achieve this timescale, overseen by weekly checkpoint meetings and monthly project boards and overseen by a programme office, the Director and the Mayor's Office.

The actual 'go-live' was a surprisingly quiet and smooth affair. There was no major traffic impacts on the Inner Ring Road, there were no surprises (Table 13.8).

Table 13.8 Summary of lessons learnt: implementation

- A coordinated and well managed implementation can deliver in a short timescale provided there is a strong team in place; this means a project can be delivered within the typical Mayoral or Government term of office
- There are many activities that can occur in parallel, for example preparing legislation, designing the operating model, soft market testing, stakeholder consultation, trials of technologies to mitigate risk, on-street surveys for infrastructure to confirm feasibility and cost, development of the cost and revenue model, estimation of volumes and preparing an implementation plan. Many of these elements are interlinked and interdependent
- Planning has to be carried out meticulously to address the interdependencies and the overall timescale
- The programme has to be treated as a series of phased milestones – for example, an approach that adopts an overall programme with subsidiary work streams and projects; we also found that for the major components having a focused 'project manager per milestone' is more likely to result in success

13.6 Expansion, other schemes and the opportunity for a 'shared service'

The London Congestion Charge zone was expanded in 2007 to include the Western Extension. However, this part of the scheme was cancelled following the new Mayor being elected in 2008. It ceased operation in Dec 2010. The Western Extension demonstrated that the existing scheme was scalable using a single zone, single charge model.

Figure 13.11 illustrates the originally intended scope of the Western Extension. Note that the central route from Edgware Road, Park Lane to Vauxhall Bridge was an uncharged through route.

The existing scheme was also used as an operational platform for the LLEZ in 2008, the area for which is illustrated in Figure 13.12. The existing congestion charging cameras and back office systems were adapted to also process information for the LEZ covering a wider area.

The same platform is to be used to deliver the Ultra-Low Emission Zone (ULEZ) in 2020; however, the current Mayor intends to implement the ULEZ more quickly – in 2019, subject to consultation.

Both the Western Extension and LEZ involved the adoption of a new camera architecture, where more processing is devolved to the roadside and which makes the system more scalable. Figure 13.13 illustrates the change in architecture.

Technically, there is no real limit to the expansion of the operation that is feasible. Cameras, customer channels and back office processes can be scaled to meet the operational need. It is also technically possible to share operational services between different schemes as most customer channels and back office processing functions can be made 'common'. The primary questions concern marginal costs and benefits and transport/charging policy.

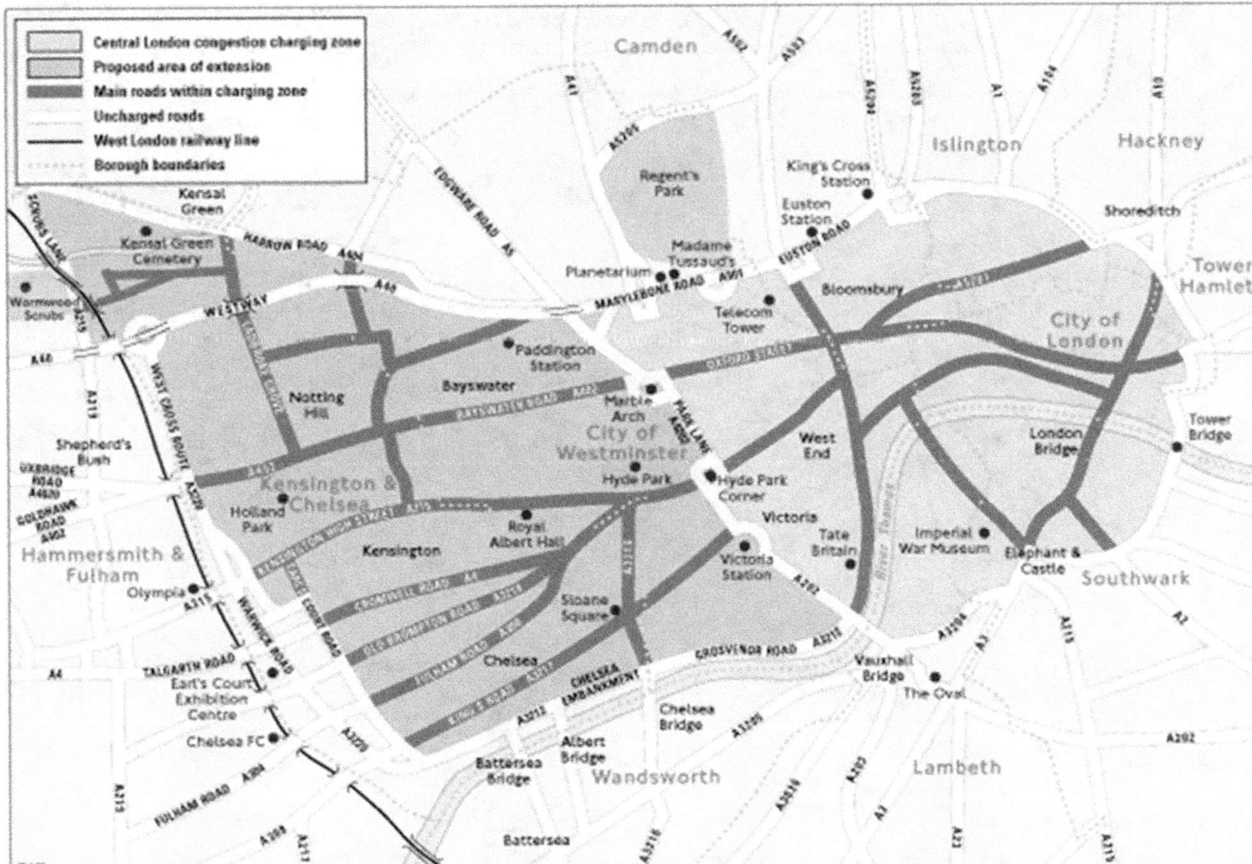

Figure 13.11 Western Extension – proposed scope 2004
[Source: TfL]

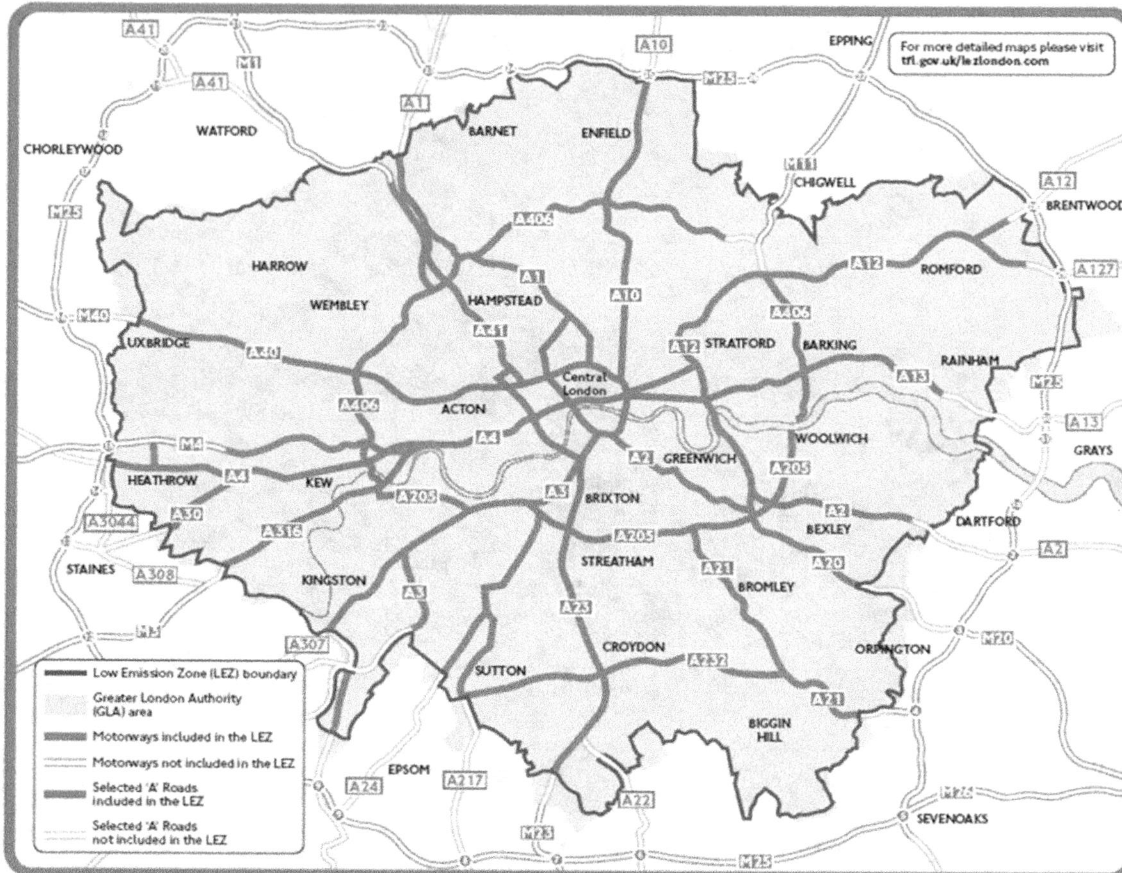

Figure 13.12 Area of the London low emission zone in 2008
[Source: TfL]

Central scheme (2003)
- 688 standard CCTV cameras
- Analogue point-to-point fibre connections
- Data-centre-based ANPR
- 27% enforcement sites with dual fibre connections available in the event of disaster recovery (DR)
- Difficult and expensive to add or move detection sites

Current scheme in 2015
- 639 integrated ANPR cameras
- Private internet protocol network based on digital subscriber line technology
- 100% of enforcement sites now available in the event of DR

Figure 13.13 Camera architecture in 2007
[Source: Harpur-Evans Consulting]

Other UK cities had also planned their own charging and low emission schemes through the former UK Government's Transport Innovation Fund (TIF) programme. One might have expected them to seek a shared scheme with London to reduce implementation and operating costs. However, this was not the case, and none of these schemes were progressed beyond the very early stages:

- Cambridge: the proposed TIF scheme was never sufficiently developed to determine how it might be implemented and operated; had it gone forward, a shared scheme with London may have been possible. It was cancelled in 2010
- Cardiff: Welsh Assembly Government proposed in 2003 to introduce a charging scheme [13] to combat the 3% annual rise in commuter traffic (some 50% above the average commuter growth) into the city, but this did not move forward
- Edinburgh: planned from 2002 onwards (before TIF was envisaged) but was rejected by a referendum in Feb 2005. While the scheme was still being developed (pre-referendum), the city decided it would procure its own separate operation and system instead of sharing facilities with London; this probably made the business case more challenging given the proposed £2 daily charge per vehicle (in comparison with London's £5) as it is unlikely that implementation and operating costs per transaction or trip would have been significantly lower than for London
- Manchester: planned a TIF scheme from 2006 onwards, but it was rejected by a referendum in Dec 2008. While the proposal was still being developed, the city took the approach of undertaking its own procurement and implementation instead of sharing facilities with London or other cities. During its development, the Association of Greater Manchester Authorities took the view that 'Manchester had a quite different requirement from London'.

Table 13.9 summarises the lessons learned from expansion of the congestion charging platform, and indicates the potential for further expansion.

Table 13.9 Summary of lessons learnt: expansion

• The platform established for congestion charging can also serve other purposes such as Low Emission Zones, Ultra-Low Emission Zones or Zero Emission Zones (Zero Emission Zones are currently being planned in the UK by Oxford, Brighton and Hove, Norwich, Nottingham. Oxford and Brighton already have Low Emission Zones for buses.)
• The unit costs per transaction can be reduced through a shared services approach to back office, payment channels and enforcement
• However, for multiple reasons, to date, transport authorities have decided not to share facilities with London and this has therefore reduced the opportunity to defray operating costs across multiple schemes

Other cities such as Birmingham, Leeds and Reading also planned congestion charging schemes under TIF, but none had sufficient political support to move forward.

There have also been a series of tolled or charged crossings in the United Kingdom with similar operational requirements:

- The Highways Agency started to plan its 'Freeflow' project for the Dartford Thurrock River (Thames) Crossing in 2010. It decided to implement its own system and operations even though there had been an opportunity to share facilities with London. Its procurement awarded a contract to Emovis (formerly SANEF UK) to operate Freeflow
- Mersey Gateway Bridge decided to procure its own tolling operations – this was linked to the scheme being funded through the Private Finance Initiative in which the private sector is encouraged to take on operational risks (such as toll revenue) as a part of the overall deal. In this case, the Merseylink Consortium won the tender and contracted Emovis (formerly SANEF UK) to operate the tolling.

In practice, the potential for a shared scheme has not materialised, though there could be some economies of scale if Mersey shares some facilities with Dartford; this is likely to be primarily driven by the supply chain. This contrasts with the operating model in Ireland where a central national clearing system was deliberately set up so that all toll roads could be used by a driver using a single interoperable account (Table 13.9).

13.7 What about national road pricing?

13.7.1 What are the arguments?

The argument for national road pricing in the United Kingdom is driven mostly by:

- The acceptability of 'payment per use' being applied to the highway – we already pay for what we use for food, water, power, public transport, taxis – so why not roads?
- Whether the more sophisticated charging policy which road pricing enables is worth the additional operational costs, in comparison with the current 'per use' equivalent – fuel tax – which is very efficient to collect. Being able to charge based on vehicle type and when the journey is made offers significant benefits over just applying an annual 'flat rate' for miles driven. However, we may

need to consider a ‘flat rate’ approach as an intermediate or alternative step – see for example Chapter 18.

- What the impact is on the motorist and whether there are alternative transport modes with sufficient capacity at affordable costs to allow someone to travel.
- The Treasury’s (the UK Ministry of Finance) attitude to the revenues collected – to date they have used the term ‘cost neutral’ – but it is not clear they mean cost neutral to the motorist as the cost of collection will inevitably be higher than for fuel tax. But fuel tax revenue is projected to dwindle as more vehicles use electricity or are hybrids – and as conventional-fuel vehicles get more and more fuel-efficient. It is clear that taxation of motorists will need to adopt a different model if the same revenues are to be collected as set out by the Royal Automobile Club (RAC) Foundation and Institute for Fiscal Studies [14].
- The marginal benefit delivered for the investment made – it may not be cost-effective to charge on every road.

13.7.2 Is it feasible?

National Road Pricing is technically feasible. If we started with just the motorway network where over 20% [15] of total vehicle miles are driven, it has been estimated that there are approximately 800 motorway intersections and junctions. This could require of the order of 1,600–2,000 camera sites, depending on where they are located and the configuration of slip roads, to measure distance driven and detect entry and exit location. This is only ten times the scale of the London scheme in terms of vehicle detection sites.

Regarding customer payment channels, there is now much greater take up of digital channels than in 2003, so payment facilities for the occasional user as well as regular users should be much easier to implement at lower unit cost per transaction. Given the better quality of DVLA’s registration data and greater opportunity to have account based payment, it should also be easier to accurately bill UK drivers for their usage and enforce against non-payers.

We conclude that a national scheme is eminently feasible for the United Kingdom.

13.8 Conclusions

The experience of the London scheme has been a significant one for us and it has been a benchmark for road pricing in the United Kingdom and globally. The lessons learnt are applicable for many other schemes around the United Kingdom and worldwide and include:

- To succeed, it needs substantial political commitment, leadership and a focused delivery team
- There needs to be a programme of complementary transport and highway measures to provide additional public transport capacity and make the scheme and road network easier to use
- Softer measures are as important as ‘hard’ measures to achieve success
- There needs to be stakeholder buy-in even if there is also significant opposition
- Scheme costs are likely to become optimised over time
- Small city-only schemes are likely to have a relatively high cost to implement and operate by themselves – but even these can be successful. The key to

making schemes cost-effective is to enable sharing of common elements on a national basis between cities, toll roads and crossings

- Given how highways are currently funded and the shift towards use of electric and hybrid vehicles, the arguments for national road pricing are more compelling.

References

[1] Scrutiny by the Greater London Authority Transport Committee headed by John Biggs, December 2002, Greater London Authority.
[2] ROCOL, "Road Charging Options for London: A Technical Assessment, Report" HMSO, London, UK, March 2000.
[3] Mayoral Manifesto, Ken Livingstone's Mayoral Manifesto, 17 April 2000.
[4] Greater London Authority Act 1999. Available at https://www.legislation.gov.uk/ukpga/1999/29/contents (accessed 16 October 2017.
[5] Greater London Authority, Hearing London's Views, July 2000.
[6] Greater London Authority, A Draft Transport Strategy, Mayors Transport Strategy, January 2001.
[7] Government Office for London, A Transport Strategy for London, HM Stationery Office 1996.
[8] Greater London Authority, Final Mayors Transport Strategy, July 2001.
[9] Transport for London, London Congestion Charging Scheme Order, February 2002. Available at http://content.tfl.gov.uk/the-central-london-congestion-charging-scheme-the-consolidated-scheme-order.pdf (accessed 16 October 2017).
[10] London Local Authorities Act 1996. Available at https://www.legislation.gov.uk/ukla/1996/9/contents/enacted (accessed 16 October 2017).
[11] The Road Traffic Act 1991. Available at https://www.legislation.gov.uk/ukpga/1991/40/contents (accessed 16 October 2017).
[12] An Introduction to Public Procurement, see http://webarchive.nationalarchives.gov.uk/20110601212617/http:/www.ogc.gov.uk/documents/Introduction_to_Public_Procurement.pdf.
[13] Cardiff City Government, Keep Cardiff Moving, A White Paper, May 2003.
[14] Johnson, P., Leicester, A., and Stoye, G. (2012). Fuel for Thought: The what, why and how of motoring taxation, Institute for Fiscal Studies and RAC Foundation, London.
[15] DfT (2016). Road Traffic Estimates Great Britain, 2015, UK Department for Transport, May 2016. Available at https://www.gov.uk/government/statistics/road-traffic-estimates-in-great-britain-2015 (accessed 15 October 2017).

Further reading

[1] Richards, M. G. (2006). Congestion Charging in London: The Policy and the Politics. Basingstoke, UK and New York. Palgrave Macmillan.
[2] Smeed, R.J. (1964). Road Pricing: The Economic and Technical Possibilities. London: HMSO.

Chapter 14

The Swedish congestion charges – lessons learnt

Ida Kristoffersson and Maria Börjesson***

14.1 Introduction

At the time of writing, 2016, the congestion charges in Sweden are celebrating their 10-year anniversary. A time-of-day-dependent toll cordon was introduced around the inner city of Stockholm in 2006, followed by the introduction of a similar toll cordon in Gothenburg in 2013. Several papers explore various aspects of the two systems, e.g. cost–benefit analysis [1,2], traffic effects [3,4], acceptability [5,6], equity [7,8], forecast accuracy [9,10], the political process [4,11], elasticities [12,13] and lessons learnt [14,15]. The purpose of this chapter is to provide a forward-looking summary of what we have learnt in 10 years.

The Stockholm and Gothenburg systems are similar, time-dependent toll cordons, designed to reduce congestion in the most critical bottlenecks. The effects and adaptation strategies have proven to be similar in the two cities: for work trips, the travellers' adaptation strategy is mainly to divert to public transport. For other trips, the adaption strategies are more heterogeneous, including destination choice, cancellation of trip, route choice, etc. The charging systems are socially beneficial in both cities, meaning that the ex-post cost–benefit analyses that were performed after introduction of congestion charging [1,2] showed in both cities that the benefits (primarily travel-time savings, but to some extent also safety, noise, pollution and CO_2 reductions) exceed the governmental costs (operating costs and fuel tax decreases) of the system. Travel-time savings are larger in Stockholm and therefore the net social benefit is also higher in Stockholm compared to Gothenburg.

Although the charging systems are similar, the cities of Stockholm and Gothenburg are different. Gothenburg is less than half the size of Stockholm. It is also more sparsely populated than Stockholm and has therefore a considerably lower public transport market share. The congestion levels are also lower and concentrated only around the highway hub north of the inner city. The objective of the charges also differs between the cities: The main objective of the Stockholm system, when it was introduced, was to reduce congestion and improve the local

*VTI Swedish National Road and Transport Research Institute, Sweden

**VTI Swedish National Road and Transport Research Institute, and KTH Royal Institute of Technology, Sweden

environment. The main objective of the Gothenburg charges was to collect revenue to finance a large infrastructure package. These differences have had an impact on the traffic effects, political process, public acceptability and forecast accuracy. Hence, the comparison of the effects of Stockholm and Gothenburg systems helps to understand the transferability of the effects of congestion charges between cities with different features and of different sizes.

The support for congestion charges is now accepted among many decision-makers in Sweden, although they were extremely controversial just 10 years ago. Before 2006, the Green Party was the only party in favour of congestion charges, whereas all established parties in the city of Stockholm and Gothenburg are now in favour. In the public debate, they are often justified not only by congestion in metropolitan areas, but also as a solution to reduce climate emissions, financing of new infrastructure and reduction of local air pollution. The question is then to what extent we actually expect congestion charges to contribute to solve these problems? What pitfalls are there when using congestion charges as a policy measure? In this chapter, we discuss these questions drawing upon the great experience of congestion charging during the past decade in Sweden.

14.2 System designs

The Stockholm charging scheme was introduced in January 2006, and is designed as a toll cordon around the inner city (Figure 14.1). Charges are time-dependent and levied 6.30–18.30 on weekdays. Vehicles are charged when crossing the cordon in both directions. Since January 2016, a charge is also levied on the Essinge bypass, which is the central part of the E4/E20 motorway shown with a solid line in Figure 14.1.

Since 2006, the charge had ranged between 1 and 2 EUR[1] per passage, depending on time of day; but in January 2016, the charging levels were increased to range from 1.1 to 3.5 EUR per passage (Table 14.1). Hence, the charge increased 75 per cent in the peak but only 10 per cent in the off peak. The maximum cost for a day also increased from 6 to 10 EUR.

The Gothenburg charging scheme consists of a cordon with two spurs or 'antlers' (see Figure 14.2). Vehicles are charged when crossing the cordon in both directions. Charges are time-dependent and levied 6.00–18.30 on weekdays. The system uses the same technology as in Stockholm – automatic number plate recognition (ANPR).

Since the introduction of the charging system on 1 January 2013 and to the end of 2014, the charge ranged from 0.8 to 1.8 EUR, but from 1 January 2015, the charge ranges from 0.9 to 2.2 EUR per passage (Table 14.2). Hence, the charge increased 22 per cent in the peak but only 13 per cent in the off peak. Vehicles are charged when crossing the cordon in both directions. In Gothenburg, a multi-passage rule states that if passing the cordon more than once within 60 min, only

[1] We will use the conversion rate 1 EUR = 10 SEK throughout this chapter.

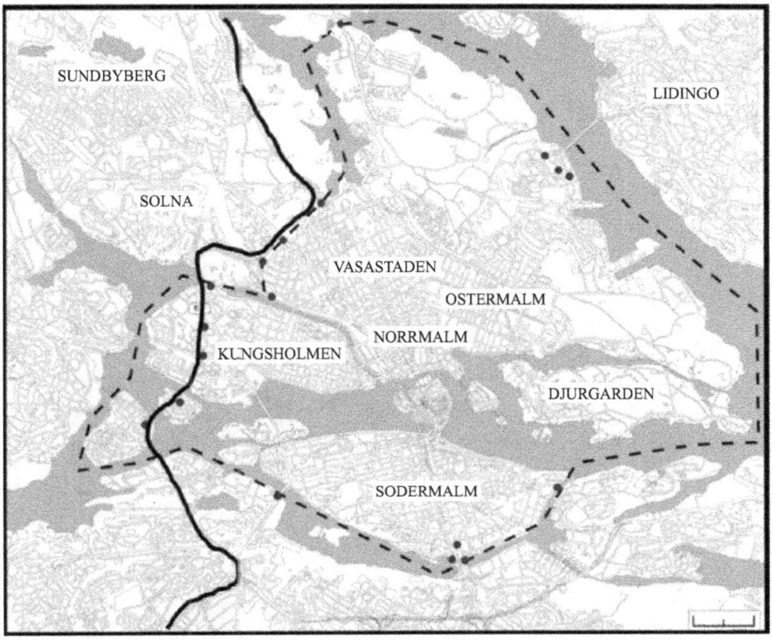

Figure 14.1 *The Stockholm congestion charging cordon (dashed line) and E4/E20 motorway (solid line) for which the central part is called the Essinge bypass*
[Source: Eliasson et al. [3]]

Table 14.1 *Charged amount in Stockholm depending on time of day before and after the charge increase*

Stockholm Time	Inner city cordon Introduced 2006	Inner city cordon Introduced 2016	Essingeleden Introduced 2016
06.30–06.59	1.0	1.5	1.5
07.00–07.29	1.5	2.5	2.2
07.30–08.29	2.0	3.5	3.0
08.30–08.59	1.5	2.5	2.2
09.00–09.29	1.0	1.5	1.5
09.30–14.59	1.0	1.1	1.1
15.00–15.29	1.0	1.5	1.5
15.30–15.59	1.5	2.5	2.2
16.00–17.29	2.0	3.5	3.0
17.30–17.59	1.5	2.5	2.2
18.00–18.29	1.0	1.5	1.5
18.30–06.29	0	0	0

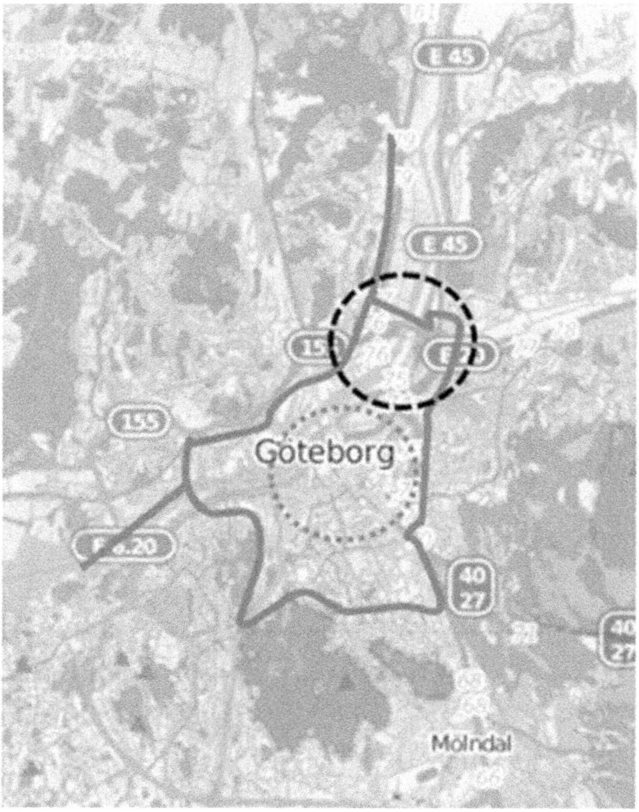

Figure 14.2 *Gothenburg with the toll cordon (solid line). The highway hub where the main bottlenecks are located is depicted with a dashed circle and the inner city with a dotted circle*

Table 14.2 *Charged amount in Gothenburg depending on time of day before and after the charge increase*

Gothenburg Time	Cordon Introduced 2013	Cordon Introduced 2015
06.00–06.29	0.8	0.9
06.30–06.59	1.3	1.6
07.00–07.59	1.8	2.2
08.00–08.29	1.3	1.6
08.30–14.59	0.8	0.9
15.00–15.29	1.3	1.6
15.30–16.59	1.8	2.2
17.00–17.59	1.3	1.6
18.00–18.29	0.8	0.9
18.30–05.59	0	0

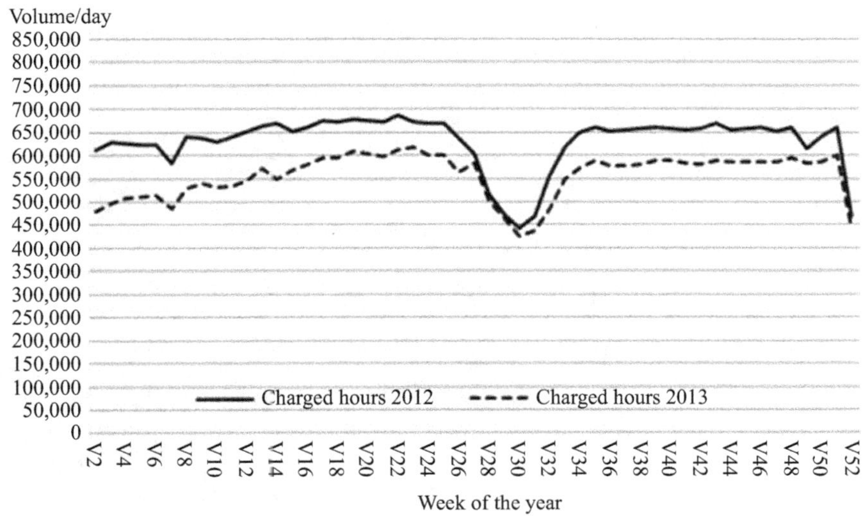

Figure 14.3 The number of vehicles passing the cordon during charged hours with and without charging

the highest charge has to be paid. The maximum cost for a day is 6 EUR, which is the same as at introduction.

The Swedish Transport Agency is responsible for the payment system and the same system is used for both Stockholm and Gothenburg congestion charges.[2] This payment system is fully automated, i.e. no manual handling is necessary. There are different ways for the user to pay the charge. The default payment is that an invoice for all passages during each month is sent to the owner of the vehicle, but vehicle owners can also sign up for electronic invoice or direct debit. In Sweden, 25 per cent pay the congestion charge via direct debit (Source: The Swedish Transport Agency). However, many of the drivers pay via e-invoice, which has also a very low transaction cost (simply tick a box in the e-bank system).

14.3 Traffic effects

14.3.1 Traffic volume across the cordon

In Gothenburg, the reduction in traffic volume across the cordon during charged hours stabilized at 12 per cent after approximately 8 months (Figure 14.3). The adjustment was slower in Gothenburg than in Stockholm, where the reduction in traffic volume across the cordon stabilized at 22 per cent after 1 month [16].

Figure 14.4 shows that the reduction in traffic volume over the cordon in Gothenburg remains almost constant throughout the charged time period, although

[2] Sweden also has two bridges with infrastructure fees – the Motala and Sundsvall bridges – for which the same payment system is also utilized.

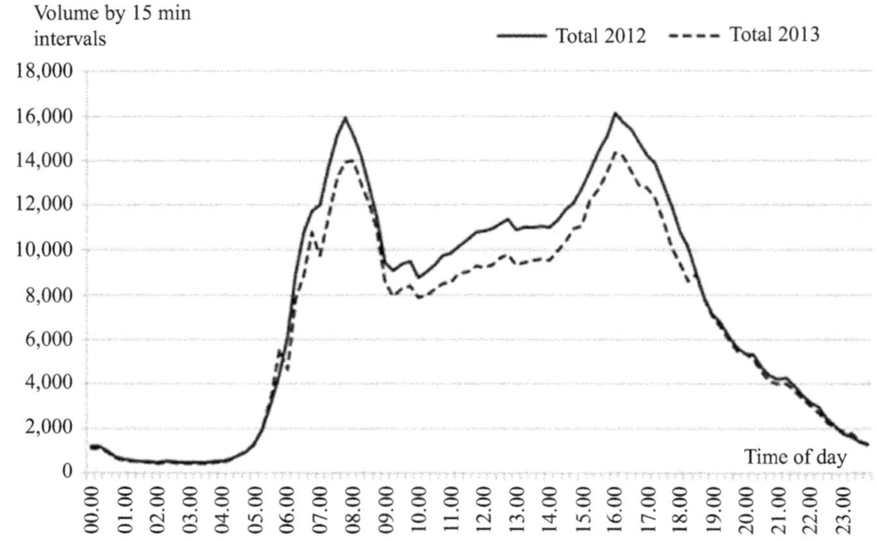

Figure 14.4 Traffic volume over the cordon in Gothenburg by 15 min intervals

the charge varies. This was the case also in Stockholm, with 22 per cent decrease over the charged hours in spite of the higher peak charge. Outside the charged hours, traffic volume remains largely unchanged in both Stockholm and Gothenburg. There is just a small reduction in the evening, presumably because these trips would have been charged in the morning going in the other direction. Spikes in the traffic volume in the morning and evening, just before and after the charge increases or decreases, indicate small shifts in departure time in Gothenburg. Similar spikes were observed in Stockholm in the first years after the introduction [17], but vanished later. Otherwise, there are no clear signs of departure time shifts.

Table 14.3 shows that the congestion charge's price elasticities

$$E_{x\to y} = \frac{\mathrm{Log}(D_y) - \mathrm{Log}(D_x)}{\mathrm{Log}(P_y) - \mathrm{Log}(P_x)},$$

where x refers to the initial state, y to the new state, D is demand, P is price level and E is elasticity. The price elasticities are higher in the off peak than in the peak in both Stockholm and Gothenburg. The relative volume reduction is constant across the day in spite of higher peak charge. Table 14.3 gives an off-peak elasticity for Gothenburg that is 1.7 times higher than the corresponding peak elasticity. For Stockholm, off-peak elasticity is 1.8 times higher than the peak elasticity at introduction.

The classic argument for the difference between off-peak and peak elasticities is that work trips are less elastic compared to trips with other purposes. However, an inspection of the car driving cost elasticities on driving distance from the Swedish and Danish national transport models [18] deepens the picture: they show

Table 14.3 Gothenburg and Stockholm congestion charge elasticities in peak and off peak at introduction

	Stockholm		Gothenburg	
	Peak	**Charged off peak**	**Peak**	**Charged off peak**
Traffic volume across the cordon without charges (veh/h)	38,934	30,119	62,516	45,984
Traffic volume across the cordon with charges (veh/h)	30,380	22,735	55,848	40,860
Real average trip cost excluding the charge (EUR)	1.95	1.95	2.78	2.78
Real average charge (EUR)	1.29	0.72	0.65	0.37
Elasticity	***−0.49***	***−0.89***	***−0.53***	***−0.93***

that non-work trips are not in general more elastic. One reason for the higher off-peak elasticities could be that already a small amount has a substantial effect at reducing traffic volume. Another reason could be that non-work trips are more elastic in dimensions relevant for congestion charging, such as destination choice, departure time choice and trip chaining.

14.3.2 Traffic volume in the inner city

One important objective in Stockholm and Gothenburg is lower traffic volumes and improved air quality in the inner city. In Gothenburg, the implementation of congestion charges coincided with other measures aimed at improving the public transport system, primarily by introducing new bus lanes. These measures reduced the capacity for driving and parking in the inner city, and have thus reduced the traffic volume over and above the reduction from congestion charges. The average reduction is 9 per cent during charged hours. During non-charged hours, however, at which the reduction across the cordon is 2 per cent, the average reduction on the city streets is still 6 per cent. These numbers indicate that the measures additional to the congestion charges reduced car traffic by 4–6 per cent, suggesting that the reduction arising from the congestion charges is 3–5 per cent.

A reduction in traffic volume within the inner city of approximately one-third of the reduction across the cordon is close to what was observed in Stockholm (where the reduction of traffic in the inner city was estimated at 8 per cent).

The reduced car traffic volume in the inner city of both Stockholm and Gothenburg implies that emissions from car traffic, such as nitrogen dioxide (NO_2) and particulates have been reduced. These reductions, however, have not been observed directly in air quality measurements, mainly due to substantial variation in air quality across the years arising from variations in weather conditions.

14.3.3 Traffic volume on roads bypassing the inner city

In the design process, it became clear that, due to the topology of the transport network, an unavoidable side effect of the charges in Gothenburg would be traffic increases on some circumferential roads due to route choice effects. Indeed, for some of the circumferential roads in Gothenburg, the traffic volume increased substantially. While these increases were an unwelcome effect for inhabitants living close to these roads, they do have enough capacity to manage the new traffic volumes. The Stockholm system did not have any substantial effects on route choices due to the natural water barriers between the inner city and the suburbs. The only route choice effect of significance that was deemed a potential concern before the charges were introduced in 2006 was traffic diverting to the Essinge bypass (Figure 14.1). But the increase in traffic volume on the bypass turned out to be marginal [16], probably because the Essinge bypass works both as a complement and as a substitute to the roads to and from the Stockholm inner city [14].

14.3.4 Travel times

Figure 14.5 shows the arterial routes and the links where travel times were observed before and after the introduction of congestion charges in Gothenburg. Travel times were also observed on selected links inside the toll cordon and on the relevant bypasses. The links where the travel times were observed are sorted into one of four different categories: inner arterials, outer arterials, inside cordon and bypasses. Inner arterials are the links closest to the highway hub as depicted in a darker colour on the map in Figure 14.5. The outer arterials category includes the links shown in grey on the map. Streets inside the cordon and the bypasses on which travel times were measured are not depicted in Figure 14.5. Their exact location can be found in the travel-time report [19].

Figure 14.6 shows the relative increase in travel times in the morning peak hour 7.00–8.00 relative to free-flow travel time,[3] before and after the introduction of the congestion charges in Gothenburg for the four categories of links. The observed travel times are averaged over all weekdays within 5 weeks in September and October 2012 (before) and 2013 (after). A relative increase in travel time of 0 per cent corresponds to free-flow travel time, and 100 per cent corresponds to a travel time twice the free-flow travel time.

The error bars in Figure 14.6 illustrate the average between-day variation in travel time for the six 10-min intervals within the peak hour 7.00–8.00. The relative increase in travel time for the days when the best and worst decile of the travel-time distribution were observed is identified for each six 10-min interval. The error bars show the best and worst decile, averaged over the six 10-min intervals within the peak hour 7.00–8.00.

[3] Free-flow travel time is defined as the tenth percentile travel time observed between 06.00 and 06.09 over 5 weeks in September and October 2012 and 2013.

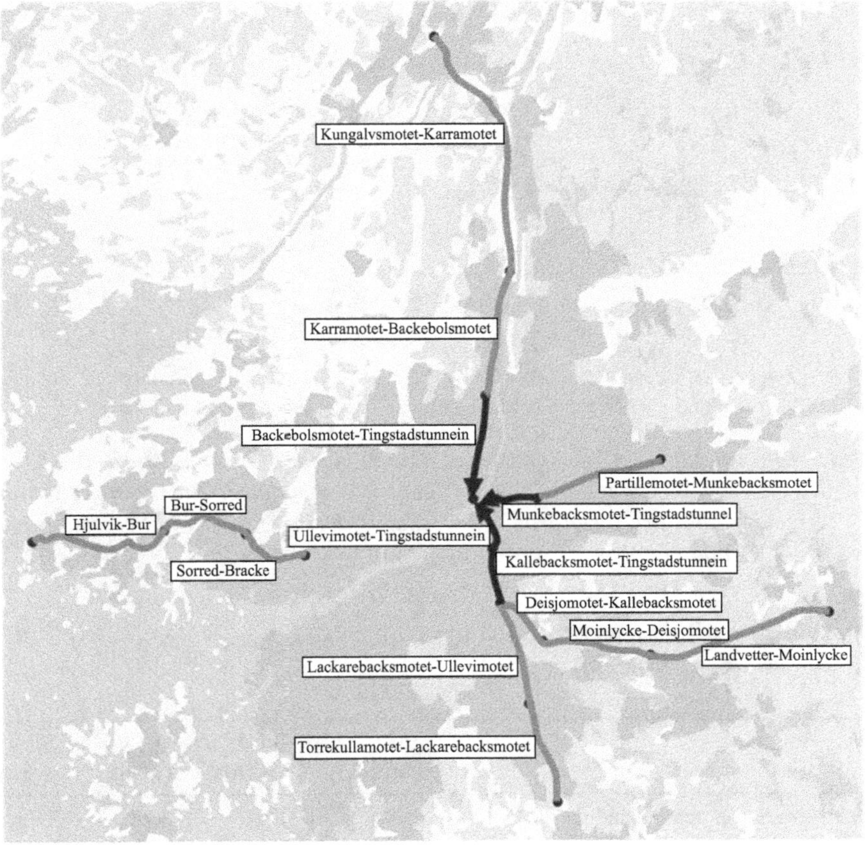

Figure 14.5 Arterial roads (shown in grey) where travel times were observed 06.00 to 10.00 before and after the introduction of charges in Gothenburg. Inner arterials are depicted in black

Figure 14.6 shows that prior to charging, congestion in Gothenburg was limited to the innermost arterials in the morning peak. Congestion on these links is significantly reduced by the charges. However, the travel-time savings are still limited because the average travel time on these links is only approximately 5 min in the morning peak. Travel-time reductions on outer arterials and inside the cordon are minor, but this is expected due to the low initial congestion levels on these streets. Travel times on bypasses remain unchanged even though volume increased on these roads, showing that bypasses have enough capacity to handle the volume increases.

In Stockholm, travel-time reductions occurred on a much larger part of the network compared to Gothenburg. Large travel-time reductions even occurred in the suburbs on streets that were not charged since the blocking back of queues reduced. The reason for more widespread travel-time improvements in Stockholm is higher initial congestion levels on a larger part of the network.

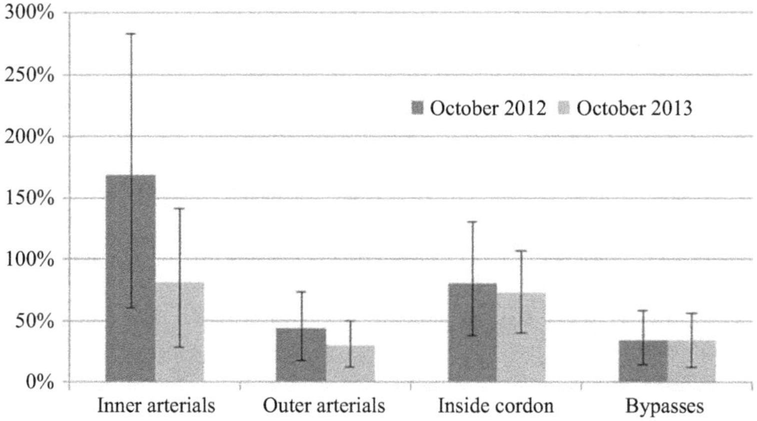

Figure 14.6 Relative increase in travel time in morning peak hour 07.00–08.00 for different categories of links in Gothenburg. The observed travel times are taken from all weekdays during 5 weeks in September and October

Table 14.4 Long-term elasticities of congestion charging in Gothenburg for the years 2013–2015

	2012 (without)	**2013 (with)**	**2014 (with)**	**2015 (with)**
Real average trip cost excluding the charge (EUR)	2.78	2.78	2.78	2.78
Traffic volume across the cordon adjusted to 2005 levels w.r.t. external factors (veh/h)	52 597	46 855	47 581	47 525
Real average charge (EUR)	–	0.51	0.50	0.59
Elasticity charged hours	–	***−0.69***	***−0.60***	***−0.52***

14.3.5 Long-term effects and effects of increased charging levels

The short-run elasticity computed for Gothenburg in 2013 is −0.69 (Table 14.4). This figure is almost identical to the elasticity in Stockholm the first year after the introduction of charges, −0.70 [12]. The similarity of these elasticities is particularly interesting given the large difference in public transport market share, which is approximately 75 per cent for peak-hour trips across the cordon in Stockholm compared to approximately 25 per cent in Gothenburg.

Differences between Stockholm and Gothenburg do, however, appear when considering long-run elasticities. Whereas the Stockholm elasticity steadily increases over the years from −0.70 at introduction in 2006 to −0.86 in 2011 [12], the Gothenburg charging elasticity declines from −0.69 in 2013 to −0.52 in 2015. The 2015-elasticity computed for Gothenburg is not entirely comparable with the previous elasticities and the Stockholm elasticities up to 2011, because of the increases in the charging levels in January 2015. However, also the 2014-elasticity before the charge increase for Gothenburg (−0.60) is lower than the one observed for 2013. Furthermore, the off-peak elasticity is also declining over the years [13] even though the charge increase in January 2015 was minimal in the off peak (from 0.8 to 0.9 EUR).

The discrepancy between Stockholm and Gothenburg regarding the direction of the trends in the elasticities might be caused by differences in city structures and transport systems. Gothenburg is much smaller and less dense compared to Stockholm, with most workplaces not situated in the city centre. For this reason, congestion is more local in Gothenburg and occurs in the highway hub north of the city centre. This is the main reason why the motorways are charged. Furthermore, public transport is not as extensive as in Stockholm. Together, this means fewer ways to adapt to the charges in the long term. Some drivers might even have tried public transport but switched back to car.

Table 14.5 shows that the elasticities of the increase in congestion charge in Gothenburg and Stockholm are substantially lower than the elasticities at first introduction. Charges were increased mainly in the peak (an increase of only 0.1 EUR in the charged off-peak hours during midday), so when it comes to the effect of the increase, we focus on the peak elasticities. In Gothenburg, the peak-period elasticity is −0.16 for the increase compared to −0.53 at first introduction. Also in Stockholm, the peak-period elasticity for the increase is lower than at first introduction: −0.24 compared to −0.49.

14.4 Adaptation strategies

To analyse adaptation strategies in Gothenburg among the drivers who changed their behaviour in response to the congestion charge, a two-wave panel travel survey was conducted in March–April 2012 and in March–April 2013 [20]. The respondents were randomly sampled from residents in the Gothenburg region,

Table 14.5 Gothenburg and Stockholm elasticities of the increase in charge in 2015 and 2016, respectively

	Stockholm peak	Gothenburg peak
Traffic volume across the cordon in peak without charge increase (veh/h)	30,898	56,609
Traffic volume across the cordon in peak with charge increase (veh/h)	29,315	56,258
Real average trip cost excluding the charge (EUR)	2.41	2.78
Real average charge (EUR) without charge increase	1.37	0.63
Real average charge (EUR) with charge increase	2.31	0.77
Elasticity	***−0.24***	***−0.16***

Table 14.6 Total number of trips by mode and purpose in OD pairs affected by the charges – before and after introduction of congestion charges

	Total 2012	Total 2013	Change	Change in per cent
Work trips				
Car	101,000	92,000	−9,000	−9
Public transport	49,000	61,000	12,000	24
Cycle	14,000	8,000	−5,000	−36
Other trips				
Car	124,000	115,000	−9,000	−7
Public transport	79,000	73,000	−6,000	−8
Cycle	8,000	7,000	−1,000	−13

commuting in OD (Origin-Destination) pairs affected by the charges. Approximately 3,000 individuals participated in both waves, reporting all together 17,000 trips.

Table 14.6 shows the total number of trips by mode and purpose in OD pairs affected by the charges, calculated based on the survey and scaled to total levels. The share of car trips priced off the road was approximately the same for commuting and discretionary trips (9 and 7 per cent, respectively). This is consistent with the finding that the total percentage reduction remained approximately constant over the day (where the bulk of commuters travel in the peak and the bulk of discretionary trip drivers travel in the off peak). The reduction of car trips is smaller among the respondents than the 12 per cent observed over the cordon, but the respondents are not a random sample of all drivers across the cordon.

The adaptation strategies, however, differ between commuting and discretionary trips. The 9,000 commuters who chose not to drive apparently switched to public transport, increasing by 12,000 trips. Among discretionary travellers, the main adaptation strategies are changing destination and reducing trip frequency. In fact, the number of public transport trips even reduced. The congestion charges have not had any measurable impact on the market share for cycling, in fact cycling trips reduced. This is likely to be due to a period of frost and colder weather in early spring 2013[4]. This also suggests that the main reason why the number of public transport trips for commuting increased more (12,000) than the decrease in car work trips (9,000) is that some commuters diverted from cycling to public transport. The drop in discretionary public transport trips in affected relations might also be due to the cold weather in March 2013. Liu *et al.* [22] find that public transport use declines and that travellers tend to make shorter trips (trips in affected relations are relatively long) when the temperature drops below −2°C.

[4] Cycling levels, even daily variations, are very sensitive to weather conditions in Sweden [21]. The weather in March 2013, all 31 days were frost days and the average temperature −0.8°C, but March 2012 had only 7 frost days and average temperature 5.7°C.

The adaptation mechanisms observed in Stockholm are very similar to those observed in Gothenburg: commuters diverted to public transport and discretionary travellers adapted in other ways [12]. The cycling levels did not increase in response to the charges in Stockholm either, although Stockholm has experienced a steady trend increase in cycling levels over many years [23].

Since the respondents were not sampled randomly, the sample is fairly small and conducted under different weather conditions in the two survey years, the total increase in public transport trips according to the travel survey is unreliable. Moreover, the traffic counts for the Gothenburg public transport system are unreliable because it is optional to tap the travel card when entering buses and trams for travellers with yearly or monthly tickets, who are in the majority. To assess how the public transport travel volume has changed, the most reliable data source is therefore ticket sales.

Since it is mainly commuters who switch to public transport, and they almost exclusively use monthly and yearly tickets, we only regard the sales of monthly and yearly tickets. The price of single tickets changed substantially between 2012 and 2013, so single ticket sales are not comparable between the years. The monthly and yearly ticket sales increased by 7.5 per cent. This number is slightly higher than the estimate based on the two-wave travel survey; according to this, the public transport trips increased by 5 per cent in charged OD pairs.

However, the sales of monthly and yearly tickets increased by around 2 per cent (between 1 and 3 per cent) yearly over several years prior to 2012, due to population growth and various marketing campaigns (both continuing during 2013). Together, these numbers suggest that the increase in public transport use due to the congestion charges is within the range of 4.5 to 6.5 per cent. This is similar to the effect of charges on public transport use in Stockholm, where the charges have been estimated to increase the number of passengers in public transport by 4.5 per cent [16].

14.5 Revenues and system costs

The congestion charge in Gothenburg generated a gross revenue of 72 million EUR during its first year. This is close to the 76 million EUR generated by the Stockholm system in 2013, although Gothenburg is less than half the size of Stockholm and the charged amounts are lower. This implies that a considerably larger proportion of the population of Gothenburg, compared to Stockholm, pays congestion charges. One of the main reasons is that the share of public transport trips is lower than in Stockholm, partly due to a different land-use pattern.

The revenue of 72 million EUR is lower than the Sampers[5] forecast from 2009, which predicted a gross revenue of 93 million EUR [24]. There are two main reasons for the under-prediction of revenues. First, the analysts made the

[5] Sampers are the national transport model in Sweden used by the Transport Administration for prediction and investment evaluation.

assumption that the traffic across the cordon would increase by 1.2 per cent per year during 2008–2013. Due to the sharp increase in fuel prices and the economic downturn in this period, however, the traffic decreased by on average 0.7 per cent per year. Second, the multi-passage rule resulted in more uncharged passages than predicted.

The late payment penalties, amounting to 8 million EUR in 2013, partly made up for the revenue shortfall. But even including those, the revenue was still 13 million EUR below the target in the first year. However, the interest rates have also fallen to a lower level than predicted, implying that the construction costs reduced as well. It is unclear who takes the risk if the revenue still does not cover the project cost after 25 years as planned, but most likely the repayment period will then be extended.

The operating cost of the Gothenburg and Stockholm systems, including costs for maintaining the technical system, customer service and invoicing, was 24 million EUR in 2013. The operating costs have declined over the years: it was 22 million EUR solely in Stockholm the second year of operation of the system. As explained by Hamilton [25], one of the main cost drivers of the Stockholm system was the requirement of an extremely high resilience, essential because of the high political risk. The payment systems of the Swedish congestion charges have now been centralized and are fully automated. Hence, there is probably limited scope for further reductions in the operating cost in the near future.

14.6 Model predictions

The design of a congestion charging system is not an easy task and heavily dependent on local conditions, such as where bottlenecks are located and route choice possibilities. In both Stockholm and Gothenburg, a number of suggested systems were rejected because the transport model forecasts indicated that they actually created problems in the transport systems. Cities interested in implementing congestion charging should therefore carry out model-based studies to evaluate different designs of the charging system.

Results from Eliasson *et al.* [9], Almroth *et al.* [26], Engelson and van Amelsfort [27] and Saifuzzaman *et al.* [28] show that in the Stockholm case, with extensive spillback queues, a dynamic assignment model is necessary for capturing the full travel-time reductions of congestion charges. In Gothenburg, on the other hand, where spillback queues are not a large problem, a static assignment model can predict the travel-time reductions with high accuracy [10].

In a network where congestion charges apply, the value of time and driving cost per kilometre determines whether the driver will take a detour or pay the charge. The topology of Gothenburg (as distinct from e.g. Stockholm) implies large possibilities to avoid charges by detouring for drivers in many OD pairs. For such areas, the forecast of the effect on traffic across the cordon and on the circumferential roads is highly sensitive to the assignment parameters relating to travel distance, travel time and charge.

A particular problem for the Gothenburg model was the multi-passage rule, stating that trips passing several gates within 60 min only have to pay once. This means that the total travel cost for a given route is not the sum of the travel cost of the links included in the route, which is a standard assumption in route choice modelling. A hierarchical route choice, applying a continuous value of travel-time (VTT) distribution, was therefore developed to forecast the effects of the Gothenburg charges. The hierarchical route choice model proved successful resulting in high accuracy of the prediction of the route choices [10]. The hierarchical route choice method can be applied to congestion charging systems resulting in non-additive link attributes, especially when there are few discrete charge levels in the network.

The continuous VTT distributions were estimated from stated preference data from the Swedish value of time study [29]. However, these VTT distributions were adjusted in the ex-ante analyses, since the predicted route choice effects were otherwise judged to be too large based on experience from Stockholm. Given the adjusted distribution, we find that the route choice in general was predicted with fairly high accuracy, indicating that a VTT distribution from stated preference data should not be directly applied in route choice modelling.

In both Stockholm and Gothenburg, almost all work trips that adapted switched to public transport and the transport model was able to predict this effect with reasonable accuracy [9,10]. However, off-peak traffic adjusted in more heterogeneous ways in both cities, such as trip chaining, change of destination and cancellation of trip [4,9]. This led to less accurate predictions of off-peak traffic with underestimation of the reduction in other car trips.

14.7 Cost–benefit analysis, equity effects and company cars

The congestion charging systems are socially beneficial (congestion and environmental costs reduce more than the operating costs of the system) in both Stockholm and Gothenburg, generating a net benefit of 65 million EUR per year in Stockholm [1] and 20 million EUR per year in Gothenburg [2]. However, even though the charges are socially beneficial on a system level, benefits and costs can still be distributed such that the system is regressive, i.e. favouring high-income individuals.

The investment cost in Gothenburg was from a financial cost perspective recovered in slightly more than a year. In Stockholm, the investment costs were recovered in financial terms in 3.3 years. The longer time in Stockholm is due to the substantially larger investment cost. When the Gothenburg system was introduced it could just use all the system development from Stockholm, which made the system implementation substantially cheaper. From a social cost perspective, the Gothenburg system had recovered costs in less than 4 years and the Stockholm system after 4.5 years.

Although the Gothenburg charges have a positive social benefit, the sums that are redistributed are substantially larger than the net benefit. The relative efficiency of a pricing scheme can be calculated as a ratio of the total net social benefits and the total revenue collected from the congestion charges. In the Gothenburg case, the relative efficiency is 20/77 = 0.26 [2]. This can be compared to the Stockholm system, where the relative efficiency was 65/80 = 0.81 [1]. Hence, the sum of money that is redistributed in the population compared to the net social benefit is substantially larger in the Gothenburg case than in the Stockholm case. The large redistribution of money compared to the social benefit is one argument for carefully considering the redistribution effect.

However, to understand the distribution effects of the Swedish company car exemption and the importance of the specific design of the system, the company car legislation is crucial. After congestion charges were introduced in Stockholm, the tax court made the decision that congestion charges should be seen as part of the operating costs of the car, and therefore should be included in the 'taxable benefit value' of a company car. This means that for drivers travelling for private purpose (including commuting) using a company car, the employer pays the charge. Company cars amount to about 20 per cent of all passenger cars crossing the cordon in Gothenburg and 39 per cent of all passenger car crossings in Stockholm. This implies that the effects of the charges are smaller than optimal and lead to negative distribution/equity effects of the charges (since it is almost only drivers in the highest income segment who have access to company cars with fringe benefits). This was disregarded by the congestion charging system designers and policy makers – but it has had substantial effects on the distributional profile of the charges.

The Gothenburg distribution analysis shows that the congestion charge is regressive [2] for several reasons. First, even low-income individuals are highly car dependent in Gothenburg due to the relatively low public transport share (about 25 per cent in charged OD pairs). Second, workers in the highest income class have considerably higher access to company cars. Third, low-income individuals have lower values of time, implying that they adapt more frequently to the charges and therefore take a larger share of the adaptation costs.

Because most residents in Gothenburg suffer a net loss from the charges, and because the distribution of the direct effects of the charges are regressive, the spending of the revenue is decisive for the total effect on equity. However, the revenue is spent mainly on a rail tunnel, which primarily gives benefits to commuters from surrounding municipalities in the regions. This could be one important reason behind the low public-support for the charges. In Stockholm, on the other hand, the larger travel-time gains resulted in a less negative consumer surplus; taking into account heterogeneity in values of time and network effects, it might even have been slightly positive [30]. Public support is also much higher in Stockholm.

14.8 Public support

When the decision was made to carry out a 'congestion charging trial' in Stockholm, it was met with great resistance: In 2004 and early 2005, 40 per cent of Stockholm citizens stated that they would vote yes to congestion charges. Support fell to 36 per cent just before the introduction of the charges. Once the trial started, however, support increased to 52 per cent. The trial ended on 31 July 2006 and was followed by a referendum in September at the same time as general and local governmental elections were held. In this, 53 per cent of Stockholm citizens voted for keeping the charges. A poll in December 2007 showed a 66 per cent support for the charges. In August 2009, a poll showed that 74 per cent were supporting the charges and in May 2011, a poll encompassing most of Stockholm County showed a support for the charges of over 70 per cent.

When the charges were introduced in Gothenburg, no referendum was planned. However, after the introduction of the charges in 2013, the public resistance resulted in a petition for a referendum, arranged by the tabloid newspaper *Göteborgs-Tidningen*, which was signed by 90,000 citizens of the City of Gothenburg. This petition triggered a decision in the City Council (5 September 2013) to arrange a consultative referendum in conjunction with the general election on 14 September 2014, although several parties were against it. The question in the referendum was formulated as: 'Do you think that the congestion tax should continue after the 2014 election?' A total of 57 per cent of the population voted 'No'. Given the result of the referendum, an investigation regarding alternative financing methods (e.g. public–private partnership or exploitation charges) of the West-Swedish package was performed [31], but no alternative were judged to be able to replace the congestion charges' part of the West-Swedish package within a reasonable time frame. Furthermore, none of the analysed alternative financing methods would have equivalent effects on the local environment and congestion and therefore require investments additional to the West-Swedish package to achieve similar effects. Therefore, since the referendum was only consultative, the congestion charges remain.

A key question is then, why the referendum in Stockholm resulted in a majority in favour of the charges while the result in Gothenburg was the opposite. The support for congestion charges has always been lower in Gothenburg than in Stockholm, at least in part due to the lower public transport shares and lower congestion levels, implying that a larger share of the population is negatively affected by charges and that fewer drivers receive any significant benefits in terms of travel-time reductions. A further key difference between the cities is that in Gothenburg, the charges are justified mainly as a tax collection instrument needed to finance an investment package, which was not the case in Stockholm. Even if the decision-makers have tried to market the charges as a measure to improve the environment and congestion, some voters might have seen through this argument.

As in Stockholm [12], the support for the charges dipped just before their introduction, but has increased thereafter. The share of positive respondents increased from 33 to 50 per cent according to a survey conducted by Börjesson *et al.* [6]. A poll in the autumn of 2014, just after the referendum, even showed a 50 per cent majority in favour of the charges among respondents residing in the municipality of Gothenburg.

Possibly, the lower support in the referendum was due to the increasingly negative opinion of the West Link (see the section below). In the run-up to the referendum, the opinion of the West Link declined. The public debate also increasingly focused on arguments against the West Link. In the poll, the respondents could separately state their support for the congestion charges and the West Link, but in the referendum, there was only one single response, not allowing a separate statement about the West Link. It is thus possible that some voted 'No' in the referendum to protest against the West Link, although being in favour of the charges. The opinion in favour of the West Link dropped also over the summer in 2012, which could also be due to the intense media debate questioning its benefits, costs and cost efficiency just after the criticism from the National Audit Office. While the respondents within the municipality of Gothenburg are more positive to congestion charges than the respondents outside, the support for the West Link is substantially lower in the municipality of Gothenburg. Moreover, the referendum result showed that also within the municipality of Gothenburg, the residents of the city centre are more positive than the residents outside the centre. There was a clear majority, 55 per cent, in favour of the charges in the city centre of Gothenburg, but a clear minority, 39 per cent, in favour outside the city centre.[6]

In fact, in the spring of 2016, the support for the charges had declined in Gothenburg as opposed to Stockholm. This might be a result of the increase in the charging level in January 2016. This increase had no visible effect on the travel times and was motivated by the financing of the West Link. In summary, our findings suggest that the focus on spending revenues on infrastructure investments with low social returns did not help to build public support for the charges, particularly not in the long run.

14.9 Political support

Before congestion charges were introduced in Stockholm in 2006, the support was as low among political parties in both Stockholm and Gothenburg as among politicians in most other cities around the world. In less than 10 years, however, congestion charges have become completely uncontroversial and accepted by all political parties in Stockholm and by all well-established political parties in Gothenburg. This is not primarily due to the increased public support in Stockholm, but rather because congestion charging has come to play a key role in the negotiations of the national grants for transport investments.

Prior to the Stockholm charges, the local and regional politicians of Stockholm of all parties were greatly concerned that Stockholm would receive fewer national infrastructure grants if they introduced congestion charges. They feared that Stockholm would be forced to use the revenues from the congestion charges and thereby miss out on national grants. This issue was solved by an agreement between the national government and the region settled in late 2007, according to which Stockholm would receive a major transport investment package, 50 per cent funded by the revenues and 50 per cent by the national government. This was a turning point for the political support, and since then all political parties in Stockholm have been in favour of the charges.

Before the Stockholm agreement was settled, most of the major transport investments in Sweden were funded by national grants. But at the outset of the

[6] Source: http://goteborg.se/wps/portal/invanare/kommun-o-politik/sa-kan-du-paverka/folkomrostning-om-trangselskatt/preliminart-valresultat/!ut/p/b1/04_Sj9Q1NDYxNDazsDA004_Qj8pLLMtMTyzJzM9LzAHxo8ziQw0NAi2cDB0NLMxdTAwcg839Qg0s3I3cnQ2BCiKBCgxwAEcDQvr9PPJzU_Vzo3IsAAtvwPQ!/dl4/d5/L2dBISEvZ0FBIS9nQSEh/.

preparation of the national investment plan 2010–2021, the Government declared that investments receiving regional co-funding would be prioritized as well as cost-efficient investments and investments in urban regions. The Transport Administration handed over a proposal for the national investment plan to the Government on 1 September 2009. Local and regional politicians in Gothenburg, however, felt that their region, in spite of being the second largest city in Sweden, did not receive their fair share of the national infrastructure grants. In particular, they argued, they were given substantially less than the Stockholm region.

In the spring of 2009, a small group representing the national Government, the regional governments (Region Västra Götaland and Region Halland), the Gothenburg Region Association of Local Authorities (GR[7]) and the local government of the City of Gothenburg begun negotiating an infrastructure package to be co-funded by the region and thereby squeezed into the plan at the last minute. This process, involving a small number of negotiators from each the stakeholders, was informal, closed and lacked authoritative or official instructions. The outcome of these unofficial negotiations was announced in late August of 2009. It resulted in a broad political coalition in the Gothenburg City Council stating its support for a West Swedish Agreement, partly funded by congestion charges (28 October 2009).

The largest investment in the package is the West Link (2.0 billion EUR), which is an 8-km-long rail link including a 6-km-long tunnel under central Gothenburg. The package also includes a 0.4 billion EUR road tunnel and 0.5 billion EUR for public transport improvements introduced jointly with, or shortly after, the charges. The latter are used to strengthen existing bus and commuter train lines with increased service frequency and separate bus lanes.

The total budget of the West Swedish Agreement is 3.4 billion EUR, of which 1.7 billion is accounted for by the national Government, 1.4 billion by revenue from the charges (of which 33 per cent are paid by citizens in the city of Gothenburg), 0.125 billion by the City of Gothenburg, 0.1 billion by the regions surrounding Gothenburg[8] and 0.75 billion by realization of increased land values.

The Government planned to establish the national plan in early spring of 2010. At the expense of some other projects previously included in the proposed plan, the West Swedish Agreement was included in the final national plan adopted by the Government on 1 April 2010. In the hurried and informal negotiation process, the stakeholders suggested investments that were not properly evaluated. When at a later stage they were evaluated, it became clear that the cost efficiency of the West Link was very low, the benefit–cost ratio being 0.45 [32]. A further problem recognized by the National Audit Office is that for the capacity of the West Link to be realized, additional investments are needed, roughly corresponding to the cost estimate for the tunnel itself. Hence, in the current cost-benefit analysis, all benefits are included but only approximately half its costs.

However, because this agreement was negotiated between many stakeholders and political parties, it could not be modified. This illustrates the problems associated with large infrastructure packages having broad political support, in

[7] A co-operative organization uniting 13 municipalities in western Sweden.

[8] Västra Götalands and Hallands län.

particular when resulting from hurried and closed negotiations. This has been discussed in the context of the Norwegian road tolls, primarily used to finance large infrastructure investments [33,34].

There may be several reasons why investments with low-cost efficiency were suggested in the first place. One possible reason is that it is difficult to assess the benefit of an investment without a proper evaluation. Another potential reason is that each stakeholder takes the total benefit of an investment into account but only factors in its own share of the cost. For instance, the City of Gothenburg carries only 4 per cent of the total cost of the Agreement (0.125 out of 3.4 billion EUR).[9] The politicians in the City Council are thus likely to consider the total benefits of the Agreement but only a small share of its cost.

All political assemblies concerned were involved in this hurried and informal process, albeit most of them at a late stage. However, there was a huge information and communication deficit to the public during this process and no public involvement. This had consequences in the election of 2010, to the extent that a new political party, Vägvalet, was elected into the City Council of Gothenburg, resistance to the congestion charge being its single issue.

The result of this is that it is now established in Sweden that investments co-financed by money from congestion charges from the region are prioritized over investments in other regions even if the social benefits are low. The result is that also in Stockholm, focus of investment from congestion charges is increasingly directed towards the use of revenues for co-financing more infrastructure investments with low social benefit.

14.10 Lessons learnt and recommendations for other cities

The charging systems in Sweden show that congestion charges can be an efficient (socio-economically beneficial) and effective policy measure for combating urban congestion. Furthermore, the technology of the Swedish charging systems has proven to work well, with high accuracy of correctly identified vehicles using the video technique with ANPR.

The case of Gothenburg demonstrates this measure is not only less efficient if initial congestion levels are low, but also less efficient in the long run: the effects are declining in the long run. In Stockholm, the effects have increased over the years. The difference between the cities in this respect could be a result of the lower density city structure and high car dependence in Gothenburg. From this perspective, congestion charges are likely most successful in cities where congestion levels are high and where there exist good alternatives to driving. Table 14.7 summarizes lessons learnt for cities thinking of introducing congestion charging.

There is a case for reducing health damaging emissions in large metropolitan areas such as Stockholm. Hence, extending the Swedish systems by a differentiation of the charging levels such that vehicle owners whose vehicles emit more health-damaging pollutants pay higher charges might be a good policy. Exempting

[9] The City of Gothenburg carries 17 per cent of the cost if the third of the charge revenue paid by the citizens of Gothenburg is included.

Table 14.7 Lessons learnt for cities thinking of introducing congestion charging

Congestion charging	Lessons learnt
Objectives	Congestion charges do reduce congestion and local air pollution. This should therefore be the objectives for introducing charges
Design	The design of a congestion charging system is a difficult task and there is a high risk of moving congestion around in the network or causing new congestion. A carefully calibrated local transport model should therefore be used in the design process to reject bad designs. For highly congested cities, this model needs to be dynamic, if there are spill-back queues in the system
Exemptions	Exemptions have a large impact on the equity and efficiency of the charging system and should be carefully analysed in advance. Exemptions might also drive operation cost of the system
Operating costs	A fully automated payment system is a key factor for keeping operating costs down
Revenues	If congestion charges are part of a larger investment package, care should be taken to the social benefit of the investments in the package
Extensions and increases of amounts	The effects (price elasticities) of increased charges and extensions of the system cannot be expected to be as large as the effects at introduction
Short-term effects	Initial congestion levels are of key importance for the resulting travel-time gains of congestion charging
Long-term effects	City density and public transport market shares are important factors for long-term success of congestion charging
Equity	Level of car dependency among low-income citizens is very important for the resulting equity effects
Public support	Framing of the charges as a green policy reducing congestion is important for public support
Political support	The power over the revenues, the system design and the role in the negotiations on national infrastructure grants, seem to be critical for political support

alternative fuel vehicles in the first years of the Stockholm congestion charge did have a large effect on the car fleet in Stockholm [35].

In Sweden, however, we find no other cities with congestion levels high enough to warrant congestion charging in the coming decade. Congestion charges are an expensive policy for pricing externalities and there is a high risk of negative distribution effects in cities with low public-transport usage. Therefore, for less congested cities, local air pollution is best dealt with by appropriate parking pricing, increasing taxes on company cars and the deductibility of commuting costs.

When considering congestion charging, it is of utmost importance to take care in the design of the system, using to this end a carefully calibrated transport model. Design details such as company car taxations seem to have very large effects on equity and also on the efficiency of the system. These impacts have proven to be larger than expected in Sweden and need to be considered more in the future.

In Stockholm, the peak-hour charge was increased from 20 to 35 SEK in January 2016 and in Gothenburg from 18 to 22 SEK in January 2015. The effects of the charge (price elasticities) observed after the increase in charging levels in Gothenburg and Stockholm are substantially lower than the elasticities at first introduction. This indicates that dynamic price adjustments are not as effective as expected. Therefore, it is not as interesting to strive for a charge that varies between days, months or seasons.

The experiences from Stockholm and Gothenburg indicate that congestion charges must be introduced as part of a bigger investment package deal and the negotiations of the national infrastructure grants. Broad political consensus has proven to be essential for such deals. However, large investment packages formed by broad political consensus have resulted in inefficient investments that cannot be modified and improved. This is a real concern since the possibilities of a more efficient utilization of existing infrastructure and capacity in the age of digitalization and automation have never have been larger.

References

[1] J. Eliasson, 'A cost-benefit analysis of the Stockholm congestion charging system', *Transportation Research Part A: Policy and Practice*, vol. 43, no. 4, pp. 468–480, May 2009.

[2] J. West and M. Börjesson, 'The Gothenburg congestion charges: CBA and equity', *CTS Working Paper 2016:17*, 2016. Technical report. http://www.transportportal.se/swopec/CTS2016-17.pdf (accessed 09 October 2017).

[3] J. Eliasson, 'Lessons from the Stockholm congestion charging trial', *Transport Policy*, vol. 15, no. 6, pp. 395–404, 2008.

[4] M. Börjesson and I. Kristoffersson, 'The Gothenburg congestion charge. Effects, design and politics', *Transportation Research Part A: Policy and Practice*, vol. 75, pp. 134–146, May 2015.

[5] J. Eliasson and L. Jonsson, 'The unexpected "yes": Explanatory factors behind the positive attitudes to congestion charges in Stockholm', *Transport Policy*, vol. 18, no. 4, pp. 636–647, Aug. 2011.

[6] M. Börjesson, J. Eliasson, and C. Hamilton, 'Why experience changes attitudes to congestion pricing: The case of Gothenburg', *Transportation Research Part A: Policy and Practice*, vol. 85, pp. 1–16, Mar. 2016.

[7] I. Kristoffersson, L. Engelson and M. Börjesson, 'Efficiency vs equity: Conflicting objectives of congestion charges', *Transport Policy*, vol. 60, pp. 99–107, Nov. 2017.

[8] J. Eliasson, 'Is congestion pricing fair? Consumer and citizen perspectives on equity effects', *Transport Policy*, vol. 52, pp. 1–15, Nov. 2016.

[9] J. Eliasson, M. Börjesson, D. van Amelsfort, K. Brundell-Freij, and L. Engelson, 'Accuracy of congestion pricing forecasts', *Transportation Research Part A: Policy and Practice*, vol. 52, pp. 34–46, Jun. 2013.

[10] J. West, M. Börjesson, and L. Engelson, 'Accuracy of the Gothenburg congestion charges forecast', *Transportation Research Part A: Policy and Practice*, vol. 94, pp. 266–277, 2016.

[11] J. Eliasson, 'The role of attitude structures, direct experience and framing for successful congestion pricing', *Transportation Research A*, vol. 67, pp. 81–95, 2014.

[12] M. Börjesson, J. Eliasson, M. B. Hugosson, and K. Brundell-Freij, 'The Stockholm congestion charges—5 years on. Effects, acceptability and lessons learnt', *Transport Policy*, vol. 20, pp. 1–12, Mar. 2012.

[13] M. Börjesson and I. Kristoffersson, 'The Swedish congestion charges: Ten years on – and effects of increasing charging levels', *CTS Working Paper*, 2017:2. Jan. 2017. Technical report. http://www.transportportal.se/swopec/CTS2017-2.pdf (accessed 09 October 2017).

[14] M. Börjesson, K. Brundell-Freij, and J. Eliasson, 'Not invented here: Transferability of congestion charges effects', *Transport Policy*, vol. 36, pp. 263–271, 2014.

[15] J. Eliasson, 'So you're considering introducing congestion charging? Here's what you need to know', *OECD Discussion Paper No. 2010-4*, 2010.

[16] J. Eliasson, L. Hultkrantz, L. Nerhagen, and L. S. Rosqvist, 'The Stockholm congestion charging trial 2006: Overview of effects', *Transportation Research Part A: Policy and Practice*, vol. 43, no. 3, pp. 240–250, Mar. 2009.

[17] I. Kristoffersson, 'Impacts of time-varying cordon pricing: Validation and application of mesoscopic model for Stockholm', *Transport Policy*, vol. 28, pp. 51–60, Jul. 2013.

[18] J. Rich and C. O. Hansen, 'The Danish National Passenger Model–model specification and results', *European Journal of Transport and Infrastructure Research*, vol. 16, no. 4, pp. 573–599, 2016.

[19] The West-Swedish Package, 'Restider i Göteborg – En jämförelse mellan 2012 och 2013 (Travel times in Gothenburg – A comparison between 2012 and 2013)', Sep. 2013. Technical Report. https://www.trafikverket.se/contentassets/6db12b4b4a844fa587b9e78a35ac7659/rapport_restider_goteborg_2012_och_2013.pdf (accessed 09 October 2017).

[20] The West-Swedish Package, 'Förändrade resvanor: Trängselskattens effekter på resandet i Göteborg (Altered travel behaviour: The congestion charge and its effect on travel in Gothenburg)', Nov. 2013. Technical Report. https://www.trafikverket.se/contentassets/2eca6e5528ea4f78bdd2dd5f64d269a8/resmonster/rapport-respanel-i-goteborg.pdf (accessed 09 October 2017).

[21] S. Berglund and L. Engelson, 'Nätverksutläggning för cykel.', Centre for Transport *Studies, KTH Royal Institute of Technology, 2014:12*, 2014.'Nätverksutläggning för cykel (Network assignment for bicycle)', *CTS Working paper* 2014:12, June 2014. Technical Report. http://www.transportportal.se/swopec/CTS2014-12.pdf (accessed 09 October 2017).

[22] C. Liu, Y. O. Susilo, and A. Karlström, 'The influence of weather characteristics variability on individual's travel mode choice in different seasons and regions in Sweden', *Transport Policy*, vol. 41, pp. 147–158, 2015.

[23] M. Börjesson and J. Eliasson, 'Chapter 10 The Benefits of Cycling: Viewing Cyclists as Travellers rather than Non-motorists', in John Parkin (ed.) Cycling and Sustainability (Transport and Sustainability, Volume 1) Emerald Group Publishing Limited (Bingley, UK), pp. 247–268, 2012.

[24] The West-Swedish Package, 'Första året med Västsvenska paketet (The first year with the West-Swedish package)', 2014:3, May 2014. Technical Report. https://www.trafikverket.se/contentassets/2109d7678fa74783b431d595e23d9ab5/rapporter/rapport_forsta_aret_med_vastsvenska_paketet.pdf (accessed 09 October 2017).

[25] C. J. Hamilton, 'Revisiting the cost of the Stockholm congestion charging system', *Transport Policy*, vol. 18, no. 6, pp. 836–847, Nov. 2011.

[26] A. Almroth, S. Berglund, O. Canella, *et al.*, 'Further development of SAMPERS and modeling of urban congestion', *CTS Working Paper 2014:10*, May 2014. Technical report. http://www.transportportal.se/swopec/CTS2014-10.pdf (accessed 09 December 2017).

[27] L. Engelson and D. van Amelsfort, 'The role of volume-delay functions in forecast and evaluation of congestion charging schemes, application to Stockholm', in *European Transport Conference 2011*, 2011.

[28] M. Saifuzzaman, L. Engelson, I. Kristoffersson, and A. de Palma, 'Stockholm congestion charging: An assessment with METROPOLIS and SILVESTER', *Transportation Planning and Technology*, vol. 39, no. 7, pp. 653–674, 2016.

[29] M. Börjesson, M. Fosgerau, and S. Algers, 'Catching the tail: Empirical identification of the distribution of the value of travel time', *Transportation Research Part A: Policy and Practice*, vol. 46, no. 2, pp. 378–391, Feb. 2012.

[30] M. Börjesson and I. Kristoffersson, 'Assessing the welfare effects of congestion charges in a real world setting', *Transportation Research Part E: Logistics and Transportation Review*, vol. 70, pp. 339–355, 2014.

[31] WSP, 'Finansieringsalternativ för Västsvenska paketet (Alternative financing of the West-Swedish package)', Jan. 2015. Technical report. https://goteborg.se/wps/wcm/connect/ce62faa8-7415-439e-b563-6e8a9675a8d5/Bilaga_3_WSP_Alt_Finans_VP_Slutrapport_150128.pdf?MOD=AJPERES (accessed 09 October 2017)

[32] A. Mellin, J.-E. Nilsson, and R. Pyddoke, 'Underlagsrapport till RiR 2011:28: Medfinansiering av statlig infrastruktur', Nov. 2011. Technical report. http://www.riksrevisionen.se/PageFiles/14666/underlagsrapport_2011_28.pdf (accessed 09 October 2017).

[33] O. Larsen and K. Ostmoe, 'The experience of urban toll cordons in Norway: Lessons for the future', *Journal of Transport Economics and Policy (JTEP)*, vol. 35, no. 3, pp. 457–471, 2001.

[34] P. Ieromonachou, S. Potter, and J. Warren, 'Norway's urban toll rings: Evolving towards congestion charging?', *Transport Policy*, vol. 13, no. 5, pp. 367–378, 2006.

[35] J. Whitehead, J. P. Franklin, and S. Washington, 'The impact of a congestion pricing exemption on the demand for new energy efficient vehicles in Stockholm', *Transportation Research Part A: Policy and Practice*, vol. 70, pp. 24–40, Dec. 2014.

Chapter 15

Moving from conventional tolling installations to open road tolling

*Bjarne Olav Tveit**

15.1 Introduction – new challenges

The tendency today is clear: to move from traditional single-lane, barrier controlled toll collection systems to multi-lane free-flow tolling (Open Road Tolling, ORT).

The consequences of this transition are substantial: better and more efficient solutions in terms of increased traffic throughput, reduced number of traffic incidents and fewer disturbances to the customers. Another important advantage is the fact that ORT will significantly contribute to reduced pollution and fuel wastage from vehicles queueing at toll booths. However, certain issues and design features need careful attention and discussions in the early design stages in order to enable the full potentials of the ORT concept without creating negative impacts that could potentially overshadow the benefits.

Traditional toll plazas have some rather distinct features, such as:

- They are quite visible.
- Toll attendants classify vehicles on the spot and disagreements over classification are settled when they occur.
- Payment and enforcement is carried out on the spot.
- Vehicles are only permitted to pass when the toll staff and the customer have agreed on the payment issue.
- When the exit barrier closes behind a vehicle, the passage and all related issues are history and no further actions are required.

The main challenges of going free-flow are

- Toll plazas become less obtrusive; they blend more into the scenery and are hence less visible to the customers.
- Vehicle detection, identification and classification need to be automated.
- Introduction of post-passage tolling.
- Changes to the payment process and payment products.

*Department of Electrical and Electronics Engineering, University of Newcastle-upon-Tyne, United Kingdom

- Implementation of post-passage enforcement of non-payers.
- Legal issues.
- Impact on system design.
- Changes to the operational organisation.
- Consequences for roadworks and civil works.

It should be noted that the chapter does not provide a comprehensive technical description of an ORT system. Nor does it discuss the pros and cons of ORT systems; it only presents and describes some of the important issues that need to be resolved if the decision has been made to move into ORT.

15.2 Unobtrusive toll plazas

Contrary to a traditional toll plaza like the one shown in Figure 15.1, from the M50 near Dublin, an ORT tolling point is not very visible to the customers and in most cases customers are likely to pass ORT tolling points without even noticing them. A photo of a modern tolling point from Göteborg, shown in Figure 15.2, will highlight some of the differences.

Particularly, in cases where customers are used to conventional tolling system with the mandatory stop to pay, the introduction of non-stop tolling must be carefully planned and prepared from a Customer Information perspective. The last thing we wish to see happening is that customers stop also in the new situation trying to find somewhere to pay the tolls.

As a consequence of free-flow tolling, the need for information will increase

- drivers must be warned they are about to enter a toll road,
- drivers must be informed about alternative *toll-free* roads,

Figure 15.1 Toll plaza on the M50 near Dublin

Figure 15.2 Free-flow tolling point in Göteborg

- drivers must be informed about the vehicle classes and the toll fees,
- drivers must be informed about how, when and where to pay,
- drivers must be urged not to stop after passage as there will be nowhere to pay at the tolling point.

Information campaigns should be prepared and launched in local and national newspapers, TV networks and all other media available. Additionally, information desks in the local environment and direct mail to the most affected customers and transport companies are all useful strategies that should be considered.

15.3 Vehicle detection

The process of detecting passing objects that are chargeable is the first important task of an ORT system. The process must be automated and the precision of the detection and the filtering of all *non-chargeable* objects are of paramount importance to the total system as revenue, system confidence and all system Key Performance Indicators (KPIs) are based on this number: the total number of vehicles detected through the system for any given period of time.

A suitable detection device must be selected which is ideally unaffected by weather and traffic conditions and so must the algorithms for detection and filtering. Vehicle detection is usually achieved through one or several detection devices, typically video systems, laser scanners, different types of radars.

In-surface, embedded sensor devices are rarely found suitable for ORT systems.

15.4 Vehicle classification

Some, maybe most of the ORT systems worldwide do not include a vehicle classification system. The reason for this is usually that the operator from a cost–benefit perspective finds this investment too expensive. However, depending on the overall objectives of the tolling system, vehicle classification must be considered. The reasons for this may be many; some of them are presented below:

- the public perception of *fairness* suggests that large vehicles should pay more than small vehicles.
- small vehicles take up less space than big vehicles and should pay less.
- small vehicles emit less pollution and noise than big vehicles and should pay less.
- small (and light) vehicles cause less damage to the road surface than big vehicles and should pay less.
- Public transport vehicles carry more people and should have special rates.

We shall discriminate between *direct measurement* classification and *declared class* classification:

- Direct measurement classification is derived from vehicle measurements in real time as the vehicle passes the tolling point
- Declared class classification is based on class information fetched from some database based on the identity of the passing vehicle.

Identification may be made by an OBU (On Board Unit, or the 'tag') installed in the vehicle or the vehicle's registration number. The database may be a Customer Account database, a national vehicle register or similar.

It is imperative that vehicle classification is correct and provides the necessary predictability and repeatability for the customers to have confidence in the system. From the operational perspective, wrong classification or even failure to classify a passing vehicle will cause considerable manual work and cost to correct. Classification parameters should be related to distinct vehicle characteristics that are easy to identify, measure and document.

15.4.1 Direct measurement classification methods

Vehicle classes for tolling are often based on dimensions (length, width, height), number of axles, vehicle weight (gross vehicle weight, e.g. above/below 3.5 t) or type of vehicle (e.g. bus or truck) or a combination of these.

In ORT, vehicle weight is typically converted to dimensions, using other vehicle features as mentioned above. In determining the correct class, the length is the most distinguishing dimension, but also vehicle height is a major contributor for a correct class.

Other features are used in order to further improve the classification performance. These are described in the following.

As an example, consider the following class definition, which is quite typical:

- Class 3: Small truck or van with two axles.
- Class 4: Truck with two or more axles.

It would seem quite difficult to discriminate between these two vehicle classes. To fulfil the classification rule in the example, additional parameters to axles must clearly be considered and appropriate values derived to enable discrimination between the two mentioned classes and other classes. Typical parameters to add would be *vehicle length* and/or *vehicle width* and/or *vehicle height*. More complex classification rules to be added could be *height over the 1st axle*, the *number of windows* of the vehicle and other even more sophisticated rules.

One important guideline is that it must be possible to determine the class of a vehicle by direct readings or measurements, not by inference (e.g. inferences of the type *if a vehicle is longer than 10 m it has 3 axles* should clearly be avoided.).

The work of identifying parameters and determining the optimal values is usually complicated and could potentially take a long time. Some typical classification parameters are presented in Table 15.1; combinations of these parameters are normally used for the purposes of direct class measurement.

Where required, the classification will apply all, or combinations of, available data such as trailer detection, length, width, height, number of axles, position of the axles and height over the first axle to calculate the correct vehicle class.

Some general recommendations are

1. The fewer classes the simpler and easier to implement.
2. Select class delimiting values in numerical regions where few (if any) vehicles are found
3. If necessary, concentrate on the predominant classes where the majority of vehicles are found and reduce the precision requirements on the less frequent classes.

Table 15.1 Typical classification parameters

Vehicle dimensions *(height, length, width)*	Vehicle *lengths* can be measured as **total length** (i.e. tractor + trailer), or tractor and trailer separately, as relevant
	Vehicle *width* can be used to separate cars from motorcycles
	Vehicle *height* can be used to separate large vehicles from smaller vehicles. (Using height measurement as a classification rule implies that the measuring sensors must be capable of filtering away objects mounted on the roof racks of smaller vehicles.)
Height over first axle	Rough form of vehicle profiling. Normally used to further improve the classification results, especially for vehicles of the same length
Number of axles	Number and position of axles is normally used to further improve the classification results especially for vehicles of the same length
Vehicle profiling	Mainly the variance of height with the length of the vehicle

Alternatively, focus could be on the vehicle classes with the highest fees to secure maximum revenue.

15.4.2 Declared class classification methods

Some systems permit the vehicle class to be read from the OBU but this requires that the Contract Provider has implemented the business rule of *one vehicle/one OBU*, which strangely enough seems a difficult decision to make.

Another alternative, found in some systems, is to fetch vehicle class data from some reliable data source, e.g. a national vehicle register or a Customer Account database. The link to the data source could be the OBU Identifier or the vehicle registration number (see the next section). This method will however require online access to vehicle data, which is not implemented in all countries.

Another challenge would emerge with a high volume of international traffic: the exchange of vehicle data between countries. This presents not only technical issues, but also administrative and potentially also legal obstructions.

Yet, another alternative to vehicle classification would be achieved if the Transport Administrations could define a vehicle licence number syntax to include vehicle class details as part of the licence number. But this seems almost impossible to even discuss with the appropriate authorities.

However, due to limitations and discrepancies in the mentioned alternative strategies for vehicle classification, any vehicle data fetched from OBU or some back office source may at best be used for comparison with the measured vehicle class.

The argument against reading vehicle class from the OBU is that different Toll Operators may have different classification schemes which are not compliant to the class information stored in the OBU or in the database linked to the OBU ID. In such cases, the back office must include some kind of class mapping applications to ensure correct tolling at the local tolling point.

15.4.3 Summary

To summarise, the recommended general principle for acquiring correct vehicle classification data is to use roadside classification equipment and perform vehicle classification in real time as the vehicle passes the tolling point.

15.5 Vehicle/vehicle owner identification

All passing vehicles will be allocated to one of the categories:

- positive identification,
- possible identification,
- no identification.

Business rules must be developed to manage these three categories.

Figure 15.3 Photo displaying an OBU mounted on a vehicle wind screen Courtesy: Kapsch TrafficCom

Vehicles may typically be identified by a number of different techniques:

- ALPR: Automatic Licence Plate Recognition. (See also Chapter 11).
- OBU: On Board Unit. An example of correct installation of an OBU has been presented in Figure 15.3.
- Vehicle fingerprinting: Identifying a unique set of characteristics for all the vehicles passing. This technique is best applied in networks where the majority of the vehicles are frequent visitors since manual verification of each vehicle is necessary upon first passage.

If the vehicle is equipped with an OBU, the OBU ID (read by the roadside transceiver during passage) will provide the link to relevant vehicle and owner data from a Customer Database.

If, on the other hand, the vehicle is *not* equipped with an OBU, other approaches must be selected; the most common is to identify the vehicle by its registration number and look up for the vehicle owner in some (national) vehicle register.

ALPR systems are capable of reading most licence plates and even determine the nationality of the vehicle with good and constantly improving accuracy. So, one important requirement for implementation of an ALPR-based solution is that relevant vehicle databases exist for the vehicles registered in your country and ideally also in the neighbouring countries if needed. Alternatively, one could consider solutions based on pre-registered customers to increase the number of known vehicles.

For enforcement purposes, it is imperative that the ALPR photos provided with each passage are accepted as legal proof of passage. There are different strategies and solutions to this, such as cryptographic hashing and overlaying appropriate metadata on the photo, e.g. time and date of passage, tolling point ID, direction of passage and other relevant information (Figures 15.4 and 15.5).

Figures 15.4 & 15.5 *Examples of video images captured of vehicles passing an ORT tolling point. Passage metadata are superimposed in the top line of the photo. Please observe that the photo on the right (Figure 15.5) has been cropped to comply with privacy act: no identification of the driver and front passenger. Courtesy: Kapsch TrafficCom*

Due to the inherent discrepancies in the OBU and particularly in the ALPR identification technology, the number of *possible ID* and *no ID* will vary from almost negligible to significant. Other reasons for wrong or no identification are the impact of bad weather, when it may become difficult or even impossible to read a vehicle registration number, incorrectly installed OBUs and dirty/unreadable vehicle number plates.

Vehicles falling into the last two categories will require manual image processing for possible identification. Although successful in general, it must be understood that not even this last effort will necessarily produce all positive identifications: some vehicles simply cannot be identified. More about this can be found in a later section.

15.6 Enforcement

There are two groups of non-paying customers:

- those who fail to pay,
- those who actively try to evade payment.

The most frequent non-payment situations are

- passing through a tolling point with dirty licence plates.
- passing through a tolling point with a non-working or malfunctioning OBU.
- deliberately passing through a tolling point undetected, i.e. too close to the vehicle ahead (tailgating).
- deliberately driving on the kerb of the tolled road to actively avoid detection.
- where the tolling point is only in one driving direction, on a road with one carriageway in each direction, deliberately changing into the oncoming carriageway to avoid detection.

The Business rules must be able to manage both non-paying customer groups and non-payment situations.

The possibility of efficacious and cost-effective enforcement is of crucial importance to Toll Chargers. It becomes an even bigger issue for toll roads that have international traffic: how to identify and enforce foreign vehicle owners? This single question represents the core question in the decision to go free-flow or not, depending on the volume of international traffic on the tolled infrastructure and the magnitude of the fees.

Important differences in the legislation in the different European countries offer in principle no immediate solutions to the issue of cross-border enforcement across Europe. This fact has led the EU Directorate-General for Mobility and Transport to state the following:

> The current challenge of cross border enforcement and exception handling seems to have no political short term solution and must therefore be solved through bilateral agreements between countries/regions if and where possible.

Bilateral agreements require willingness and commitment among the involved parties; difficult enough to achieve on their own, and on top of that they raise important and difficult legal issues. Once again, the answers to these questions could take a long time to sort out and should definitely be completed by the time of tender.

15.7 Legal issues

There is no physical barrier to stop a vehicle from driving through an ORT tolling point without paying. The enforcement of unknowing or deliberate non-payers thus becomes an issue of post-passage identification and invoicing based on the acquisition of relevant passage data captured at the tolling point. This raises many important questions to the national legislative bodies, the most important being:

1. What kind of documentation is required to satisfy the requirement for legal proof of passage?
2. Will this documentation violate the national privacy acts, and if so, is it possible to amend the national legislation to accommodate the new requirements?
3. Who shall be invoiced, the vehicle owner or the vehicle driver?
4. What is the approved legal process of issuing payment notices, surcharges and evasion fines?
5. Should the money collected by the tolling system be classed as taxes or as user fees/charges?

These decisions will have significant impact on the Business rules, the system architecture and payment means.

The following ORT features will impact on legislative issues and need to be discussed and resolved before a tender is issued:

1. Passages are associated with personal data captured upon passage, based either on OBU identification or on vehicle registration number identification and video images. The important question is: *how and for how long may such data be stored?*
2. The passage transaction and video images must be retained until payment has been recorded or a formal complaint has been processed. In this case, the important question is: *is there a conflict between the complaints process and the privacy act?*
3. Payment data may be received from customers before the passage has been made. This may also raise legal issues that must be addressed.

15.8 Changes in payment process and payment products

As payment will no longer take place at the time of passage, new payment products must be introduced to accommodate the new payment situation. Normally, most of the existing payment means will disappear as they are usually designed for the *stop-and-pay* situation previously in force.

For ORT-systems, post-payment and/or prepayment products are both quite common and so is the method of direct charging upon passage (often referred to as *pay-as-you-go*). The latter product requires a credit card link to the customer account.

Operators do not have to establish a comprehensive range of payment means. The above mentioned products are quite versatile and may be used as a basis for *return trips, weekend subscription, off-peak subscription* and other suitable combinations of calendar days, time of day and the number of trips permitted.

Interoperability agreements between different Operators and Contract Providers could enhance the payment process and product range significantly.

Another trend to be observed is the co-branding with e.g. Parking companies, Petrol stations and Ferry lines for whom it is also of interest to introduce cashless services.

In general, the new payment process will include the following steps:

- Post-process collected passage transactions.
- Compile passage records (for a predefined period of time, e.g. 1 month).
- Calculate fee and issue invoice/statement to the vehicle owner.
- Receive payment.
- Manage complaints.
- Manage late payments, reminders and the enforcement process.

15.9 Impact on system design

15.9.1 Introduction

The absence of staff at the tolling point creates the unavoidable uncertainty of whether passing vehicles are detected and correctly classified or not. This uncertainty must be eliminated for several reasons, the most important being to ensure the credibility of the system towards the public and to ensure the highest possible revenue for the owners of the infrastructure.

System availability and precision/performance are crucial features to the operation of the system and such features must be monitored and validated continuously throughout the lifetime of the system. To some extent, monitoring may be achieved through a comprehensive set of internal system monitoring functions, but in the end, the most important parameters can best be evaluated by statistical analysis.

A general modular overview of an ORT system is shown in Figure 15.6; it should be observed that the figure does not specify whether the modules are running in the Roadside System or in the Back Office System.

The different modules serve the following purposes (Table 15.2).

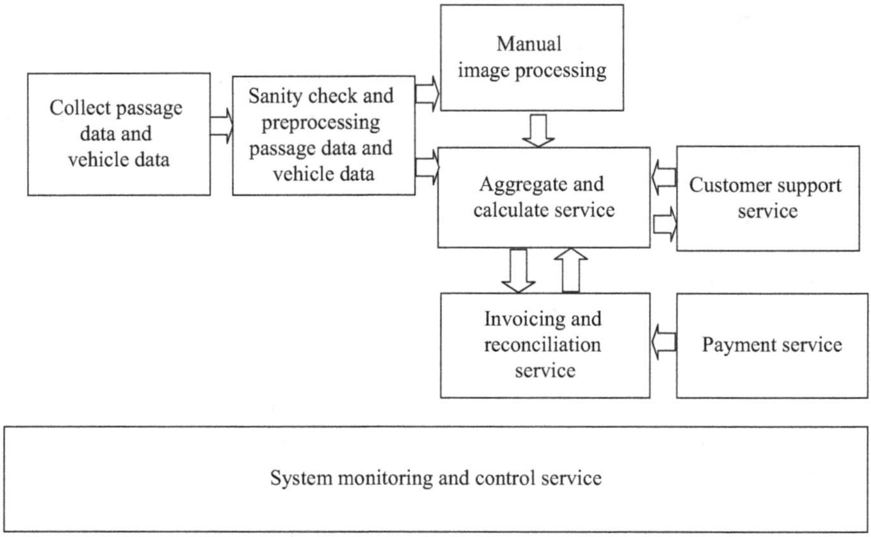

Figure 15.6 General overview of important modules in an ORT system

Table 15.2 Purpose of the important modules in an ORT system

Collect passage data and vehicle data	Collecting vehicle classification and identification data, and passage data such as toll point ID, lane ID and time stamp
Sanity check and preprocessing	Processing the different raw data checking whether a passage is valid or not and whether manual inspection is needed
Manual image processing	Manual inspection and correction of collected video images where the EFC (Electronic Fee Collection), ALPR or other systems have failed to determine correct vehicle identification/registration number, further to check possible tailgating, vehicle class deviations (measured vs. declared)
Aggregate and calculate service	Aggregate valid passages in the vehicle owner's (Customer) account
	Calculate the correct fee in accordance with appropriate Business Rules
Customer support service	Service to support the vehicle owner/customer in queries regarding passages and payments, changes in account information or complaints
	The Service may also include a Customer's web service where passage and payment information can be inspected by each customer
Invoicing and reconciliation service	Generate and issue invoices to the customers
	Receive payment from the Payment Service to perform the reconciliation and clearing of the accounts
Payment service	Receive and validate payments from the customers
System monitoring and control service	Monitor all system components and issue warnings and alarms as appropriate
	Report on Service Level Agreement (SLA) Key Performance Indicator (KPI) values
	Enable automated and manual control of system functions

15.9.2 Asynchronous reception of passage and payment data – individual accounts

In the traditional tolling systems, accounts were associated only with registered customers, and cash payments needed no follow-up after the passage had been made.

In an ORT system, transactions can no longer be closed at the time of passage.

Passage and payment transactions happen at different times, asynchrously. For this reason, passage information and payment information must be stored separately and remain open until passage and payment can be reconciled and cleared. The system must be designed such that it enables post-processing of passage transactions, issuance of invoices to identified vehicle owners, complaints processing and management and finally payment reconciliation.

15.9.3 Data storage capacity

As the basis for invoicing, passage data, metadata and video images need to be stored at the roadside, and eventually transmitted to the back office computer. Data storage needs to be fully redundant at all times and, depending on the chosen Business Model the passage data must be available in the back office computer in near real time although this latter feature is not very common.

From this, it follows that data storage facilities and data communication bandwidth normally shall need to be expanded considerably compared to the old system.

15.9.4 Monitoring of system performance

Operators need confirmation that the automated tolling system has the required availability and provides correct and repeatable data for a large number of vehicles.

This is of importance firstly to satisfy the public by proving availability and predictability, and also for the operator to ascertain that the correct vehicle owners are debited the correct amounts and that correct revenue is secured.

In the traditional tolling installation, the system quality figures are constantly observed by attendants and both availability and performance is rectified immediately if not satisfactory. In ORT systems, this is difficult to achieve: there is nobody on-site and no one will report any sort of errors or anomalies.

The easiest parameter is availability: system monitoring and error messages should normally be able to provide satisfactory information here. More complex is the identification of performance indicators regarding vehicle detection, classification and identification rates. But also in this field, it will be possible to come up with some quite good system monitoring and control applications.

Critical KPIs must be identified and subjected to continuous monitoring and statistical analyses to ensure that the system runs with satisfactory performance and at satisfactory service levels.

As there will no longer be staff present at the toll plaza, the roadside equipment and systems must be monitored remotely. Important questions to be discussed in this context are who shall monitor and service the installation and a practical work share between the Toll Charger and the Systems Supplier.

15.10 Changes to the operational organisation

15.10.1 Customer support

It is the justified requirement of all customers to easily and unconstrained be able to obtain relevant information about passages made, debts accumulated and payments received. However, as payment will become a more abstract process in an ORT environment, there will be an increase in the need for customers to receive information or guidance from the Toll Charger which in turn will increase the demand on customer service functions and facilities.

A significant offload for the customer service staff will be achieved if a website could be developed where customers can look up their own passages and payments and monitor the movements on their accounts.

15.10.2 Transaction control

Classification systems are by no means absolutely correct and neither is OBU nor are ALPR processing systems. In order to correct such faults, manual corrective work is required. There will be situations where the system cannot identify or classify a vehicle with the required confidence. In such cases, trained staff must manually inspect the passage transactions and images to determine the correct identity and/or vehicle class.

The amount of manual work required for transaction control should not be neglected as it will be of some magnitude and also strongly dependent on the weather situation and the robustness of the system. As an example, approximately 1%–3% of all OBU readings will probably be incorrect and 3%–5% of all ALPR readings will be incorrect at the best of times; in bad weather ALPR readings may give correct read rates of less than 60% and in the worst cases obviously down to 0 although this happens very rarely.

In summary, transaction control will always require manual work to ensure the required transaction quality and revenue. At times of adverse weather, a lot of manual work is needed and this places some challenges on how the organisation is built.

15.10.3 Money management

As all passages are cashless, there will no longer be the need for cash management.

This eliminates the need for strong rooms, counting rooms and money handling equipment, bringing cash from toll plaza to the bank, the entire change money situation and cash distribution to the different lanes.

15.10.4 Summary

Staff will not necessarily become redundant; they will only need to be re-trained to execute new and different tasks. Re-organisation of human beings is always difficult and time consuming. The entire change process must involve all staff affected by the change and should be initiated well in advance of the transition.

15.11 Road works/civil works

Moving into ORT implies a removal of the toll plaza as a physical construction and ideally a reduction in the number of lanes to match the number of lanes on the associated road network. Behind this lies the rationale that the need to spread traffic flow into individual payment lanes no longer exists and that traffic flows at its best capacity when uninterrupted. Also, toll plaza constructions such as canopies, administration buildings, lane islands and protective rails may normally be removed.

When installing the new equipment, usually gantry-based, it is important to reduce the lane closure times to a minimum. Hence, available work space should be allocated for the supplier to be able to carry out as much of his installation work as possible in this area and spend minimum time to erect the installation across the running lanes.

Examples of ORT installations are presented in Figures 15.7–15.9 to illustrate the differences in sizes and complexity. Hopefully, this will give some indications of the amount of road works and civil works that could be needed.

Figure 15.7 Roadside installation in Sydney Courtesy: Kapsch TrafficCom

Figure 15.8 *Roadside installation in Sydney Courtesy: Kapsch TrafficCom*

Figure 15.9 *Roadside installation in Singapore Courtesy: Kapsch TrafficCom*

15.12 Conclusion

There are no answers to the issues addressed in this chapter. The objective is to identify some important issues and to emphasise that thorough discussions and clear decisions are required in order to succeed in the implementation of an ORT system. The solution to some of the issues could be of a generic nature whereas others will require local adaptations.

Chapter 16
GNSS-based tolling: standards and implementations

Norbert Schindler and Erich Erker***

16.1 Overview of major GNSS tolling projects in Europe and worldwide

16.1.1 Introduction

Since the turn of the century, the use of satellite-based tolling has been spreading throughout Europe. By 2013, more than one million tolling On Board Units (OBUs) using Global Navigation Satellite System (GNSS) technology have been in use in Germany, Slovakia, and Hungary. More recently, nationwide truck tolling systems have been implemented in Russia and in Belgium – which have more than doubled the number of satellite-based OBUs being used for tolling. Singapore is presently preparing to transition its Electronic Road Pricing system to GNSS-based technology, where it is anticipated that another million new GNSS-based OBUs will be installed into passenger cars using this urban tolling scheme – as described in Chapter 6.

In Oregon and in California, solutions for road pricing using satellite technology have been developed as well, which is described in detail in Chapter 18.

There have also been three failed attempts to introduce GNSS-based tolling – in the United Kingdom, the Netherlands, and France, which we briefly address in an effort to explore the lessons to be learned from these cases.

16.1.2 The Swiss distance-based 'LSVA' system

Switzerland was the first country to introduce a kilometre-based scheme for all trucks on its entire road network in 2001, known as LSVA.[1] The tax replaced the flat-fee vignette levied on all domestic and international trucks above 3.5 t. Each travelled kilometre, regardless of road category, is charged. An OBU (Figure 16.1)

*GNSS Consulting, Austria
**Kapsch TrafficCom, Austria

[1] In German: **LSVA** (Leistungsabhängige Schwerverkehrsabgabe), which roughly translates to 'performance-related heavy vehicle duty'.

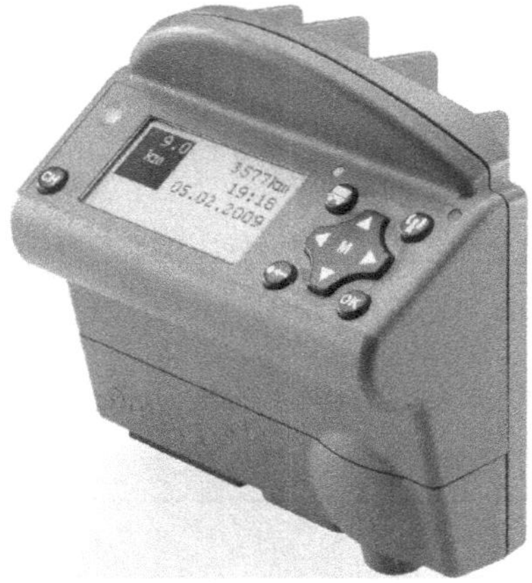

Figure 16.1 The second-generation windshield-mounted OBU for Switzerland

is mandatory for all trucks registered in Switzerland; foreign trucks can use a manual booking system available at each border crossing [1].

Initially, 55,000 trucks were equipped with an OBU which has the purpose of logging the number of kilometres travelled and writing this information on a chipcard. The OBU does not calculate the fees and does not perform any data transmission. The installation of the Swiss OBU must be performed by an authorised automotive workshop.

The primary data source of the OBU is the truck's tachograph, which serves as the legally recognised source of the distance travelled. The Global Positioning System (GPS) has been shown to measure the distance more accurately and is therefore used to verify the exact distance travelled. A microwave Dedicated Short Range Communication (DSRC) interface is built in to detect border crossings, effectively activating and de-activating the units when they enter and leave the country, respectively. A movement sensor detects whether the OBU has been disconnected without authorisation when the vehicle is in motion. The Swiss scheme was the first system to implement GNSS technology in an OBU for tolling, although it was only for verification purposes.

Even though the principle of counting kilometres is simple, the Swiss OBU is rather complex. It consists of the following components:

- Tachograph connection
- Power connection
- GPS receiver and internal GPS antenna
- Trailer sensor
- Chipcard reader
- Motion sensor
- DSRC interface (based on the standard CEN TC278 at 5.8 GHz)
- Physical interface for programming and data access

- Internal battery
- A set of light-emitting diodes on the back which indicates the OBU status – which is also visible to enforcement officers from a distance – to verify whether the truck is paying the correct amount.

In 2009, a new generation OBU was introduced to Switzerland [2]. New features of this OBU include:

- DSRC interoperability with Austria (CEN TC278) and with Italy (UNI[2]).
- Bluetooth interface
- Increased data security

16.1.3 German 'Toll Collect' system

On 1 January 2005, after more than a year of delay, Germany started its 'LKW MAUT' (truck toll) system for trucks over 12 t on its motorway network – more than 12,000 km in length. Two years after the system was launched, a number of first-class roads were also added to the tolled road network in areas where a dramatic increase in truck traffic was observed once the motorways were tolled [3]. After 11 years of operation, more than one million OBUs have been installed. Vehicles without OBUs use the manual booking system which was made available since the overhead required for the installation by an authorised automotive workshop could have been considered discriminatory for occasional users.

Unlike Switzerland, where the total distance travelled is measured, tolling in Germany is based on tolled road sections, each having a corresponding fee. A digital map of all roads in the tolled network is stored in the German OBU. Based on the GPS signal, the OBU can recognise each tolled section the vehicle has travelled. The section and the cumulative fee for the trip are visible on the liquid crystal display of the OBU. In instances where toll-free parallel roads are very close to the tolled roads, supporting gantries have been installed at the roadside for short-range communication to ensure that the OBU recognises the correct road section.

16.1.3.1 Fat client vs. thin client approach

The OBUs in Germany store all recognised road sections, apply the corresponding fees (which also depend on the vehicle category and emissions class), and make the entire toll calculation. The German OBU is therefore a so-called 'fat client' or 'intelligent OBU', since the entire digital map and toll tariff structure is stored on the unit. The result of the toll calculation is sent to the tolling back office via the GSM network, if not periodically during the trip, at least at the end of each trip. Compared to the storage of data on smart cards inserted in the Swiss OBU, the use of GSM to transfer tolling data to the back office was a significant advancement which has been implemented in each European GNSS-based tolling system thereafter. When changes are made to the road network or the applicable fees, the new data can be updated to the units via GSM data communication.

[2] Italy, in contrast to all other EU member states, has its own microwave standard at 5.8 GHz for electronic tolling: UNI-10607.

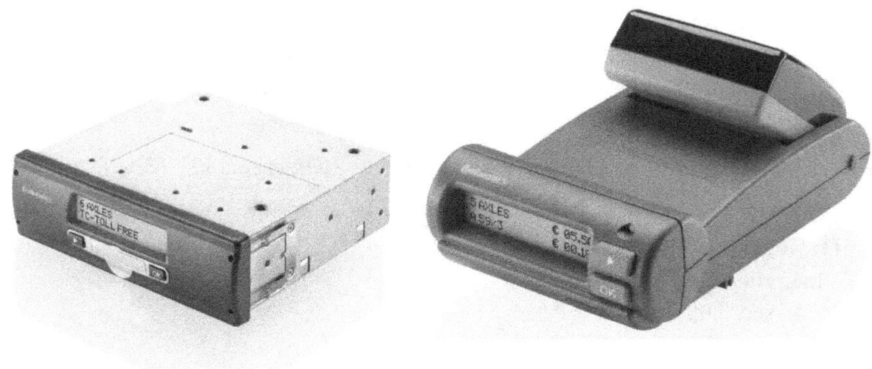

Figure 16.2 The German DIN-slot OBU (left) and dashboard-mounted OBU (right)

By contrast, as will be described later in the chapter (in Section 16.4.4), a 'thin-client' OBU architecture results in the deployment of an OBU that sends ***all*** of the position information of a vehicle's entire trip to the tolling back office, regardless of whether the vehicles travelled on a tolled road section or not. The entire road charge calculation is then performed in the back office [4].

In the German approach, a fixed connection to the vehicle's tachograph is used to verify the calculation made based on GPS positioning information. In combination with the built-in gyroscope, it allows for section recognition when GPS signals are not available, using the technique of 'dead-reckoning' [5].

Two OBU types have been made available (Figure 16.2), the DIN-slot OBU (having the form factor of a car radio) and a dashboard-mounted OBU (since a DIN slot might not be available). A majority of the installed OBUs are dashboard-mounted.

The German OBU consists of the following components:

- Tachograph connection
- Power connection
- GPS receiver and connection to an external GPS antenna
- GSM module for SMS communication
- Gyroscope
- External DSRC interface (based on the standard CEN TC278 at 5.8 GHz)
- External infrared interface (for enforcement purposes)
- Physical interface for programming and data access
- Internal battery

16.1.4 The planned lorry road user charge in the United Kingdom

Since 1994, the Department for Transport had been researching a means of implementing road user charging in the United Kingdom. In 2000, fuel tax protests forced the government to deal with the increasing inflows of foreign Heavy Goods Vehicle (HGV) hauliers using the British road network 'for free', since they would fill up their fuel tanks outside of the United Kingdom and thus avoided paying the high fuel duties. It was considered unfair that British hauliers paid the UK fuel tax while foreign vehicles did not [6]. This led to the attempted procurement of the Lorry Road User Charge (LRUC) in 2004, with the intention to levy a distance-based fee for HGVs from 2008. Since the goal of the UK scheme was to offset the high fuel tax paid by the UK hauliers, a complicated reimbursement scheme was planned for those vehicles which tanked up their vehicles in the United Kingdom.

At the time, the nationwide truck tolling schemes in Switzerland and Austria were already in operation – and the German system was about to be launched as well. In fact, the German truck tolling system was the de facto reference for the LRUC scheme, as it demonstrated the ability of using a GPS-based system for identifying distance-based road usage on a comprehensive road network – without the need for the costly installation of roadside infrastructure. The LRUC scheme would have placed a charge on motorways, trunk roads, and other roads – with different tariffs, depending on the time of day.

The 'LRUC' tender was probably the most complicated issued for a distance-based charging scheme in Europe. The procurement was broken down into three individually tendered lots: the charge data system (to identify the distance-based charge to be levied), the enforcement system (to ensure compliance of all HGVs to the scheme), and the central system (for user registration, billing, etc.). This division into three parts would have made the system implementation very difficult to manage, since it was not fully clear where the responsibilities for each of the three system suppliers would have been within the scope of the overall system implementation. It was also not clear who would have taken the responsibility of integrating these three subsystems.

To make matters more complicated, there would have been three components to the LRUC scheme:

1. The main scheme: Primarily for UK-registered vehicles, as well as for foreign-registered HGVs frequently visiting the United Kingdom, which would most likely have been based on GNSS technology.
2. The occasional user scheme: Primarily for foreign-registered lorries visiting the United Kingdom infrequently that would be equipped with a 'low-use OBU' that could have been based on DSRC.
3. The fuel duty repayment scheme: To rebate fuel duty to UK operators and to foreign hauliers that purchased fuel in the United Kingdom [7].

The estimated cost for installing and operating the LRUC scheme for 10 years was estimated at £4 billion (the equivalent value of approximately € 5.5 billion at the time). Some studies indicated that the generated annual revenue would have been less than the actual operation costs [8].

The procurement process led to the selection of pre-qualified vendors in each of the three lots. The 'proof of concept' of all lots were already being performed by many of the chosen vendors when, in 2005, the procurement process was cancelled completely – primarily due to the growing criticism of the system total costs. The termination of LRUC caused major financial setbacks to many leading tolling technology suppliers that participated in the various stages of the procurement procedure. Even worse, the tender cancellation sent shockwaves across Europe – leaving the impression that the failure of the LRUC scheme was somehow related to the 'high cost' of satellite-based tolling technology (and not due to the complexity of the scheme as such). The misperception that a GNSS-based approach to tolling was too expensive was clearly disproven in 2010, when Slovakia successfully implemented a GNSS-based system in less than 1 year – demonstrating how this technology could be highly cost-effective (see Section 16.1.6).

16.1.5 The planned tolling scheme of the Netherlands for all vehicles on all roads

The Netherlands was among the very first countries in Europe to consider the implementation of a comprehensive tolling approach on the public road network. Since 1988, various proposals for road pricing policies were under evaluation that considered distance-based fees to be differentiated based on time of day and location. Because Holland is a very densely populated country, the proposed tolling schemes needed to address the ever-growing problems of congestion on motorways during peak travel hours. Due to the pressure placed on the political leadership by influential interest groups, none of the proposals were ever pursued. In 2004, however, there appeared to be broad political support for a nationwide tolling scheme involving all vehicles on all roads. A national platform was founded called 'Anders Betalen voor Mobiliteit' (ABVM, which translates to 'Paying Differently for Mobility'). The committee behind the platform, consisting of government organisations and interest groups, recommended a kilometre-based charge ('kilometerprijs') for all vehicles, replacing the fixed cost of vehicle ownership. This recommendation was widely accepted, as the 'pay for use' principle was considered fair and would finally tackle traffic congestion problems – since the charge for travel at peak hours would be significantly higher. It was promised that the new scheme would not collect more revenues than the existing ownership-based fees for vehicles. Furthermore, these annual fees of approximately €7 billion would be ear-marked for infrastructure development rather than being passed on to the general treasury.

As a first step, feasibility studies were procured in 2005 during a 'market consultation' with technology vendors to determine whether technology was sufficiently mature and reliable – and also affordable – for such a large undertaking, in which every vehicle would be needed to be equipped with an OBU (Figure 16.3). The industry participants almost unanimously recommended a satellite-based approach. Although satellite-based OBUs at the time had a cost well above €200, some of the vendors estimated for the production of 8 million units, economies of scale could bring the OBU cost down to approximately €100. During the market consultation, a number of vendors were able to confirm the initial estimates that the overall investment needed for building the tolling system would cost at most € 2.2 billion and that the operational costs would not exceed 5% of the revenue generated [9].

In 2007, feasibility trials were performed with commercial suppliers to demonstrate that the technology was ready for the challenge of this ambitious project.

Figure 16.3 The vision of the Dutch 'kilometerprijs' in all passenger cars

Eleven (!) technology suppliers participated in the trials, each offered their own OBU solution – though only one would eventually supply its OBU to a national tolling system (in Slovakia, which launched its tolling system in 2010).

With the fall of the of the government coalition in 2010 (due to differences within the coalition about the deployment of troops in Afghanistan), the Dutch parliament decided to halt the project indefinitely – despite the significant progress made and the wealth of resources already invested. In the end, without clear political leadership, most of the parties represented in parliament feared that the subject of kilometre-pricing was too controversial when new elections were imminent. See also Chapter 20.

The Dutch approach was highly ambitious; charging every vehicle on every road category would have been an enormous challenge, both technically and politically. Ten years later, no other country ever considered such an all-encompassing tolling approach. However, even in 2007, it was demonstrated by the leading technology vendors at the time that the implementation of such a tolling scheme would have been technically possible.

16.1.6 Slovakia's innovative 'Myto' system

With the launch of its nationwide tolling system on 1 January 2010, Slovakia set a new milestone for electronic toll collection. For the first time, the use of GNSS-based OBUs for tolling was mandatory. Since the new generation 'plug and play' OBU could be mounted by the driver on the windshield within a few minutes, expensive installation by an authorised automotive workshop was no longer necessary.

Also unique was the fact that the greatest part of the tolled road network consisted of first-class roads – approximately 2,000 km – in addition to a mere 400 km of motorways and expressways [10]. With the exception of Switzerland, all previous nationwide tolling schemes (most of which used microwave-based technology) were applied to motorways and expressways. The Slovak Authorities decided to toll all the major routes from the very beginning, since:

- the motorway and expressway network at the time was relatively small (only 400 km),
- a large proportion of the heavy vehicle traffic uses the first-class road network, and revenue was needed for the maintenance and development of these roads as well as for the motorways,
- first-class roads exist alongside most of the motorways, and tolling only the motorways would have resulted in traffic diversion and thus significantly less toll revenue,
- it was necessary to eliminate any motivation for heavy vehicles to avoid paying tolls and divert to parallel roads, thus ensuring that these vehicles remain on the motorways rather than causing increased congestion and road damage (and consequently higher maintenance costs) along the parallel roads.

With this new approach in Slovakia, a paradigm shift was witnessed in tolling: rather than tolling vehicles on specific roads (typically motorways), ***specific*** groups of vehicles (in this case buses and trucks above 3.5 t) could be tolled on ***any*** road category.

In 2014, the Slovak tolled road network was extended to a total of 17,762 km of motorways, expressways, first, second, and third class roads (Figure 16.4). Thanks to the inherent flexibility of satellite-based tolling technology, this massive extension of more than seven (!) times the original size could be implemented in a few short months [11]. In the neighbouring countries of Poland and the Czech Republic, where microwave-based technology was chosen, extensions of the tolls to lower road categories were very limited. When the construction of a microwave gantry is necessary for every new tolled road segment (along with installation of new power and data connections at each gantry), the cost of extending a tolled road network becomes prohibitive. On smaller road categories, a tolled segment is typically about 1 km in length. One need only imagine the time and cost required for the installation of more than 17,000 gantries across all of Slovakia!

The costs of the mandatory OBU, which were covered by the toll operator (and not the end user), were significantly less than the OBUs used in Switzerland and Germany. From this moment on, new generation GNSS-based OBUs would have sensors and communication modules fully integrated. They can be easily mounted on the windshield, requiring only a power connection – typically by means of the cigarette lighter socket (see Figure 16.5). A tachograph connection was no longer needed, thanks to the improved position accuracy of the OBU hardware and advanced positioning algorithms that reliably detected tolled road sections with an accuracy exceeding 99.8% from the very beginning.[3] The new OBU included integrated GPS, DSRC, and GSM communication modules and antennas.

[3] The Slovak Road Authority (NDS) required a 'Functional Test of Concept' within 20 days of the contract award, to prove that all requirements were technically met. Nearly 20,000 km of test driving with 10 vehicles yielded a result of 99.86% correctly detected toll sections.

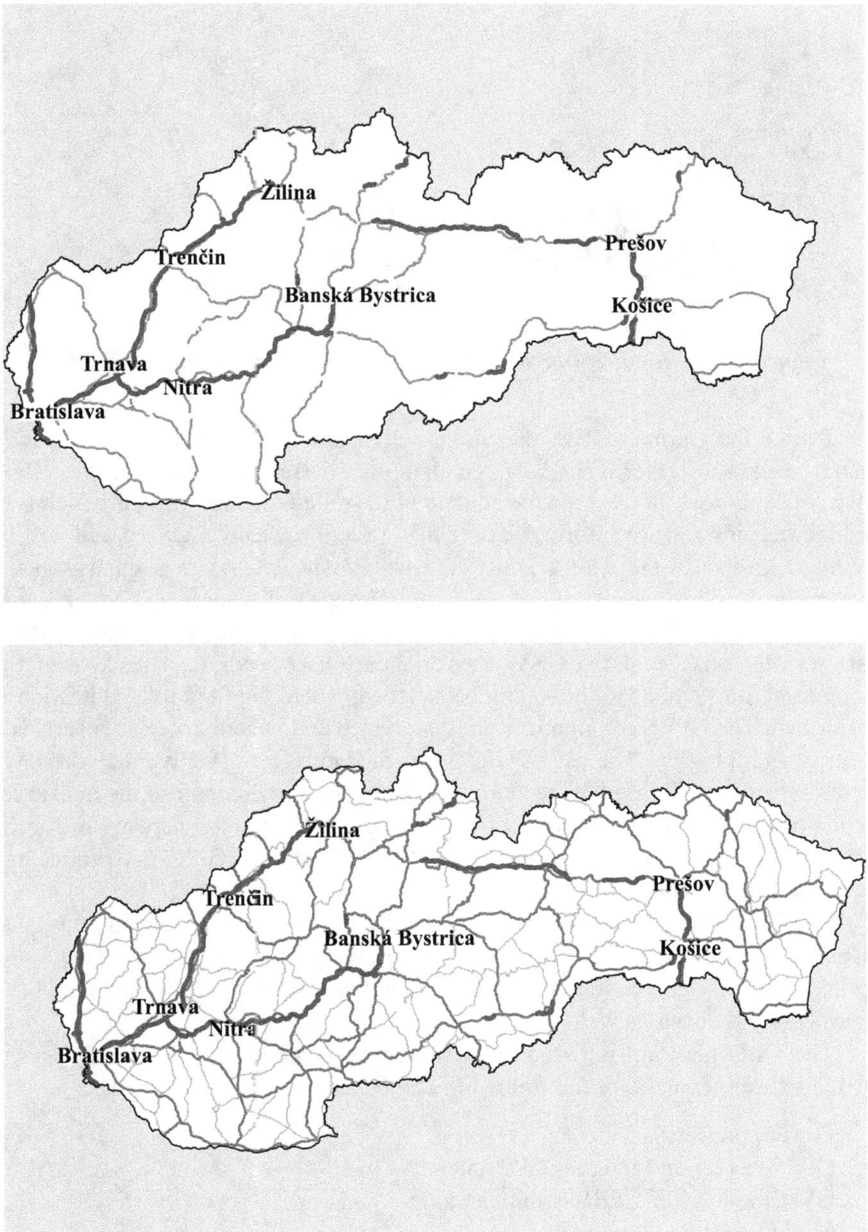

Figure 16.4 The Slovak tolled network from 2010 (top) and from 2014 (bottom)

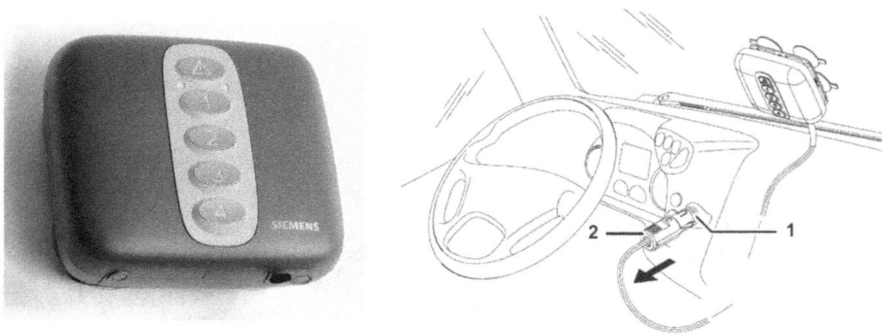

Figure 16.5 The first Slovak 'plug and play' windshield-mounted OBU

Unlike the German OBU, the Slovak OBU does not store tariff information that is needed to perform the cost calculations – these are realised in the tolling back office. It does, however, store all data of the tolled road network, producing an audible beeping tone to inform the user that a section was recognised and will be charged, similar to the way a DSRC tag beeps when it passes a gantry on the motorway. In other words, the *fat client* OBU approach chosen in Slovakia was 'less fat' than the German OBU. Using the General Packet Radio Services (GPRS) data transfer service of the GSM network, the OBU sends data packets of the recognised toll sections to the tolling back office, where the final toll calculation is performed. The GPRS communication is encrypted and watermarked to ensure data security and integrity. The use of GPRS was far more cost-effective than the SMS data services initially used in the German toll system, and guaranteed the transfer of tolling data within the GSM network. Updates of the tolled road network or for the OBU application software could be sent to the OBU via GPRS – without any interruption to the tolling application [12].

Since the OBU is mandatory in Slovakia, there is no need for an additional manual booking system, thus further reducing the overall system costs. Due to the optimisation of OBU communication with the back office, GSM costs have been significantly reduced as well.

The OBU platform is based on open source technologies and interfaces; the OBU hardware consists of the following components:

- Power connection
- GPS receiver and internal GPS antenna
- GSM module for GPRS communication
- Motion sensor
- DSRC interface (based on the standard CEN TC278 at 5.8 GHz)
- Internal battery

16.1.7 The French *écotaxe* system

In the framework of environmental and transportation policies, French authorities prepared the *écotaxe* by which trucks should pay a fee when travelling on

Figure 16.6 Two of the Hybrid OBUs developed for France: Autostrade (left) and Siemens (right)

15,000 km of national roads which were previously free of charge. The GNSS-based system, to be launched in 2013, was intended to be compliant with the European Commission's European Electronic Tolling Service (EETS) Decision and thus open to European Tolling Service Providers. The design of the *écotaxe* system constituted the first major steps towards electronic tolling implementation according to EETS (see Chapter 10). Four Toll Service Providers (TSPs) were granted the right to offer their services from the very beginning, each providing their own OBUs to their clients.

By 2014, all trucks in France above 3.5 t would have been equipped with new Hybrid OBUs (Figure 16.6), which would use GNSS technology for the automatic payment of tolls on the new tolled road network. These mandatory OBUs also contained a microwave interface, allowing for the automatic payment at all the existing toll plazas of the concession-based motorways already subject to tolls (which were approximately 8,000 km in length). More than 800,000 new OBUs were to be distributed to the trucks driving in France. By making the new system interoperable with its traditional tolling schemes, and by involving the TSPs during the test period of the system, France was the first country to attempt to implement a tolling environment based on the principles of EETS [13]. This would have been a very significant advancement of EETS, since the efforts of the European Commission to establish EETS by the end of 2012 were unsuccessful.

Four TSPs offered their clients the new hybrid OBUs to replace existing microwave tags. The new OBUs, from three different technology suppliers, were equipped with DSRC modules that worked on all the microwave infrastructure installed throughout France – and also complied with the Spanish VIA-T standard [14]. New telematic features using satellite positioning were also developed and are still in use, even though the GNSS functionality was never used for tolling purposes.

As previously witnessed in the United Kingdom and in the Netherlands, the lack of political willpower in France to follow through with new distance-based tax for road use led to the termination of the *écotaxe* system. After a weekend of protests by approximately 1,000 people in Brittany in October 2013, the scheduled launch in January 2014 was 'suspended indefinitely'.[4] Unlike other 'failed' tolling

[4] http://www.france24.com/en/20131029-france-ecotax-brittany-farmers-food-climate-taxes-protests

projects, however, the French tolling system was already fully implemented! Even though the technological implementation met all specifications, a newly elected government decided to scrap the project – at the expense of billions of Euros of public money that had already been invested in the fully operational system.

Nevertheless, this implementation brought significant advancements to electronic tolling in the European context. The theory of allowing independent TSPs to operate in national tolling domains was finally set into practice, breaking down barriers which would eventually benefit neighbouring Belgium a few years later. By allowing multiple service providers to operate in the same charging system, multivendor interoperability was ensured from the beginning. This prevented the main subcontractor (for the design, build, and operation of the new system on behalf of the French Toll Charger) from using closed systems with proprietary interfaces. Furthermore, the *écotaxe* implementation demonstrated how a new GNSS-based approach to tolling could be introduced to an extensive infrastructure-based tolling environment, without any inconvenience to the road user. The Hybrid OBUs that were introduced in France were able to perform tolling transactions both in the DSRC-based tolling systems (by communicating with the existing roadside infrastructure) and in the new GNSS-based environment of a very large newly tolled road network [15]. These new OBUs were in fact the first 'EETS-compliant' OBUs to reach the market and would eventually be used throughout Europe by the first EETS TSPs. In this regard, everyone in the European Union could benefit from the significant investment made in France that basically gave birth to EETS.

16.1.8 Hungary's 'HU-GO' system

Hungary was among the first European countries to procure a nationwide truck tolling scheme, but cancelled the procedure during the tender phase (twice). Nevertheless, the Hungarians were determined to launch a comprehensive tolling system of their own, and eventually became the third country in the European Union to launch a nationwide truck tolling system using GNSS technology. Since July 2013, tolls are charged for all trucks above 3.5 ton 6,500 km of designated sections of the public road network. The distance-based fees can be paid either by manual ticketing or by using the positioning information of satellite-based fleet management devices offered from one of many Toll Declaration Operators (TDOs) registered in Hungary [16]. These basic tracking devices act as 'thin-clients' and automatically transmit all positioning data of the travelled routes to the TDOs. The TDOs then determine, through their own individual algorithms, whether their clients' vehicles travelled on a tolled road section.

Unique to the Hungarian approach is that these toll declarations must be forwarded to the national Toll Charger within just 15 minutes, in the form of an 'electronic ticket' for each toll section. In other nationwide electronic truck tolling schemes, toll transactions are either deducted from the users' prepaid account or invoiced at the end of the month for registered post-paid users. The Hungarian HU-GO system deploys such a strict time constraint since it is based on the principle of user self-declaration. In contrast to other schemes, it is the responsibility of the user in Hungary (as opposed to the national toll operator) to ensure that the toll road sections travelled are declared and paid for – either in the form of prepaid tickets that are manually booked for an entire route, or by the declaration of electronic 'tickets' en route which authorise the National Toll Operator to deduct the corresponding amount of the ticket from a prepaid account. While this approach allows enforcement officers to catch violators 'just in time', it is demanding both for users who opt for the electronic approach and the service providers who need to ensure that the tolls are declared in time – even when there is no GSM coverage and/or poor GPS reception. Unlike similar tolling schemes in other countries, the HU-GO system does not (yet) rely on the use of microwave interfaces for enforcement (since DSRC modules are not included in fleet management devices). Instead, camera-based technology is used to read the number plates of the passing vehicles and determine whether those vehicles are compliant with the toll system.

Since the HU-GO system places the responsibility for correct toll declaration on the users, it has proven to be extremely cost-effective. By enabling 20 Service Providers to use more than 50 different tracking devices to determine the distance-based fees, the implementation of the Hungarian system took less than 3 months! This record-breaking time for implementation was something that most experts in the field of tolling would not have thought possible. Equally surprising was the low implementation cost of the HU-GO system, which is described in detail in Chapter 17. As remarkable as this achievement is, it is important to realise that such a quick implementation was only possible because the Hungarians could take advantage of the existing infrastructure already in place for the time-based vignette system which had already been in operation for many years. Thanks to the prior deployment of the highly innovative electronic vignette system, the camera-based automatic number plate recognition (ANPR) technology was already in place and could be used to monitor compliance to the new truck tolling system as well. By avoiding the use of DSRC microwave components for enforcement, and relying only on ANPR technology, the cost of building and operating the system could be kept to a minimum.

The Hungarian approach is unique among all nationwide tolling systems in Europe. An advantage of this approach is that the costs for building and operating the toll system can be kept to a minimum, while passing the responsibility of accurate toll charging on to the TDOs in the form of required Service-Level Agreements (SLAs). Rather than defining strict technical requirements with respect to the hardware functionality of the OBUs, like the other GNSS-based tolling schemes in Europe, the Hungarian Authorities have defined functional SLAs which must be met by the TDOs. Therefore, the TDOs are required to take the risk of ensuring that the strict requirements of the HU-GO system are kept – especially with respect to the level of accuracy required in reporting the declared toll sections used by vehicles within the time limitation of 15 minutes.

The Hungarian implementation of tolling is similar to the EETS-oriented approach, in which a number of TSPs provide electronic tolling services to their respective clients. The Hungarian TDOs, however, work primarily as fleet management operators and provide the additional service of performing electronic toll

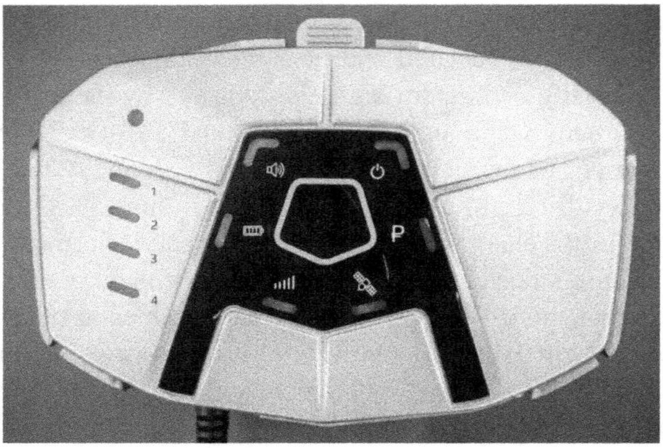

Figure 16.7 The Russian 'PLATON' OBU that uses GLONASS and GPS

declarations using the GPS tracking device that is already installed in the vehicle for fleet management. Furthermore, the service fee granted to the TDO for the declaration of electronic 'tickets' by use of tracking devices is a fixed fee defined by the Hungarian Road Authority as a fixed percentage of the toll revenue generated by the user. TDOs typically need to provide other services (such as fleet management) to generate enough revenue to cover their operation costs. Most significantly, the burden of ensuring correct toll-declaration lies on the end user: the truck driver (or truck haulier company) must ensure that the tracking device in the vehicle is always working correctly (e.g. has a good GPS signal) and transfers the positioning data within 15 min (e.g. has sufficient GSM coverage) – at least 95% of the time. Consequently, a large number of users prefer to pay the HU-GO fees manually.

In any case, Hungary successfully eliminated the long-lasting misperception that the implementation of satellite-based tolling system was inevitably a high-risk, time-consuming, and cost-intensive undertaking. In fact, by taking advantage of the innate flexibility of GNSS technology – and using technologies that were already in place – Hungary demonstrated how a cost-effective tolling system could be quickly implemented, generating more than €500 million in annual tolling revenues.

16.1.9 Russia's 'PLATON' system

The Russian tolling scheme is by far the largest satellite-based tolling system to date. Launched on 15 November 2015, it subjects an estimated two million trucks above 12 t to distance-based fee on more than 50,000 km of Federal Roads. A centrepiece of the tolling system technology is Russia's own satellite positioning system, GLONASS (GLObal NAvigation Satellite System[5]). The OBUs deployed in the Russian scheme (Figure 16.7) use GLONASS in combination with GPS to identify a vehicle's passage on segments of the tolled road network. The use of

[5] In Russian: ГЛОНАСС: Глобальная навигационная спутниковая система (Globalnaya Navigazionnaya Sputnikovaya Sistema)

OBUs is not mandatory; a manual ticketing scheme is also provided. Due to the vast size of the country, the GNSS-based tolling approach in Russia has faced some technical challenges, such as the varying density of mobile network coverage (for the transfer of data from the OBU to the back office) and extreme climatic conditions. The need for reliable and accurate satellite positioning in higher latitudes has placed new demands on the robustness of a GNSS-based solution; GLONASS was especially designed to offer good coverage in these northern regions. By combining the positioning information of GLONASS and GPS, the overall availability of the GNSS system can be increased significantly [17].

Unlike most of the GNSS-based solutions deployed in the European Union, the new PLATON system does not use DSRC for enforcement – or for interoperability with existing toll roads that deploy microwave-based electronic tolling. Furthermore, the PLATON OBUs do not collect data outside of the Russian Federation. Not only is the transmission of data via GSM communication limited within Russia; the integrated security access module that supports data transfer from the OBU to the back office within the Russian Federation is proprietary. Consequently, GNSS-based OBUs from other countries cannot be used for tolling in the Russian system.

The sheer scale of the Russian tolling system, spanning 11 time zones, is unlike any distance-based charging scheme implemented before. With Russia being the largest country in the world, it is only logical that this new tolling scheme was developed based on use of multi-constellation GNSS. GLONASS positioning signals can be received reliably throughout the country; the satellite receivers integrated into the OBUs also use GPS signals to increase the level of reliability and accuracy of the positioning data – which can be particularly useful in challenging environments, such as urban canyons or mountainous regions. Another challenge faced by the Russian solution is the ability to operate in extreme weather conditions: the system must be operational at temperatures ranging from −50 to +50 °C.

The PLATON scheme, which was developed using open-source software, is a dual system (like in Germany and in Hungary) using both a manual ticketing system and an electronic system – with a windshield-mounted OBU (Figure 16.7). These OBUs act as thin-clients, sending all positioning data to the back office, where it is determined whether the vehicles have driven on the tolled road network, and the corresponding fees are deducted from the prepaid accounts of the users.

16.1.10 The Belgium 'Viapass' system

In April 2016, Belgium became the fifth country to introduce a new nationwide distance-based road charge (for trucks above 3.5 t) based on satellite-positioning technology. Initially implemented on 6,700 km of roads, the Viapass system has the potential to include up to 50,000 km of roads in the future. The 'plug and play' OBUs deployed in Belgium do not differ significantly from the windshield-mounted units already used in Slovakia. However, Belgium set a very important new milestone in the world of tolling since, for the first time, users of the new tolling system could choose among different EETS TSPs from the first day of operation. Thanks to the advancement of EETS in the development of the ill-fated French *écotaxe* system, Belgium became the first country to bring EETS into practice. In other words, multiple TSPs offer their clients' tolling services on the territory of Belgium, using their individual OBU-based solutions (Figure 16.8).

Because Belgium consists of three essentially autonomous regions, there are actually three different Toll Chargers: for Brussels, Flanders, and Wallonia. The 'Viapass' system actually emerged from an interregional agreement made in 2011 in which it was decided to implement a reform to the road tax system that included the introduction of a satellite-based kilometre charge with a mandatory GNSS-based OBU. Each of the three regions defines its individual toll policy, publishes its own Toll Doman Statement (which sets down the rules for EETS TSPs to be accepted), and determines its own tolling tariffs (Figure 16.9). While Flanders and Wallonia primarily toll their motorway network, Brussels charges for every single

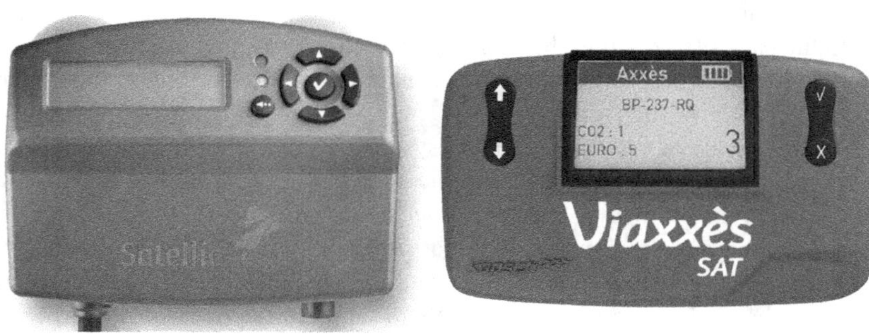

Figure 16.8 The two OBUs initially available for Viapass – Satellic (left) and Viaxxès (right)

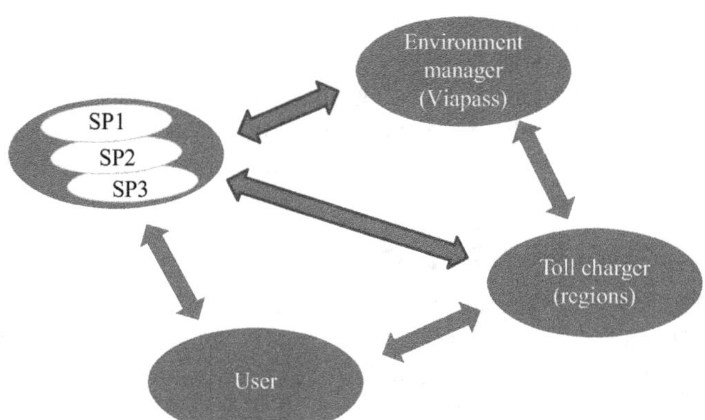

Figure 16.9 The EETS approach in Belgium with multiple service providers (SP), a Toll Charger in each of the three regions, and Viapass as the coordinating environment manager

kilometre travelled – and has declared a higher tariff than the other regions, up to 65% higher for large trucks. In this sense, the Belgian tolling system has a 'congestion charge' already built in. Following the EETS Directive 2004/52/EC, Flanders and Brussels also split the toll into 'infrastructure' and 'external' costs by dividing the overall charge – which is actually a tax. Wallonia, on the other hand, charges a road-user fee with added value added tax (VAT) since the motorways that levy the fee are road concessions. For trucks paying for their trips in Belgium, these differences were not visible for the first year of operation: the fees on the highways cost the same in all of Belgium, while the cost of driving in Brussels is higher (except for the highways).

Unique to the tolling approach in Belgium is the definition of more than 100,000 toll areas, each of which can (theoretically) have its own toll model based on the time of day, road category, vehicle category, and even the direction of travel. However, when Viapass was first launched, the method of tolls charged on the motorways was very similar to the section-based approach implemented in Germany, Slovakia, Hungary, France, and Russia.

The TSPs are also given flexibility in some of the details of their technical implementation. For example, the Toll Charger accepts the final toll charge calculated by the TSP, regardless of whether the TSP has chosen a *fat client* or a *thin client* approach. In fact, when the system was launched, one of the first two TSPs stored the geographic data and determines the tolling fees directly on the OBU, while the other TSP chose to deliver all position information to the back office, where the algorithms for map-matching and calculating the tolling fees are implemented. Unlike the EETS approach chosen in France, there is no common map or geo-model made available by the Toll Charger. Therefore, each TSP needs to take the responsibility for implementing an accurate geographic model.

At the time of writing, six new EETS Providers are registered and are engaged in the approval process to become TSPs in Belgium. Thanks to the fact that the Viapass tolling scheme was designed and implemented according to the EETS Directive of the European Commission, international tolling interoperability has become a reality.

16.1.11 Common themes in all GNSS-based tolling systems

At the turn of the century, Switzerland courageously led the way of tolling heavy good vehicles on all roads – based on tachograph measurement, but taking advantage of electronic tolling technologies. Germany soon followed the Swiss example of tolling trucks, yet with a much more ambitious approach: using satellite positioning technology to detect a vehicle's passage on tolled road sections, by installing OBUs in trucks above 12 t. The innovative German approach to electronic tolling set the de-facto standard for GNSS-based tolling in Europe: the automated detection of tolling events using positioning information, combined with the transmission of the toll data via GSM to the tolling back office. Although the costs of the German Toll Collect system were substantial – both the capital expenditures (Capexs) and the operational costs being quite high – it was a wise decision to use GPS technology rather than to invest even more in the construction of roadside equipment (RSE) along more than 12,000 km of motorways. Since its launch, the German system generates between €4 and €5 billion of revenues annually – significantly more than the initial investment costs.

In 2010, Slovakia established what has now become the de-facto standard in GNSS-based tolling in Europe, in particular for (future) EETS providers:

- Mandatory windshield-mounted OBUs
- Integrated (rather than external) antennas in the OBUs
- GSM/GPRS transmission of data to a proxy server
- Updates of OBU software (and, for *fat clients*, geo-data) over the cellular network, without interruption of the tolling system operation
- DSRC (CEN TC278) for tolling enforcement and interoperability with microwave-based systems

The French *écotaxe* solution (even though never operational) was implemented according to these criteria, as was the *Viapass* system in Belgium. The Hungarian approach to tolling is an exception, however, as it is based on the self-declaration of users – either with a ticketing scheme or through the use of GPS tracking devices that were already in use for fleet management. Time will tell whether the Hungarian approach will become a role model for other countries, but at least in Bulgaria (where a tender procedure is ongoing at the time of writing) the notion of using existing fleet management devices for electronic tolling has been seriously considered.

The French tolling solution was defined according to the EETS Directive of the European Union, and it was only logical that the nationwide scheme in Belgium would follow suit. With the launch of the Belgian system, we have witnessed the dawn of EETS – a defining moment in the transformation of tolling in Europe. Electronic tolling is now being offered as a service, and the deployed tolling technology enables this service to be implemented in all of the unique tolling environments throughout Europe [18].

GNSS has now become the technology of choice for nationwide tolling schemes in Europe. Countries that already invested in DSRC-based solutions, such as the Czech Republic and Poland, are seriously planning to 'upgrade' to a satellite-based approach, in order to cover a much larger tolled road network – and not be limited by the need for roadside infrastructure being constructed along all new tolled roads. The cost-effective expansion of the Slovak tolled road network from 2,700 to 17,000 km in a matter of months was an ideal demonstration of the level of flexibility that satellite-based tolling solutions can offer. Table 16.1 provides an overview of all nationwide GNSS-based tolling solutions implemented in Europe.

As will be described at the end of this chapter, the use of GNSS technology in tolling has already become a driver for the deployment of other value-added services that can take advantage of the track and trace features made available by the satellite-based tolling architecture.

Table 16.1 Comparison of all existing GNSS-based tolling systems in Europe

Country, roads tolled	Launch date	Vehicles	Mandatory OBU?	OBU type	Unique aspects
Switzerland, all of the roads are charged	1 January 2001	Trucks >3.5 t	Only for trucks registered in Switzerland	Fixed tachograph connection	GPS is used to confirm distance measured by the tachograph DSRC is used for enforcement
Germany, 14,000 km (mostly motorways)	1 January 2005	Trucks >12 t*	No, manual declaration is also possible	DIN-slot or dashboard-mounted	First distance-based scheme using GPS to identify toll road sections DSRC is used for enforcement
Slovakia, 17,700 km (most of the roads)	1 January 2010	Trucks, buses >3.5 t	Yes, with self-installed OBUs	Windshield-mounted OBUs	First mandatory OBU, tolling includes motorways and first, second, and third class roads DSRC is used for enforcement
Hungary, 6,500 km (motorways + first class)	1 July 2013	Trucks >3.5 t	No. GPS tracking or manual declaration	Mostly basic GPS tracking devices	Fleet management tracking devices can be used for self-declaration of tolls DSRC is not used
Russia, 50,000 km (motorways + first class)	15 November 2015	Trucks >12 t	No, also manual declaration possible	Windshield-mounted OBUs	GLONASS and GPS can be used simultaneously DSRC is not used
Belgium, 6,700 km (mostly motorways + all of Brussels)	1 April 2016	Trucks >3.5 t	Yes, with self-installed OBUs	Windshield-mounted OBUs	Designed to enable multiple EETSs providers from the beginning. All roads in Brussels are charged DSRC is used for enforcement

*Since 1 October 2015, also trucks >7.5 t are subject to tolls in Germany.

16.1.12 Summary of failed projects

Although there are several successful European toll systems based on GNSS technology in operation today, we have also witnessed a few examples of failed projects. In the case of France, an innovative and complex tolling system was fully implemented – and paid for – only to be cancelled a few months before the planned system launch. There are lessons to be learned from these failures, also with respect to technical implementation aspects.

The satellite-based tolling systems developed in Europe, whether they were actually launched or cancelled, have had the following issues to deal with:

- Difficulties with a 'big bang' approach, especially with short implementation timeframes. This problem is addressed in Chapter 7 with respect to some tolling initiatives in the United Kingdom.
- Political issues (lack of public support), which are also discussed in Chapters 5 and 7.
- Consideration of the business case, which is of paramount importance.

In light of the failed projects (United Kingdom, Holland, France), authorities or agencies planning to implement tolling schemes should be wary of designing and tendering overly ambitious projects.

16.1.12.1 United Kingdom: LRUC

The tender for the LRUC in the United Kingdom was most likely cancelled because there was no convincing business case: all heavy goods vehicles would be charged with a road usage tax, but domestic trucks would get a refund since they already pay the high fuel tax in the United Kingdom. This would have resulted in a huge administrative effort, consuming a significant portion of the revenue that the new system could generate.

16.1.12.2 Netherlands: ABVM

The plan to establish a distance-based charge for all vehicles on all roads in Holland was extremely ambitious. The 'over-design' of the system was highly problematic. Even though the break-up of the government coalition led to this tender being cancelled, it seemed doubtful that such an all-encompassing system could have been cost-effective since the detailed design requirements gave little room for technology vendors to develop and implement an efficient solution. After the tender cancellation, no Dutch government has since attempted to relaunch a similar tender.

16.1.12.3 France: Écotaxe

The fact that a very costly and fully operational system was terminated just months before the 'go-live' date is an especially painful demonstration of political failure. Without public acceptance, even a system that met all technical requirements and cost the public billions of Euros can fail. From a technical point of view, the French écotaxe was highly complex: interoperability was required with 18 existing

Toll Chargers, as well as with the existing microwave-based systems in Spain and Portugal, and 4 TSPs would have offered the service of tolling the new road network. Although very ambitious, the project was fully implemented (with some delays) prior to cancellation.

16.2 Technical challenges of GNSS in tolling

In order for European GNSS-based tolling systems to be successfully implemented and operated, a number of technology issues have required attention. The most relevant technological challenges have been

- GNSS reception
- time to first fix (TTFF)
- jamming and spoofing

Although there is still room for improvement in addressing each of these issues, the respective risks have been sufficiently mitigated by technological improvements, as described in the paragraphs below. Workarounds have, of course, also been devised to meet the required performance targets set forth by the individual tolling systems.

In this section, we review the expected Key Performance Indicators (KPIs) and Service Level Agreements (SLAs) in existing (and planned) tolling systems and the challenging task of measuring them. These issues are especially relevant for future EETS TSPs.

16.2.1 Introduction

As illustrated in the examples above, GNSS-based tolling offers a level of flexibility that other tolling technologies can never come close to. Since no RSE is needed for charge determination,[6] large road networks can be covered easily, irrespective of the availability of power and communication infrastructure. The behaviour of each vehicle can be determined with the resolution of a few metres and a few seconds.

However, there is a price to pay for this level of flexibility – not only that GNSS-based OBUs are more expensive than simple microwave tags. GNSS tolling solutions need to address a number of technological challenges, in particular with respect to the accuracy of positioning data and the availability of a mobile communication link.

Given the numerous ways that GNSS-based tolling technologies can be deployed, there are also challenges in standardisation. On the one hand, there is the aim to minimise the range of implementation options; on the other hand, stakeholders do not want to give up on flexibility – each system is designed based on a different set of rules, regulations, and objectives.

[6] Some RSE is typically needed for enforcement purposes and, in cases where GNSS reception is lacking, for localisation augmentation (as described in Section 16.2.2.6).

16.2.2 Challenges of GNSS technology

Satellite-based tolling solutions can only be accurate when sufficient GNSS signals can be received to reliably determine the position of vehicles on the tolled road network. In some real-life scenarios in road traffic, the required level of accuracy cannot be readily guaranteed. GNSS performance can be impaired by the movement of the vehicle, obstacles in the line of sight to the satellites, weather conditions, and even environmental conditions in the Earth's magnetosphere, ionosphere, and thermosphere caused by the sun and by solar winds [19].

16.2.2.1 Accuracy vs. availability

The correct determination of road usage clearly increases when the accuracy of the underlying Position, Navigation and Timing (PNT) solution is improved. In practical systems, however, the availability of any PNT solution at all (i.e. the GNSS track and trace data) is of equal importance. In other words: it is better to have an inaccurate trace than none at all. High sensitivity receivers, for example, might deliver a less accurate trace when GNSS signals are of inferior quality (e.g. due to reflection) but these could nevertheless be used for the PNT solution. As long as positioning information can be obtained, when at least four satellite signals are available (or even less, when state-of-the-art GNSS modules are used), a GNSS-based tolling solution would likely make use of this signal (as opposed to having no positioning data at all).

16.2.2.2 GNSS reception

When applying GNSS technology to tolling, there are always issues with satellite signal reception. When an OBU operates in an open sky environment, the satellite signal reception is only limited by the way in which the OBU is physically integrated into the vehicle. This depends on

- the GNSS antenna orientation,
- objects in front of the antenna,
- the type of windscreen (e.g. metallisation),
- the gap width between antenna and windscreen,
- deposits in front of the antenna, such as ice, snow, or water.

All of these have an impact on reception quality. However, GNSS-based solutions used on the road must also operate in a variety of challenging scenarios:

- **Urban canyons**
  High buildings on both sides of the street result in only few satellites having line-of-sight conditions. Since those satellites in view are aligned along the street axis, large lateral errors in position will result. Other satellite signals are either blocked or subject to multipath propagation. When there are high buildings on only one side of the street, a considerable offset of the measured vehicle trajectory can result, due to reflections (multipath issues are described below).

- **Wet canopy**
  Rural roads traversing areas with dense vegetation can result in GNSS reception being heavily dependent on the weather conditions. Wet foliage can cause strong signal attenuation.
- **Indoor and tunnels**
  Vehicles sometimes navigate in locations without GNSS reception (e.g. garages and tunnels). This scenario can only be countered by alternative positioning technologies or by the development of unique algorithms during system design.

16.2.2.3 Multipath issues

In addition to buildings, natural slopes near the roadside, or steep valleys in mountainous regions, can lead to reflected signals. Therefore, the resulting pseudoranges are too large, leading to large systematic errors in the PNT solution. This problem has actually increased with the emergence of high sensitivity GNSS receivers. Reflected signals usually experience larger propagation attenuation, but are still available. Therefore, the availability of PNT data improves while accuracy might actually deteriorate.

The consequence of these effects depends on the circumstances: a large building on one side of the road leads to a lateral offset of the GNSS trace, for example. This is of little consequence for determining the travelled distance, but the passage of a defined road cross-section might be left undetected. Of course, the detection algorithms can be modified to compensate for the offset. Therefore, in a well-designed system, these effects can be mitigated while the GNSS data is still useful.

16.2.2.4 Time to first fix

When an OBU is powered on, it takes time for the GNSS module to deliver the first position fix. The device must scan the GNSS spectrum for available satellite signals, select the most promising ones, and then calculate a PNT solution from those signals. This usually takes about 30 to 40 s, but may take up to a few minutes under adverse conditions. Modern GNSS modules store satellite ephemerides[7] for a given time (e.g. 2 weeks), which improves the TTFF considerably. But in cases when the equipment has been deactivated for a longer period, or if the location reactivation is far from the last measured fix ('tow and carry'), this information becomes useless.

Newer generations of GNSS receivers offer functionality for reconstructing the path before the first fix, using buffered pseudo-ranges. Assisted GPS is only of limited use in this case; the cellular network module must first power itself up, register with the nearest base station, and download the ephemerides; this results in delays similar to an unassisted TTFF.

[7] Ephemerides are accurate data about the satellite orbits. The position of each satellite can be calculated from this, which is a necessary prerequisite for finding a PNT solution.

16.2.2.5 Jamming and spoofing

The potential of disrupting GNSS signals through spoofing and jamming have been a major concern since the launch of the nationwide truck tolling system in Germany. Jamming is the generation of signals that result in a deterioration or failure of GNSS reception. Considering the fact that GNSS signals are rather weak, it is relatively easy to jam the signals in the same frequency range – in the vicinity of a GNSS receiver. GNSS signals can also be unintentionally jammed due to faulty equipment located near the receiver.

Spoofing, on the other hand, is the intentional generation of misleading or 'forged' GNSS signals. Since spoofing requires considerable technical knowhow and expensive equipment, it is not yet considered to pose a significant threat to the practical operation of tolling systems. Furthermore, it is relatively easy for the operator of a GNSS-based tolling solution to detect spoofing since gaps and jumps in the timing solution can be detected.

Jamming and spoofing of GNSS signals are illegal in most countries. Widespread attacks can be detected with relative ease, and severe penalties can be enforced. One documented case of GNSS jamming took place when North Korea impaired GPS reception in large parts of South Korea using high power interference signals.[8] Small-scale attacks are of limited economic impact while the risk of being detected and paying high fines can potentially dwarf any potential benefits of the perpetrator. The installation of a jamming device within the vehicle in order to avoid paying road charges can be detected with relative ease. New generation tolling OBUs can detect signal manipulations when this condition continues for a given time span. In such a case, the OBU automatically switches into a non-operational ('red light') mode – when the user can see that the OBU is not operational, and alarms are transmitted to the toll operator.

16.2.2.6 The remedy: inertial navigation and other sources of location information

Most of the problems described above can be mitigated by using other kinds of location information to support and verify the GNSS data. When GNSS reception is strong, additional sensors are calibrated using the PNT solution, or an integrated solution incorporating all available sensor data is used.

The most common technologies for improving position location are:

- Accelerometers and gyroscopes: modern GNSS chipsets support the input from such sensors or even integrate the sensors on the chip. If GNSS positioning fails, it is still possible to trace the vehicle's position.
- Vehicle data: with in-vehicle networks improving, sensor data from the vehicle become more readily available for PNT solutions. For example, wheel ticks give information about the circumference speed of each wheel, which can be very useful to improve the measured vehicle trace.

[8] *See e.g.* http://gpsworld.com/massive-gps-jamming-attack-by-north-korea/ *from 8 May 2012*

- Localisation augmentation: DSRC beacons can be installed in areas having limited GNSS reception. There is a dedicated standard suite (ISO 13141 plus the respective test standard ISO 13140[9]) covering this application. In order to minimise the additional costs of roadside infrastructure, the definition of tolling detection points should be considered carefully when drafting tolling regulations.
- Cellular network data:
  - The cell ID identifies the cellular network base station to which the OBU (i.e. the terminal) is currently connected, but this information is not always available to toll system operators.
  - Direct location using cellular network signals: the triangulation of signals from the nearest base stations is already used in smart phones, e.g. to determine indoor location, but is not typically available for external applications such as electronic tolling.
  - Unfortunately, cellular network data is usually not available to dedicated tolling devices. This might change in future when mobile network providers decide to offer this service.

Fortunately, through the use of multi-frequency by GLONASS and Galileo, GNSS signals are more robust against jamming. Furthermore, in 2018, Galileo will implement signal authentication services, which will make spoofing virtually impossible since the GNSS receiver will only accept signals transmitted directly from the Galileo satellites [20].

16.2.2.7 Satellite-based augmentation systems

In many parts of the world, regular GNSS satellites are complemented by Satellite-Based Augmentation Systems (SBAS). In Europe, the European Geostationary Navigation Overlay Service (EGNOS) is used; in Northern America, the Wide Area Augmentation System (WAAS); in Asia, the System for Differential Correction and Monitoring from Russia, GPS Aided GEO Augmented Navigation (GAGAN) in India and the Quasi-Zenith Satellite System (QZSS) in Japan. They all follow similar concepts of measuring the quality of PNT data at reference stations and deriving correction data These data are transmitted from dedicated satellites, typically in geostationary orbits. These augmentation systems were primarily devised for applications in aviation systems. It turns out that they are of limited value in the road sector for at least the following reasons:

- Satellite visibility: the position of geostationary satellites is quite low on the horizon for areas that are not close to the equator. Therefore, especially in challenging environments (such as urban canyons) the chance of good signal reception is relatively small.

[9] ISO 13141:2015, *Electronic fee collection – Localisation augmentation communication for autonomous systems*, ISO 13140-1:2016, *Electronic fee collection – Evaluation of on-board and roadside equipment for conformity to ISO 13141 – Part 1: Test suite structure and test purposes;* and ISO 13140-2:2016, *Electronic fee collection – Evaluation of on-board and roadside equipment for conformity to ISO 13141 – Part 2: Abstract test suite*. Available at www.iso.org.

- Local factors determining the PNT data quality: while the correction information provided by SBAS is of a global nature (e.g. ionospheric corrections), the quality of PNT data in the road sector is largely determined by local influences, such as reflections, multipath phenomena, and local signal deterioration (e.g. wet canopy).
- GNSS-based OBUs are usually mounted on or close to the windshield. The GNSS antenna is therefore oriented towards the driving direction. If a vehicle in the northern hemisphere is heading in a northeast or northwest direction, for example reception is almost impossible since geostationary satellites are only in view in the south.

16.2.3 Issues with the communication link

Since GNSS-based tolling systems typically rely on mobile communication networks for sending the relevant data to and from the OBUs, this technology represents a critical part of the tolling infrastructure. Therefore, the issues described in this section are highly critical for successful GNSS-based toll system operation.

16.2.3.1 Network coverage

Although the cellular network coverage has improved significantly over time, there are still locations with limited cellular reception. Therefore, tolling solutions need to be designed to ensure that they are still functional in areas having an unreliable communication – without any loss of toll data. For example, an OBU can store the tolling data generated in areas without adequate GSM availability and transmit the data when the network coverage is available. In fact, an OBU's memory can serve as an ideal backup facility, storing the tolling data for a given time so that they can be retrieved on request (e.g. for disaster recovery).

16.2.3.2 Bandwidth

The communication bandwidth of modern cellular networks is usually more than adequate for tolling purposes. There may be cases where the available bandwidth might become an issue, for example when a fall-back to GPRS communication is necessary. A GNSS-based toll system should therefore be defined in a way that minimises the data volume required via the cellular network link: both for charge data in uplink communication, and for map or tariff updates in downlink communication (for *fat client* solutions).

16.2.3.3 End of journey

When a vehicle terminates a trip in an area without cellular network reception, for example, when entering a garage, the last data relating to the trip cannot be transmitted to the tolling back office. This data must be stored in the OBU and can only be transmitted after the next trip is started. This can be a problem when there are time limitations for issuing invoices or for monitoring traffic.

16.2.4 Issues with power supply

GNSS OBUs still have high power consumption compared to other tolling technologies, such as 5.8 GHz DSRC tags. It is not possible to operate a GNSS-based

OBU for extended periods without an external power connection. Since users of tolling OBUs are not necessarily motivated to ensure that the OBU is always working properly, they may not recharge a battery-operated OBU when needed. This leaves two options:

- the OBU is permanently integrated into the vehicle, having a fixed connection to the car battery.
- for quick self-installation (e.g. for occasional users), the OBU can be connected to the cigarette lighter plug. This connection is not always reliable, however, since the plug may already be in use, the connecting jack might be pulled out accidentally, or the cable might break due to permanent bending and stretching.

In both cases, there is the need for a rechargeable backup battery – not only to buffer gaps in power supply, but also to ensure the basic functioning of the OBU so that the user is always compliant to the tolling scheme. In most GNSS-based tolling systems, OBUs are required to operate autonomously for at least a few hours.

In any case, an important design goal of GNSS-based OBUs is to keep the power consumption as low as possible, since:

- the vehicle's battery cannot be drained during long time spans of inactivity;
- a backup battery in the OBU can be costly and have a limited lifetime, and therefore should be as small as possible – and easily exchangeable.

This implies the need for sophisticated power management in the OBU, including low power design and an elaborate concept of maintaining different activity levels of the OBU (such as a sleep mode and a suspend mode).

16.2.5 Issues with maps

Since the GNSS receiver only delivers a series of position and time fixes, it is necessary to maintain and use map data even in the most basic toll systems. At least the borders of the toll domain must be known, so that data is only collected when necessary; this will minimise communication costs and also protect the user's privacy. Therefore, it is advisable also for thin-client solutions to act accordingly when passing a 'geo-fence'. Otherwise, potentially high (and completely unnecessary) data roaming costs can be the result.

16.2.5.1 Updates

In practical implementations like the German truck tolling system, each section of the tolled road network is detected as a basis for calculating the road user charge. For large road networks subject to tolls, regular modifications of the section data could be required (e.g. due to road construction and maintenance). Keeping all the map information up to date can be challenging, especially when this data is required in the OBU for a *fat client* solution (as is the case in Germany and Slovakia, for example).

16.2.5.2 Coordinate reference systems

Although reference systems may be of minor importance in most systems, toll system designers must be aware that there are inconsistencies between the coordinate systems used in different parts of the world, and even between the various GNSS constellations. These differing reference systems ('geographic datum') can lead to additional errors.

16.2.6 Key performance indicators (KPIs) and service-level agreements (SLAs)

In real world tolling environments, it is important to measure system performance in the implementation phase as well as during system operation, using KPIs and SLAs. Depending on the business architecture, the measurement of certain parameters is selected. Most typically, the correct charging rate is the most important parameter – usually measured in per cent. The number of 'false positives' is usually measured as well, to ensure that no charges are assigned to a user without justification, since overcharging obviously leads to reduced user acceptance. The ISO standard 17444 defines performance metrics and strategies for measuring these parameters (in part 1 and part 2, respectively).[10] If a toll is levied as a tax, rather than a charge, the requirements for system performance become even more stringent and critical, since taxation must be fair and accurate.

16.2.6.1 Measurement

If a KPI was set on rarely occurring events, it would be difficult to measure. For example, requirements for false positives have been observed in the range of 10^{-6}. To get reliable data about such a KPI, a very large sample size is required. The generation of large amounts of samples can be realised during system operation, but not during the implementation phase (i.e. during acceptance testing). In performance measurement, there is also the problem of reference values (a.k.a. 'ground truth'). For ambitious KPIs that are close to 100% (or to 0%, depending on the target), it is extremely difficult to find reliable reference values to compare the system performance with.

16.2.6.2 KPIs on PNT information

When a toll system architecture assigns the measurements of vehicle traces to one entity, and the evaluation of those traces to another entity, it is important to measure the quality of the traces for KPI and SLA measurements. This measurement is a very complicated task, due to the wide range of parameters influencing the performance of a PNT subsystem. (More on the separation between the Toll Charger and TSP is described below.)

[10] ISO 17444-1:2012, *Electronic fee collection – Charging performance – Part 1: Metrics*, and ISO 17444-2:2013, *Electronic fee collection – Charging performance – Part 2: Examination Framework*, download available for both on www.iso.org

Ongoing work in standardisation tries to resolve these issues with a set of standards and specifications, but it will take some time until a standardised prescription of position performance metrics and their measurement is available.

The European Cooperation in Science and Technology (COST) has initiated a COST Action called SaPPART (Satellite Positioning Performance Assessment for Road Transport). Quoting the SaPPART website: 'This Action brings together experts in GNSS, intelligent transportation systems (ITS) and mobility to address the open issues and guarantee the success of the standardisation for underpinning certification initiatives. The Action will propose a unified framework for the definition and the assessment of performance for the GNSS-based positioning terminals'. On the SaPPART website (www.sappart.net), a white paper can be downloaded that analyses the topic of PNT metrics for the road sector.[11]

The CEN/CLC/TC 5 is part of the European standardisation organisation CEN. The Task Committee TC 5 deals with all topics related to space technology. Working Group 1, CEN/CLC/TC 5/WG 1, for navigation and positioning receivers for road applications, has recently released the first part of a standard suite for performance metrics on PNT for the road sector.[12]

Also in this case, the issue of reference data (ground truth) is a problem. Even expensive high-precision GNSS receivers can measure the wrong trace very precisely, being prone to the same influences (such as multipath issues) as the actual system.

16.3 Commercial and political obstacles in implementing tolling systems

The commercial and political process of introducing a nationwide tolling system often follow a typical pattern in which the public authorities begin to gather information about how toll systems are operated – both technically and commercially. These authorities often issue detailed technical requirements that reflect preferences for certain types of technology or even specific suppliers. It is also typical for technical trials to be initiated in which vendors can, at their own expense, demonstrate how successful their solutions can be deployed for electronic tolling. It can be vital for technical suppliers to participate in such demonstrations – not only to enter the competitive process, but also to 'educate' the tendering authority. As pointed out in the previous section, the tolling requirements issued by public authorities strongly influence feasibility, the business case, and chances of success in general. Since most technology suppliers have already worked in the field for many years, they can provide a very valuable source of knowledge for the authorities.

[11] http://www.sappart.net/wp-content/uploads/2014/08/White-Paper_SaPPART_sept15.pdf: 'SaPPART White Paper – Better use of Global Navigation Satellite Systems for safer and greener transport'.

[12] EN 16803-1:2016, *Space – Use of GNSS-based positioning for road Intelligent Transport Systems (ITS) – Part 1: Definitions and system engineering procedures for the establishment and assessment of performances*, available from standards.cen.eu.

Once a tender is published, legal claims are often made against the tender procedure by some of the suppliers (who believe that certain tender requirements are discriminatory towards their solutions) – thus prolonging the tender process. After a tender has been awarded, legal claims are almost always made by competing suppliers who were not selected. These legal battles can easily delay the planned start of system implementation by at least 1 year. To make matters worse, public debates are often launched by various stakeholders, using all available media channels to support their specific interests.

16.3.1 Long lead times

It is obvious that the procedure of establishing a nationwide tolling system, as described above, can take several years and require considerable investments from all involved parties. Consequently, the political leadership of the tendering authority can change during the course of the tender process. Shifts in political power often result in cancellation of the tolling projects, as most prominently witnessed with the French *écotaxe* system. The economic loss of billions of Euros in France was staggering.

16.3.2 Privacy concerns

The protection of the drivers' privacy has been a major concern with the planned introduction of tolling schemes. In a public debate, it is difficult to convince citizens that their privacy is well protected, and the recent advances in big data technology have shown that those concerns are well justified. While it is undeniable that tolling systems collect a lot of privacy-relevant data, new European regulations on data protection will become more stringent.[13]

Those designing tolling systems therefore need to consider data privacy issues in all stages of project planning, ensuring that only relevant data is collected – and discarded as soon as it has fulfilled its purpose, thus impeding any connections made between people and location data. Chapter 18 describes how privacy issues are addressed in Oregon.

16.3.3 Increase in transportation costs

In many parts of the world, road infrastructure is traditionally perceived as a free commodity. The notion of suddenly charging for the use of roads can seriously influence public perception. Hauliers, for example, seek to protect their profit, and the fear of a negative economic impact is common.

Since the development and maintenance of road infrastructure requires significant funding, clear communication is needed for road users to understand the necessity of their contribution to the required investment. Convincing the public can be a very demanding task, and working with various media outlets is essential

[13] Regulation (EU) 2016/679 of the European Parliament and of the Council of 27 April 2016 on the protection of natural persons with regard to the processing of personal data and on the free movement of such data, repealing Directive 95/46/E

to raise awareness of the true benefits of tolling (e.g. fair distribution of costs, cost transparency, or environmental benefits). See also Chapters 5 and 7.

Road construction projects by private entities, or those based on private or public–private partnerships, are far better in this respect since the public can easily comprehend that a private entity which invests in the construction of road infrastructure will need to charge for the use of those roads.

16.3.4 Fear of failure and over-specification

Past tolling projects have either failed or have faced serious struggles due to technical design decisions. As a reaction to this, tendering authorities have attempted to specify solutions to a great level of detail. However, the more technical detail the authorities prescribe in the tender, the less flexibility is given for potential suppliers to reuse their proven solutions – thereby increasing cost efficiency and reliability. When tolling systems are over-specified, technology suppliers have limited potential for optimising their solutions, even for newly developed components.

16.3.5 Lack of a sound business case

Another pitfall in the decisions made by public authorities may be triggered by the fear of public pressure through bad press. In order to avoid the opposition of certain interest groups, tendering authorities may be tempted to restrict the tolling fees among certain groups of road users. Granting too many exceptions can lead to a significant loss in the toll revenues, as well as increased system complexity and cost, in which case the overall business case may be questioned: considering the large investment needed for installing a new tolling system, and taking also the operational expenses into account, there may be little revenue left which would justify the overhead of such a complex system in the first place.

16.4 Trade-offs in the definition of tolling systems

As we have seen, there is a very wide spectrum of how GNSS-based tolling systems can be implemented. But there is no 'free lunch'. Here are a few of the trade-offs:

- the number of vehicles vs. the size of the tolled road network
- flexibility vs. complexity
- cost of implementation vs. cost of operation
- thin vs. fat client

We will now briefly look into these aspects and try to suggest a sound balance, using the advantages given by state-of-the-art GNSS technology.

16.4.1 Number of vehicles vs. size of the road network

It is obvious that for a very large number of vehicles driving on a small road network (a single bridge in a city, for example) it is advantageous to deploy OBUs (or tags) that are as inexpensive as possible, while the investments in RSE are relatively low. On the other hand, for large road networks with a low tolled traffic density (such as the truck tolling scheme in Russia), RSE is the major cost driver and should thus be kept to a minimum. Especially in rural areas, the lack of power and communication infrastructure can significantly increase the implementation costs of RSE.

In this context, the complexity of the road network must also be considered. If the toll roads consist of highways having long sections with well-defined entries and exits, it is relatively easy to cover this road network with infrastructure-based solutions (e.g. DSRC or ANPR). If a lot of surface roads having frequent intersections are part of the tolled road network, the implementation of a GNSS-based solution will be far more cost-effective. Since the Austrian and Czech truck tolling schemes were implemented using DSRC technology, for example, the extension of the tolling system to first-class roads has been very limited due to the high cost of the microwave infrastructure that would be required for these roads.

16.4.2 Flexibility vs. complexity

The trade-off between flexibility and complexity is familiar to a system architect in any field, and is of particular relevance to electronic tolling systems. On the one hand, there is a large legacy base, and new systems are frequently designed to incorporate new concepts and ideas. On the other hand, the implementation schedules are often extremely challenging. Therefore, toll solutions suppliers would be well advised to provide a flexible solution which can be easily adapted to evolving customer requirements. Unfortunately, the development of a fully flexible solution would require a very high amount of resources that are rarely available for a single project implementation.

It is therefore advisable to start with a flexible and basic system architecture that can be easily extended and modified. The system concept should allow for the addition of functionality, forming the basis for long term system development. The design of a successful product line forming the basis of a comprehensive solution is one of the biggest challenges in system architecture.

Unfortunately, the resulting product line must also support considerable scalability, since the system should be efficient for very small numbers of vehicles (e.g. 50 vehicles for a trial) as well as for large scale system having several hundred thousand or even millions of users.

16.4.3 Cost of implementation vs. cost of operation

The chosen solution must also represent the optimum balance of upfront investments (Capex) and costs for operating the system (operational expenditure, i.e. Opex). This is all the more critical since potential operators usually shy away from a higher Capex even if the resulting solution would offer a lower Opex and thus provide the overall cost optimum for the duration of the project.

Obvious contributions to the Capex are the installation of RSE and data centres, but also the production and initial distribution of OBUs. In GNSS-based systems, as also with DSRC-based systems, the major cost drivers with respect to

Opex are communication and maintenance costs, points of sales, and the billing, and clearing processes.

16.4.4 Thin vs. fat client

The 'client/server' environment is well-known in system architecture. In the tolling domain, the server is the central system (proxy), while the OBU takes the role of the client. In general, two main variants of this architecture exist: *fat client* systems perform a lot of processing in the OBU (e.g. charge object recognition, tariff determination), while *thin client* OBUs transmit the pure PNT data to the central system, where the data is further evaluated.

Both approaches have distinct advantages: the *fat client* makes efficient use of the processing power in all of OBUs (of which there are typically hundreds of thousands), enables easy implementation of real-time user feedback (e.g. current tariff), while the data payload for charge communication is small. It is also easier to dissipate privacy concerns of the road users and authorities since PNT data generally never leave the OBU. The price for that is the complexity in OBU software, resulting in reliability issues and the need for more frequent software updates, and the need for keeping the frequently changing map data current in all the OBUs. Considerable extensions of the tolled road network also result in large updates.

Thin client OBUs make available all PNT data in the central system that can be used for other purposes (e.g. traffic management, fleet management), the software is lean and easy to maintain and changes in map data or tolled road network can be accommodated easily with changes mostly in the central system. These advantages are counterbalanced by the higher data payload for charge communication (which can be reduced using sophisticated compression algorithms). The system performance relies more heavily on the available communication bandwidth, usually resulting in the need for more buffer memory. Depending on the applicable legislation, it can be difficult to ensure and certify appropriate protection of road user privacy.

16.5 Trends in GNSS-based tolling systems

This section looks more deeply into the state-of-the-art GNSS technology and how technological advancements can affect future implementations and standards in electronic tolling.

16.5.1 Integration into cooperative intelligent transportation systems (C-ITS)

New cars will soon be equipped with electronic equipment that facilitates communication among cars and to roadside infrastructure (V2X, i.e. Vehicle to Vehicle and Vehicle to infrastructure). This new domain of vehicle communication, also known as cooperative ITS (or C-ITS), has been tested for a number of years and should soon become commercially available. The standardisation of C-ITS communication is also gaining momentum.

Vehicles that are equipped with 5.9 GHz microwave communication can exchange data with ITS stations at the roadside, also for electronic tolling. Could the cost distributing and installing tolling OBUs to all vehicles become obsolete if C-ITS becomes universally available? Several challenges remain before this can become reality:

- Standard for C-ITS stations have not yet been defined
- Setting priorities for competing services at a single ITS station has yet to be solved
- Certification of the C-ITS stations must be possible to ensure that the resulting toll is correct.

Nevertheless, the use of 5.9 GHz microwave communication for C-ITS applications offers enormous potential in reducing costs. Furthermore, the use of pre-certified equipment and trusted sensor data can offer a great advantage with respect to non-repudiation issues.

16.5.2 Multi-constellation GNSS

Today, there are already two complete GNSS systems (GPS, GLONASS) in place with worldwide coverage. The European Galileo system launched its first open services in 2016. By 2018, Galileo will have a complete constellation in place. China has begun operation of its Compass/Beidou system, and the Indian GAGAN and Japanese QZSS systems are now also fully functional.

The advantage of this multitude of constellations for GNSS tolling systems is the increased availability of satellite positioning signals. This leads to significant advancements in challenging environments (such as urban canyons) since any additional visible satellite could make the difference between having a valid location measurement or not. A great improvement in signal robustness can be observed as well, since GPS has suffered from irregularities in the past. Nevertheless, there is a price to pay for multi-constellation capability: new GNSS receivers require higher performance, which results in higher cost and power consumption. Some minor issues with respect to the different coordinate reference systems must also be dealt with (see also Section 16.2.5.2).

16.5.3 Fusion of GNSS signals with inertial sensors and vehicle data (CAN bus)

In the future, when 100 or even more GNSS satellites will roam the skies above us, there will be situations when GNSS is still not capable of providing useful PNT fixes. The most obvious example is tunnels, but also parking garages or deep urban canyons will remain problematic. For this reason, modern GNSS modules integrate other data to improve the positioning accuracy. Inertial sensors such as accelerometers and gyroscopes can be connected or are even integrated on chip thanks to Micro-Electro-Mechanical Systems technology.

Existing systems such as the German truck tolling require a connection to the tachograph, which provides a legally certified data source with respect to the distance travelled. This idea is taken further by incorporating wheel ticks: signals giving the revolutions of each wheel of a vehicle. This can be used not only for distance measurement, but also to determine changes in direction from the differences between the inner and outer wheels. The integrated traces derived from such sensor fusion have shown considerable improvements compared to using GNSS alone. Such data are often also available on the CAN bus,[14] but not easy to access – since this would require a complex OBU installation. Hopefully, standards will eventually be established that allow easy access to this information in the future.

16.5.4 Separation of Toll Service Providers and Toll Chargers

In simple terms, the Toll Charger is the organisation representing the road owner in terms of toll collection. The TSP is responsible for contracts with and relations to the road user, e.g. billing, OBU distribution, and dispute settlement.

Until a few years ago, these roles were usually taken on by the same entity – simply called toll operator. The need for separating these roles emerged, initially in countries having several private road operators (such as France) where it turned out to be problematic for road users to have separate business relationships with all the local operators. Companies eventually emerged that acted as brokers between road users and operators. This separation of responsibilities was also incorporated into the EETS legislation. The basic principles of this effort (one OBU, one contract, one bill per road user) strongly implied the existence of TSPs acting independently of local Toll Chargers.

16.5.5 Interoperability and EETS

The emergence of more local, regional, and national tolling schemes increases the need for the emergence of TSPs. For these TSPs to operate efficiently, it is crucial that their OBUs interact flawlessly with all the RSE of each tolling domain, and that the interfaces to the toll collection data centres work according to agreed standards.

Fortunately, and as described in other parts of this book, a comprehensive set of standards has been developed to achieve these goals. Since interoperability issues will become increasingly relevant in the future, the design of new tolling systems will need to take standards and norms into account.

16.5.6 Use of smartphones in electronic tolling

In recent years, it has been suggested that smartphones can be used instead of OBUs for road usage schemes – both for trucks and for passenger cars. The European tolling schemes are designed to secure the revenue streams from the vehicles obliged to pay, with virtually no chance for road users to evade those charges. The distance-based fee is not voluntary and, with the exception of the use of fleet management devices in Hungary, the technical equipment used to calculate the fees

[14] Controller Area Network (CAN bus) is a vehicle bus standard designed to allow microcontrollers and devices to communicate with each other in applications without a host computer.

must meet very strict requirements. In some schemes, such as Germany and Hungary, self-declaration is made possible by means of roadside terminals, call centres, and the internet – and here smartphones can be easily used as means of electronic payment. However, the use of smartphones to detect the use of tolled road sections would be far more complex – and costly.

Windshield-mounted GNSS-based OBUs, which are used in Slovakia and Belgium, for example, are cost-effective as they are designed for the specific task of accurately detecting the travelled toll road sections and reliably transferring this data to the back office to secure payment of the road usage. A central idea behind the European road usage charging systems is that the truck drivers should not be burdened with the issues of declarations, payments, and handling of devices. A mandatory OBU can be installed within minutes by the driver who attaches the device to the vehicle's windshield, connects it to the cigarette lighter, and sets the vehicle category. From then on, everything is done automatically: the OBU knows when to turn itself on and off, when to record trip information, and when to send and receive data from the back-office server using the cellular (GSM) network.

A tolling OBU differs from a smartphone in that it cannot be tampered with; you cannot open it up, remove the internal battery, or change the SIM card. An attempt to cut off the power will set off an alarm on the OBU and immediately notify the back office of potential non-compliance. The data being transferred to and from the OBU cannot be tampered with (due to encryption and, in many cases, a built-in Security Access Module) nor can any data be 'lost'. Most importantly, there are no excuses for an OBU not functioning properly: you don't need to 'boot up' the device, install an application, or make sure that the batteries are charged. Enforcement officers can control user compliance by reading the data directly from the OBU – even when it is turned off. In short, the truck tolling systems in Europe are closed systems with the sole purpose of ensuring full compliance to the scheme and thus securing the revenue stream. Of course, there is great potential for other value-added services, particularly those that could increase traffic safety and provide real-time traffic information, but these have yet to be realised in any of the European schemes. A smartphone, on the other hand, could not easily meet such rigid requirements. There would be countless versions of hardware and software out in the field and no guarantees that the correct version of the tolling application is operating as it should be. Even the basic operating systems could differ significantly among the various smartphones on the market, not to mention the potential of other 'apps' operating in parallel that could impair a smartphone in its ability to detect the use of tolled roads, even causing the road usage charge 'app' to crash (along with the entire smartphone). An 'open platform' software could be potentially manipulated or sabotaged, viruses could be spread, and the proper functionality of the tolling application be manipulated in ways to avoid paying the obligatory fees. GNSS-enabled smartphones are thus hardly a suitable substitute for GNSS-based OBUs when it comes to mandatory fee collection; at least as long as it is not a 'friendly user' scenario (i.e. when the driver is highly motivated to make sure that the tolling application works).

Studies have been made to demonstrate the potential for using smartphones that are already being used in areas where millions of users will be potentially obliged to pay distance-based fees. For example, in the Minnesota trials in the USA, users received a discount when using a smartphone: they paid only 3 cents per mile, rather than the obligatory 5 cents per mile. In the case that their smartphone crashed or ran out of power, they would pay the full price. The default payment was to be measured by the odometer readings. Since regular technical vehicle inspections are not mandatory everywhere, it is not certain how to secure regular and reliable readings from the odometer [21].

In the European road usage charge schemes, there are typically 'only' a few hundreds of thousands of users in each country having a nationwide scheme. Once there are millions of users in each toll domain, the cost of producing and distributing OBUs to all users is a factor to be considered. A European OBU might cost a fraction of the price of a smartphone, but the fundamental question being asked (and being explored by various trials) is whether you can realistically offset the capital costs of a system by allowing users to deploy their own devices for tolling. There are many issues to be solved, of course, yet progress is being made in a number of studies (in particular in the United States). Based on the years of experience made in Europe with GNSS-based tolling, a key success factor for the introduction of an 'open system' (in which the drivers are offered a number of technology choices for tolling) would be to ensure that not only capital investments could be saved, but also that operational costs would not end up being higher than that of a closed system. The 'bring your own' approach would probably only work if users had very strong incentives that their devices work properly (and high penalties when they don't).

16.6 Using GNSS-based tolling systems for other purposes

16.6.1 Missing trader fraud

Missing trader fraud, i.e. the theft of VAT, has become a growing concern within the EU since the movement of goods between EU member states is VAT free. An organisation committing this crime charges VAT on the sale of goods and absconds with the VAT rather than handing it over to the state's tax collection authority, for example. Or the payment of VAT could be illegally avoided by declaring that goods were exported to another member state when they were not.

At the beginning of 2015, an 'Electronic Public Road Trade Control' system was launched in Hungary to enforce VAT laws. By using this new system, the actual route of the goods can be tracked since transport-related data is registered in a central electronic system before the transport commences. Thus, the actual delivery of goods to and from specific warehouses where import and/or export has been declared can be monitored. Such a detailed monitoring system requires a high level of technological complexity and, one might think, a significant investment. However, since Hungary had already introduced GNSS-based electronic tolling system on all of its major transit routes in 2013, the new Electronic Public Road

Trade Control system took advantage of this tolling infrastructure to track the movement of all transport vehicles [22]. The required capital investment was so low that it paid for itself within a few weeks. In fact, the VAT surplus during the first half year of system operation (compared to the same period in the previous year) approached €500 million, which is approximately the income generated by nationwide truck tolling scheme on the tolled road network of approximately 6,700 km over the period of an entire year.

16.6.2 A Multipurpose GNSS-based OBU – with data privacy

Information provided by an OBU based on satellite technology can be the basis for individual or global services. Aggregation of information may open up a range of new services that can serve the public interest. A multipurpose GNSS-based OBU (Figure 16.10) captures road usage data from every vehicle and transmits the data to the back office for further processing. Consequently, the OBU – which until now has been the key component of satellite-based Electronic Toll Collection systems – can also form the centrepiece of other mandatory and optional GNSS-based services. Since data privacy is typically required for various location-based services, state-of-the-art security technology is deployed to ensure protection of the data during transmission and while it is stored in the central computer system ('back end').

A GNSS-based tolling OBU can operate completely on its own, establishing a communication link to the system back end. After many years of deployment experience, the robust, cost-effective, and user-friendly GNSS platform for distance-based road user charging can be the basis for more diverse and demanding applications, particularly in domains in which state regulations require the monitoring of position data.

The demand for tracking and tracing of particular groups of vehicles has been steadily growing, particularly within the European Union. As indicated above, the Hungarian authorities demonstrated how easily it could establish an electronic trade control system to combat cross-border tax fraud. This is but one example of the new opportunities that are now unfolding in which the technological platform established for satellite-based tolling systems can provide a completely new

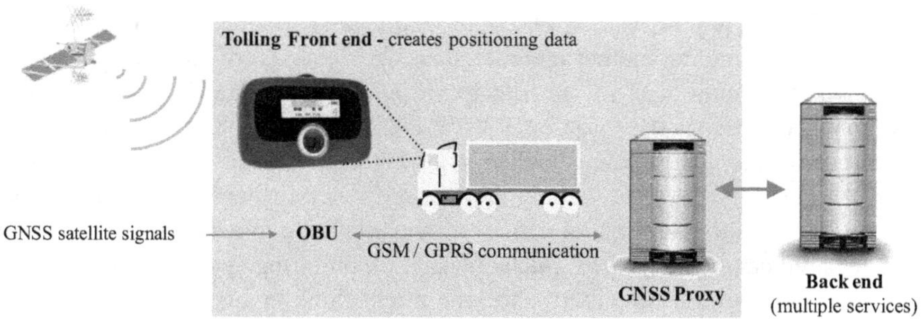

Figure 16.10 A common platform for multiple positioning services

generation of services and applications in the road transportation sector and beyond. Other services that could use the infrastructure of GNSS-based systems include the obligatory tracking of dangerous goods and livestock [23].

16.7 Lessons learned from existing GNSS-based tolling systems

By now there is a wealth of experience with GNSS-based tolling systems throughout Europe, yet with each new tender, the approach towards defining and implementing such systems is virtually re-invented. What can be learned from the existing systems, what are the best practices, and how can they potentially fit into new standards?

1. We have witnessed the implementation of several GNSS tolling systems. The requirements for each of these projects vary so widely that large parts of the technology had to be newly developed. This drives up the costs and lead times, and increases the risk of substantial delays or even complete failure.
2. Technology vendors can build a solid base of their GNSS solutions by establishing a highly adaptive solution to mitigate these effects. This can be challenging, however, considering the short amount of time typically given between contract award and system launch.
3. Most existing systems have been devised to start operation with a large road user base and considerable road network coverage – a so-called 'big-bang approach'. From the technology and implementation perspective, it would be advantageous to slowly increase the size and complexity of the system, starting with a few vehicles and gradually scaling the service. Since the development overhead for the GNSS technology and IT infrastructure is basically the same for both small and large numbers of users, the contractual structure between the tendering authority (typically the Toll Charger) and the vendor (typically the Toll Operator) could consider a guaranteed initial number of users and take into account a later increase of users so that the Capexs can be distributed among a larger number of OBUs.
4. Jamming and spoofing are hotly debated topics in the GNSS arena that are also addressed in virtually all GNSS-based toll system tenders. Until now, the risk of jamming and spoofing have been minor for tolling purposes since the achievable economic gain is relatively small, while the risks for the perpetrator are substantial – high fines are issued for non-compliance with the tolling scheme and for the illegal sabotage of GNSS signals. Nevertheless, Galileo will introduce signal authentication services from 2018, to ensure that the GNSS receivers will only accept direct signals from Galileo satellites and not from other sources (i.e. from the perpetrators of spoofing and meaconing[15]).

[15] Rebroadcasting a GNSS signal maliciously, accidentally, or to improve reception but causing misreporting of a position.

5. At least for truck tolling systems, a dedicated OBU is difficult to replace – for example by a smart phone. The necessity for reliable operation far outweighs the potential cost savings of using smart phones (as described in detail in Section 16.5.6). Once a truck tolling system has been introduced, hauliers are primarily interested in the smooth and reliable fulfilment of their tolling obligations – and not having to worry that their tracking devices are working properly.
6. Although most of the implemented systems so far favour a fat client solution, there is a growing trend towards thin client architectures. With the steady improvement of mobile communication services, the advantages of making frequent changes in map information centrally (at the proxy server) can potentially outweigh the higher data payload of transmitting track and trace data from the OBU to the proxy. Also, considerable progress has been made with respect to communication availability and round-trip delays in GSM networks.

The advantages of using GNSS-based tolling for a comprehensive nationwide tolling scheme, as described in this chapter, are indisputable. The flexibility of GNSS technology allows for all roads to be tolled, thus making it possible to easily address the phenomenon of traffic diversion and toll evasion that typically occur when motorways and expressways are tolled. There are no limitations to the development of a toll tariff structure that can discourage vehicles from using specific roads or entering certain zones or areas, reduce the risk of traffic congestion, minimise the risk of road accidents (e.g. during adverse weather conditions), or to reduce vehicle emissions in sensitive areas – also in real time, for example during a smog alarm. Even traffic control can become viable when tolling fees are adjusted in real time.

With the ready availability of accurate position data, the door is wide open for value-added services to be provided in addition to tolling. Although additional services may not generate additional revenue, they create added benefits for the toll road users that can help to increase the acceptance of a tolling scheme. We have already witnessed how new services are being offered in combination with GNSS-based tolling infrastructure, creating additional revenue for TSPs: in Hungary, fleet management services are combined with electronic tolling services. With the recent interest in the EETS, where the number of registered EETS providers has more than doubled in less than 1 year, it would be only logical that new value-added services will be offered in combination with tolling services.

Looking into the future, new GNSS tolling systems are to be expected: in Sweden, a new initiative ('The Nordic Way') is pursuing this course, and Switzerland plans to implement its own EETS as a true GNSS domain without any dependence on the tachograph – at least for foreign vehicles. Bulgaria has issued a tender for the implementation of a GNSS-based nationwide tolling scheme, with the intention to support both the Hungarian approach (in which local fleet management providers can offer toll services) and the EETS approach, where foreign TSPs can offer their services in Bulgaria from the moment the new tolling scheme goes live. Poland and the Czech Republic are also considering the implementation of GNSS-based tolling systems since an extensive expansion of the tolled road network with their existing DSRC-based solutions is not feasible.

There has also been a mounting interest for GNSS-based tolling in South America and other parts of the world, where widespread road networks having low traffic density and the lack of infrastructure begs for satellite-based solutions. Another notable trend is the development of GNSS tolling solutions for passenger cars: Singapore is already leading the way with its ERP-II program (Chapter 6), and a number of States on the West Coast of the United States are implementing GNSS-based solutions for charging specific groups of vehicles for their road usage (see Chapter 18). We are therefore witnessing a paradigm change in electronic tolling, where no longer specific road networks but rather groups of vehicles (such as trucks and electric cars) can be targeted for paying distance-based fees. This trend will inevitably lead to a large increase in the installation base of GNSS-based OBUs, which can potentially reduce the costs of these units thanks to the economies of scale, thus making the deployment of satellite-based tolling technology less expensive and therefore even more attractive.

In summary, the future for GNSS tolling looks brighter than ever. For the technology vendors in this domain, it has been extremely demanding to survive in this challenging market – with only a few GNSS-based tolling schemes being installed over the past decade, and many failed projects in between. But many new opportunities lie ahead, and we can look forward to navigate these new paths, with the guidance of increasingly more accurate and cost-effective satellite-based equipment.

References

[1] Balmer, U. (2003). Demand Management by User Charging – the Swiss Experience. In *Proceedings 10th World Congress on Intelligent Transport Systems and Services*, Madrid. ERTICO (ITS Europe).

[2] Benz, S., and Hoffstetter, B. (2005). Introduction of the 2nd Generation on Board Units in the Swiss LSVA Systems. In *Proceedings 5th European Congress on ITS*, Hannover. ITS Germany.

[3] Rottinger, F., Blum, M., and Jestädt, T. (2013). Quick and Clean – Expanding Toll Operations to Include 1,100 Additional km of Roads. In *9th ITS European Congress*, Special Interest Session 18, 8 June 2013, Dublin.

[4] Erker, E., and Schindler, N. (2005). A Third Weigh, Etc., H3B Media, Vol. 4, No. 4, pp. 16–19.

[5] zur Bonsen, G., Ammann, D., Ammann, M., Favey, E., and Flammant, P. (2005). Continuous Navigation: Combining GPS with Sensor-based Dead Reckoning, GPS World, Vol. 16, No. 4, pp. 47–54.

[6] Blythe, P. T., Knight, P., and Walker, J. (2000). The Technical and Operational Feasibility of Automatic Number Plate Recognition as the Primary Means for Road User Charging, The Journal of Navigation, Vol. 54, pp. 345–353.

[7] Walker, J. (2011). The Acceptability of Road Pricing. RAC Foundation. http://www.racfoundation.org/research/economics/road-pricing-acceptability.

[8] McKinnon, A. C. (2006). Government Plans for Lorry Road User Charging in the UK: a Critique and an Alternative. *Transport Policy*, Vol. 13, pp. 204–216.

[9] Geurs, K. (2010). The Dutch National Kilometre Charge: Impacts on the Dutch Car Market and Environment. In *ERSA 50th* Congress, Special Session on Road Pricing, 19–23 August 2010, Jönköping, Sweden.

[10] Bobošík, M. (2013). Hidden Benefits, Tolltrans 2013, Traffic Technology International, pp. 78–79. UKIP Media.

[11] Schindler, N. (2014). Making the Slovak Tolling Network the Largest in the EU with the Use of GNSS Technology. In *Proceedings 10th ITS European Congress*, Helsinki. ERTICO (ITS Europe).

[12] Schindler, N. (2009). Fourth Generation On Board Unit for Cost-Effective Automated Electronic Tolling. In *Proceedings 16th World ITS Congress*, Stockholm. ERTICO (ITS Europe).

[13] Erker, E., Schindler, N., and Tschulik, P. (2013). EETS: From Theory to Reality, ITS International, Nov/Dec, pp. 31–32. Route One.

[14] Schindler, N. (2013). Hybrid Electronic Tolling Solution for France. In *Proceedings 20th World ITS Congress*, Tokyo. ERTICO (ITS Europe).

[15] Schindler, N. (2013). On Board Unit for the European Electronic Tolling Service. In *Proceedings 9th ITS European Congress*, Dublin. ERTICO (ITS Europe).

[16] Diani, F., and Schindler, N. (2015). The Future is Now, Thinking Highways Tolling Review, H3B Media, Vol. 8, No. 3, pp. 30–33.

[17] Schindler, N. (2012). Combined Use of GLONASS and GPS in Electronic Tolling. In *Proceedings 19th World ITS Congress*, Vienna. ERTICO (ITS Europe).

[18] Tschulik, P., and Schindler, N. (2014). New Hardware Platform Supporting Interoperability of European Electronic Tolling System. In *Proceedings 10th ITS European Congress*, Helsinki. ERTICO (ITS Europe).

[19] Hegarty, C., and Kaplan, E. D. (2006). Understanding GPS, Second Edition, Boston and London: Artech House Inc. ISBN: 9781580538954.

[20] European GNSS Agency (2017). GNSS Market Report, Issue 5. European GNSS Agency. ISBN 978-92-9206-032-9. www.gsa.europa.eu/2017-gnss-market-report.

[21] Schindler, N. (2015). A Tale of Two Worlds, Thinking Highways Tolling Review, H3B Media, Vol. 7, No. 4, pp. 4–11.

[22] Schindler, N. (2015). Multiple Use of GNSS On Board Units: Electronic Tolling and other Mandatory Services. In *Proceedings 22nd World ITS Congress*, Bordeaux. ERTICO (ITS Europe).

[23] Diani, F. (2012). EGNOS2ROAD: Assessing Benefits for Road Applications. In *Proceedings International Week on Space Applications*, Toulouse. Centre National d'Etudes Spatiales.

Chapter 17

HU-GO: the Hungarian distance-based electronic toll system

*Zoltán Varga**

17.1 The Hungarian tolling and RUC (road user charging) environment

17.1.1 History of tolling and RUC in Hungary

The history of Hungarian toll collection goes back to the Middle Ages. The transport infrastructure was tailored to that level of economic development, but certainly the paved road network was much smaller than in these days. The value of different transportation infrastructure elements like roads, bridges and ferries determined the price of their usage which was collected mainly on the spot as a one-time tax payment. The generated tax income was the major financial source to cover the investment, operational, maintenance and any additional cost related to the infrastructure element.

On the legislative level, the first act about the public roads and tolls of the year 1890 regulated the imposition of tolls. This article is very much comparable to the current existing regulations as the major principles have not changed, although the current legislation can consider the environmental impact as well.

In the first decades of the twentieth century, these direct toll items were slowly replaced by indirect charging (public taxes) and finally the usage of the transportation infrastructure was considered as a welfare service.

The lack of funding of the necessary transport infrastructure development forced the Hungarian government to develop the concessionaire system, which was finally implemented in 1996. In this step-by-step approach, the different highway sections, depending on their development level and the different toll collection methodology and tariffs, were integrated into the Hungarian tolling system. This undeveloped system was a kind of patchwork that created even more problems, whilst not solving the financing needs of the transportation infrastructure.

The transport expert's recommendation was clear based on the experience gained in the mix of the different pricing and systems: one common toll policy, a free-flow technology and a centrally monitored, managed, time-based road user

*Toll Service PLC, Hungary

charging (RUC) scheme would solve the occurred problems in Hungary. Finally, all the highways were included in this highway sticker (vignette) model between 2000 and 2004.

The customer needs and the technology development pushed the state-owned operator to further develop the system by implementing new pricing categories to differentiate the vehicle categories and increase the service level of the sticker purchases by introducing the first mobile phone text message (SMS) purchase option in Europe. On top of this, the RUC network extension on the major transit routes and on the public road network resulted incremental charge income and so incremental road fund for the central budget as well.

By implementing all the above-mentioned changes, Hungary was in 2008 the first European country where the all-electronic-tolling (AET) was created without toll plazas and physical stickers for all vehicle categories. Even though multilane free-flow distance-based electronic tolling systems for commercial vehicles were implemented earlier in some European countries, like Austria and the Czech Republic, for the personal vehicles they were still using (until today!!!) the physical sticker on the windscreen.

Hungary is a landlocked country in the heart of Europe and plays a very important role in terms of operating some of the major east-west transit routes. This was clearly visible by analyzing the electronic purchasing data, as the nationalities of the vehicles are also submitted. The analysis showed that at least 50% of the commercial traffic was non-Hungarian registered vehicles. This fact already highlighted the need for the implementation of a distance-based tolling system as well, in line with tolling system implementations in the neighboring countries (see Chapters 9, 10 and 16).

17.1.2 The need for implementation

The central budget needed additional income to fulfill the financing need of the road fund. This gap could not be covered by the time-based RUC scheme any more with the EC limitations defined in the 1999/62/European Commission Directive (Eurovignette).

The very extensive international usage of the Hungarian road infrastructure also pushed the government to find the best possible solution to stop the rapid wear-out, the devaluation and the complete destruction of the Hungarian road network.

In 2010, the Hungarian government announced the mid-range economic program, where the implementation of the distance-based free-flow electronic toll system was projected for 2013. The government was looking for a solution which was easy to install, highly secure, flexible, reliable, nondependent on suppliers, easy to adapt to local requirements and truly interoperable.

The respective ministry and authorities had already spent years to evaluate the opportunities of development of the road tolling infrastructure; finally, all the circumstances indicated that only a free-flow Global Navigation Satellite System (GNSS) tolling system would be flexible enough to support the local tolling requirements.

The first attempt at tendering was unsuccessful by the beginning of 2013, as the winning bidder of the completed tendering procedure refused to sign the delivery contract because of very short and unchangeable deadline. In this situation, a very pragmatic and logical way of thinking in terms of the system development was needed to result in a real sustainable solution, which was never approached before.

The major issue in the development was to identify all those existing technologies which could speed up a country-wide toll system implementation compared to previous examples and make it very cost effective. On top of these expectations, the limited time availability encouraged the Hungarian experts to further develop the known GNSS tolling system solutions. With these innovative and cost-effective developments, the HU-GO system created a new chapter in road tolling history (Figure 17.1).

As a result, the HU-GO distance-based free-flow electronic toll system for commercial vehicles has been implemented and started to operate on the 6,500 km highway and trunk road network of Hungary on July 1, 2013.

The HU-GO solution utilizes the already applied Global Positioning System (GPS) positioning technology, providing a solid basis for other intelligent transportation system (ITS) and electronic payment solutions for a much more reasonable price than all other competitors on the market. On top of the unique technology, the Hungarian solution supports the preparation of the required legal, operational and enforcement environment in advance of the technical installation of the system.

17.1.3 Hungary—the home of innovations

Hungarians have an enviable record of innovation, as inventors of the ballpoint pen, the safety match, the telephone exchange, the electric train, binoculars, soda water, the carburetor for stationary engines, the turbo generator and the Rubik's Cube. Similar innovation was applied in the HU-GO tolling system in that all elements of the system existed and were used previously but no one considered linking them in the way the HU-GO system developers did—including the use of fleet management GPS tracking devices for tolling.

17.2 The innovative approach

17.2.1 Prevailing circumstances of implementation

The development of an open toll declaration platform with the integration of the already properly working telematics service and fleet management providers delivered the solution to implement the Hungarian electronic distance-based tolling system within 2.5 months.

It was obvious that according to the planned toll policy, with 6,501 km of toll road network in the country, no other tolling technology than GNSS should have been used. A centrally maintained and regularly updated road database provides the information to the digital roadmap of the country, where virtual tolling sections are identified. At the start of the system, there were 2,243 tolling sections, precisely measured and priced in line with the pricing matrix.

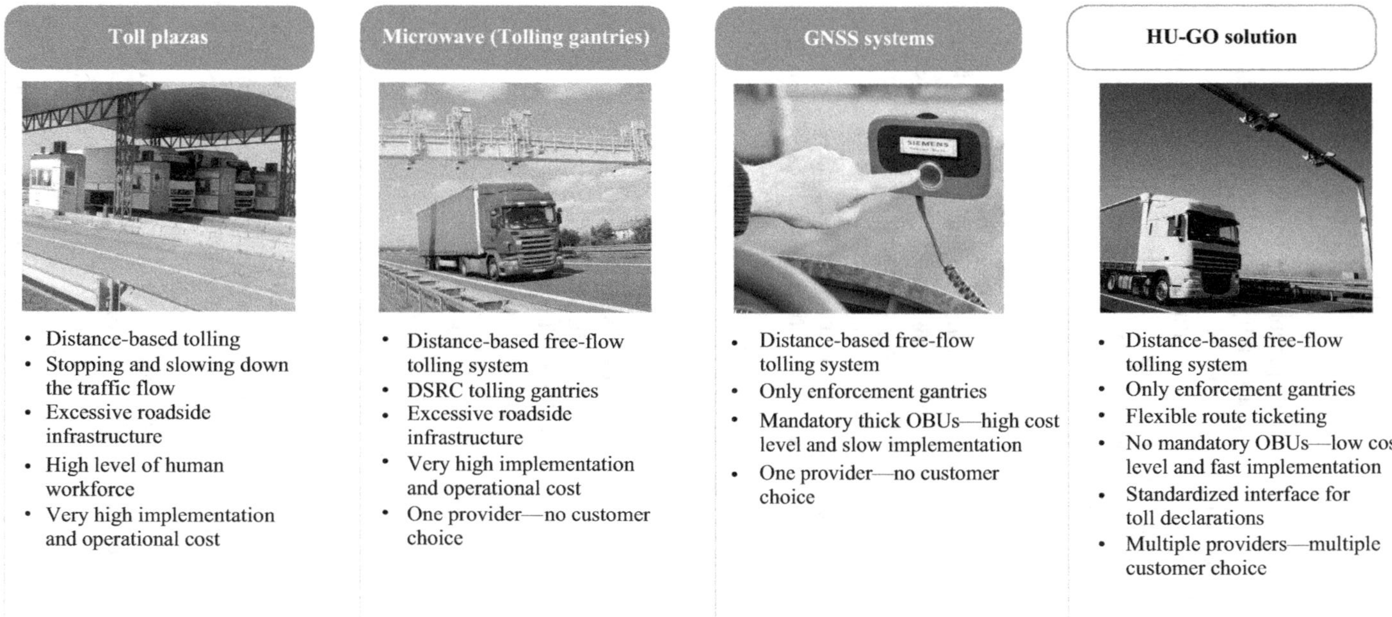

Figure 17.1 A comparison of HU-GO with other kinds of road tolling systems as of 2013

According to the business scheme, the telematics service providers were identified as potential toll declaration operators (TDO) of the HU-GO system, if they were able to fulfill all the specified legal, financial and technical requirements, like service levels, financial liability and proper map-matching. This open TDO platform was and is still open for any domestic and international telematics service provider to join. The technical and financial conditions of an audited TDO were set up to secure the continuous and safe operation of the toll system, by providing sound tolling transactions and reliable tolling data for enforcement purposes. This TDO set up enabled also to dramatically decrease the one-time investment and the continuous operational cost of the HU-GO system. The On Board Units (OBUs) are provided, distributed and maintained by the TDOs, although the digital roadmap of Hungary is provided and maintained centrally. If a vehicle already had a built-in device, no further investment was needed to declare the toll, as that particular equipment could be used for the tolling transactions.

The legal framework provides clear and equal treatment also in terms of the TDOs, as the remuneration of their services are centrally controlled and maintained on an equal and fair way and the opportunity to join the system is continuously open for any new applicant.

Beside the common TDO platform, the flexible route planning and ticketing module was also necessary to be integrated, because of the high number of transit road users and the rapid system start. By prepurchasing an individually planned route ticket, the user can declare and pay the toll in advance without any complicated registration process. The minimal vehicle data requirements, like license plate number, nationality and emission category, the number of axles and the basic vehicle measurements provide enough information to the HU-GO system to securely plan the best possible route, taking into consideration those road limitations, like height or gross weight that would influence the vehicle's path.

Even though the future of the online services were obvious for the system developers, the customer and payment behavior still showed a need for personal contact at the point of sales. There was no question that the previously existing retail Point of Sales (POS) network had to be used for distributional purposes. In Hungary, this network mainly consists of the majority of petrol stations, retailers on the border stations and the customer service points of the Hungarian RUC and tolling system operator, the National Toll Payment Services PLC (NTPS). Obviously, the ad hoc users, who are mainly purchasing route tickets, are using these point of sales, but some of the registered users are also managing their account balance at these points. A few months after the start of the system, the postpaid facility was also opened for those customers who qualified themselves on the credit rating procedure.

The technical solution on its own is a basic requirement, but not enough to successfully implement a tolling system. The legislative and operational background, plus the tolling policy development, makes the whole system complete. The basic rule in the Hungarian tolling is the self-declaration methodology that legally transfers the responsibility to the road user. Within that legal framework, the user can make his own choice on how he wants to declare the toll, mainly depending on his road usage habits and on his financial circumstances. This gives

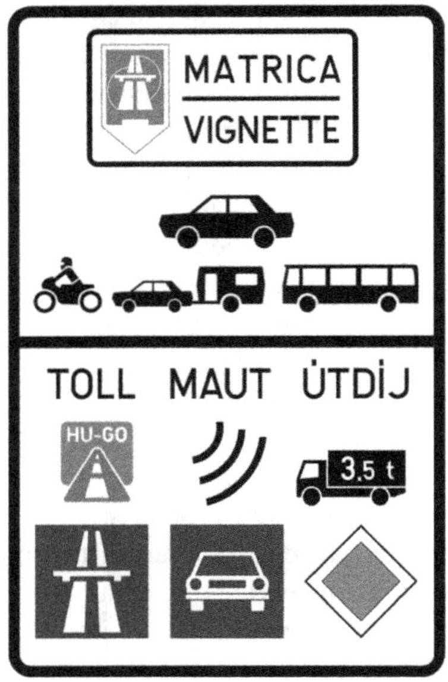

Figure 17.2 A traffic sign summarizing the Hungarian road user charging rules, policies and systems

the user freedom to adjust the toll declaration and payment to his transportation needs within a predefined legal framework.

The previously applied time-based e-vignette user charge system still exists for motorcycles, automobiles, camper-vans and coaches and their trailers, and cargo vehicles with a maximum permissible gross weight of 3.5 t. These vehicles are authorized to use the Hungarian highway and motorway system with the pre-purchased road user permits, the so-called e-stickers or e-vignettes. The RUC rate depends on the category of the motor vehicle and the validity periods. The category of the vehicles depends on the number of passengers and the weight of the vehicle. The following traffic sign (Figure 17.2) clearly summarizes and explains the whole Hungarian RUC and tolling rules, policies and systems.

17.2.2 The unified open operational platform

The HU-GO system is designed in a way that every external interface (Figure 17.3) ensures online communication with secure, standardized message packets, in order to avoid message or data loss and to log all message exchanges. Continuous acknowledgment will also be ensured during message exchanges, meaning that the recipient must send an acknowledgment of the successful receipt of all the messages (handshake). The HU-GO system does not limit the number of interface connections; any number of satellite systems can be connected if the proper device is used. The majority of the OBUs in Hungary are using the satellite signals of GPS as the most common provider, although there are OBUs available for European Electronic Toll Service (EETS), equipped with GLONASS and GALILEO receivers as well.

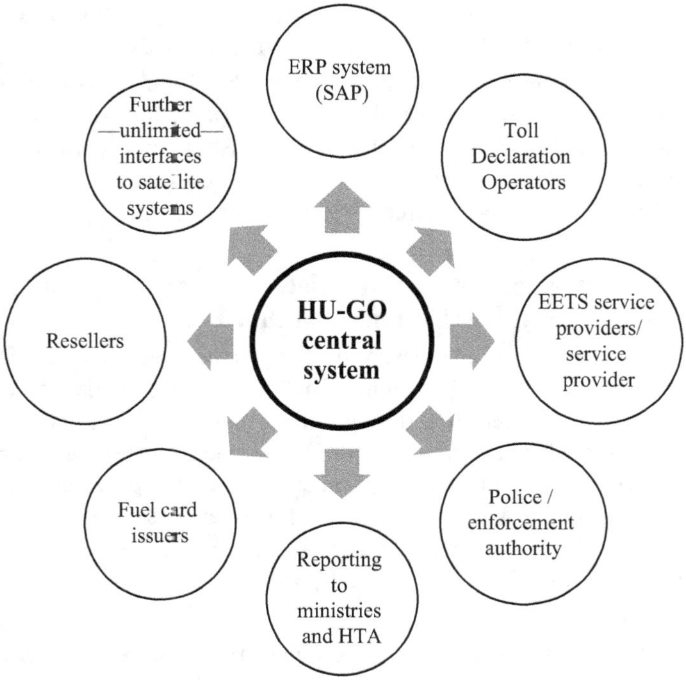

Figure 17.3 Conceptual overview of the HU-GO system's external interfaces

17.2.2.1 The interface between the electronic toll system and the ERP system of the service provider

The purpose of the interface is to ensure communication between the Enterprise Resource Planning (ERP) system implemented by the service provider and the HU-GO system through a standardized, secured channel. In Hungary, the NTPS is providing two different services. On the one hand, NTPS is the Bound Service Provider (BSP) and on the other hand acting as a Toll Charger (TC) as well. These two activities require a complete separation in terms of operation and in financial records as well. So, the general ledger data of the BSP and the TC are entered into the ERP system through a separated data connection in financial and Information Technology (IT) terms as well.

17.2.2.2 The interface between the electronic toll system and the processing systems of Toll Declaration Operators

The purpose of the interface is to ensure data connection between the processing systems of TDOs and those of the BSP. The interface is used for

- transmitting data required to compile Toll Declarations to the BSP's system;
- confirming the OBU for the BSP on the basis of the identifier of the OBU to be entered during registration;
- providing the BSP with the identifiers of installable OBUs the TDO wishes to put on the market, but are not yet sold.

17.2.2.3 The interface between the electronic toll system and additional EETS and other service providers

The purpose of the interface is to ensure standardized and predefined data connection between the TC and any service providers (including EETS Service Providers) that allows for message exchange established in accordance with the applicable standards. It is important to note that the data connection between the TC and the BSP must also be ensured through this interface.

17.2.2.4 The interface between the electronic toll system and the systems of the enforcement authority

The purpose of the interface is to ensure data connection between the TC's system and the Enforcement Authority's systems. The TC transmits the data on potentially unauthorized road users identified during enforcement support tasks to the Enforcement Authority for further actions. Additionally, the Enforcement Authority can request from the TC's system to clarify and supplement the previous transmission. Furthermore, to perform a more efficient enforcement task the IT subsystem is operational on the Toll Charger's side as part of the HU-GO system to control the Mobile Data Collection Vehicles' physical position data.

17.2.2.5 The reporting between the electronic toll system and the systems of bodies specified in legislation

As a part of the HU-GO system, multiple interfaces must be established to ensure data connection between the systems of the BSP, the TC and bodies specified in legislation.

17.2.2.6 The interface between the electronic toll system and the systems of fuel card issuers

The primary purpose of the interface is to ensure data connection between the systems of the BSP and those of certain fuel card issuers that previously concluded contracts with the BSP. On the BSP's side, the option to pay tolls using certain fuel cards as one of the noncash payment solutions must be ensured. The IT solutions used for authentication, administration and settlement during the payment process must be implemented using standardized interfaces.

17.2.2.7 The interface between the electronic toll system and bank systems

The primary purpose of the interface is to ensure data connection between the systems of the BSP and those of banks that previously concluded contracts with the BSP. On the BSP's side, bank card payment as one of the noncash payment solutions must be ensured for both personal (POS) and online (Virtual Point of Sales, (VPOS)) purchasing. Online payment must be established in a way that allows it to be integrated and embedded into the BSP's online services designed to support Online Customer Service functions.

17.2.2.8 The interface between the electronic toll system and reseller systems

The primary purpose of the interface is to ensure data connection between the systems of the BSP and the systems of resellers that previously concluded contracts with the BSP, and data exchange performed during reseller activities, required for sales, settlement and authentication through a secure channel.

17.2.3 The toll declaration methodology

The major rule in the Hungarian electronic tolling system is that the user is responsible for the toll declaration. By transferring this responsibility to the user means that several options need to be provided by the operator. These options make it easy for users to find the best and most convenient toll declaration method suited to their personal or business needs.

Toll declarations regarding the use of tolled roads can be made in the following ways (Figure 17.4):

1. With a prepurchased route ticket for a preplanned route;
2. Via a fixed or installable OBU provided typically by the fleet management service provider companies (hereinafter: TDO) who are in contractual partnership with the BSP;
3. Via an OBU provided by the EETS Service Provider or other Service Provider within a contractual framework. (This toll declaration method will be available after signing the cooperation agreement with the possible Service Providers.)

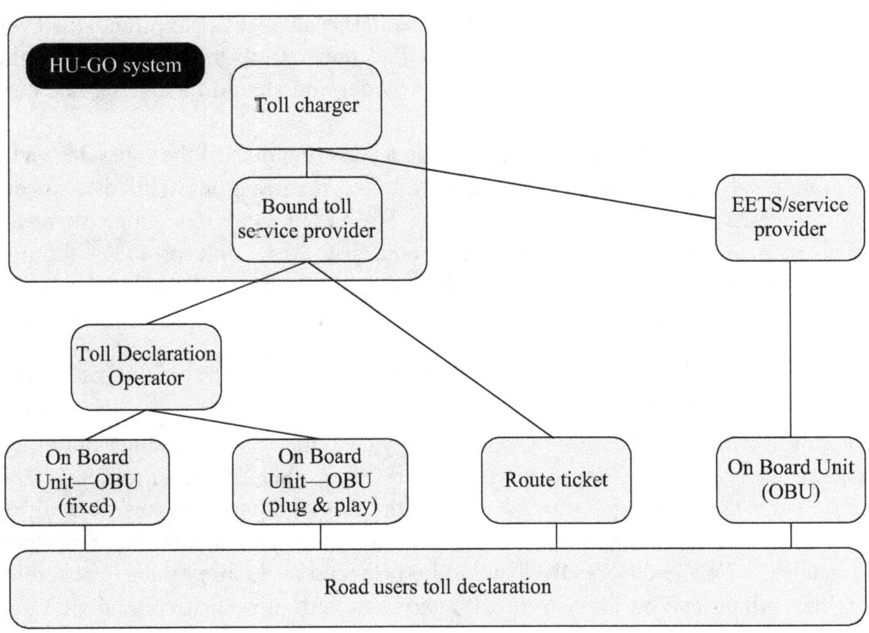

Figure 17.4 HU-GO system architecture

17.2.3.1 Route ticket

Toll declaration can be obtained by purchasing a so-called route ticket. In this case, users can plan their route (the HU-GO system allows users currently to identify one departure and one destination point and a maximum of four points en route) and obtain the road usage without registering, but providing the necessary vehicle information like registration number, country code, environmental class, number of axles, authorized gross weight, truck stop notice and vehicle height for the entire route. Based on the information provided by the user, the Bound Toll Service Provider's system determines the tolled road sections and generates Toll Declarations which are then sent to the TC. The TC, based on the previously received data, determines the toll charge to be paid by the user for the preplanned route.

The route tickets can be purchased a maximum of 30 calendar days in advance, but at the latest just prior to entering the toll road. The purchased route tickets are valid for 2 days from the time specified by the user; they grant authorization to use the tolled road of the previously defined route and direction only. Only registered users can prepurchase route tickets. It must be stressed that users can neither deviate from the preplanned route, nor modify purchased route tickets within the validity period. This service offers benefits especially for the ad hoc users who use the Hungarian toll roads very rarely.

17.2.3.2 The use of Toll Declaration Operators

It is much more convenient for frequent users to perform their toll declarations via OBUs provided by the TDOs. Currently in Hungary, there are 22 audited and operational TDOs with valid contracts. This solution does not impose any limitations on road users regarding the toll payment either in space or in time. However, the precondition for using this solution is that the road users must have a valid registration with the Bound Toll Service Provider and should have a service contract with one of the TDOs as well.

In this case, the TDO continuously collects position data of the route driven by the contracted user and determines the used tolled road sections. This data is continuously transferred to the BSP. The Bound Toll Service Provider supplements the tolled road section data with the registered user data and provides the toll declaration to the TC. The TC determines the toll to be paid and provides this information to the BSP who charges it to the user account.

17.2.3.3 Toll declaration with the help of a service provider or EETS service provider

In addition to the toll declaration methodology described above, another option for frequent users can be to register with an EETS or other Service Providers. This solution would make it possible to declare the toll in other countries as well, by using the same equipment. This is mainly an option for international transport companies; a DSRC-GNSS OBU should be provided to them by their new service provider, which can be used in the Hungarian system as well. Potential EETS or other Service Providers need to have a valid contract with the local TC to be able to service their customers.

Figure 17.5 Modules of the HU-GO system

The HU-GO system was designed to handle toll declarations provided by EETS or other Service Providers to the TC through standardized communication channels, calculating the toll to be paid and performing the related administrative procedures like invoicing, customer care, claim management.

17.3 Modules of the HU-GO system

17.3.1 The central system

The core system of HU-GO (Figure 17.5) is the central module where all the necessary dates are stored and managed. This module provides information for the back office applications as well. The major input sources of the central module are the sales and toll declaration (circles left) and the enforcement modules (circles right).

The sales module includes the flexible route ticketing solution, the GNSS toll declaration opportunity and the European Union (EU) required EETS conformity. This is very important, especially considering that the new EC Directorate of Transport highlighted a program where in all EU countries one common, distance-based e-tolling scheme should be in place within the near future.

The enforcement module collects and organizes the collected enforcement data from the fixed and mobile data points, validates the data in the central module and provides penalty data for the Hungarian Police. The HU-GO system handles with equal accuracy the local and international license plates, including the nationality and the vehicle category. This is very important, especially considering that approximately 50% of the toll income is generated by foreign registered vehicles.

17.3.2 The sales module

The sales module provides purchasing solutions to any type of transportation need, from huge international fleet users to local small one-truck entrepreneurs. The great variety of purchasing options in combination with the different toll declaration methodologies makes HU-GO a very flexible road tolling system.

Usually, the toll system operators do not take too much care on the customer service level, as the related legal environment forces customers to use the provided technology with all its pros and cons.

It was obvious at the implementation phase of the Hungarian system that the previously applied sales methodology, namely the existing retailer network, is a well-developed system and with certain extensions can serve the new system requirements. On top of this, the further development of the online purchasing channels opened up an additional sales facility for the customers.

By technical means, the sales module differentiates registered and non-registered customers. It is worth mentioning that only the HU-GO road tolling system provides toll declaration and payment facility without customer registration. The nonregistered customers are eligible to purchase route tickets, rather like a train ticket. As a train ticket is also purchased in advance, before the trip starts, for a previously defined route and direction, the route tickets are working in the same way. For a purchase like that, only basic vehicle data is required.

17.3.2.1 Payment methods

The regular payment solutions like cash, debit or credit cards, fuel card and also direct bank transfer are used in the HU-GO system. For online payment, only debit or credit cards can be used.

The system was originally started as a prepaid system. This approach was related to the fast implementation need. Obviously, the qualified users can apply for the postpaid facility, as this is not a technical issue, but more a financial risk issue that needs to be very carefully evaluated by the Toll Service Provider. Those qualified customers who already had a credit line with the major fuel card providers were in the position to maintain their account balance with the postpayment facility provided by the fuel card provider. On the other hand, the Bound Toll Service Provider established the direct post payment facility for the qualified customers.

There are several administrative online options available for registered users:

- Redeeming route tickets purchased as a registered user
  Registered users can redeem purchased route tickets before the start of the validity period. When redeeming the route ticket, the full amount of the pre-paid toll will be credited to the customer account.
- Administration concerning invoicing

In the HU-GO system, invoices are created with electronic signatures in electronic form (pdf) and kept in the road user's account in the HU-GO system for 5 years. During the invoice retention period, electronic files can be searched, viewed and if necessary downloaded at any time.

- Balance query
  The actual balance of the prepaid accounts and the outstanding amounts on the postpaid accounts can be checked with this facility. The user can also select different reports by account or even by vehicle.
- Tolling transaction query
  Similar to the balance query, it is possible to check all tolling transactions that have already been recorded in the system.
- General complaint reporting

  Any type of customer complaint can be reported trough the customer interface. From the service provider point of view, all complaints are registered, managed and followed by the complaint management system. This ensures that all customer complaints are properly handled and followed-up until resolved. All related workflows are also monitored and the required measures are set.

On top of the online customer account management facilities, the call center and the customer service points are also available for the users to manage their accounts.

17.3.3 The enforcement module [1]

From the enforcement point of view, the HU-GO system features mature ANPR technology with 3D scanning to collect, maintain and store all the vehicle information collected on 101 fixed enforcement gantries and 45 mobile data collection vehicles, on the whole tolled road network and around the clock.

The enforcement methodology and processes are also very much determined by the legislative background. The penalization process is based on the so-called "Objective Responsibility" that makes the workflow much easier and more flexible primarily for the domestic registered vehicles. In case of the "Objective Responsibility," always the registered owner or user bears all responsibility related to any type of traffic offenses.

In these enforcement cases, there is no need to physically stop the vehicles on the road. The officially registered owner or user will be approached afterwards by the respective authority to collect the penalties.

As the Hungarian authorities do not have access to international vehicle databases, foreign road users need to be stopped in case of a penalty, either on the road or at the latest at the border station when leaving the country. As the system is configured to store all penalty cases for 24 months, the enforcement activities can also be done retrospectively, even if the user has a valid toll declaration at the time of the control—so the user can be penalized on the spot for a past offense.

The ANPR software is designed to read all vehicle license plate types in the world in any traffic speed. In addition to the toll collection, this data can be used for other purposes, like congestion charging, traffic monitoring, speed and journey time measurement, bus lane and traffic light enforcement, parking or access control and many others. The ANPR engine reads license plates from many image sources remarkably fast and with high recognition accuracy. It offers country-independent recognition not only of the Latin characters, but also the Arabic, Cyrillic, Chinese, Korean, Thai and many more, as well as reflective, nonreflective, personalized and special interest plates. The ANPR's strength comes from its ability to automatically recognize virtually any license plate of vehicles in moving traffic, regardless of country, state or province, plate or character type, and adapts to environmental changes for 24/7 operation. It is fast and straightforward to use with any hardware, compatible with any image source (analog/digital/still images/video streams) and supports both Windows and Linux platforms. Its development libraries allow flexible integration through their software development kit and in server mode it is able to perform parallel processing and handling of unlimited camera inputs.

This technology is used on the variable sensing and monitoring system installed on a single, fixed detection point (i.e., traffic gantry or bridge) for accurate surveillance and data gathering. The standard list of components includes: radar, laser, overview camera, DSRC antenna and the ANPR software.

The additional on-board processing unit intelligently computes all measured and detected data; marks each vehicle-related event with a timestamp, location and lane identification; bundles the gathered data in an encrypted package; and finally sends it to a predesignated central location.

The enforcement gantries detect all passing vehicles using a laser trigger. The TÜV Rheinland Technical Inspection Association-audited 98.5%+ automatic recognition accuracy attests high detection rate even during heavy traffic, limited visibility and at speeds of up to 250 km/h (155 mph). The purpose-built hardware provides secure data retention and continued functioning offline for at least 5 days, IP-based communication, efficient data compression and upload. Each necessary data set is bundled into a single event package for the central data server where all monitoring and component management can also be executed.

The Hungarian developer of this single-gantry solution aimed to provide an answer to the basic challenges that continue to afflict the latest generation of road side equipment on the data collection points, the so called traffic endpoints. What are these challenges?

As soon as the first gantries had been installed in the recent past, the practical reality proved that many of the mundane but everyday events on the road were too complicated for them to handle. Useful motion triggering was almost impossible to achieve without installing additional inductive loops into the road surface. Lane changes resulted in a loss of data that was supposed to relate to a particular vehicle. Semitrucks with different license plates on the front and back lead to confusing results and misidentifications.

The HU-GO system fixed-gantry solution is responding to the system integrators' number one concern: device calibration and synchronization is an onerous and time-consuming process with difficult-to-pin-down parameters. The HU-GO gantry provider solved this conundrum by developing and synchronizing all gantry elements in-house and delivered them preset as variable and connectable modules assembled on individual customer demand, designated in the first round for toll collection, but with the option of utilizing the gantry features in weigh-in-motion measurement, speed enforcement, traffic surveillance or combined with other data collection capabilities.

The central data server is the ultimate solution for managing complex traffic-related systems for law enforcement, toll collection, border control, tariff, tax and statistics. It is an ITS with a combined data server; middleware and graphical user interface (GUI) that gathers information from several endpoints and makes them available for various end user applications with a wide selection of premade modules.

The operators of the central data server manage the processes through a dedicated graphical interface while the data server delivers solid, secure connections between monitoring points and client applications.

The server receives standard, customizable independent data packages from various endpoints that are managed by a central server with standard security technology for establishing an encrypted link between a server and a client. The fast IP traffic inflow and outflow is assured by xml or binary communication. The internal structure of the system is characterized by data redundancy through high-availability replication, clustered storage, highly efficient image storage and dynamic hardware scalability without maximum limits. Simultaneous end user applications are managed with standard interface and software development kit. The GUI allows high effective remote operation and it reflects detailed conditions in real time. The user-friendly display provides maps, statistics and various fast, flexible searches with preset automation and export functions.

17.4 Implementation and operational challenges

17.4.1 The time frame

After several unsuccessful attempts to implement distance-based electronic tolling in Hungary, the willingness of the government was unquestionable. The operation, maintenance and development processes of the road network were suffering because of the lack of necessary funds. It was almost impossible to properly operate the road network; the maintenance of the infrastructure was very much slowing down and there were no new developments in the pipeline. Every month of operation within the old RUC and financing scheme increased the risk of the faster deterioration of the road infrastructure.

In such a situation, in 2012, the preparation started, considering all the known international experiences, especially in Central European Region. On behalf of the Hungarian government, the State Motorway Management Company was in a position to manage the tendering and later the implementation activities. But after an unsuccessful tendering process, there were only five and half months left to start the electronic toll collection in Hungary, and the central budget was already planning to use the increased toll income for the second half of the year. The unexpected time pressure created such a situation that the idea and the essence of the current HU-GO system was born. After the necessary modifications, a special public procurement tender was announced to make the "impossible" happen, to avoid the risk of losing central budget income. On April 16, 2013, the delivery contract was signed with the winning consortium, to deliver the required electronic toll collection system on more than 6,500 km of toll road network of Hungary, within 2.5 months.

17.4.2 Internal and international communication

One of the key success factors of the incredibly fast system implementation was the well prepared communication strategy. The first and most important issue was to establish an unquestionable relation with all the existing Hungarian haulers' associations by making them clearly understand the Hungarian government's aim to implement the HU-GO system. The clear and unquestionable list of benefits transferred the associations' leaders and later their members to absolutely supporting the idea of the HU-GO system and related pricing policy implementation.

In the first step, it was necessary to make it clear that the increased toll rate was not directly affecting the haulers themselves, but was transferrable to the customer. So the increased toll rates don't have any direct influence on the transportation market and the competition. The transport and hauler companies were supported by their associations to communicate and describe the new toll rate as a separate item on their transportation invoice. This transportation cost increase impacts the final customers at the end of the day, but the real financial effect is hardly measurable on the consumer price level.

As the HU-GO tolling system clearly affects all users equally, the impacts of the system implementation either on the local or the international users were equal too. This fact also supported the Hungarian haulers' association to be helpful in the international communication of the toll system changes to their international partners.

The total Hungarian toll income is generated 50% by the Hungarian and 50% by the international toll paying customers.

17.4.3 Major stakeholders—the involved organizations

17.4.3.1 Toll charger

The TC is the legal entity designated for charging and collecting the tolls as regards the Toll Domain and supports the Enforcement Authority.

The TC in Hungary is the NTPS. Its primary task is to receive Toll Declarations regarding the usage of tolled road sections from Service Providers and EETS Service Providers (including BSP) and determines the toll to be paid for tolled road sections.

17.4.3.2 Bound service provider

With respect to the Toll Domain, the entity designated on the basis of a legislative provision that provides Electronic Toll Service and is bound to enter into contract with any road user required to pay toll, if so requested.

The Bound Toll Service Provider in Hungary is the NTPS. Its primary task is to accept the toll declarations regarding the use of tolled road sections submitted by road users required to pay toll in a contractual framework, then to forward these declarations to the TC and ensure the settlement of the toll payment determined by the TC.

17.4.3.3 Enforcement support agency

This agency, as an integrated part of the TC, performs enforcement support tasks regarding the Toll Domain. The Enforcement Support Agency in Hungary is the NTPS. Its primary task is to record and evaluate road usage data, use them to detect unauthorized road users and transfer these data to the Enforcement Authority.

17.4.3.4 Enforcement authority

This authority performs the enforcement activity on the Toll Domain. The Enforcement Authority in Hungary is the Hungarian Police. Its primary task is to take enforcement action against road user suspects identified by the Enforcement Support Agency.

17.4.4 EETS compatibility

The HU-GO system has been developed in line with the EETS requirements and this is the only system operating with multiple OBUs and multiple toll declaration modes at the same time. The system is technically adopted to receive toll declarations from any type of GNSS tolling device if the service provider enables his own system to provide the tolling data in the necessary format to the HU-GO system.

HU-GO's strength is focusing on the server-to-server communication, instead of forcing the Service Providers to use unified OBUs to manage device-to-server communication locally on the respective Toll Domain. By setting up one common platform supporting the server-to-server communication and data flow, approved and accepted by all TCs, would speed-up the activities of all participants, as they will not be forced to invest in high cost equipment and deal with the different local technical requirements, regulations and approval processes.

17.5 New opportunities for utilization of the toll system

On the basis of the core modules of the HU-GO system, additional transportation and customer account management systems are developed. These systems utilize either the HU-GO system infrastructure or the HU-GO system generated data to increase the efficiency of data utilization. The examples listed below, although not all directly linked to the tolling industry, show how to use the tremendous amount of traffic-related data that is generated daily on such systems.

17.5.1 The National Mobile Payment services [2]

The National Mobile Payment Plc. is a fully state-owned joint stock company that has been established on October 15, 2012, based on the Act CC of 2011 on the

National Mobile Payment System. Its tasks and operating conditions are governed by this Act and its implementation regulation, Government decree No. 356/2012 (XII. 13). In market and social terms, the National Mobile Payment Plc. is an innovative organization of the twenty-first century that as a state-owned coordinator provides public services in the field of technology and facilitates the uniform mobile payment system in Hungary.

17.5.1.1 National strategic goals and tasks of the National Mobile Payment Plc.

- Facilitation of the uniform mobile payment of public services:
  The most important task of the National Mobile Payment Plc. (in line with Act CC of 2011) is to provide simple, convenient, cost-effective and safe access for the public to Hungarian public services through mobile payment by establishing a high-quality public service and mobile payment culture.
- Making mobile payment accessible for increasingly wider sections of the Hungarian public:
  The strategic aim of the National Mobile Payment Plc. is to make available mobile payment of public services not only for the potential customers with financial knowledge but to motivate sections of the society who are less skilled in finance, to use the mobile payment option. It endeavors to provide secondary motivation by involving resellers and integrating more public services, by exploiting the potential of technological innovation, ensuring the interoperability of public services as well as targeting younger generations and so their families.
- Positioning mobile payment as a realistic payment method in order to reduce the proportion of cash payments:

  This assists society and consequently facilitates the change of its financial culture, the application of mobile payment, the improvement of payment compliance and aims at future generations through young citizens.

- Improvement and provision of mobile and electronic payment services connected to domestic public transport:
  Through its developments and services, National Mobile Payment Plc. as a strategic partner of the National Automated Fare Collection Platform facilitating its mobile and electronic payment services is able to contribute significantly to the provision of services for people using public transport services in Europe and to the successful development of interoperable solutions for the smart public transport systems of Hungary.
- Development and provision of mobile payment services linked to the "Good State" concept:
  In connection with subparagraph, "extension of the electronic administration" of the Magyary Program, the National Mobile Payment Plc. identifies the development and introduction of electronic mobile payment public administration services as strategic tasks within the framework of the Public Administration and Public Service Mobile Payment Platform project.

17.5.2 The National Electronic Ticketing Platform (NETP) [3]

The Hungarian government's strategic program aims to promote the renewal of public transport and supports the implementation of the common electronic ticketing platform. The strategic goals of the program are to provide a unified electronic transportation card system with authentic personalization and at the same time to implement a personalized subsidy and discount management system, both regulated by the State.

The National Electronic Ticketing Platform (NETP) public transport ticketing and fare collection system central module includes basic data clearing and subsystems, the national transport card system, the related mobile application modules and the central sales registry and data transmission modules providing interoperability between the different transport-related ticketing systems.

One of the major outcomes of the NETP project as a common transportation ticketing platform is the real transparency. All related participants, like public service providers, their owners, the municipalities and all other institutions, will have a clear and transparent view and control of the related income. This will greatly contribute to optimizing the cost structure of the public transport sector.

17.5.3 The Electronic Public Road Trade Control System (EPRTCS) [4]

The Electronic Public Road Trade Control System (EPRTCS) was launched on January 1, 2015.

The essence of the system is to strengthen the market positions of compliant market players, to make the circulation of the goods more transparent, to eliminate frauds related to food products and to eliminate tax evaders. By using EPRTCS, the actual route of the delivered goods can be tracked, because the transport-related data must be registered in a central electronic system before delivery.

Some of these data (product weight, value and registration number of vehicle) can be modified up to the arrival of the cargo.

The obligation to submit data applies for all road transport of goods using vehicles subject to road toll payment, like vehicles exceeding 3.5 t of maximum gross weight. It is obligatory to report:

- Import of goods from any member states of the European Union to Hungary
- Export of goods from Hungary to any member states of the European Union
- Delivery of goods subject to first taxable domestic sale if it is not for an end-user

Domestic products subject to reverse taxation are not exempt from the obligation to submit data.

However, certain goods are exempt from the data submission obligation, e.g., relief supplies or vehicles participating in disaster relief, or road transport in the context of collecting freight if the quantity/value of the given product does not exceed the limit specified in the law, and in the case of transporting certain excise goods specified in the law. Nonrisky goods with a net value not exceeding 2 million

HUF[1] and with a weight not exceeding 2.5 t are exempt from the data submission obligation.

In addition to submitting the required data, a deposit must be provided and maintained for risky food and other products throughout the business operation. The amount of deposit is 15% of the net value of the products registered in EPRTCS in the course of 60 days prior to the submission of data. The following entities are exempted from the deposits:

- Qualified tax payers
- Tax payers, in business at least for 2 years, free of public debt

17.5.4 Weigh in Motion

The consortium of the National Transport Authority and the National Toll Payment PLC was assigned to implement the National Weigh in Motion (WiM) control system. The main goal of the Project was to implement a countrywide, complex controlling system which prevents and eliminates overloaded transport. The WiM was built on the already existing complex transportation enforcement system used for the HU-GO electronic toll enforcement, utilizing the already-existing technology- and system-related cooperation. Besides the main goal of the project, major positive effects are expected in the following fields:

- Improvement of road safety
- Increased protection of the road network, the national property
- Improving the efficiency of the EPRTCS
- Increasing the convenience of public road users

The WiM is based on the physical and application level IT infrastructure that supports the automatic measuring, data processing and fining processes without stopping the traffic and with the help of the different interfaces provides valuable data to all relevant authorities. The WiM and the related project consists of 89 fixed control points on the Hungarian road network, all the necessary equipment to perform on the spot enforcement activities in case of traffic diversions and the relevant legal and operational background.

17.6 Facts and figures

HU-GO facts and figures in a nutshell are shown in Table 17.1

17.7 Summary and conclusions

The HU-GO system brought a completely new view of the "Tolling World" with its innovative and unique technological solutions, with the open GNSS-based

[1] Hungarian Forint. There are currently (June 2017) 308, 274 and 353 forints per EUR, USD and GBP, respectively.

Table 17.1 HU-GO facts and figures

In operation	Since July 1, 2013
Number of average monthly transactions	58 million
Total toll income in the first 3 years of operation	1,700 million EUR
Total toll system operational cost, including 30% of TDO and retailer remuneration	100 million EUR
Initial investment	100 million EUR
Return on initial investment	2.5 months

tolling platform. The HU-GO solution utilizes the already applied GPS positioning technology, providing a solid basis for other ITS and electronic payment solutions for a relatively low price.

The Hungarian government spent years evaluating the opportunities to develop the road tolling infrastructure in Hungary; the conclusion was that only a free-flow GNSS tolling system would be flexible enough to support the local tolling requirements. The limited timescale forced the creative development of existing GNSS system solutions. With these innovative and cost-effective developments, the HU-GO system created a new chapter in road tolling history.

The past 3 years of experience of the system operation clearly shows the planned and expected income generating ability of the system. The Hungarian toll collection model also proved that the operation of a countrywide common tolling system is an efficient way of collecting toll income, which has exceeded 1,700 MEUR in the first 3 years. The annual income generating level of HU-GO is eight times higher than the previously applied e-vignette system, for the commercial vehicle segments.

The HU-GO system has also created a unique platform to implement additional services. By collecting the transportation "Big Data" of Hungary, a multipurpose ITS solution has been developed that has provoked international interest.

The HU-GO system is fast, easy and inexpensive to implement, creates numerous additional utilization opportunities within the transportation sector, increases traffic safety, provides excellent traffic control measures, immediately generates toll income, eliminates tax frauds and increases the tax income. There is no question that the future of the transportation sector is determined by such complex ITS solutions providing extensive information on one platform and at an affordable cost level!

References

[1] ARH (2017). "The Image recognition company", http://www.arh.hu/index.php/en/index.html. [Accessed 08 August 2017].

[2] National Mobile Payment company (2017) "National Mobile Payment news", https://www.nemzetimobilfizetes.hu [Accessed 08 August 2017]. National Mobile Payment is a 100% subsidiary of the National Toll Payment Services.

[3] National Mobile Payment company (2017) "Presentation of NEJP", http://www.neszip.hu/articles/view/a-nejp-bemutatasa [Accessed 08 August 2017].

[4] National Tax and Customs Administration (2017) "A brief introduction to EKÁER", https://ekaer.nav.gov.hu/ [Accessed 08 August 2017].

Chapter 18

West Coast distance charge programs: an open market as the gateway to implementation in the United States

*James M. Whitty**, **

18.1 Introduction

A new approach for road pricing has emerged in the United States: use of an open commercial market, robust and competitive, to do the work of collecting road revenues. The Pacific Coast states, working in succession—Oregon then California then Washington, have put together the elements of a privately operated distance charge system that could eventually replace volume-based fuel excise taxes as the nation's primary road funding mechanism.[1]

For nearly a century, several US states have charged heavy vehicles for distance traveled,[2] combining a tax rate with a weight factor among other characteristics, but charging light vehicles for distance traveled has long been a conundrum. Collecting taxes from a multitude of taxpayers tends to be an expensive endeavor, and light vehicle owners pay fairly small amounts for road use. With millions of light vehicles on the road, how could a government afford the operating costs to accurately and frequently collect small amounts from each one?

Formal research on light vehicle distance charging in the United States began early in the first decade of the twenty-first century. Fifteen US states put together a

*Partner of D'Artagnan Consulting LLP, USA

**Much of what I rely upon in this text comes from my own first-hand knowledge and observations during 15 year working in the field of US road usage charge projects, especially, but not exclusively, in Oregon where I led the entire effort from the start to implementation of an operational road usage charge program, including visioning, policy development, political development, system architecture and design, testing, public-facing communications throughout the nation and internationally and legislative strategy and engagement. I have included references where possible, but it is difficult to find third-party references for this personal knowledge.

[1] For purposes of this chapter, *distance charge*, or *road usage charge*, is defined as a set charge, fee or tax for the use of public roadways based on distance traveled by a subject vehicle and, as such, is one basis for road pricing along with time, location, place, vehicle type and others.

[2] For heavy vehicles, Oregon, New Mexico, New York and Kentucky currently have a road use tax rate that combines charging for distance traveled with a weight factor and other characteristics.

special, pooled-fund research arrangement with the US Federal Highway Administration [1], and the state of Oregon passed legislation directing policy development and the nation's first pilot programs for a road user fee for light vehicles.

Over the past dozen years, the United States has experienced eight formal trials for charging by distance traveled. Four of the trials—Puget Sound Regional Council in Washington State, University of Iowa, Nevada and Minnesota—constituted research efforts primarily focused on assessing public attitudes toward various methods of collecting distance traveled and location data. Only the Oregon pilots and road usage charge program, and the California road charge pilot, tested central elements for an actual operational distance charge collection system. The aspirational structure of Oregon's program, California's pilot and similar, nascent efforts in Washington and Hawaii, including the prudent decision to lead with policy in directing these efforts, is the central focus of this chapter (Table 18.1).

Table 18.1 History of distance charge tests, pilots and programs in the United States, 2005 to 2017

- 2005–07. The Puget Sound Regional Council conducted the *Traffic Choices Study*, an 80% federally funded, congestion pricing pilot in Seattle [2]
- 2006–07. The state of Oregon conducted the 75% federally funded *Road User Fee Pilot Program (RUFPP)*, a pay-at-the-pump distance charge technology and payment system with an area congestion pricing pilot as an added feature [3]
- 2008–10. The University of Iowa conducted the *National Evaluation of a Mileage-Based Road User Charge*, a 100% federally funded, multicity test of public attitudes about mileage reporting technologies [4]
- 2009. The Nevada Department of Transportation conducted a technical distance charge test and motorist survey leading to design of a pay-at-the-pump field test [5]
- 2011–12. The state of Minnesota conducted a state-funded, mileage reporting technology field test featuring a dedicated smartphone app [6]
- 2012–13. The state of Oregon conducted the *Road Usage Charge Pilot Program (RUCPP)*, a state-funded, open system technology and payment system involving not only Oregon residents but also residents of Washington State and Nevada [7]
- 2015 and ongoing. The state of Oregon implemented the state-funded, fully functioning, operational *Road Usage Charge Program* (branded ORe*GO*) in accordance with legislative directive, with an open system and choices of state or commercial account management for volunteer payers of a distance-based charge, the first US distance charge program for light vehicles with actual payments [8]
- 2016–17. The state of California conducted the 9-month, state-funded *road charge pilot program*, testing an open system with state and commercial mileage data collection and seven different methods of reporting mileage data and simulated payments [9]

18.2 Context for distance charge activities in the United States

Across the world, policymakers, academics, economists and advocates often have very different reasons for pursuing road pricing policies. Some advocate for congestion pricing, wanting to improve upon commendable schemes in London, Stockholm and Singapore (see Chapters 6, 13 and 14, respectively) to bring time of day and location pricing elements into road travel in hopes of managing congestion by giving a price signal to drivers. Others see charging for distance traveled on roads as a way to achieve greater, or at least sustainable, revenues for maintaining and improving the functionalities of road networks.

Although extensively researched in the United States a decade ago [2,3,10], congestion charging has not advanced as quickly in the United States as has a simple, flat charge on distance traveled. Both approaches to road pricing have strong political challenges to weather before going forward. An efficient congestion pricing scheme, like Singapore's new scheme (Chapter 6), uses location detection technologies installed within vehicles rather than roadside equipment. Given the sensitivities to privacy protection in the United States, location-aware mileage-reporting technologies are currently not viable for required use in a mandated road pricing system in the United States but *do* seem viable as a *choice*, along with nonlocation-aware options, for mileage reporting in a distance charge system. The US experience shows that putting road charging and congestion pricing policies together in one road pricing legislative proposal may be too politically challenging for the first quarter of the twenty-first century.

Distance charging for light vehicles has advanced politically in the United States for essentially two reasons. First, road revenues are not meeting expectations. The fuel efficiency of the US light vehicle fleet is steadily improving. With the nation's funding for roads heavily dependent upon the volume-based fuel excise tax, expected fuel tax revenues are declining and will continue to do so. According to the US Energy Information Administration, light vehicle energy use in the United States will enter a permanent and steep decline by 2020. Volume-based fuel excise taxes will follow a similar path downward (Figure 18.1) unless legislatures adopt regular and large rate increases, an unlikely scenario given the unpopularity of fuel taxes among voters and potential impact on low-income drivers.

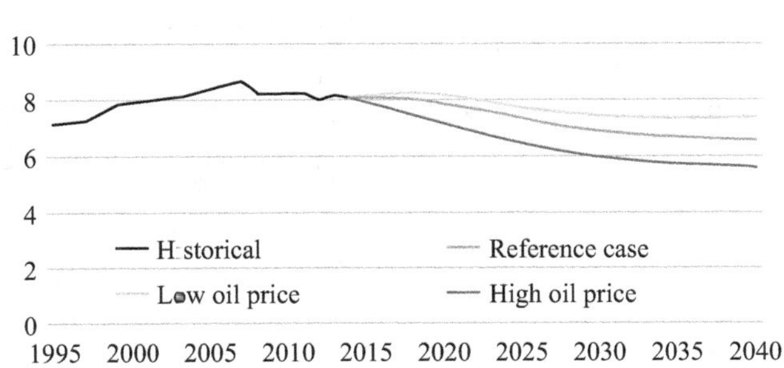

Figure 18.1 *US light-duty vehicle energy use in three cases (million barrels of oil equivalent per day)*

[Source: US Energy Information Administration 2014 Annual Energy Outlook (historical data) and 2015 Annual Energy outlook (projection)]

The downward trend of fuel tax revenue is driven by two factors: consumer demand for fuel-efficient light vehicles and regulations that promote reductions in fuel consumption. In the early part of this decade, the Obama administration negotiated with automakers an increase in the corporate average fuel economy (CAFE) standards for new light vehicles to 54.5 miles per gallon by 2025. Automaker efforts to reach that CAFE goal have already produced a shift of the light vehicle fleet to greater fuel economy.

The second reason for exploring distance charging is the growing unfairness of the burden of paying for road use. Figure 18.2 shows that some light-vehicle operators pay a lot per mile for road use while others pay very little. Operators of electric vehicles pay nothing at all for road use. The flat, volume-based fuel excise tax produces this inequity. For example, at a fuel tax rate of US$0.30 per gallon, the owner of a Ford F-150 (common in rural areas) pays US$167 for driving 10,000 miles while a Prius owner pays only US$60 for the same number of miles. The fully electric vehicle driver pays no fuel tax at all for road use.

Further, the cost difference for the same distance traveled produces divergent travel behavior. People driving highly fuel-efficient vehicles tend to drive more because the next mile costs them less than people driving a less efficient vehicle. Thus, drivers of fuel-efficient vehicles doubly impact the road system by driving more but paying less for doing so.

The growing unfairness of the fuel tax as a road funding mechanism was the primary motivation for Oregon legislators to pass the nation's first distance charge law in 2013 to implement a fully functioning, operational, distance charge program for light vehicles. No legislator questioned the proposition that a charge on distance traveled is more fair to road users than the fuel excise tax.

As the western states of the United States continue to experiment with distance charge pilot projects, a third reason emerges for pursing distance charge policies to

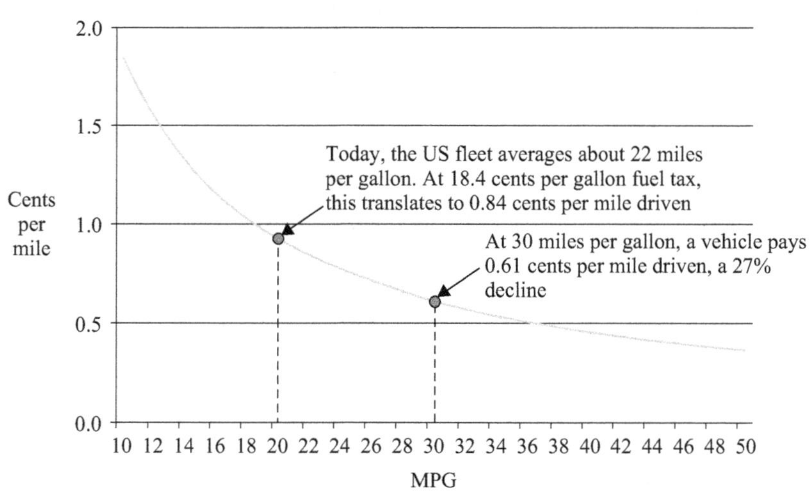

Figure 18.2 Fuel tax expressed in cents per mile as a function of fleet fuel economy

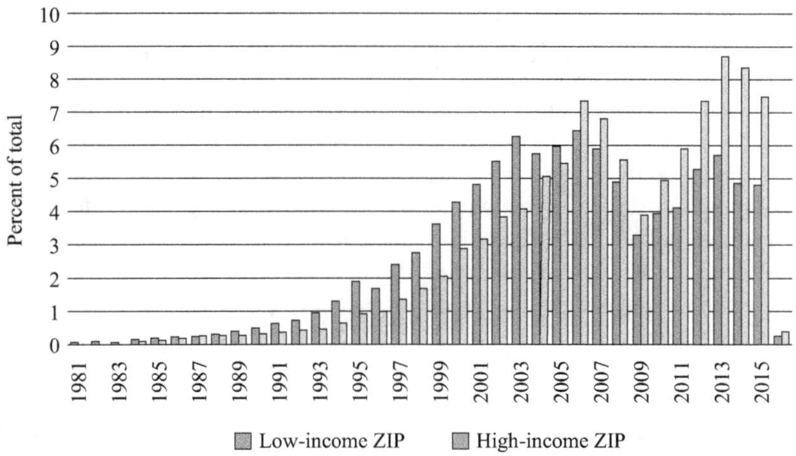

Figure 18.3 Vehicle model year in context of relative ownership by income class (ZIP means ZIP code or postal code, a code of letters and digits added to a postal address to aid in sorting mail for delivery)
[Source: D'Artagnan Consulting Paper on Income Equity before the California Road Charge Technical Advisory Committee, June 2016]

replace the fuel tax for road funding. The relative burden of the fuel tax is shifting to low-income drivers. Persons driving less fuel-efficient vehicles pay more fuel tax per mile than those driving fuel-efficient vehicles. The less affluent drivers tend to drive older cars that have low fuel efficiency (Figure 18.3). Higher income drivers, on the other hand, have the ability to, and do, operate the newer, technology-laden, fuel-efficient models. More affluent households also tend to own more vehicles, which affords the flexibility to drive more efficient vehicles for daily or long-distance travel and less efficient vehicles for shorter trips and occasional uses such as hauling goods and equipment.

The demographic groups harmed by this effect include less affluent drivers and very wealthy drivers, because both groups drive less fuel-efficient vehicles. The societal issue at hand, however, is that the mobility of the less affluent is greatly impeded by increases in the fuel tax rate while the mobility of the very wealthy is essentially unaffected. To put it another way, the next mile driven by low-income drivers is relatively more expensive than for high-income drivers and, therefore, increasing the excise fuel tax rate disproportionately decreases the mobility of the low-income driving population.

18.3 Legislative direction of distance charge policy development and technical research and development

After acquiring interest in a charge on road use as a way to meet transportation objectives, government officials will ask, "How do we start?" Some start with technical analysis, the tendency after which is to stop for lack of further direction.

By contrast, the US states along the Pacific Coast started with policy development for distance charging at the direction of each state's legislature, enabling these states to move forward successfully.

To reach the goal of implementing a mandated, fully functioning, operational road pricing system within a jurisdiction, a legislative body must enact legislation under its taxing authority. Sound policies and supportive politics are fundamental to passing legislation for any new tax program. Smart policies appeal to the public and encourage political supporters and champions to emerge within the legislative body. As the Pacific Coast examples discussed below attest, directed research yields the foundational elements for development of policies that can then become road pricing systems.

18.3.1 Oregon

Oregon's 15-year odyssey with distance charging for light vehicles began with legislative passage of House Bill 3,946 in 2001. This legislation created the 12-member Road User Fee Task Force (RUFTF) with the purpose of recommending alternatives to the existing road funding mechanisms, primarily the fuel tax, and directed the Oregon Department of Transportation (DOT) to develop and operate pilot programs to test alternatives. After examining 28 possible road funding mechanisms, the RUFTF identified charging a fee for distance traveled as the best broad-based funding alternative to the fuel tax [11]. The task force also identified congestion pricing as an aligned mechanism worthy of exploration as a potential solution for traffic management. This led to Oregon DOT undertaking its first pilot, the Road User Fee Pilot Project in 2006–07, a pay-at-the-pump model, in which volunteer participants paid a flat fee for distance traveled within Oregon and received a credit for fuel tax paid [3]. A smaller subset of participants paid a larger congestion charge for driving within the Portland Metropolitan Area during peak periods.

The purpose of the pay-at-the-pump model was to emulate the simple collection mechanism of the fuel tax and apply an immediate fuel tax credit to the per-mile charge transaction so that no motorist would pay both. Each pilot participant had a distance reporting meter installed in their vehicle. Plugged into the vehicle's on-board diagnostic port (OBD-II), these distance meters read and wirelessly reported a vehicle's distance traveled. While transportation officials celebrated the nation's first test of a distance charge for light vehicles, national feedback from academics, economists and think tanks foretold high costs, complexity, a bloated new governmental agency and transition mayhem if an attempt was ever made to implement the model. The general public expressed concern about the potential loss of privacy.

Buffeted by the feedback, Oregon DOT and the RUFTF changed course on policy objectives for a distance charge system and set aside the transactional approach to distance charge collection. At the time, mobile computing technology was changing rapidly and the next iteration of Oregon's distance charge concept shifted to an account-based approach to collection resembling the customer-friendly mobile phone industry.

To encourage evolution of mileage reporting technologies and competition among providers to reduce costs for administration, in 2011, the RUFTF recommended an open system that fostered competition by relying heavily upon commercial entities to collect the distance charge and manage the bulk of accounts. These policies led to design of the successful second pilot, the Road Usage Charge Pilot Program (RUCPP) in 2012–13, a demonstration of a potentially interoperable per-mile charge system with volunteer motorists from Oregon and the neighboring states of Nevada and Washington [7].[3]

In alignment with the policies tested in Oregon's successful second pilot—open system and private-sector data collection, tax processing and account management—the RUFTF proposed legislation to implement the new policies for an operational distance charge program, called a per-mile RUC. The state's legislature responded by enacting authorizing legislation (Senate Bill 810) [12], with strong majorities in both houses, to implement the per-mile RUC program according to a policy framework similar to the one used for the second pilot. Governor John Kitzhaber signed the bill into law in July 2013.

18.3.2 California

Movement on distance charging began in California in early 2014 following thorough research of the Oregon program. Later that year, the state's legislature passed Senate Bill 1077 directing development and testing of a voluntary distance charge program for light and heavy vehicles. Upon the recommendation of California's State Transportation Agency, Governor Jerry Brown signed the bill into law [9].

Senate Bill 1077 created the 15-member Road Charge Technical Advisory Committee (TAC) to study distance charge alternatives to the fuel excise tax and recommend design of a pilot program to California's Transportation Agency. The California Transportation Commission appointed 15 members to the TAC to undertake a year-long policy development process. The TAC guided design for the development, implementation and evaluation of a pilot program to assess the potential for a distance-based revenue collection system for the state's roads and highways as an alternative to the fuel excise tax system.

In developing recommendations for the design of the pilot program, the California law directed the TAC to consider factors such as availability, adaptability, reliability and security of methods for recording and reporting highway use. The law also compelled the TAC to consider privacy concerns for use of location data and to recommend protection for all personally identifiable information collected in reporting roadway use. Further, the law required the TAC to consider ease and cost of administration by the government and of recording and reporting roadway use by motorists as a factor in pilot program design, as well as effective methods of maintaining compliance [9].

[3] Each state imposed its own rate for distance traveled by participants internal to the state. Oregon's rate was 1.56 cents per mile; Washington's rate was 1.87 cents per mile; Nevada's rate was 1.19 cents per mile.

18.3.3 Washington State

The Washington State Legislature has directed five phases of research on distance charging since 2012. The first 2 years constituted an investigation phase for examining the potential of a distance charge to replace the fuel excise tax as the core revenue mechanism for road funding. Under the auspices of the Washington State Transportation Commission, the RUC Steering Committee performed a feasibility study for distance charging that addressed policy, technical, legal and administrative issues. In 2013, at the direction of the legislature, the steering committee considered various alternative approaches for a distance charge system and decided that a flat rate, per-mile charge would best serve the state. In 2014, the steering committee examined potential impacts of a distance charge, developed a concept of operations, recommended a policy framework and guiding principles (Table 18.2) and conducted a business case analysis, again as per the direction of the legislature. Among the guiding principles were *consumer choice* and *open to competing vendors* [13].

Table 18.2 Common principles [3,7–9,11,13,14]

Principle	State		
	Oregon	**California**	**Washington**
User pay	√	√	√
Open system	√	√	√
User choices	√	√	√
Private-sector administration option	√	√	√
Transparency	√	√	√
Cost-effectiveness and efficiency	√	√	√
System flexibility, adaptability and scalability	√	√	√
Reliability, security and technological feasibility	√	√	√
Simplicity of compliance and ease of use	√	√	√
Ease of administration	√	√	√
Enforceability	√	√	√
Privacy protection	√	√	√
Accuracy and system performance	√	√	√
Fairness and equity	√	√	√
Minimal nongovernmental burden	√	√	√
Integrate with other state policies	√	√	√
Accountable oversight		√	√
Interoperability		√	√

In 2015, the Washington Legislature directed the steering committee to look outward to review distance charge research in other states and coordinate with them in pursuing federal funding for distance charge pilot projects.

With passage of Engrossed Substitute House Bill 2524 in 2016, the Washington Legislature authorized updating the business case. The legislature also directed design of a distance charge pilot program to confirm the pilot's purpose and need and determine the location, number and the nature of the participant pool, the concept for testing and pilot duration.

Oregon, Washington and California have many common, fundamental and guiding principles for designing a distance charge pilot program. These are contained in Table 18.2. The states had a few noncommon principles that are not listed. California's Road Charge TAC, coming third in time, identified additional noncommon principles in the pilot's evaluation criteria.

18.4 Pacific Coast distance charge programs

The states of Oregon, California and Washington worked from similar polices and structure to develop their research efforts for distance charging for light vehicles. Each state relied upon establishing an account for each motorist to pay a charge for miles driven on public roadways within the state. Each state also relied upon motorists to report miles driven. Though voluntary, the Oregon, California and Washington programs could become the basis for a mandatory distance charge applied to all light vehicles.

18.4.1 Oregon's road usage charge program

Following passage of the nation's first law in 2013 creating a program for distance-based charging for light vehicles, Oregon's DOT worked toward the following objectives for the new program based on the provisions of the law,

> Create a sustainable road usage charge market that is simple and easy for payers, flexible and encourages evolution of mileage reporting technologies and business systems into effective, affordable, convenient and attractive options for the motoring public.

Oregon's RUC Program, branded as ORe*GO,* is a primarily state-funded, statewide, fully functioning, operational system with no termination date [15]. Volunteer owners of light vehicles[4] electronically report mileage driven and pay a charge quarterly of US$0.015 per mile traveled,[5] automatically receiving a credit for fuel tax paid. The per-mile charge applies to miles driven on public roads within the geographic boundaries of the state of Oregon. The number of volunteers for the program hovers around 1,000.[6]

In preparation for procurement of private-sector entities for account management and program operations, Oregon held a national RUC Summit in November 2013. Attending were 38 prospective commercial account managers (CAMs),

[4] For purposes of ORe*GO*, "light vehicle" means a motor vehicle of 10,000 pounds or less with at least four wheels in contact with the ground.

[5] The rate of 1.5 cents per mile is roughly equivalent to the average fuel tax rate a motorist pays to drive a light vehicle in the state: (30 cents per gallon/20 miles per gallon) = 1.5 cents per mile.

[6] By statute, Oregon's Road Usage Charge Program is limited to 5,000 volunteers. Oregon Revised Statute §367.890(2)(d).

mileage reporting device providers, data collection, accounting and other related service providers from firms around the world. Oregon legislators and agency executives sought to impress upon the world's vendors that creation of an open commercial market to support the distance charge market in the United States was a wise business opportunity. Importantly, representatives from 15 state governments also joined the gathering. Oregon wanted to encourage other states that policy development, researching and testing distance charge systems were worthy endeavor.

By all measures, Oregon's RUC Summit was a success, generating a buzz in a nascent commercial market and excitement among the states. The next year, the state of Oregon created the conditions for a distance charge open market, which materialized, and was formalized, in the execution of market-based contracts with several vendors to provide services for the new program in a competitive environment. Further, the summit encouraged formation of the Western RUC Consortium (WRUCC), now branded as *RUC West*, a research partnership among state departments of transportation. During the summit, the consortium held its first steering committee meeting which all governmental summit attendees witnessed. Fourteen state DOTs are now members of RUC West [16].

Oregon's RUC Program, branded as ORe*GO,* launched successfully on July 1, 2015. The majority of the participants—73%—chose CAMs providing location-aware mileage reporting equipment. The remaining 27% chose the government account management option only providing basic mileage reporting with no location awareness [17].

Before year's end, ORe*GO* proved itself an end-to-end technical success. The program established program rules and standards,[7] procured and certified three vendors (two as CAMs and another to provide the state account manager option), which enrolled volunteer participants, installed mileage reporting equipment in vehicles, collected metered use data, invoiced vehicle owners their per-mile charges with credits for fuel tax paid, accepted payments and forwarded them to ORe*GO* which deposited the revenue in the state's treasury; in essence, all that needs to be done.

In 2016, Oregon sought and received federal grant funding under the Surface Transportation System Funding Alternatives (STSFA) Act to enhance ORe*GO* [18]. This infusion of US federal funds will allow Oregon to develop and offer more mileage reporting options than currently offered (discussed further below). This funding allows Oregon to foster greater access to the market of CAMs by refining and streamlining its technology and account manager certification process for easier market entrance and exit and to align with national standards. Oregon will also explore financial interoperability with Washington, evaluate compliance and increase public awareness.

[7] ORe*GO* is designed with an open architecture that uses standard functional requirements and interfaces that are fully accessible to the marketplace.

18.4.2 California's road charge pilot program

A state-funded, statewide, 9-month (July 2015–March 2016) endeavor, the California road charge pilot program (CRCPP) tested distance traveled data collection and billing processes for a distance charge set at US$0.018 per mile traveled within the state of California [9]. The pilot included a cross section of 5,000 light and heavy vehicles owned by volunteer participants and fleets. Most of the participants were California residents but a few were nonresidents from neighboring states. Participants used a simulated payment process that included a mock credit for fuel tax paid. Unlike the Oregon program, CRCPP participants were not required to make actual payments.

The California pilot launched successfully on July 1, 2016. Within 6 weeks, over 5,000 volunteer vehicles enrolled in the program. To investigate the effects of distance charging on out-of-state drivers, the CRCPP accepted the participation of nonresidents in the pilot but they were limited to choices of a time permit or automated, nonlocation-aware devices for reporting distance traveled. The CRCPP enrollment process divided prospective volunteers into demographic groups to assess how a distance charge affects different parts of the citizenry.

Table 18.3 indicates the spread of mileage reporting preferences in the California pilot. The strong preference for location-aware methods of mileage reporting mirrors the ORe*GO* experience. The CRCPP's 35% preference for nonlocation-aware methods is higher than ORe*GO*'s 27%, perhaps because of the availability of more nonlocation-aware options.

The CRCPP was technically successful. Although it did not feature actual payments, it offered more distance reporting options than ORe*GO*. Like Oregon, California's program procured private-sector account managers to operate data flow and account management for the program with two vendors certified as CAMs

Table 18.3 California Vehicle Enrollments by mileage reporting method (This table was prepared based on data compiled internally for the CRCPP mid-pilot)

Location-aware methods	65%
• Plug-in device	53%
• Smartphone	12%
Nonlocation-aware methods	35%
• Plug-in device	7%
• Smartphone	8%
• Odometer read	11%
• Mileage permit	4%
• Time permit	2%
• Telematics	2%
• Other	1%

and another vendor to provide the state account management (SAM) option. During the initial enrollment period, 82% of light vehicle participants chose a CAM and 18% chose the SAM.[8] California enrolled volunteer participants, installed or accessed mileage reporting equipment, collected metered use data electronically and manually, invoiced per-mile charges with mock credits for fuel tax paid and accepted mock payments. California's pilot was a research project and potentially a step toward an operational program; as such, actual payments were not considered necessary to demonstrate viability of the distance charge concept to policymakers.

The California DOT sought and received federal grant funding under the STSFA Act in 2016 to enhance the state's distance charge investigation. The federal infusion of funds allows California to investigate a pay-at-the-pump/charging station method for paying a distance charge, explore organizational structures for road charging and improve outreach to the public [19].

18.4.3 Washington State's road usage charge pilot program

In 2016, Washington's legislature allowed use of federally allocated funding for the Washington State Transportation Commission to design a year-long, distance charge pilot program in five geographic regions of the state. As a US federal grant recipient under the STSFA program, Washington will launch its distance charge pilot in early 2018 for 2,000 light vehicles. The federal grant funding covers 8 months of design and pilot preparation (stage one) followed by 12 months of pilot operations (stage two) [20].

With elements similar to the Oregon and California programs, Washington's pilot demonstrates an open system with private-sector account managers and motorist choices for mileage reporting from among several options. Washington's pilot goes a step further by testing international interoperability with participants from the City of Surrey and Greater Vancouver, British Columbia, Canada and actual financial interoperability with Oregon's program, a neighboring state.[9] Like California, the Washington pilot simulates distance charge and fuel tax credits. This pilot also explores distance charging as a replacement for the fuel tax as a road funding mechanism.

18.5 Commonalities and differences among the Oregon, California and Washington distance charge programs

Oregon's extensive, decade-plus work on distance charging for light vehicles gave birth to public policies that proved much more politically acceptable than earlier efforts. These policies formed common foundational elements for the Oregon,

[8] These ratios are based on data compiled internally for the CRCPP.

[9] Although Washington State enrolled resident participants in Oregon's 2012–13 Road Usage Charge Pilot Program, actual financial interoperability was not attempted.

California and Washington distance charge efforts. There are differences as well. In California and Washington, policy development bodies sought to expand upon Oregon's achievements and to fold in additional, specific policies to accommodate local sensitivities and preferences.

18.5.1 An open market

Analysis of Oregon's authorizing legislation shows the elements of creating an open commercial market to manage the collection of distance charges. This law specifically requires the Oregon DOT to partner with private-sector entities for collection of metered use data, tax processing and account management [21]. Further, methods for collection of data, including technology used in methods, must adhere to an "open system" [22].

> As defined in Oregon's law, "open system" means an integrated system based on common standards and an operating system that has been made public so that components performing the same function can be readily substituted or provided by multiple providers. [23]

By adopting the requirement for an open system, the legislature made it clear that the state's distance program would not rely upon the proprietary systems and technologies of a single vendor. Even further, the legislation requires the program to provide payers of the distance charge the opportunity to *select a method from among multiple options* for collecting and reporting the metered use by a subject vehicle [24].

Read together, the law's provisions requiring an *open system, private-sector data collection and selection of a data reporting method from among multiple options* show the Oregon legislature's intent to form an open commercial market for collecting data for distance charges. On the other hand, there is no indication the legislature intended to proscribe a *government option* for distance charge collection. Rather, government collection, should it exist at all, must exist as an alternative to, and alongside, commercial market collection.

As per the legislative directive, the state of Oregon adopted administrative rules, established performance standards for data flow and operations, formed a government oversight and auditing system to manage private-sector account managers and issued a request for qualifications in May 2014.[10] With these steps, Oregon took a large leap toward establishment of an open market for metered use data collection, per-mile charge processing and account management. Four private-sector vendors submitted proposals and met the first two stages of certification requirements sufficient for each to sign an umbrella agreement with the state. The umbrella agreement was unique in that Oregon negotiated a *market price* that each vendor had to accept and a *market contract*, negotiated with all the vendors, that each vendor had to sign in order to enter the Oregon distance charge market. Only three of the vendors agreed with financial terms and met qualifications by

[10] The state of Oregon also issued a request for proposals for private operation of the state government account management option. One firm won this award.

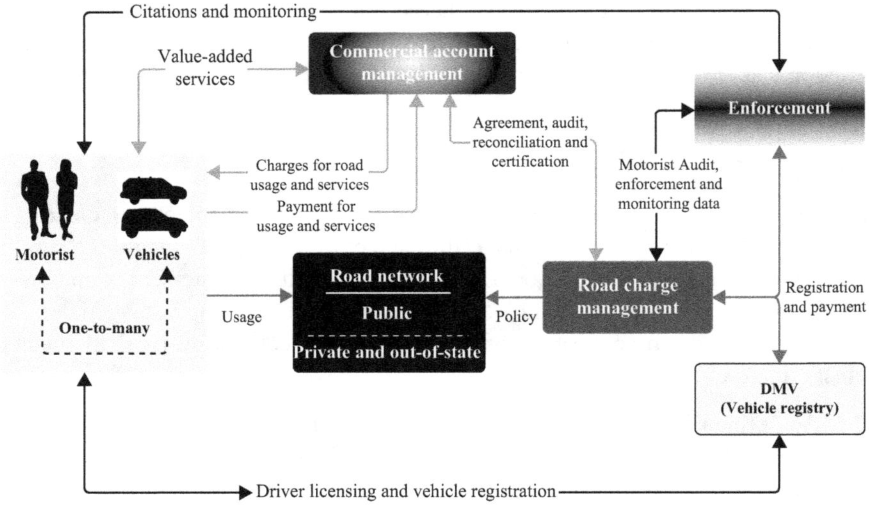

Figure 18.4 Open system overview with commercial account managers

completing the next three stages of the certification process. By doing so, these firms proved they could meet Oregon's established standards and requirements to become CAMs for ORe*GO*. One of the firms also completed certification to provide a state government account management option. The Oregon program permits CAMs to sell value-added services along with data gathering, billing and account management services for the distance charge. See Figure 18.4.

California's Road Charge TAC's design of its road charge pilot program also embraced the open-market approach to collect a distance charge and use of commercial account management [14], as did Washington's Road Charge Steering Committee. These states' policy bodies recommended that their pilots test an open system design and offer motorists a choice in mileage reporting methods and account managers. Each state offered pilot participants commercial account management, as per choice of each participant, while Oregon and California also offered SAM as an additional option.

18.5.2 Distance traveled reporting methods

Oregon law requires the Oregon DOT to establish methods for reporting vehicle distance traveled. In doing so, the Oregon DOT must consider accuracy, security, resistance to tampering, ability to audit compliance and privacy options. The Oregon DOT must provide persons liable to pay the distance charge, the vehicle owner or lessee, the opportunity to select a metered use reporting method from among multiple options. At least one of the methods for collecting and reporting distance traveled must not use vehicle location-aware technology [25].

In Oregon's May 2014 procurement for distance charge account management services, Oregon asked the market to propose mileage reporting methods and technologies for certification as meeting Oregon DOT's established technical standards. After a lengthy contracting and certification process, Oregon DOT certified the mileage reporting methods offered by three firms for use in the program. These certifications included demonstration of capability for the metering technology to store road use data and report the data on a daily basis through a standard mileage message format. The location-aware metering devices also demonstrated whether miles were driven within Oregon or elsewhere.

The devices Oregon certified as automated mileage reporting methods included two types of electronic distance meters used in usage-based automobile insurance. Plugged into a vehicle's on-board diagnostic port (OBD-II),[11] these distance meters read and wirelessly report a vehicle's distance traveled and, for most vehicles, fuel consumption. One type has an embedded global positioning system (GPS)[12] chip that can determine general location of distance traveled to record whether miles are driven on public roads within the state or somewhere else. The other type has no GPS chip and simply wirelessly reports all miles driven. Both types of distance meters have proprietary features for other optional, value-added applications, depending upon the vendor.

Reporting of fuel consumption via the distance metering devices has been somewhat problematic because data available for this purpose is not consistent among vehicles. For some vehicles, OReGO defaults to using a formula to estimate fuel use rather than actual fuel use.[13] These issues will fade away beginning with the 2019 model year when California's new OBD-II regulations standardize electronic reporting of fuel use.

Rather than simply adopting Oregon's standards limiting automated mileage reporting using two types of OBD-II devices (with and without location awareness), California ambitiously augmented Oregon's automated reporting methods with additional technologies and also expanded mileage reporting offerings to methods (or proxies) not requiring the use of technology [9,14].

California's augmentation of automated reporting:

- *Factory installed telematics.*[14] Factory installed telematics and OBD-II devices have access to the same vehicle data. Only some newer vehicles

[11] Only light vehicles since 1996 have the requisite OBD-II port available for this purpose.

[12] GPS is an acronym for the US global positioning system which is a part of the Global Navigation Satellite System.

[13] OReGO's *formula* for estimating fuel use is the US Environmental Protection Agency's fuel economy estimate for a given make/model divided into the number of miles driven.

[14] Telematics, or native wireless telematics, is a wireless communications system built into vehicles that communicate with the internet in order to display real-time traffic and travel information within a vehicle and provide online driving performance information. Telematics systems first appeared in 2012 in high-end luxury vehicles but are now available in other vehicle models. Motorists access a telematics system through the same interface as the vehicle navigation system, but not all navigation systems include native wireless telematics. Many are simply stand-alone computer systems preloaded with navigation data. If a vehicle has telematics, the vehicle owner's manual will have live data features on the navigation system. *Summarized from California Road Charge Pilot Program website. Available from:* https://www.californiaroadchargepilot.com/account-manager/participant-faq/.

currently have this capability but telematics will likely become standard equipment in the future. The telematics option for the California pilot is not location aware.[15]

- *Smartphone app.* This method requires installation of a certified smartphone app that measures distance traveled from a specific vehicle, with or without location awareness, when the smartphone is powered-on within the vehicle.

California's additional reporting methods not requiring distance reporting technologies:

1. *Time permit.* Motorist purchases a permit for a given period of time during which the motorist may drive an unlimited distance. The price for the time permit is set at a high level.[16] No distance reporting is required for this method.
2. *Mileage permit.* Motorist purchases a specific number of miles in advance after an official odometer reading by a designated third party or a smartphone photo app with additional odometer readings taken at regular intervals.
3. *Odometer reporting.* Motorist self-reports distance traveled quarterly and periodically verifies reports via manual reading by a designated third party or via smartphone app that records an odometer image for a specific vehicle.

Oregon's federal grant funding obtained in 2016 will allow ORe*GO* to adopt some of the additional mileage reporting options explored in California. These mileage reporting options will not rely upon access to the OBD-II port including telematics, smartphone apps and a manual mileage reporting option for participants.

The Washington State RUC Steering Committee recommended mileage reporting methods for the Washington pilot similar to those of the CRCPP, except for the time permit and telematics. Oregon has requested each of the ORe*GO* CAMs to consider proposing for certification distance reporting via telematics and smartphone apps but, because the program is operational, the decision to propose a new mileage reporting method for ORe*GO* is left to the marketplace.

Table 18.4 summarizes the commonalities and differences among the Oregon, California and Washington distance charge pilots and programs.

[15] Telematics currently do not provide a cost-effective way to report distance traveled by location.

[16] For the California Road Charge Pilot Program, the time permit rate is determined by multiplying the road charge rate of US$0.018 per mile by 25,000 miles traveled per year, scaled proportionately to the term of the permit. The 25,000 annual miles traveled figure is set at the 95th percentile for California light vehicles. In Oregon's second pilot, the annual miles traveled figure was set at 35,000, at the 98th percentile, for a year-long time permit.

Table 18.4 Mileage reporting methods in US West Coast distance charge pilots and programs [7–9]

Mileage reporting method	Payment type	Oregon (OReGO)	California (CRCPP)	Washington[a] (WRUCPP)
1. Time permit: purchase covers specific period of time; no mileage reporting	Prepay	Tested in 2012	Yes	No
2. Mileage permit: purchase of a specific number of miles	Prepay	No	Yes	Yes
3. Odometer reporting: third-party or smartphone photo verification	Postpay	Yes[b]	Yes	Yes
4. Automated reporting: OBD-II plug-in device (with or without location awareness)	Prepay or post-pay	Yes	Yes	Yes
5. Automated reporting: factory installed, in-vehicle telematics (without location awareness)	Postpay	Yes[c]	Yes	No
6. Automated reporting: smartphone app (with or without location awareness)	Postpay	Yes[c]	Yes	Yes
7. Automated reporting: smartphone app connected to OBD-II plug-in device (switchable[c])	Postpay	Tested in 2013	No	No
8. Pay at the pump	Postpay	Tested in 2006–07	Planned for 2017–18	No

[a]The mileage reporting methods for the Washington State Road Usage Charge Program were determined at a meeting of the Washington Road Usage Charge Steering Committee on December 13, 2016, in Olympia, Washington.
[b]US Federal STSFA enhancement.
[c]"Switchable" means the use of location information can be switched on and off by the vehicle user.

18.6 Other Western States' distance charge pilots and authorities

18.6.1 Colorado road usage charge pilot program

In late 2014, the Colorado DOT, without legislative direction, announced development of a small-scale, distance charge pilot demonstration for light vehicles, funded by US federal research funds allocated to the state. The state recruited 100 volunteers to participate in a distance charge pilot over 4 months with elements similar to those of Oregon's second pilot program, including providing motorists with choices for mileage reporting [26]. Participants tested several mileage reporting methods already proven in Oregon and California. Invoices for a distance charge of US$0.012 per mile were illustrative and, as such, not for payment and included a mock credit for fuel tax paid.

The intent for the Colorado pilot was to evaluate issues related to potential implementation, determine the feasibility of distance reporting technologies and obtain feedback for system design. The Colorado pilot launched in December 2016 and ended in April 2017.

18.6.2 Hawaii road usage charge pilot program

As a US federal grant recipient under the STSFA program, the state of Hawaii's pilot will explore the potential for a distance charge to replace not only state fuel taxes but also federal and county fuel taxes and other flat road fees such as inspection fees and state and county automobile registration fees and weight fees [27]. All relevant state and county agencies will be involved.

Owing to the singular geographical characteristics of the state of Hawaii—no borders with another governmental jurisdiction—and the state's practice of collecting an annual odometer reading for every vehicle, Hawaii's approach to distance charging has unique elements relative to its counterparts in the western United States. The statewide Hawaii pilot will offer both manual and automated reporting options and provide illustrative invoices for owners of light vehicles.

Development and operations for the pilot will occur over five phases. Phase 1 involves policy development by a select stakeholder support group throughout the 36 months of pilot program. This stakeholder group will determine essential operational features for the pilot.

Manual distance data reporting of odometer readings occurs in phase 2. Pilot operations will begin (phase 3) by accessing distance data already reported during the annual vehicle safety inspection and mailing illustrative billings to over one million motorists.

The state plans to prepare automated reporting options in early 2018 (phase 4), with third-party vendors certified to provide automated distance reporting options and illustrative billings later in 2018.

The pilot will commence automated reporting operations later in 2018 (phase 5). The pilot will recruit up to 2,000 volunteer motorists to engage in automated reporting of distance traveled with location-awareness optional. Which automated reporting options the pilot offers in phase 5 will depend upon policy direction. Nevertheless, the automated reporting options in Hawaii's pilot will likely operate as an open system with open market standards similar to those adopted by Oregon, California and Washington to foster competition in the distance charge marketplace and provide motorists with choices for mileage reporting. The Hawaii pilot should conclude operations in late 2018.

18.6.3 Utah

Pursuant to passage of House Bill 362 by the Utah legislature in 2014, the Utah DOT is actively researching the policy implications and systems for distance charging. The legislation also authorizes development of a small distance charging pilot program.

18.7 Key issues for distance charge programs in the United States[17]

Two congressionally established surface transportation commissions in 2007 independently explored the policy issues surrounding transition from a road revenue mechanism based on fuel consumed to one that charged motorists for the amount of road use [28,29]. At the same time, the general public weighed in with concerns about Oregon's first pilot test for a fee on distance traveled. Issues emerged that were, after a period of reflection and investigation, addressed by redesign of Oregon's distance charge concept for a second pilot, legislation establishing the OReGO program and similar design, structure and policies adopted for the California and Washington pilots. The open commercial market for collecting distance data and charges and managing accounts opened a path for resolution of key issues listed in Table 18.5.

18.7.1 Issues essentially resolved or substantially calmed

18.7.1.1 Complexity of implementation and difficulty of operations

The weight and complexity of a government creating a new tax payment system from scratch to manage millions of distance charge payers seems insurmountable. The solution was to move a substantial amount of the burden away from the government. Outsourcing some or all account management to private-sector entities with expertise in electronic data collection and account management relieves the government from actually operating the data and payment collection mechanisms to instead regulate and audit the market, a much less daunting task.

Table 18.5 Key implementation issues

Implementation issues	Resolution status/path
Complexity of implementation	Essentially resolved
Difficulty of operations	Essentially resolved
Simplicity of compliance	Essentially resolved
Scalability	Essentially resolved
Flexibility for policy adaptations	Essentially resolved
Evolution of technologies and business systems	Essentially resolved
Privacy protection	Substantially calmed
Security of data	Continually managed
Interoperability among state systems	Toward resolution
Nonresident driving	Toward resolution
Equity by income, geography and vehicle type	Determined by legislature
Enforcement	Determined by legislature
Cost of start-up and administration	Determined by legislature
Transition	Determined by legislature

[17] Much of this section is based on my personal knowledge of the development and external communications for the Oregon, California and Washington road usage charge programs. There are few if any published references regarding this material.

18.7.1.2 Simplicity of compliance

Understandably, citizens judge a system through a personal lens. If offered only one way to do something—such as Oregon's pay-at-the-pump model in its first pilot—people tend to find error or inconvenience and, accordingly, judge their ability to comply with the system's requirements as difficult. Offering motorists choices of account manager and mileage reporting methods permits each motorist to select the simplest and easiest manner of reporting and paying the distance charge according to the individual's needs and preferences. Governments regulating the market can improve the ease of compliance through carefully crafted service agreements with certified CAMs.

18.7.1.3 Scalability and flexibility for policy adaptions

As the states experimented with distance charging, analysts at the national level showed concern that any single state's system would not be replicable or interoperable with other state systems nor would that system prove adaptable for other, related, road pricing policies [28–30]. In response, the US Pacific Coast states tested an open system design with established and published, but not identical, common standards. Additionally, Oregon developed contractual provisions requiring vendors to accept alterations to policy applications and a vendor certification process that supports easy upgrading of mileage reporting methods. Accordingly, an open market distance charge system with provisions supporting adaptability for policy needs and jurisdiction variations will be scalable and flexible because it will be founded as an open market relying upon private-sector account managers who are obligated to make adjustments and accept opportunities as directed by policymakers.

18.7.1.4 Evolution of technologies and business systems

Technology evolves continually leaving most government entities struggling to keep business systems and technologies updated. Like all government systems, distance charging is vulnerable to selection of closed system[18] technology for mileage reporting that would quickly go out-of-date and operational business systems that would stagnate and be hard to change. An open market solution avoids governmental selection of the mileage reporting technology or the operational business systems but rather allows an open commercial market to provide them. The government's role should be to ensure private-sector vendors have access to a flexible and swift certification process, probably operated by more than one third-party private-sector entity. An open market supported by an efficient certification process ensures the continual evolution of mileage reporting technologies and operational business systems toward least cost and greatest efficiency.

18.7.1.5 Privacy protection

As interest in light vehicle distance charging became serious in the United States, protection of personal privacy emerged as the most critical policy issue. Sensitivity about privacy is strong enough in the United States that complete resolution of this

[18] A "closed system" is proprietary and which only one provider is, in practice, able to support.

issue may not be possible, at least in the short term. Nevertheless, the Oregon approach shows that adoption of policies and laws protecting privacy can have a calming effect sufficient to allow legislation for a distance charge to pass state legislatures. Policymakers in California and Washington similarly understand the necessity of protecting privacy in a distance charge system.

In its 2013 legislation, Oregon addressed the privacy issue in three ways. First, by allowing a motorist to select between government and private-sector entities, the motorist *chooses* the entity which will manage his or her data collection; thus, giving control of which entity manages sensitive personal information to the motorist. Second, the motorist has the opportunity to *choose* the mileage reporting method and associated technology with at least one option not involving location awareness, such as global positioning system technology, which, although not necessarily indicative of privacy invasion, many citizens perceive as having high risk for it. Finally, and most importantly, extensive statutory provisions require the government and the private-sector account managers to ensure all motorists' personally identifiable information will be protected from disclosure.

Oregon statutes prohibit the Oregon DOT, its certified private-sector account managers and their contractors from disclosing personally identifiable information used or developed to report metered use of a vehicle or for administrative services to collect the per-mile charge [31]. "*Personally identifiable information*" means any information that identifies or describes a person. There are exceptions for the vehicle's owner, a financial institution involved with payment, employees of the department, a police officer pursuant to a valid warrant or an entity the owner expressly approves to receive the information.

Oregon negotiated the privacy protection provisions for the RUC program with the American Civil Liberties Union, a privacy watchdog group. A key negotiated provision was a requirement to destroy the records of location and daily metered use within 30 days after completion of the later of payment processing, dispute resolution or noncompliance investigation. Exceptions include traffic management and research (if the personally identifiable information is removed), monthly summaries of metered use and retention by a CAM when the vehicle owner consents.

In addition to these statutory requirements, the department created an administrative rule granting the vehicle owner the following additional rights pertaining to personally identifiable information:

- The right to inquire about the accuracy, status and use of the information;
- The right to examine the information;
- The right to request corrections where there is error and
- The right to request to delete location and daily metered use data that has not been destroyed within the required time frame [32].

California law specifically prohibits the California Transportation Agency from taking actions to "disclose, distribute, make available, sell, access or otherwise provide for another purpose,[19] personal information or data collected through the

[19] This implies another purpose *other than operation of the Road Charge Pilot Program.*

pilot program to any private entity or individual." Exceptions include authorization by court order of a civil case or a subpoena issue for a criminal case, a search warrant or in the aggregate when all personal information is removed [33].

California's Road Charge TAC added additional detail to the state's statutory policies directing protection of privacy for the pilot by adopting the Road Charge Pilot Program Privacy Policy. The California Transportation agency adopted this privacy policy as an attachment to the pilot participation agreement. It informs the participant about the nature of information collected during the program. More importantly, it imposes the legal requirement to protect all participants' personal information upon all the entities performing data collection, tax collection and account management services on behalf of the state of California.

In Washington, authority for protection of privacy is contained in the state's constitution, article I, section 7, which states, "No person shall be disturbed in his private affairs, or his home invaded, without authority of law." The appellate courts of the state of Washington have interpreted this right to privacy as creating the right to nondisclosure of intimate personal information and the right to personal autonomy. These provisions are likely to apply to collection of a distance charge from motorists but the reach of this provision is not yet known.

18.7.2 Issue requiring continual management

18.7.2.1 Security

Security of personal information requires a government implementing and operating a distance charge system to manage data security according to best practices, either national or international. As information security in the digital age is continually under assault, best practices require continual monitoring and upgrading. Governmental entities must ensure that application of these best practices occurs not only by government but also by its CAMs. The California Road Charge TAC recommended adoption of specific data security measures based on industry standards for online financial-grade transactions for authentication, authorization and encryption [14].

18.7.3 Issues for later resolution

While open system design enabled solutions for the many issues raised by citizens and analysts, other issues are not so easily resolved. Some issues will reach resolution only through cooperation among states.

18.7.3.1 Interoperability

Motorists in the United States travel freely through states without impediment. A distance charge system, for one state, must be able to integrate with other states' distance charge systems. Because electronic toll systems in the nation interoperate only on a regional basis, if at all, analysts express concern about whether distance charge systems would be hampered by a similar limitation [28,29]. Through application of an open system with identical or closely similar standards and requirements, programs of neighboring states would easily interoperate. While individual peculiarities will make some states' requirements unique, the requirements important to interoperability—like electronic communication of data—can be the same. Judging from the structure of the distance charge systems in place in the Oregon and California programs, interoperability is well under development in the western United States. Whether other states adopt a similar open system approach will be settled at a later date.

18.7.3.2 Non-resident driving

Under the fuel tax system, nonresident drivers pay for the roads within a state in which they travel when they refuel within that state. If drivers of light vehicles do not refuel within a state despite traveling within that state, they do not contribute to the upkeep of that state's roads.[20] If one state has a distance charge while a neighboring state does not, the situation remains the same for nonresident drivers in the distance charge state if the fuel tax is retained and applied to them. If the neighboring states both have distance charges, the extent to which they are interoperable can resolve the question. As such, this issue falls into the category of *later resolution*.

18.7.4 Issues determined by legislatures

Some issues can only be resolved through further action by legislatures as they make the final policy choices for a mandatory distance charge system.

18.7.4.1 Equity by income group, geography and vehicle type

Whether a distance charge system will be considered fair to payers will be determined by legislatures as they apply a mandate and a rate structure to those liable to pay the charge, the vehicle owner or lessee. The perception of fairness is always in the eye of the beholder. Some owners of highly fuel-efficient vehicles will regard paying the same rate as those with less efficient vehicles as unfair. Rural drivers may want to pay a lower rate because they view themselves as driving more and longer distances for essential services. Drivers of less fuel-efficient vehicles may believe every vehicle should pay the same rate for the same burden placed on the road system. Less affluent drivers may want a lower rate. Application of a distance charge and setting the rate structure is a conundrum only a legislature can resolve.

18.7.4.2 Enforcement

The enforcement measures in place for the Oregon distance charge program are limited. Because these measures apply to volunteers rather than mandatory payers,

[20] In contrast, reporting of fuel use by heavy vehicles for purposes of redistribution of revenues among the lower 48 US states and the Canadian provinces is possible through the International Fuel Tax Agreement (IFTA).

the program applies a light touch in enforcement situations. Payment, on the other hand, is required for billed amounts. ORe*GO* volunteer participants sign an agreement to pay the distance charge in lieu of the fuel tax. The ORe*GO* CAMs have responsibility to pursue payment should delay occur.

If a participant cheats the system by tampering with mileage reporting equipment or false reporting, Oregon statute requires the government agency to act. Detection of cheating is not necessarily simple to determine. For example, the technology used in the ORe*GO* program for mileage reporting can be unplugged from the diagnostic port causing gaps of data flow into the system. Fortunately, California's regulatory upgrade of the on-board diagnostic port (OBD-II) will eliminate this data gap beginning with 2019 light vehicle models.

More needs to be done on enforcement measures for distance charge programs. As states move closer to mandatory application, enforcement policies and practices will get more attention. Legislatures will enact serious enforcement regimes once a mandatory distance charging system is in place with sizable numbers of payers.

The California Road Charge TAC recommended checking for anomalies in mileage data for the CRCPP, as Oregon does for ORe*GO*, but does not recommend rigorous enforcement activities for the volunteer participants.

18.7.4.3 Cost of administration

Perhaps the chief difficulty of implementing a distance charge for light vehicles is the cost of operations. Any new tax system is necessarily costly at the start. Few payers at the beginning means smaller revenues to absorb high start-up costs as well as high operational costs for any system designed for a large number of payers. As the number of payers grows, this ratio will improve. Government tax programs, on the other hand, are not known for easy improvements to gain efficiencies. Encouraging distance charge payers to access CAMs within an open competitive system animates market forces fostering innovation and cost efficiencies, especially when sharing collection costs for value-added services.

Setup costs for a distance charge system can be affordable. The only experience with implementing such a system is Oregon's operational ORe*GO* program. The Oregon DOT spent US$6.5 million to implement the state and CAM-based system over a 2-year period. California spent just under $5.0 million to procure, set up, test certify and operate its pilot system for over 5,000 vehicles for 9 months, including the range of reporting methods for light and heavy vehicles.

Although operational, Oregon's RUC program includes a relatively small number of volunteers. The most recent research on the operational costs for a large-scale distance charge system comes from the state of Washington. As Table 18.6 indicates, the higher the number of payers, the more affordable a distance charge system is to operate. Legislative action and time can yield an affordable distance charge system.

Washington's financial model compares the cost of a state-run distance charge system versus a hybrid system operated partially by the state and partially by CAMs. This model found that the hybrid system was slightly less expensive than

Table 18.6 Operational costs for distance charging[21]

Number of payers	Relative operations cost
250,000 payers	16% to 18% of revenue
1 million payers	8% to 10% of revenue
6 million payers	4% to 6% of revenue

the government-run system when applied to one million payers or less but larger applications clearly favors the hybrid approach with the government-only system at 6% of revenue and the hybrid system at 4% of revenue. The commercial market's ability to achieve operational efficiencies leads to reductions in operational cost. The larger the number of payers in the distance charge system, the greater the cost efficiencies the private sector can gain.

18.7.4.4 Transition

Transition from the fuel tax to a distance charge as the central feature of road funding has been a conundrum for US thought leaders. They cite political, institutional, technical and technological challenges requiring development of a strategy to guide transition to the new system [28,29].

Notwithstanding these concerns, shifting the bulk of responsibility from an entirely government operated system to one operated through an open commercial market moves the burden to the most creative element of the economy, making transition not only feasible but easier. Even so, for the advantages of a commercial market to reach its potential in this sphere, legislatures must apply a distance charge mandate to a sizable number of motorists to enable the market to form and prosper.

Developing a strategic plan for full implementation is the final stage of development for distance charging. Once a state has engaged in sufficient research, policy development and testing resulting in development of strong institutional knowledge embedded within the responsible government agencies, strategic planning for transition to a mature, mandated distance charge system can begin. A strategic plan must indicate the end state—what the program looks like when it is finished—and how transition will occur. Necessary to these determinations are recommendations to legislatures for (1) which road taxes will be replaced—for example, fuel tax, vehicle registrations fees or license fees, if any, (2) which vehicle types will be subject to the charge and when and (3) the rate structure.

18.8 Impact of the US Federal STSFA grant program on distance charging

The United States Congress actively supported the states' investigations into user-based alternative revenue mechanisms by including in the Fixing America's

[21] Estimated by the Washington Road Usage Charge Steering Committee.

Surface Transportation (FAST) Act passed in December 2015 [34], a 5-year US$95 million grant program to the states called the STSFA.[22] The STSFA grant program provides funding to US states to conduct pilot programs demonstrating user-based alternative revenue mechanisms that use a user fee structure. In late summer 2016, the United States DOT announced awards of grant funds to eight STSFA proposals for distance charging in the first year of funding, five to individual states and two to consortia of states [34] (Table 18.7).

Table 18.7 US Federal STSFA grant funding for distance charge demonstrations

Awardee	Type	Amount (US$)
Oregon	Enhancement	$2,100,000
California	Enhancement and research	$750,000
Washington	New large pilot	$3,847,000
Hawaii	New large pilot	$3,998,000
Minnesota	Predevelopment	$300,000
Missouri	Predevelopment	$250,000
Oregon, as lead state for RUC West (11 states)	Predevelopment	$1,500,000
Delaware, as lead state for I-95 Corridor Coalition (5 states)	Predevelopment and four demonstrations	$1,490,000
Total		$14,235,000

Each state awarded STSFA grant funding committed to provide a 50/50 match of actual pecuniary or in-kind resources in a grant agreement with the US Federal government. Thus, the actual endeavors equal double the amount awarded under the STSFA grant funding.

18.9 Awards for predevelopment of new pilots

First-year grants under the federal STSFA program allowed states with new ideas and consortia of states to have 18 months of breathing room to fully develop pilot programs for distance charging.

18.9.1 Minnesota

With federal STSFA funding, Minnesota's DOT will investigate a distance charge pilot using telematics embedded in autonomous vehicles for reporting distance traveled [35]. Minnesota DOT intends to develop a partnership with MaaS[23] providers to understand the value proposition before undertaking the pilot.

[22] Section 6020 of the FAST Act.

[23] MaaS (Mobility-as-a-Service) providers offer integrated mobility services—public, private and shared—for on-demand, door-to-door, transport options for a monthly subscription fee. MaaS providers intend for MaaS to provide an alternative to privately owned vehicle travel.

18.9.2 Missouri

Missouri's DOT will use the federal STSFA funding to develop a new registration fee schedule based on miles-per-gallon for light and heavy vehicles taking into account the estimated average miles traveled [36].

18.9.3 Western Road Usage charge consortium (RUC West)

The WRUCC, branded RUC West, is a collaborative, distance charge, research partnership of 14 states comprising 38% of the land mass of the United States. The STSFA grant, awarded to Oregon DOT as lead state on behalf of WRUCC, allows the WRUCC to conduct two phases of predevelopment work for a regional pilot [37]. Phase 1A involves defining the regional system and requirements in coordination with the private sector. Phase 1B involves project planning to develop plans for communications, oversight and management, procurement and implementation. Eleven states have funded and will participate in phase 1A: Arizona, California, Colorado, Hawaii, Idaho, Montana, Nevada, Oklahoma, Oregon, Utah and Washington. Four of these states will participate in phase 1B: California, Colorado, Oregon and Washington.

Phase 2 is the pilot demonstration which, under STSFA guidance, must start no later than 18 months following award of this grant. While the WRUCC intends to seek additional STSFA funding in 2017 for pilot operations, the US federal government grant agreement obligates the WRUCC to launch the regional pilot in early 2019. California and Oregon have already committed to participation in phase 2 but other states may do so as well once their state legislatures grant approval.

18.9.4 I-95 Corridor Coalition

The I-95 Corridor Coalition is a membership organization consisting of departments of transportation from 16 states along the eastern seaboard of the United States plus the District of Columbia. The coalition is devoted to the well-being of the interstate transportation system along the lengthy corridor running from Canada to the Caribbean along the East Coast of the United States. The STSFA grant, awarded to Delaware DOT on behalf of the coalition, allows the coalition to conduct two main components, one for planning and predevelopment, including an operational concept, and the other for individual pilots in four states in the northeastern United States: Connecticut, Delaware, New Hampshire and Pennsylvania [38]. Of these states, only Delaware and Pennsylvania have a contiguous border. Vermont is participating in planning but will not conduct an individual pilot for the state. None of the states received legislative support in advance for this endeavor.

The nature of the focused individual state pilots will be determined by the operational concept developed in the first component. Each 4-month state pilot will consist of 50 light vehicles. Each state will select one or more private-sector vendors to conduct distance charge data collection and accounting activities. No actual drivers' money will change hands as the invoices will be illustrative. The pilots will deploy electronic distance meters plugged into the vehicle's OBD-II port for reporting distance traveled. A time-based flat fee may be deployed by some states as an alternative to distance reporting. Other distance reporting technologies, such as deployed in California, may also be selected for use in the individual state pilots. The timing of each state's pilot's launch will be determined by the terms of the grant agreement with the US federal government.

18.10 Planning for mandatory road usage charging in the United States

The success of the US Pacific Coast open market approach to distance charging for light vehicles proves the feasibility of the fundamental policy, systems and technological elements necessary for application of a mandate to a large group of vehicles. While a few policies remain to develop, largely concerning an enforcement regime, the next task to undertake is planning for full transition to the new distance charge system.

In 2015, Oregon's RUFTF began strategic planning for a mandated distance charge for light vehicles. The task force completed this work on September 21, 2016, by adopting a recommendation to the state legislature for enactment of legislation, effective in 2025, to

- impose a distance charge of US$0.015 per mile to resident operators of all new light vehicles with a fuel efficiency rating of 20 miles-per-gallon or better;
- offer a credit for fuel tax paid by distance charge payers;
- keep the fuel tax in place for operators of light vehicles with a fuel-efficiency rating of less than 20 miles-per-gallon and for nonresident motorists; and
- retain the voluntary distance charge program until 2025.[24]

The proposed long wait (8 years) for application of a distance charge mandate for light vehicles and the fact that nearly all necessary, operable distance charge systems already exist in Oregon, seems to indicate the late date was selected for political reasons. After all, most legislators voting to support this legislation will no longer hold office at the time of the effective date. Even so, this recommendation recognizes that distance reporting technologies embedded within vehicles (called telematics) are not currently common nor easily accessible, and a lengthy wait offers sufficient time for this desirable distance reporting technology to achieve broad-based application.

If adopted, this policy will have the effect of operators of fuel-efficient light vehicles paying more for the roads than they do now and operators of fuel-inefficient light vehicles (and nonresident drivers) paying the same as they do now. Or, to put it another way, operators of vehicles above the fuel efficiency midpoint will pay the distance charge and receive a tax credit for fuel tax paid while operators of vehicles below the fuel efficiency midpoint will continue to pay the fuel tax.

[24] This information was obtained by personally witnessing the deliberations and decision-making process of the Road User Fee Task Force at the September 16, 2016, meeting at the Oregon State Capitol, room 50.

The big question is, "What will California do?" Some policymakers in California want to revisit the pay-at-the-pump model first piloted in Oregon. Their reasoning is that a significant number of Californians do not consistently pay accounts and, therefore, a transaction-based system would work better for this group. To bolster the account-based system recently tested in the CRCPP, California may examine a hybrid arrangement where payment at the pump is a payment option but not a requirement. Following the state's successful road charge pilot program and investigation of payment at the pump, the state will have sufficient information to enact and implement a viable distance charge program beginning as early as the 2018 legislative assembly, although recent enactment of a comprehensive transportation funding package may calm the urgency to more forward quickly on a distance charge program.

18.11 Strategic engagement and political acceptance

The distance charging efforts in Oregon, California and Washington all proceeded with an extensive strategic engagement process with lawmakers and the public. No effective planning for distance charging should go forward without the involvement of lawmakers. No mandate to charge by distance traveled can go forward without the consent of the voting public. To convince lawmakers and the public of the viability and appropriateness of distance charging to fund road systems, strategic engagement with the public must take specific steps. These steps come in two stages.

The first stage involves *exploration* including policy development and technical research and development. The Pacific Coast states' distance charge efforts progressed during stage one by taking three steps. In step one, the government explains why exploring distance charging is a worthy endeavor. This entails communicating the nature of the problem being addressed. For distance charging, the explorers explain the effect of increases in vehicle fleet fuel efficiency on revenue generation from the volume-based fuel tax, the dire prospects for this revenue stream in the future and the impact of inadequate revenues on road travel. Lawmakers in Oregon, however, report a more effective rationale for convincing the public of the need to find a new revenue source for roads: the growing inequity of the fuel tax. These lawmakers' anecdotes indicate the public shows more favor toward distance charging as a replacement for volume-based fuel taxation when people understand the unfairness of one driver paying a lot for road use while another paying very little for the same use and that this unfairness is getting worse as new vehicles become more fuel efficient.

In step two, the government takes into account particular sensitivities of their citizens in order to inform the design of a distance charge system. Though similar, the Pacific Coast states' distance charge systems are not identical. Consideration of the sensitivities of the citizenry in program design put these programs on a natural pathway to resolution of issues identified by the public.

In step three, the government presents an actual application of the distance charge system in the form of a demonstration, pilot, trial or test. To achieve public acceptance, people need to experience something similar to the system proposed for implementation. In places where trust in government lags, obtaining knowledge and understanding eases fears. To aid this process, communications with pilot participants and the general public occurs throughout pilot operations.

For the second stage of strategic engagement—*achieving a mandate*—an even more extensive public outreach and education process will be necessary. Trials are one thing; a legislated mandate is on another scale. To pass a distance charge mandate, lawmakers must feel comfortable enough that the public will consent to legislative action. To achieve this level of comfort, distance charge supporters must pursue a public information campaign to dispel myths and inform the voting public of the proposed action and the rationale for taking it. The recent legislative enactments by Congress [39] and the Oregon and California legislatures [12,33] and several states' FAST Act grant applications [18–20,27,36–38] indicate such informative activities for distance charging have begun to take place in the United States.

In the 2015 federal transportation reauthorization legislation [39], the United States Congress pressed the importance of public involvement for researching user-based funding alternatives. In the STSFA program, Congress set forth as a key objective increasing public awareness of the need for alternative road funding methods and informing the public on alternative approaches. The US Federal government granted awards to Oregon and California specifically to fund public awareness and education activities to present distance charging as an alternative to the fuel tax for road funding.

As the nation's most advanced distance charging program, Oregon's public awareness and education campaign illuminates one approach to the second stage of strategic engagement for advocates seeking a distance charge mandate. Oregon's campaign proceeds, again, in three steps [18].

Oregon first seeks to *map the path to acceptance* by analyzing equity impacts and local attitudes in two ways: first, a statewide research and education tour addressing particular populations in key demographics: urban and rural, affluent and low income, education and age, and second, surveys and focus groups throughout the state assessing localized opinions of Oregon's operational distance charge program. These methods identify local differences in perceptions of distance charging, ascertain the character of objections and identify points of support. Oregon will also test proposed messages and communications tools for relevance and impact.

With a deeper understanding of the cognitive process to acceptance, Oregon officials will take the second step to *design the education* program by mapping the process and developing the education program. Oregon's education program will retool the existing ORe*GO* communications media (website, social media, blog, emails lists, presentations, etc.) to provide personally relevant and compelling messaging and develop new, interactive education media (video, online modules, games and apps) to engage a diverse range of learning styles. Oregon will also train speakers for statewide outreach. The third step is implementation and promotion of the education program statewide. This federally funded communications program began in 2017 and will continue into 2018.

18.12 Conclusion

Now known are the technologies, systems and many of the appropriate policies for adopting and implementing a fully functioning, operational distance charge system for light vehicles in the United States, alongside emergence of a supporting commercial market. As this chapter notes, many US states are headed in this direction especially in the western part of the nation. The new US federal grant program greatly assists the states in these investigations.

The next steps toward a distance charge mandate in the United States are, in essence, political. As the general public and policymakers in the United States gain more intimate knowledge of distance charge models, political support should improve (provided the public continues to accept the longstanding user pays policy for road funding). If during this process, discovery of additional helpful technological, systemic or policy measures occurs, states will adopt them and further improve the chances for adoption of a mandate for distance charging in the near term. Widespread, mandated distance charging in the United States is not that far away.

References

[1] Forkenbrock, David J. and Kuhl, Jon G., *A New Approach to Assessing Road User Charges*. Transportation Policy Research. Iowa City: University of Iowa Public Policy Center, 2002. Available from: http://ir.uiowa.edu/cgi/viewcontent.cgi?article=1005&context=ppc_transportation. [Accessed 22 July 2016].

[2] Puget Sound Regional Council, *Traffic Choices Study – Summary Report*, April 2008. Available from: http://www.psrc.org/assets/37/summaryreport.pdf. [Accessed 18 July 2016].

[3] Whitty, James M., *Oregon's Mileage Fee Concept and Road User Fee Pilot Program Final Report*, November 2007; Available from: http://www.oregon.gov/ODOT/HWY/RUFPP/Pages/rucpp.aspx. [Accessed 19 July 2016].

[4] Paul Hanley and Jon Kuhl, Public Policy Center, University of Iowa, *National Evaluation of a Mileage-Based Road User Charge*, 2011. Transportation Research Record: Journal of The Transportation Research Board, 2011. Available from: https://utcm.tamu.edu/mbuf/presentations/pdfs/Kuhl-UIowa-Eval.pdf. [Accessed 22 November 2017].

[5] Nevada Department of Transportation, *Nevada Vehicle Miles Traveled Fee Study: Program Summary*, June 2015. [unpublished].

[6] Minnesota Department of Transportation, *Connected Vehicles for Safety, Mobility, and User Fees: Evaluation of the Minnesota Road Fee Test*, February 2013. Available from: http://www.dot.state.mn.us/mileagebaseduserfee/pdf/EvaluationFinalReport.pdf. [Accessed 22 July 2016].

[7] Whitty, James M. and Capps, Darel F., *Road Usage Charge Pilot Program 2013 & Per-Mile Charge Policy in Oregon*, May 2014. Available from http://www.oregon.gov/ODOT/HWY/RUFPP/Pages/rucpp.aspx. [Accessed 20 July 2016].

[8] Oregon Department of Transportation, *Road Usage Charge Pilot Program* web page, 2016. Available from: http://www.oregon.gov/ODOT/HWY/RUFPP/Pages/index.aspx. [Accessed 20 July 2016].

[9] California Department of Transportation, *California Road Charge Pilot Program* web page, 2016. Available from: http://www.dot.ca.gov/road_charge/. [Accessed 13 September 2016].

[10] Schaller, Bruce, *New York City's Congestion Pricing Experience and Implications for Road Pricing Acceptance in the United States*, Transport Policy 17, 2010, 266–273. Available from: https://www1.nyc.gov/html/dot/downloads/pdf/schaller_paper_2010trb.pdf. [Accessed 28 November 2016].

[11] Whitty, James M., *Road User Fee Task Force Report to the 72nd Oregon Legislative Assembly*, March 2003. Available from: http://library.state.or.us/repository/2011/201109060934082/www.oregon.gov_ODOT_HWY_OIPP_docs_FinalReportP2003march.pdf [Accessed 19 July 2016].

[12] Oregon Laws 2013, chapter 781. Available from: https://www.oregonlegislature.gov/bills_laws/lawsstatutes/2013orLaw0781.pdf, 2013.

[13] Washington Road Usage Charge Steering Committee, *Briefing Book*, 9 November 2016. p 10; Available from: https://waroadusagecharge.files.wordpress.com/2016/11/wstc-ruc-sc-meeting-nov-9-briefing-book-20161104_v04c.pdf. [Accessed 14 November 2016].

[14] California Road Charge Technical Advisory Committee, *Road Charge Pilot Design Recommendations*, November 2015. Available from: http://www.catc.ca.gov/meetings/Committees/Road_Charge/Road_Charge_Pilot_Design_Recommendations_121115.pdf. [Accessed 8 August 2016].

[15] Oregon Department of Transportation, *OReGO* web page, 2016. Available from: http://www.myorego.org. [Accessed 14 November 2016].

[16] Western Road Usage Charge Consortium, *RUC West* web page, 2016. Available from: https://www.rucwest.org/. [Accessed 16 November 2016].

[17] Oregon Department of Transportation, *OReGO First Year Highlights*, 1 July 2016. p 10. Available from: https://www.oregon.gov/ODOT/HWY/RUFPP/Sept%202016%20Meeting%20Materials/ItemB2_OReGO-RUCFirstYearHighlts.pdf. [Accessed 22 November 2016].

[18] Oregon Department of Transportation, 'ODOT RUC Program Enhancements to Improve Functionality, Public Acceptance, and Interoperability'. Presented to the US Federal Highway Administration as STSFA Grant Application (Opportunity Number: DTFH6116RA00013), 2016 (unpublished).

[19] Ortega, Norma, 'California Road Charge Pilot Program Update'. Presented to the Tri-State Commission Meeting of California, Oregon and Washington State Transportation Commissions, 17 June 2016, slide 12 (unpublished).

[20] Washington State Department of Transportation, 'Washington State Road Usage Charge Pilot Project'. Presented to the US Federal Highway Administration as STSFA Grant Application (Opportunity Number: DTFH6116RA00013), 2016 (unpublished).

[21] Oregon Revised Statutes §367.806(2). Available from: https://www.oregonlegislature.gov/bills_laws/ors/ors367.html, 2015.

[22] Oregon Revised Statutes §319.900(2)(d)(A). Available from: https://www.oregonlegislature.gov/bills_laws/ors/ors319.html, 2015.

[23] Oregon Revised Statutes §319.900(1). Available from: https://www.oregonlegislature.gov/bills_laws/ors/ors319.html, 2015.

[24] Oregon Revised Statutes §319.900(3). Available from: https://www.oregonlegislature.gov/bills_laws/ors/ors319.html, 2015

[25] Oregon Revised Statutes §319.900(2)(c). Available from: https://www.oregonlegislature.gov/bills_laws/ors/ors319.html, 2015.

[26] Colorado Department of Transportation, *Colorado Road Usage Charge Pilot Program website*. Available from: https://www.codot.gov/programs/ruc/ruc. [Accessed 26 December 2016].

[27] Hawaii Department of Transportation, 'Hawaii Road Usage Charge Pilot Project'. Presented to the US Federal Highway Administration as STSFA Grant Application (Opportunity Number: DTFH6116RA00013), 2016 (unpublished).

[28] The National Surface Transportation Policy and Revenue Commission, *Transportation for Tomorrow*, December 2007. Available from: http://www.transportationfortomorrow.com/final_report/index.htm. [Accessed 3 August 2016].

[29] The National Surface Transportation Infrastructure Financing Commission, *Paying Our Way*, February 2009. Available from: http://financecommission.dot.gov/Documents/NSTIF_Commission_Final_Report_Advance%20Copy_Feb09.pdf. [Accessed 3 August 2016].

[30] Sorensen, Paul, Rand Corporation, National Cooperative Highway Research Program, *System Trials to Demonstrate Mileage-Based Road Use Charges*, October 2010. Available from: http://www.ftc.state.fl.us/documents/presentations/nchrp_w143.pdf. [Accessed 23 November 2016].

[31] Oregon Revised Statutes §319.915. Available from: https://www.oregonlegislature.gov/bills_laws/ors/ors319.html, 2015.

[32] Oregon Administrative Rules §731-090-0010. Available from: http://arcweb.sos.state.or.us/pages/rules/oars_700/oar_731/731_090.html, 2017.

[33] California Senate Bill 1077, section 3091(c), (Chapter 835, Statutes of 2014).

[34] Federal Highway Administration, August 2016 Press Releases. Available from: http://www.fhwa.dot.gov/pressroom/fhwa1648.cfm. [Accessed 16 September 2016].

[35] Minnesota Department of Transportation and University of Minnesota, Humphrey School of Public Affairs, 'Minnesota's Pre-Deployment Proposal for a User Based Fee Demonstration'. Presented at the Mileage Based User Fee Alliance Quarterly Meeting, 20 June 2016.

[36] Missouri Department of Transportation, 'Request for Qualifications: Surface Transportation System Funding Alternative Revenue Strategy'. Issued 23 December 2016.

[37] Western Road Usage Charge Consortium, 'Regional Road Usage Charge System Definition and Pilot Planning Project'. Presented to the US Federal

Highway Administration as STSFA Grant Application (Opportunity Number: DTFH6116RA00013), 2016 (unpublished).

[38] I-95 Corridor Coalition, 'Mileage-Based User Fees in a Multi-State Region'. Presented to the US Federal Highway Administration as STSFA Grant Application (Opportunity Number: DTFH6116RA00013), 2016 (unpublished).

[39] Fixing America's Surface Transportation (FAST) Act (Pub. L. No. 114-94). Available from: https://www.congress.gov/bill/114th-congress/house-bill/22, 2015.

Chapter 19
Four years of Milan's road charge: effectiveness, acceptability and impacts

Paolo Beria, Luca Tosi**, and Davide Nuccio***

19.1 Introduction

The city of Milan had introduced a form of cordon pricing to enter the city centre already in 2008. Originally, it was conceived as a pollution charge, aimed at reducing the polluting vehicles going into the most delicate area of the city, and only secondarily affecting the traffic volumes. That policy, whose effects declined over the years due to the fleet turnover, was substituted in 2012 with a genuine congestion charge, called 'Area C'. The substitution came after a city referendum in June 2011.

The chapter discusses the history of the policy and summarises the main outcomes, 5 years after implementation, in terms of traffic reduction, emissions and impact on user groups. A section is devoted to the description of the technology used and of the architecture of the system, entailing real time data exchange with the vehicle registry, the urban police and a number of payment systems. The chapter concludes with a discussion on the motivations behind the (unexpected) public and political acceptability of the policy and proposes some hints to transfer its experience to other cases.

19.2 Milan's road pricing

Milan is the second largest city in Italy and the core of the largest economic and productive area of the country. Its geography is characterised by a large plain, namely the Po river valley, surrounded by Alps in the North and West and by the Apennines in the South. Eastward, the plain ends into the Adriatic sea in Venice. The city boundary includes about 1.3 million inhabitants, but the whole metropolitan area spreads for dozens of km, especially northward, and includes, according to various estimations, 6 to 8 million inhabitants.

*Politecnico di Milano, Italy
**AMAT (Agenzia Mobilità Ambiente e Territorio), Italy

Table 19.1 Main characteristics of Ecopass [22] and Area C [4]

	Ecopass	Area C
Type of toll	Daily cordon toll, allowing unlimited entrances during the day of the first entrance	
Toll classes	5: according to engine emission class (0 to 10€/day)	3: residents, registered commercial vehicles and others (2€, 3€, 5€)
Toll payment	Single prepaid, multiple entries prepaid (40% to 50% discount), yearly pass for residents, post-paid (within the next day)	Single prepaid, multiple entries prepaid (no discount), single post-paid (+10 € surcharge if paid within 1 week, or a larger fine if later)
Primary purpose	Pollution reduction	Congestion reduction
Technology used	Automatic number plate recognition (ANPR)	
Revenues	14.2 M€/2008, excluding fines	20.3 M€/2012, excluding fines

Due to the combined effect of large population, lively economic activities and enclosed geography, the area is characterised by high pollutant concentration. Over the years, the administrations have taken general measures to cope with the problem, mainly the development of an extensive metro and urban rail network [6]. However, given the already extensive use of public transport, further significant improvements may mainly come from the evolution of the car fleet.

To push the fleet evolution, in 2007, the city implemented a policy measure, called 'Ecopass', started experimentally on January 2nd. It was explicitly aimed at introducing the 'polluter pays' principle through a daily entrance toll to be paid by vehicles entering the city central area from Monday to Friday and from 7.30 a.m. to 7.30 p.m. [5]. The toll was differentiated according to the Euro emission class of the vehicle: low emission vehicles entered for free (natural gas, LPG, electric, hybrid), older than Euro 0 vehicles were banned, while intermediate engines paid a toll ranging from 2 to 10€. Table 19.1, left column, includes the main characteristics of the toll [22]. Of course, as a secondary effect, Ecopass acted as a rough congestion charge, reducing car use in the area [4].

With respect to the city of Milan, and moreover to its metropolitan area, the priced area is very small – just 8 square kilometres – and corresponding to the XVI century city walls. Inhabitants nowadays are relatively few (90,000), but the area hosts 27% of total Milan employees (2011 census) with average densities during the working hours up to 40,000 persons/km^2. Also from the commercial viewpoint, the area is the core of Milan, with 23% of total commercial activities [2]. Figure 19.1 gives, in a glimpse, the extension of the priced area within Milan metropolitan area and its main infrastructure, while Figure 19.2 details the inner zone.

The Ecopass policy proved to be quite effective, in the short run and locally: emissions and traffic were reduced. Bedogni *et al.* [2] report – 16% traffic inside

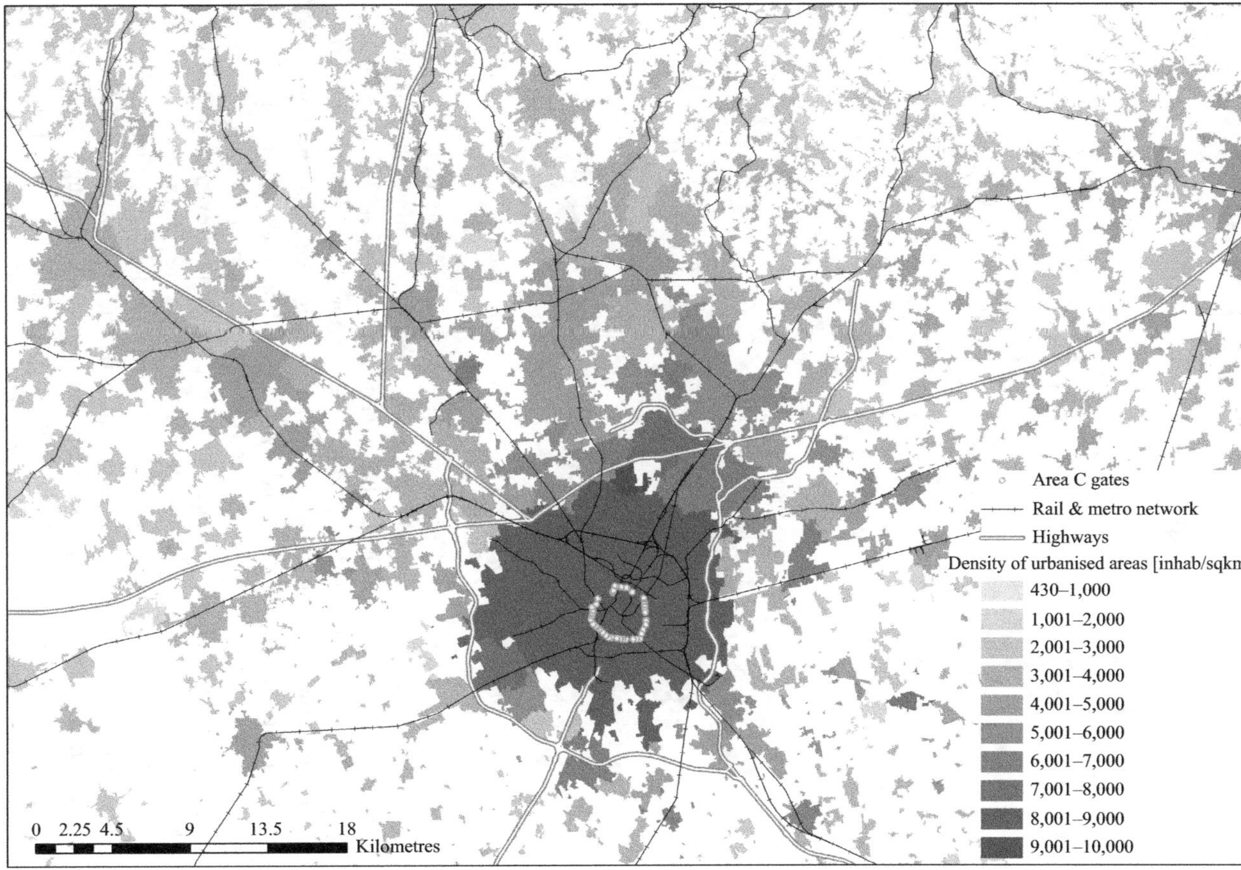

Figure 19.1 Area C gates and Milan metropolitan area
[Source: Our elaborations]

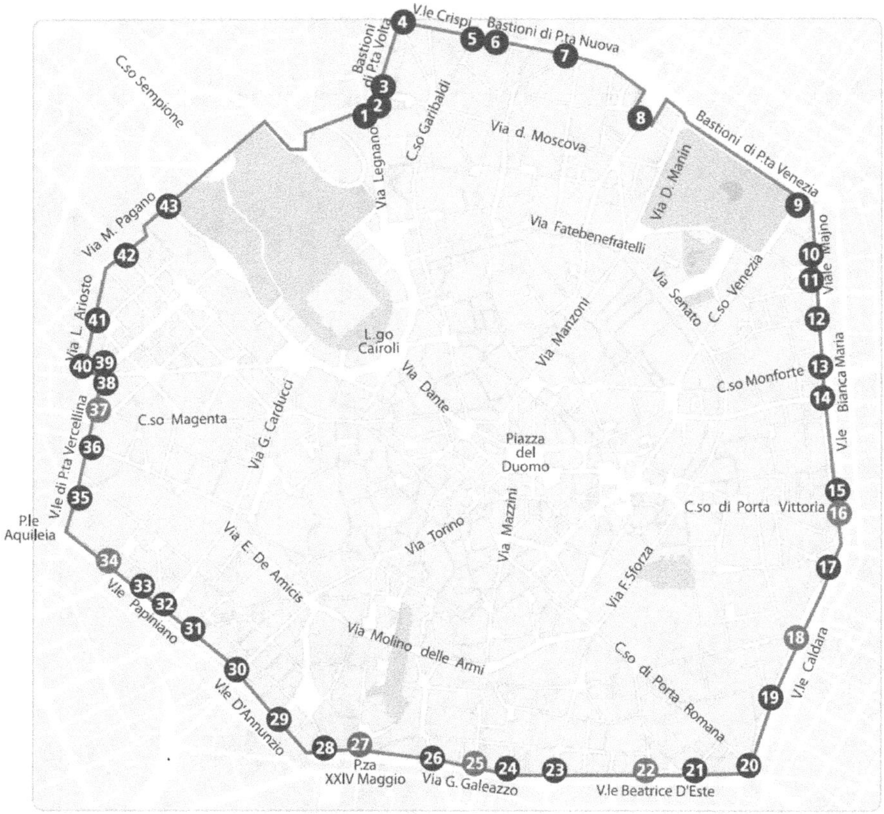

Figure 19.2 Detail of Area C zone and gates

the tolled area.[1] However, this effect decreased over time along with the fleet renewal that increased the number of exempt vehicles. For example, the total of PM10 particulates emitted inside the central area decreased from −23% during the first year [22] to −15% at the end [2]. Actually, the main mid-term effect was to stimulate the substitution of the older and more polluting vehicles: in 4 years, newer cars entering in the area increased by 478% and commercial vehicles by 1,400%, indirectly giving a benefit also elsewhere. In addition to declining effectiveness, another major problem was the lack of clarity on how revenues, initially 12.4 M€/year [22] and later reduced to 8–9 M€/year [5], were reinvested.

In 2011, Ecopass had been reconsidered at the end of the trial period. A technical–political commission was established by the right-wing city Mayor Letizia Moratti to evaluate the measure [5]. In the meantime, a group of associations and parties promoted a referendum on the topic, which took place on 12 June 2011. Quite surprisingly, before the referendum, nearly all political parties

[1] That, however, must be separated from exogenous effects such as other policies or an overall reduction of car mobility, resulting in a net −6% to −7%.

Table 19.2 Vehicle categories of Area C

	Paying categories	**Fare**
Residents	Vehicles registered as belonging to residents within the priced area	40 free, then 2€
Service vehicles	Vehicles registered as serving commercial activities within the area	3€
Others	All other vehicles	5€
	Exempted categories	
Authorised	Public vehicles (police, buses, ambulances, taxi, car sharing, etc.)	Free
Ecological vehicles	Private vehicles temporarily exempted (LPG, NG, electric, hybrid)	Free
Motorbikes	Motorbikes and 2-wheeled vehicles	Free
Unclassifiable	Vehicles not present in the database, typically foreign ones from countries without a number plate exchange agreement	(5€)

directly or indirectly supported the extension of the policy. Realistically, governing right-wing parties used the referendum to demonstrate their will to solve the pollution problem, while left-wing parties, traditionally more environmentally oriented, did not oppose [4]. We will come back to the referendum when discussing the acceptability of the measure in Section 19.4.2. In brief, the electoral result was clear, with an overall 'yes' rate of 79.12% and a good participation rate of 48%. Despite the differences, all city zones presented a positive vote, ranging from 71.7% to 89.2% [3].

The referendum, despite not being binding, provided the necessary political legitimacy to evolve the previous scheme towards a new, and possibly more effective, congestion charging scheme. The new left-wing mayor, Giuliano Pisapia, promoted a new ex-ante study analysing the possible policy extension scenarios [2] and setting the new policy goals.

The new measure was called 'Area C' to suggest the new target, namely 'congestion' reduction. While it applies exactly to the same zone and uses the same technological infrastructure of Ecopass, the charging structure was changed (Table 19.1, right column, and Table 19.2) coherently with the new goals. Older polluting vehicles, as well as trucks longer than 7.5 m, are banned, while newer ones pay an undifferentiated fare of 5€/day. Vehicles registered as belonging to the priced area obtained discounted fares: 3€/day for commercial vehicles, 40 free accesses per year and then 2€/day for the residents.

In the following section, we will discuss the effects of 4 years of Area C in terms of entrances, emissions, revenues and house prices.

19.3 Four years of Area C: consolidated effects

To show the additional effect of Area C with respect to the previous Ecopass regulation, we analyse a dataset consisting of five full years of traffic counts entering

Milan city centre. The area of Ecopass being exactly the same as Area C, as described above, we rely on fully consistent measures. The years of analysis are 2011, the last year of Ecopass and 2012–15, the first four years of Area C. We also have 4 months of 2016, used only in some elaborations.

The data measured is not limited to the number of vehicles, but we also know the type of vehicle (associating the plate number to Italian vehicle registry) and if it belongs to one of the pre-registered categories of Table 19.2, benefiting from exemption or discounts.

It is important to remember that the days of application of Area C are not always the same. In 2012, the pricing was applied for 196 days, excluding Saturdays, Sundays, the days before Christmas and between 1 and 16 January. In addition, starting in July, the state council enacted a 39-day forced suspension, later resolved in favour of the Municipality. The 2011 Ecopass and Area C after 2012 have similar application periods, around 230 days/year. Differences depend on specific days of suspension (e.g. in some years during Christmas or during public transport strikes). Normal application period is 7.30 a.m. to 7.30 p.m., except on Thursdays, when Area C is shortened to 6 p.m. This is the outcome of an agreement with commerce associations to reduce the possible impact on shopping, in recognition of special events organised by street associations.

19.3.1 Traffic reduction

To represent the entire period we use a weekly detail, as in Figure 19.3. Most of the negative peaks correspond to weeks 32–34 approx., which is August, when the city

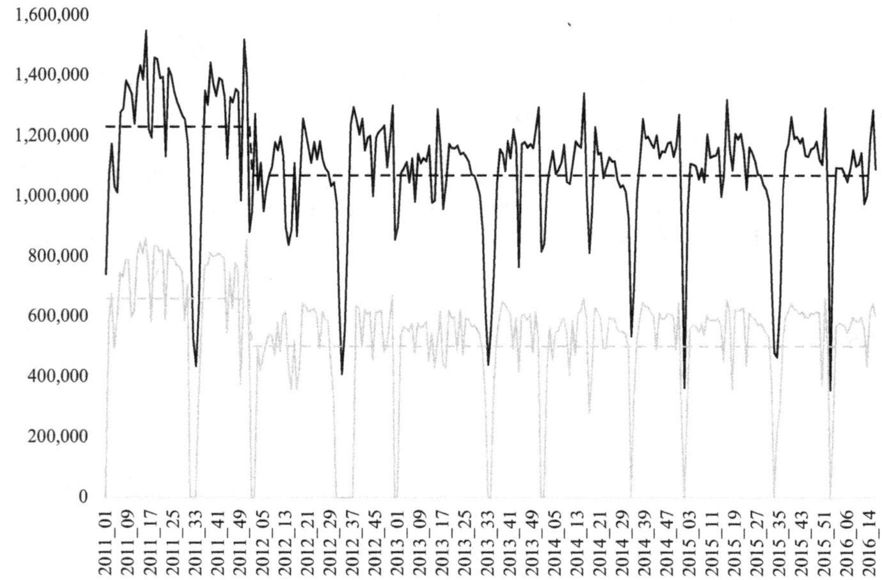

Figure 19.3 Weekly entrances in the priced area, 2011–15, whole year, all vehicles, both total (in black) and during priced hours only (grey)

empties due to concentrated summer holidays and Area C is suspended. In general, also end-June and July is an off-peak period because of school closures, as can be seen by declining trends from week 25 approx. Other low traffic periods are Christmas holidays, but this is due also to week split between years. In mid-2012 well visible is the longer suspension period between weeks 31 and 37.

The two series represented account for total entrances (black series) and for priced hours only (grey series). In both cases, we can easily notice a steep decrease exactly across 2012 New Year, when Area C came into force. Averages of 2011 and of 2012–15 roughly quantify the effect, consisting of about 160,000 vehicles less per week.

What is interesting is not only that the effect is very visible, but also that it lasted for the entire period, except a small increase in the very first months (already documented in [4]). In fact, we do not see a significant increase of entrances in the long run due to vehicle substitution or to a reconsideration of the actual burden of the fare by drivers: tens of thousands of trips have been permanently cut because considered, by the drivers, as substitutable or avoidable. We will come back to this in Section 19.4 when commenting on the acceptability of the measure. Second, the reduction in trips is almost the same in both series: there was not a significant shift from priced to unpriced hours, just to avoid the fare. Even if an entry peak is measured just after 7.30 p.m. (see Section 19.2.3), overall the total entries reduction is not modified significantly.

The total number of entrances is telling us the total effect, showing how many fewer vehicles have moved in the period. However, their interpretation might be misleading if we consider that the number of days per year in which the measure is active varies. For this reason, we estimate the average number of vehicles per day, only for the periods in which Area C was active.

Table 19.3 and Figure 19.4 show some important elements:

1. The reduction in terms of average daily entries is clear and is around 33–35 thousand vehicles, or 22%–23% of 2011 total daily figures. This effect would be even more evident, around 40%, if we limit to private cars excluding motorbikes, foreign cars, buses and other public vehicles.

Table 19.3 Average daily entrances 2011–15 (all vehicle types) and variations vs. 2011

	Days w. active policy	***Average daily entrances during restrictions***	***Variation vs. 2011 (%)***	***Average daily entrances other hours***	***Variation vs. 2011 (%)***
2011	230	149,109		51,816	
2012	196	116,255	*−22.0*	53,956	*4.1*
2013	229	114,094	*−23.5*	52,680	*1.7*
2014	226	116,227	*−22.1*	51,982	*0.3*
2015	241	114,998	*−22.9*	52,052	*0.5*
2016	76	115,881	−22.3	50,972	−1.6

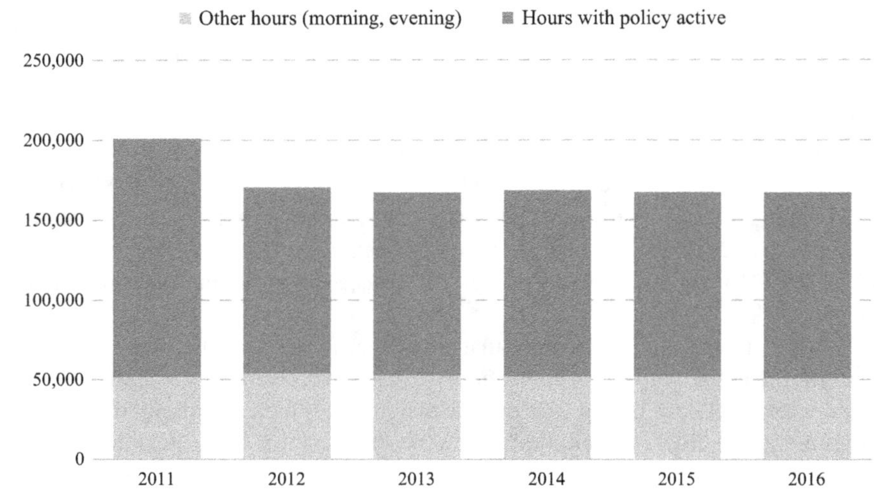

Figure 19.4 Average daily entrances 2011–15 (all vehicle types). Comparison of priced and unpriced periods of the day

2. This effect does not decline after the first year. Actually, the average entrances are slightly decreasing further.
3. The number of vehicles entering the area when there is no road pricing is rather constant: there was no shift to unpriced hours, but a reduction of car transport in favour of other areas or other modes. Apparently, there is also a small downward trend over the years.

In conclusion, we may affirm that the effect of Area C seen during the first year [4] is continuing after 4–5 years of policy application. The 33,000 trips per day reduction remain such and have not moved to other periods of the day.

19.3.2 Vehicle mix

The number-plate recognition system, used to verify the correct payment, is also a source of important information about the classes and fuels of the vehicles used.

The interpretation of Figure 19.5 requires two clarifications. The 2012 trips appear as reduced, but, as we already showed in terms of daily averages, this is not true. The lower total is instead due to about 40 days less of road pricing being active. The 2015, too, is a peculiar year, as Milan hosted the Universal Exposition, attracting about 21 million visitors in 6 months. In fact, the number of unclassified 'NC' vehicles is higher, probably belonging to foreign visitors and operators. Consequently, a more consistent comparison should mainly refer to 2011, 2013 and 2014.

Of course, paying vehicles dramatically reduced from being 62% of the total, to about 45%. Exempted ('authorised') vehicles such as buses, instead, remained basically the same and now are the majority. Residents' trips are about 9% of total and service vehicles' about 5%. In terms of total figures, they appear as slightly

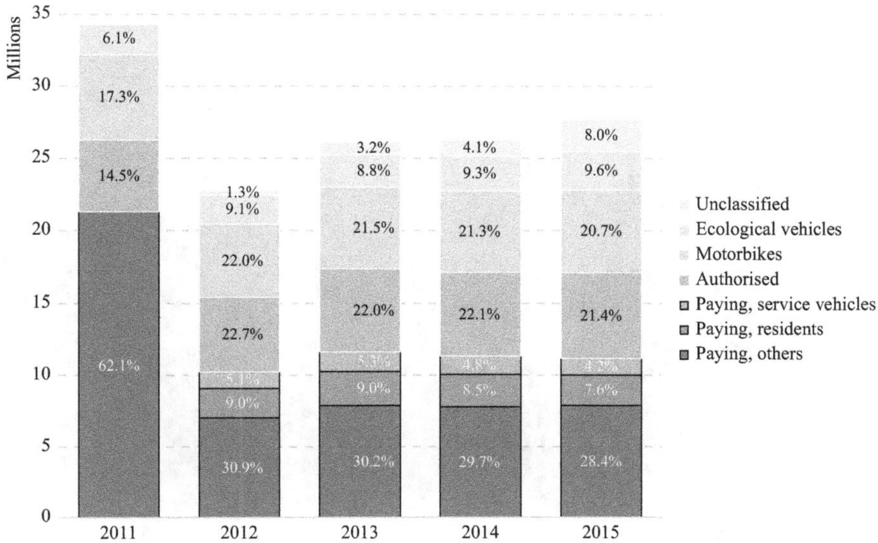

Figure 19.5 Classification of vehicles crossing the cordon during priced hours and days, 2011–15

decreasing, probably thanks to a change in residents' behaviour. Unregistered paying vehicles are around 30% and constant. Overall, compared to 2011, there are about 8 million entrances less per year.

Concerning other classes, public vehicles are not very interesting, being unaffected by the policy. Motorbikes did not increase: once again citizens did not search for an alternative private transport mode in substitution of car trips. Only 'ecological vehicles' increased, including natural gas, LPG and electric ones. The increase from 2 to 2.6 million per year is probably both due to normal vehicle substitution maybe increased somewhat by the exemption. This would be a marginally negative effect in terms of congestion, but good news in terms of emissions reduction. By the way, the exemption of 'ecological vehicles' is due to end in 2017 (originally from 2015, but extended in 2014).

We also have information about vehicle fuel and engines, as in Figure 19.6.

In the case of private cars, diesel and gasoline vehicles more than halved, with a sharper effect on gasoline ones falling from 44% to about 30%. This is because diesel vehicles, the majority of the fleet, were already heavily limited during the Ecopass period (up to Euro 3), while gasoline cars were relatively less so (Euro 0 only). Fully electric vehicles remain a very minor component of the fleet (0.3% in 2015), but entrances passed from 4,000 to 41,000/year, also thanks to an enlarging fleet of shared cars. Hybrid cars' trips rose from 132,000 in 2011 to 415,000 in 2015, and now account for 3% of the total. LPG and Natural Gas vehicles did not rise much because already very common in the fleet, and account for a significant 16% of total cars circulating in the city centre.

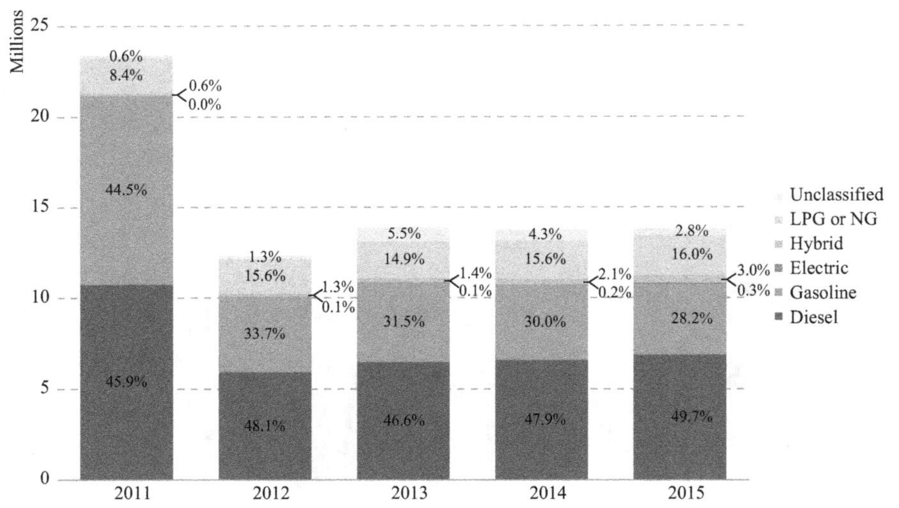

Figure 19.6 Fuel mix 2011–15 of vehicles entering during priced periods (excluding motorbikes, public vehicles, unclassified category)

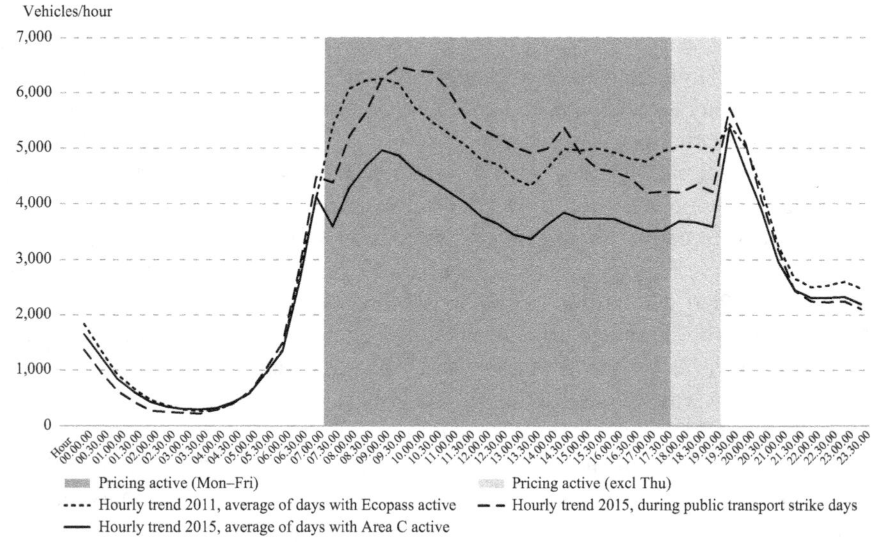

Figure 19.7 Comparison between 2015 and 2011 average entrances, per hour of the day

19.3.3 Hourly trends

The continuous, automatic measure of entrances in the traffic zone is always active and this lets us analyse the trend of entrances also outside priced hours. Figure 19.7 compares the entrances during 2011, when the former Ecopass was active, and

2015, when Area C is operative and consolidated. In addition, we separated the 'normal' days and the days in which there was a strike of public transport services and Area C was suspended.

Once again, the difference between 2011 and 2015 is clear and self-explaining. Before 7.00 a.m. and after 7.30 p.m., the two trends are almost the same. Although there is a small peak of entrance between 7.00 a.m. and 7.30 a.m. before payment is active, the figure of 2015 remains below the corresponding hour of 2011. The peak of entrances is now at 9.00 a.m., at a lower level than 2011, when the peak also started earlier, at 8.00. As soon as the pricing ends, at 7.30 p.m., there is the highest peak of entrances, but 2015 is not higher than 2011. This points out once again, that Area C does not simply push travel out of priced hours more than it was already before. Moreover, it must be noted that the congestion-effect of this counter-flow of vehicles is less problematic because entering into an area that is becoming emptier as people return home after work.

19.3.4 Emissions

Area C is not primarily intended as a pollution charging measure. In fact it does not discriminate among emissions classes except forbidding the entry to older and polluting vehicles and temporarily exempting alternative engines from the payment.

However, the net reduction of entries and travelled distances obviously induced a reduction of local emissions. The estimated [1] reduction of particulate matters (PM10) is of −18%, ammonia −42%, nitrogen oxides −18%, carbon dioxide −35%. The largest effect is related to the concentration of Black Carbon inside the area with respect to the outside: depending on weather conditions, we have seen a decrease of −52% (Sept) to −32% (Oct).

19.3.5 Revenues and reinvestment

One of the main criticisms of the former Ecopass pricing was the lack of transparency regarding the use of collected revenues. The new city administration promised more transparency on all aspects concerning Area C and some figures were actually made available, both for tolls and fines. Table 19.4 summarises the main figures available, which appear as rather stable.

Table 19.4 Main financial figures for Area C, 2012–14

[M€]	2012	2013	2014
Running costs	7.1	6.0	6.0
Revenues*	20.3	29.4	24.0
Fines†	Not available	21.4	Not available

*The variability of revenues in 2013 and 2014 depends on a change in accounting, but the actual difference is smaller.

†Fines have decreased from 760,000 in 2012 to 660,000 in 2013 and 570,000 in 2014. A precise calculation of revenues is not available because not all fines are actually paid and if fines are paid within 5 days they are reduced by 30%.

It is worth mentioning that the investment and running costs of the system were relatively low. Concerning the investment, it was reduced as gates, software and other investment of Ecopass were just adapted from the already existing traffic control centre. Also, running costs are low compared with other cities such as London[2] (and moreover because they include all costs of the traffic control centre). In fact, they amount to about 25% of revenues excluding fines (2014).

Given the importance of the redistribution [14], in 2012, the Administration spent the entire net revenues for two projects, namely €10 million to finance a public transport supply increase and €3 million to install 46 new bike-sharing stations. Later, revenues were reinvested to partially cover the running costs of Metro line 5, recently opened.

19.3.6 Housing market

The introduction of Area C is too recent to have a measurable effect on house prices, especially considering that it occurred in a period of global crisis on all housing markets. However, a couple of studies had checked the effect of Ecopass on purchase and rent prices in Milan.

Percoco [20] analyses empirically the house prices in Milan between 2007 and 2012, adopting a simple model of a monocentric city in which the housing rent is a decreasing function of the distance from the Central Business District. Actually, even if slowly changing, Milan can be defined as a monocentric city and its Central Business District has historically been matched with the city centre, making the simplification acceptable. The model assumes the variation of house prices due to the implementation of a charging system as depending on two opposite factors: the increase of private transport costs, which reduces the house prices, and the increase of quality of the environment, potentially increasing prices. In the short period analysed, and with the limitations of the case and data, he finds a negative impact of Ecopass, quantified in a decrease of 60 to 180 €/sqm of house prices inside the area, or 1.2% to 1.8% less.

Similarly, D'Arcangelo and Percoco [7] studied the rent prices. Applied to the same period and dataset, they estimate instead an increase in rent prices of about 0.75% with respect to before Ecopass conditions.

Our interpretation of these apparently counterintuitive results is that, when buying a house – a typical long-run investment – the potential lack in accessibility plays a larger role in decreasing the willingness to pay, than the benefits of a better urban environment. To the contrary, renters can more freely chose the house to rent and those opting for the expensive central locations are more positively influenced by the positive characteristics of the location itself, including the (improved) quality of the neighbourhood. In conclusion, the effect of road pricing on house prices exists, but its magnitude is limited to a few percentage points and going in opposite directions between purchase and rentals.

[2] London running costs initially were about 42% of revenues.

19.4 Technological and organisational architecture

19.4.1 The technological infrastructure

The technology used is based on a conventional OCR plate recognition. The cordon is controlled by 43 gates, each provided with a double camera. One does the plate recognition and is provided with an infrared illuminator. The second one registers the context and can be used for the fines and for general traffic control. The OCR camera populates a database of transits across the gate, provided with a timestamp, which is transmitted to the rest of the system to verify the payment. The cameras do not record the type of vehicle, as it is derived from the vehicle registry (see below).

19.4.2 The informative architecture

The system relies on a complex architecture, interfacing the cameras with the payment system, the vehicle registry and the municipal police bureau which issues the fines. Figure 19.8 illustrates the system, structured into three separated components, or flows.

A first component is the one managing the exemptions. As described before, the residents within the tolled area and the vehicles associated to commercial activities enjoy discounts and free entries. Taxis and other public vehicles are exempted. The vehicle owner must pre-register the vehicle plate number and the

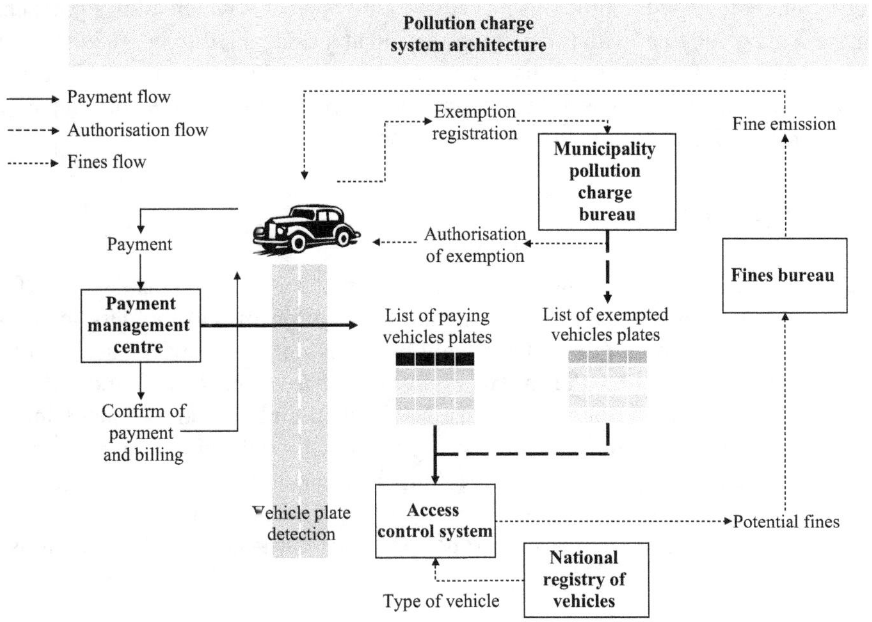

Figure 19.8 Schematic architecture of Area C system
[Source: Our elaborations]

pollution charging bureau checks if it is entitled to the discount or the exemption and populates a list of plate numbers.

A second flow involves the payments. Independently from the trip, the user purchases a transit 'token' paying the 2/3/5 € toll. The tokens are stored in a database and associated with the plate number of the purchaser. Multiple entries can be purchased in advance.

The third component of the system collects the plates from the gates, and compares them with the list of paid tokens together with the list of exempted vehicles. If the plate number does not have an active token or is not included in the exempted list, a potential fine is transmitted to the fines bureau of the municipal police. Similarly, all plates are cross-checked with the national registry of vehicles to check the type of engine and the emission class. Also in this case, if the vehicle falls out of the allowed classes, a potential fine is issued. After a check by the municipal police, the fine is issued to the user and enters in the usual credit collection process.

19.4.3 Payment systems

In order to ease the payment of the toll, a number of payment channels have been activated during the first year of operation. Tickets can be purchased at newsstands, or via website, call centre or park metres by credit card. The ATMs of one important bank also allow the direct payment. In addition, two further alternatives exist. On the one side, all 'Telepass' users, namely the automatic payment used for highways at the national scale, are active by default and the payment is done autonomously without the need to activate anything. Second, the municipality has signed agreements with numerous private parking operators which allows payment for the Area C together with the parking fee and at a discounted price of 3€. This is both convenient for the occasional users (e.g. foreigners visiting the city and not aware of the toll) but also useful to support private parking inside the cordon which experienced a decrease in patronage due to the toll.

19.5 Acceptability

Area C can be considered a successful policy: it met the goals it was designed for and, quite surprisingly, survived the natural opposition of stakeholders towards anything that used to be free and suddenly becomes tolled. We have cases of pricing policies aborted after the initial proposals (for example New York and San Francisco in the United States, and West Midlands, Edinburgh and Manchester in the United Kingdom) due to opposition. To date, the only cities adopting road pricing are London, Stockholm, Gothenburg,[3] Milan, four Norwegian cities and Valletta [4]. Other cities, such as Rome [10], have similar policies, but usually based on monthly or yearly passes and not on single entrance tolls. Having shown

[3] Gothenburg is particularly interesting in that the scheme was introduced by the Local Authorities without any consultation; citizens had the right to call a consultative referendum, which they did, and voted to remove the scheme. But the Local Authorities decided to keep it, because it leveraged additional transport funding from central government; but they stated that they would investigate other methods of traffic management to see if they could be equally effective. For more details see Chapter 14.

in the previous sections the effects of Area C and how it works, we can now move to discuss its acceptability.

19.5.1 The acceptability of pricing policies in the literature

The factors influencing or determining the acceptability of urban road pricing are numerous and broadly studied in the literature. At the root, there is the value citizens give to individual car use (independence, status) and the related frequent use, and consequently the difficulty to change the car users' behaviour. The factors are either exogenous, belonging to the environment in which the pricing is set, or endogenous, depending on how the policy is designed and presented. Table 19.5 summarises these factors, also mentioning some relevant literature on them.

19.5.2 The results of the 2011 referendum

The switch between Ecopass and Area C was preceded by a local referendum, which took place on 12 June 2011. The referendum is particularly important because it is one of the few which took place after a test period and not before the introduction of a new policy. This means that citizens already knew what would be the impact on them of the new policy, based on the previous Ecopass experience.

Table 19.5 Factors influencing acceptability of urban road charging schemes

Perception of the problem [8,11,13–15,19]
The existence and the perception of the problem (congestion, environment) by the involved citizens; the knowledge of available options to face the problems; the perceived/expected effectiveness and efficiency of the proposed measures; the effectiveness of communication campaigns
Social norms [23]
The existing mobility-related social norms, such as the attribution of responsibility to the citizens; the attitude and beliefs of citizens related to the environment and other externalities; the perceptions of freedom and fairness of the measure; equity issues in general
Socio-economic environment [12]
Numerous socio-economic variables (males vs. females; rich vs. poor; singles vs. families, etc.); the relative proportion of car users vs. transit users
Design of the scheme [8,13,14,18,19]
The features of the system (simplicity, etc.); Cordon vs. distance based charging; the level of charge; the contextual improvement of existing public transport; the promise of revenue hypothecation (typically reinvestment in transport or lowering local taxes); the existence of discounts for excessively affected categories
Dynamics [16,21]
The timing of the survey (before or after the policy is introduced): design idea, implementation phase, ex-post
Spatial aspects [9,12,17,21]
Land use; availability of alternative modes of transport

Consequently, the referendum can be seen as a field experiment providing a dataset of revealed preferences, which can be used to study which were the aggregate determinants of the vote.

Beria and Boggio [3] analyse the forces behind the vote for the Ecopass extension, with the help of a linear regression model. They run two models, one explaining the referendum result per electoral precinct (%yes) and the other the referendum result weighted with participation (%yes·%participation). The independent variables tested refer to the 286 electoral precincts and belong to all categories mentioned above, except those referring to single individuals. They were demographic (average income, proportion of women, proportion of foreign inhabitants, proportion of families with children), geographic (distance and distance squared to centre), political (political and environmental orientation of the zone), accessibility related (public transport supply, density of bus stops) and land-use related (density, share of offices, commerce, industry and services).

The question of the referendum was far from clear. The 'title' was:

> Do you want the Municipality of Milan to adopt and realise a plan of interventions to empower the public transport and the 'clean' mobility alternative to cars, through the extension to all vehicles (excluding the zero emissions ones) and the progressive broadening of the cordon charging, with the goal of halving traffic and pollution?

Followed by a list of specific measures (e.g. doubling of pedestrian areas, extension of night metro opening) to be financed by the extension of the cordon and parking pricing. However, the preceding debate made clear to the voters that the referendum was a yes/no option about the extension in space and scope of the existing Ecopass pricing [3].

As already commented, the result was very positive: 'yes' were 79.12% and participation was much higher than the usual local referenda, at 48%.[4] Nevertheless, significant differences occurred among zones of the city, as in Figure 19.9.

The results confirm the previous intuitions of the literature. The ecological and political attitude of citizens is very relevant. The mix between socio-demographic characteristics and land-use also matters: income and office density in the area are slightly positively related with the acceptance of charging measures. Poor, remote and residential districts instead suffer more from the charging and, conversely, have no direct benefits. Public transport accessibility (in terms of metro, bus and tram supply) had also a role in explaining the results, to some extent also visible in the picture, but only outside the cordon, and distance influence is U-shaped, with a minimum at about 4.5 km from city centre[5] (Milan compact built area radius is about 6–7 km) [3]. The intuitive explanation is that the more a zone is accessible by

[4] For a comment on the 'limits' of the referendum results, see [3]. For example, it was not open to citizens of the surrounding cities and was associated to four national referenda, for which large communication campaigns were conducted, like the one about nuclear power.

[5] In other words, acceptance is non-linearly related to distance from the charged zone, with nearby and further out respondents more in favour than those living around 4.5 km away from the centre.

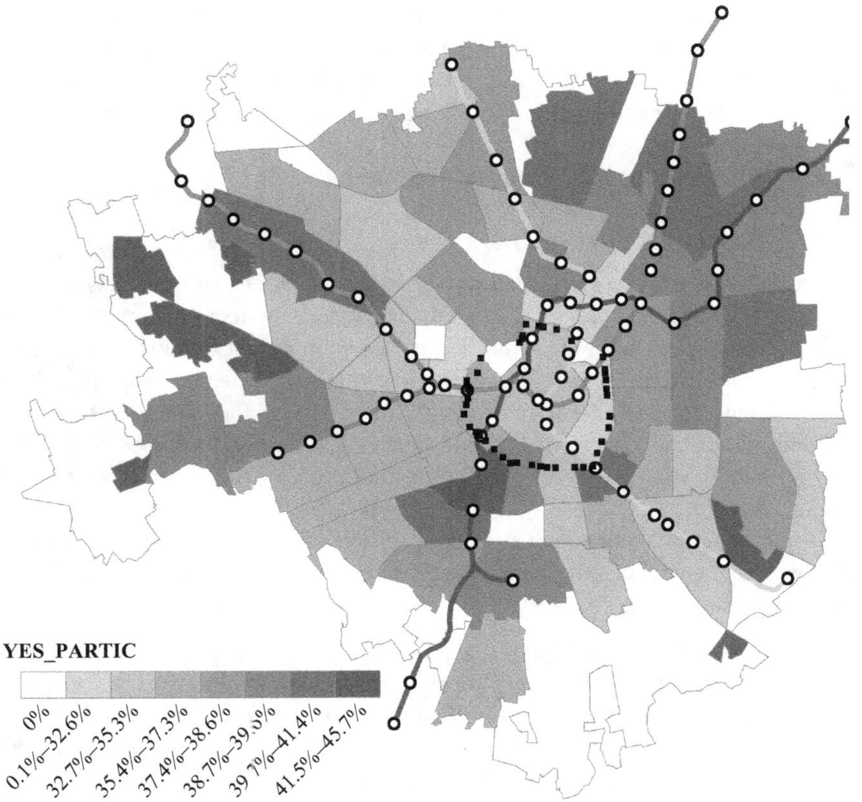

Figure 19.9 Ecopass referendum yes rate, times participation rate, by zone, with metro lines and stations
[Source: [3]]

public transport, the more the inhabitants will be prone to accept the road charging because they have realistic alternatives to the car. At the same time, the farther out a zone is, the less it will be dependent on the priced city centre.

19.5.3 Monetary impact on user groups

The local support (or non-opposition) to the road pricing is somewhat surprising, especially when compared with other cities where there was no chance even to test the policy.

Beria, in [4], gives an explanation of acceptance based on the quantification of the actual monetary burden paid by city users. He finds that only a small percentage of single cars pay, for example, more than 1€/day on average (4% of residents, 3% of other cars, 13% of service vehicles), both thanks to the discounts and to the fact that the accessibility to the priced area is, by far, much easier by public transport. In other words, out of the 1.01 million single vehicles which crossed the Area C gates during 2012, just a few thousand did it many times and consequently paid appreciable amounts of money. Put simply, car users in Milan city centre are, typically, occasional and consequently just marginally influenced by the toll. Differently from other cities, Milan citizens *already knew* such facts, because they had already been experiencing a similar pricing, Ecopass, for 5 years before.

19.6 Conclusions and lessons learned

19.6.1 Summary of results

After some years of Area C, it is interesting to draw a picture of its effects, benefits, costs and to wonder why it has been well accepted by the local community despite being an (efficiency) tax. Moreover, it is not only important to know the effects of the policy, but also, if these effects are going to last or if people adapt their behaviour, thereby decreasing (or increasing) the effect of the policy. As we have indicated above, the effects of Ecopass declined as drivers switched to less polluting vehicles. Area C is more 'stable' in this respect, with its effects remaining constant over time.

In the paper we have analysed, using official traffic counts provided by the monitoring system, four full years of application, compared to the year before the new policy entered into force. Here the main facts found are

1. Traffic reduction is of 33,000 fewer average entrances per day, or 22%, compared to 2011. These trips were not shifted to other periods of the day: unpriced hours figures did not change significantly. Similarly, alternative private modes such as motorbikes did not increase. So, trips were avoided or moved to public transport or non-motorised modes, in line with policy goals.
2. This effect has not been temporary: 2012 figures are very similar to those of the following years.
3. Now, the majority of vehicles entering the area are public (buses, taxies), authorised (police, ambulances, disabled, etc.) or 'ecological'. The two discounted categories of residents and vehicles associated with commercial activities stabilised around 13% of total entrances.
4. The minutes before and after the priced period show a peak of entrances, but this peak is not higher than during the previous Ecopass period and generally does not generate congestion as it is early or because it is counter-flow traffic.
5. Emissions are locally reduced, but compared to the entire metropolitan area, the effect is marginal. Area C is congestion pricing, and as a pollution charge, its effect is negligible except within Area C itself.
6. There is an impact on the housing market, quantified at less than 2% decrease in house prices and less than 1% increase in rent prices.

19.6.2 What can be taken from Milan's experience?

As we already have seen, the main and most surprising fact of Milan Area C is not that it reduced the trips in the priced area but, rather, that the policy has been quickly accepted by the population. Some elements of the explanation are in line with literature (communication and transparency, awareness of the problem, ecological attitude, existence of alternatives, etc.). However, others are less intuitive and could be considered as good practice for the transfer of this kind of policy to other cities.

First, the referendum, namely the democratic check of citizens' opinion out of direct lobbies' pressures on the Administration,[6] took place after a long trial period in which the population 'measured' how much they were individually impacted, positively or negatively. This direct experience is worth much more than any communication campaign to explain what will happen.

Second, it must be remembered that the acceptability of such a measure is different between occasional and regular users. In Milan, most regular travellers were already walking or using public transport to enter the city centre and cars accounted for less than 20% already before Area C was implemented. The private car is typically used for infrequent and somewhat 'special' occasions, such as a hospital visit or a day of shopping. So, a limited share of trips is actually influenced and almost all of them are occasional, resulting in small annual payments.

Also, crucial is the concession of special conditions to those users who are 'forced' to be regular: residents and logistics operators. The concession of discounts made the measure less draconian and the cost was counterbalanced by benefits in terms of liveability. By the way, both categories are not very price sensitive: logistics can transfer the small tax to the clients and residents of the centre who usually belong to the upper classes, given the house prices of the area.

Other elements can also be mentioned as good practice applied in Milan and important to be transferred to other cities:

1. payment channels are numerous and make the payment of the toll not a problem;
2. there were no season-tickets which would encourage extra and possibly unnecessary trips;
3. there was an acceptable level of transparency in policy goals and results.

19.6.3 What cannot be taken from Milan experience?

Not all elements which positively contributed to the Milan case can easily be transferred to any other city. Two of them are probably the most crucial, in addition to cultural factors probably less present in other Italian cities.

First, the trial was possible because most of the technological infrastructure (gates and control centre) was already in existence and consequently already paid for. In the absence of such infrastructure, it would be hard to justify to a suspicious public opinion the investment needed just for a trial.

Second, the area of application of Area C is small, central, extremely well served by public transport, with few residents and with a dominance of offices and shops. Not least, it is actually the weakest and most delicate area of the city, including most of the cultural heritage buildings, museums and institutions, where

[6] The owners of commercial activities and others lobbied with the usual tools (newspaper articles, protests, etc.), but citizens basically voted ignoring them.

traffic problems used to be critical in the recent past. If it had been larger (like in Stockholm) or less effectively served by public transport (like in Rome) or simply less problematic (like in Norwegian cities), the probability of resistance to the toll would have been much higher.[7]

In conclusion, there are a number of elements which proved to be important in Milan's Area C and which can reasonably be considered by other cities willing to implement a road pricing policy. On the other hand, the Milan case, as well as the few other road pricing systems working in Europe, present a handful of local conditions that cannot be ignored to successfully implement this kind of policy elsewhere.

References

[1] Bedogni, M. *Emissioni atmosferiche da traffico stradale in Area C nel periodo gennaio – dicembre 2012*. Agenzia Mobilità Ambiente e Territorio. Comune di Milano, 2013, Milano, Italy.

[2] Bedogni, M., Pulpito, A., and Tosi, L. *Valutazione nuovi scenari di regolamentazione degli accessi alla ZTL Cerchia dei Bastioni.*Agenzia Mobilità Ambiente e Territorio. Comune di Milano, 2011, Milano, Italy.

[3] Beria P., and Boggio M. 'The role of transport supply in the acceptability of pollution charging extension. The case of Milan'. Presented at 13th WCTR – World Conference of Transport Research, Rio de Janeiro, Brasil, 2013.

[4] Beria, P. 'Effectiveness and monetary impact of Milan's road charge, one year after implementation'. *International Journal of Sustainable Transportation*, 2016;10(7): 657–669.

[5] Commissione Ecopass. *Commissione Ecopass. Sintesi Conclusiva*. 2011. Available at: http://www.milanosimuove.it/wordpress/wp-content/uploads/2011/06/commissione-ecopass1.pdf, visited: 26/08/2016.

[6] Comune di Milano. *PUMS – Piano Urbano della Mobilità Sostenibile*. 2015. Comune di Milano, Milan, Italy.

[7] D'Arcangelo, F.M., and Percoco M. 'Housing rent and road pricing in Milan: Evidence from a geographical discontinuity approach.' *Transport Policy*, 2015; 44: 108–116.

[8] Eliasson, J., and Jonsson, L. 'The unexpected "yes": Explanatory factors behind the positive attitudes to congestion charges in Stockholm'. *Transport Policy*, 2011;18(4): 636–647.

[9] Eliasson, J., and Mattsson, L. G. 'Equity effects of congestion pricing. Quantitative methodology and a case study for Stockholm'. *Transportation Research Part A*, 2006;40: 602–620.

[7] In this respect, there is another application of road charging in Italy with similar goals even if at a smaller case: Bologna. There, a restricted traffic area covers the city centre – a medieval town – where only residents and other categories can enter. Occasional drivers can obtain a 'temporary permit' by paying 6€. Despite the wording (it is a temporary permit, not a toll…), the structure is rather similar.

[10] European Commission. Urban road charge in European cities: A possible means towards a new culture for urban mobility?. 2010.

[11] Fujii, S., Gärling, T., Jakobsson, C., and Jou, R.C. 'A cross-country study of fairness and infringement on freedom as determinants of car owners' acceptance of road pricing'. *Transportation*, 2004;31: 285–295.

[12] Hårsman, B., and Quigley, J.M. 'Political and public acceptability of congestion pricing: ideology and self-interest'. *Journal of Policy Analysis and Management*, 2010;29(4): 854–874.

[13] Ieromonachou P., Potter S., and Warren J.P. 'Norway's urban toll rings: Evolving towards congestion charging?' *Transport Policy*. 2016;13: 367–378.

[14] Ison, S., and Rye, T. 'Implementing road user charging: the lessons learnt from Hong Kong, Cambridge and Central London'. *Transport Reviews*, 2005;25(4): 451–465.

[15] Jakobsson, C., Fujii, S., and Gärling, T. 'Determinants of private car users' acceptance of road pricing'. *Transport Policy*, 2000;7: 153–158.

[16] Jones, P. 'Acceptability of Road User Charging: Meeting the Challenge'. In Schade, J. and Schlag, B., *Acceptability of Transport Pricing Strategies*, 2003;27–62. Oxford Elsevier.

[17] Kottenhoff, K., and Brundell-Freij, K. 'The role of public transport for feasibility and acceptability of congestion charging. The case of Stockholm'. *Transportation Research Part A*, 2009; 43(3): 297–305

[18] Marcucci E., Marini M. A., and Ticchi D. 'Road pricing as a citizen-candidate game'. European *Transport/Trasporti Europei*, 2005;31: 28–45.

[19] May, A., Koh, A., Blackledge, D., Fioretto, M. 'Overcoming the barriers to implementing urban road user charging schemes'. *European Transport Research Review*, 2010;2(1): 53–68.

[20] Percoco, M. 'The impact of road pricing on housing prices: preliminary evidence from Milan'. *Transportation Research Part A: Policy and Practice*. 2014;67: 188–194.

[21] Rentziou A., Milioti C., Gkritza K., and Karlaftis M. G. 'Urban road pricing: modeling public acceptance'. *Journal of Urban Planning and Development*, 2011;137(1): 56–64.

[22] Rotaris, L., Danielis, R., Marcucci, E., and Massiani, J. 'The urban road pricing scheme to curb pollution in Milan, Italy: Description, impacts and preliminary cost–benefit analysis assessment'. *Transportation Research Part A*, 2010;444: 359–375.

[23] Schade J., and Schlag, B. *Acceptability of Transport Pricing Strategies*. Elsevier, Oxford, 2003.

Chapter 20

Optimising use – using incentives to address traffic congestion

Rob Mouris, Jorrit Nijhuis**, and Colin Black****

20.1 Introduction

In this book, many characteristics of road pricing have been thoroughly addressed. Road pricing is broadly considered to be a viable transportation demand solution that mitigates some of the negative impacts of vehicular traffic. Transport economists have long argued that road pricing is the most efficient and most effective way of reducing congestion [1]. Road pricing is often presented as an alternative to more traditional transport policies such as large-scale infrastructure construction. Road pricing builds on the principles of behavioural economics, which argues that people are susceptible to monetary incentives and disincentives. Behavioural economics indicates that placing a monetary value on the limited network capacity (scarcity) will lead to a more equal distribution of road users over time, geography and modality, thereby optimising the use of existing infrastructure.

However, from a political and social point of view, road pricing is highly controversial and does not appeal to many voters. The general public's appreciation for road pricing as a tool for reducing congestion is low because people do not like to be inconvenienced, as Chapters 5–8 illustrate. Congestion charges have faced criticism for their regressive nature, whereby poorer motorists are charged the same tax as richer car owners, which arguably increases social exclusion. These controversies, along with privacy issues and the technological challenges associated with road pricing prior to implementation, were the main reasons that road pricing has never taken off in the Netherlands, despite continuous academic and political interest in the matter.[1]

Research suggests that using carrots instead of sticks can also be a successful method for behavioural change [2]. It tends to be more politically acceptable than using monetary disincentives. In many policy domains, there is strong evidence of

*Arcadis, Netherlands
**Rijkswaterstaat, Netherland
***Arcadis, UK

[1] The ambitious goals of the Dutch government to differentiate the road price on the base of time (rush hours), location (urban areas) and environmental characteristics of the car substantially increased the technological requirements for registration, thereby increasing the cost of implementation.

the positive effects of 'nudging' behaviour by using rewards. Existing research reveals the importance of positive incentives for promoting long-term behaviour changes and modal shift and demonstrates the potential benefits of transport demand management programmes [3] of the Dutch strategy of 'Spitsmijden', or 'avoiding peak traffic'. Several reward-based experiments have been conducted over the years within the field of transport and mobility. For instance, financial rewards have been used to promote cycling and the use of public transport, and to encourage road users to drive more safely in the Netherlands; financial incentives have been used as a strategy to reduce congestion and improve traffic flows. Tillema *et al.* [4] even go so far as to indicate that, in terms of decreasing traffic congestion, a peak-avoidance reward scheme may be preferred to road pricing.

Over the last 10 years, more than 20 reward schemes have been piloted in the Netherlands, providing a wealth of data on the impact of Spitsmijden on reducing traffic volumes and other benefits. This growing body of research inspired us to test the effectiveness of Spitsmijden as an alternative strategy to road charging. In this chapter, we will share insight about the efficacy of rewarding drivers through Spitsmijden, based on our research. This chapter will also describe the approach of successfully implementing rush-hour avoidance schemes, as well as reviewing the behavioural effects of rewarding, the effect on reducing congestion and return on investment.

20.2 Development of peak-hour avoidance initiatives

Spitsmijden schemes share many similarities with road-charging schemes, mainly with the technologies for enforcing the schemes used and the incorporation of behaviour economic theory into transport planning. That being said, Spitsmijden schemes differ from road-charging schemes in two fundamental ways. First, conventional road pricing schemes charge motorists tolls for using a specific infrastructure asset or make use of surcharges for those entering a designated zone during peak hours, as is the case with London's congestion charge (Chapter 13). Spitsmijden, on the other hand, believes in carrots, not sticks: road users are financially rewarded for avoiding certain roads or areas during select peak hours. In essence, they receive a financial reward each time they choose an alternative. Alternatives can include

1. Travel by car outside peak periods.
2. Travel by car during peak hours but making use of designated alternative routes.
3. Travel by other modes of transport (cycling or public transport) or making use of car-pooling.
4. Reducing the need to travel by working from home or holding virtual meetings.

Second, rush-hour avoidance schemes select volunteers to participate in the scheme instead of imposing a blanket charge on all road users. Participants of Spitsmijden are individually selected after monitoring their daily commute with traffic cameras

(ALPR, Automatic Licence Plate Recognition). Only road users who make daily use of a certain section of a road over a given period of time are invited to participate in a pilot scheme.

Spitsmijden was first trialled as part of a series of research initiatives on the Dutch motorway A12 between 2006 and 2008. Funded by the Transition to Sustainable Mobility foundation, these pilot projects aimed to help researchers analyse new mobility management concepts [5]. The pilot projects were designed and rolled out by a public–private partnership which included consultants, universities (such as TUDelft and VU Amsterdam) and government authorities (such as Rijkswaterstaat[2]). The research investigated whether monetary incentives are an effective way of persuading road users to avoid travelling during peak hours.

Using financial rewards to nudge behaviour proved more complex than initially expected, because people are generally averse to change, and old habits are hard to break. An individual's travel patterns become automatic and habitual, and people stop weighing the pros and cons each time they undertake a journey. Several social psychologists have highlighted the role of habits in individual decision-making processes and behavioural interventions [6]. According to Verplanken [7], habit formation occurs when behaviour is repeated in a stable context, helping to decrease the complexity of our decision-making. Research by Aarts *et al.* [8] on habits indicated that intervening on engrained, everyday habits through behavioural interventions is complex. Routine behaviour leads to a 'blind spot' or lack of interest in considering better alternatives [9]. For instance, car users with strong habits use significantly less information to make a travel choice. Policymakers who want to change routine behaviour should consider different types of interventions for breaking the cycle of a habit. We identified two powerful levers of change to break through established habits and nudge people towards reconsidering options to their default routine.

The first lever is to keep track of any big developments that are taking place or where travel plans are about to be disrupted that would require route replanning. An example would be rehousing, major road construction works and severe congestion during peak hours. These situations provide a 'window of opportunity' for transport planners because people would already be forced to re-evaluate existing habits and are generally more open to experimenting new transport options. The second lever of change involves presenting the user with a clear and well-structured alternative to their regular daily commute. By providing a pre-designed and simple alternative to their default routine, travellers will be able to easily reflect upon their route and make a more calculated decision [9].

These two levers of change form the foundations for mobility management schemes such as Spitsmijden. The levers' aim is to encourage road users to break their routine behaviour and consider better (and often more efficient) alternatives.

[2] Rijkswaterstaat, part of the Ministry of Infrastructure and the Environment, is responsible for the design, construction, management and maintenance of the main infrastructure facilities in the Netherlands. This includes the main road network, the main waterway network and water systems.

During a pilot project, volunteers are invited to review and assess alternative travel options to their routine and select the most attractive one.

Participants during the first experiment on the A12 motorway in 2006 were rewarded with either a gift (a smartphone) or cash, worth either 3 or 7 euros. This difference in the type and value of the reward allowed us to investigate which were more effective. This first experiment proved that financial incentives were very effective and delivered the anticipated outcomes. Over 340 participants stopped travelling during the morning peak on the A12 between Zoetermeer and The Hague [10]. There seemed to be little difference in success rates between the type and value of the incentive. Commuters who received a higher financial reward (7 euro) did not show significantly higher peak-hour avoidance rates (only 19% higher) than those who received 3 euros. Participants who opted to save for the gift (in this case, a smartphone) had an equal avoidance percentage as those who received the higher reward.

The first experiment provided evidence that temporary use of rewards was an effective tool to change behaviour and reduce traffic volumes. This provided the right context for rolling out the concept across two other programmes called 'Anders Betalen voor Mobiliteit' ('Paying Differently for Mobility') and the first rush-hour avoidance projects by the Dutch Highway Agency (Rijkswaterstaat) linked to roadworks (Programme Spoedaanpak & Beter Benutten). The results of these three programmes will be presented in the following sections.

20.3 Launch of the initiative

After the success of the pilot project in 2006, the Spitsmijden concept was used on a number of programmes that are still in use today. This section describes these initiatives.

20.3.1 Programme paying differently for mobility

For decades, the Netherlands has suffered from high levels of congestion. The Randstad, a highly dense area of the country that is home to the four largest Dutch cities (Amsterdam, Rotterdam, The Hague and Utrecht) and 7 million people, has suffered from the highest rates of congestion in the country. The Dutch government understood the economic costs associated with congestion on these roads, but it did not have the funds or physical space to widen any motorways in the early 2000s.

During the Cabinet Balkenende II (2003–06), plans were introduced to implement a national congestion charge to tackle these problems. A committee, the National Platform 'Anders Betalen voor Mobiliteit' ('Different Payment for Mobility') consisting of Algemene Nederlandsche Wielrijders-Bond translated: National Breakdown Service (ANWB), employers and employee representatives, the municipalities and the pro-environmental organisations drew up recommendations for the government to introduce road pricing.

The original intention for the programme 'Different Payment for Mobility' (ABvM) was to introduce a fixed charge per kilometre depending on the time of day, place, fuel consumption and pollution of the car. It was to be introduced in 2012 for lorries and then rolled out later on in the year for passenger cars. In 2018, the entire system would have been implemented using a satellite (Global Positioning System (GPS)) charging system instead of toll booths. However, the plan never came to fruition, because one of the political parties of Mark Rutte's coalition government (2010–12) withdrew its support for the scheme. According to the Christian democrats, the system would be too complicated, expensive and an infraction to the privacy of people.

Instead, the government made funding available for experimenting with price incentives and Intelligent Transportation Systems (ITS) technologies such as on-board units (OBU) with GPS and ALPR cameras (see Chapter 11 for more information). A budget of 100 million euros was released for ten pilot projects in six regions over 4 years (2008–12). These and other projects were reviewed and evaluated over time. There are three relevant reports, which are as follows:

- Resultaten mobiliteitsprojecten, Ministerie van Infrastructuur en Milieu, 14 June 2011 [11]
- Eindevaluatie Mobiliteitsprojecten, eindrapportage, MuConsult, 2 November 2013 [12]
- Meta-evaluatie Spitsmijdenprojecten, eindrapportage, MuConsult, January 2017 [13]

The lessons learnt of the latest study have been included in this article. The selected 'Spitsmijden' projects of the 2017 study were

1. Winnen van de File between Eindhoven and Den Bosch ('beat the traffic');
2. Spitsmijden 010 in Rotterdam ('avoiding rush hour Rotterdam');
3. Spitsvrij 2 between Utrecht, Hilversum and Amersfoort ('rush-hour free');
4. Wild! van de Spits (sub-projects 1 and 2) in Rotterdam ('wild! Of the rush hour');
5. Slim uit de Spits (sub-projects 1 and 2) in Arnhem-Nijmegen ('smart outside of the rush hour');
6. Spitsmijden Galecopperbrug and the follow-up Spitsmijden Galecopperbrug in Utrecht ('avoiding rush-hour Galecopper bridge') GCB2A and GCB2B;
7. Slim uit de Spits (sub-project 3) between Ede and Grijsoord ('smart outside of the rush hour').

In Figure 20.1, these projects are visualised geographically. Table 20.1 contains the main features of these projects. The results of the projects are elaborated in Section 20.5.

20.3.2 Improving the value of existing road infra (Programme Spoedaanpak & Beter Benutten)

Between 2008 and 2011, the Rijkswaterstaat widened highways at 30 bottlenecks and renovated a number of steel bridges around the country. Without construction mitigation measures, these roadworks ('Spoedaanpak' or translated: 'immediate approach') would result in increased congestion. The Minister of Infrastructure and

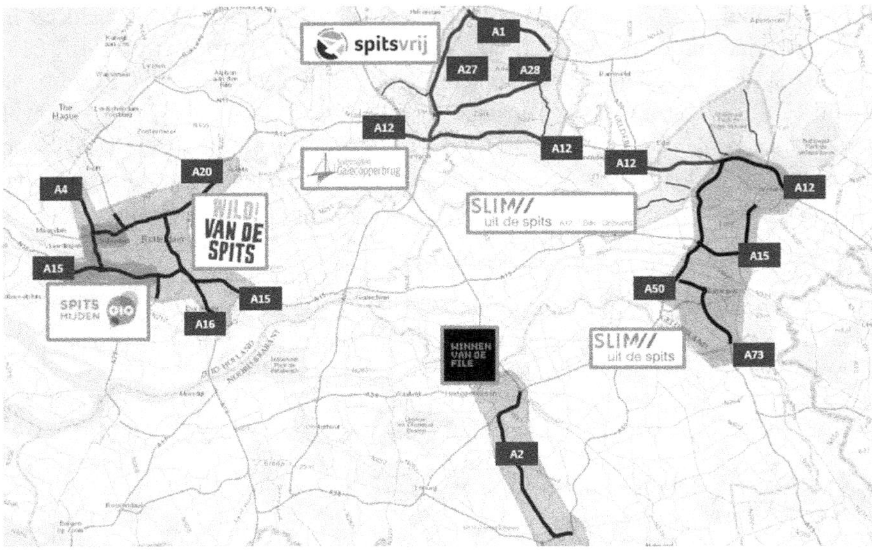

Figure 20.1 Spitsmijden projects [13]

the Environment made an agreement with the House of Representatives that the congestion was not to exceed an increase of 10% which was a great window of opportunity to roll out Spitsmijden across all projects to minimise congestion.

There was also a second window of opportunity to roll out Spitsmijden across the Netherlands: in the 'Beter Benutten' ('Optimising Use') programme [14], the Dutch national government, regional authorities and businesses have been working together to improve road, water and rail accessibility in the busiest regions. The programme had a budget of 1.1 billion euros in the period up to 2017. In the follow-up programme, the emphasis is primarily on measures to enable the traveller to arrive at their destination quicker and smarter. The first objective was to reduce congestion by 20% in the most congested areas of the Netherlands in 2014, using a variety of strategies. The next phase of the programme (2014–17) aims to achieve 10% shorter journey times from door to door in the busiest areas.

Investing in new roads could result in less congestion. That is beneficial to economic development, but asphalt alone is not enough to increase accessibility between and within cities because budget and space are limited. The *Beter Benutten* programme also aims to improve quality of life in Dutch cities, something that is at odds with building more roads. Optimising existing roads has therefore become the main strategy for reducing congestion and improving road accessibility, by

- Improving the existing roads. Implementing small solutions which collectively help to improve traffic flows. For example, an extra lane at traffic junctions or small adjustments of traffic lights.
- Customised solutions for each region. The government and businesses are working together to improve accessibility in the 12 busiest urban regions,

Table 20.1 Project's most important features [13]

<table>
<tr><th>Project</th><th>Winnen van de File</th><th>Spitsmijden 010</th><th>Spitsvrij 2</th><th>Wild! van de Spits 1</th><th>Wild! van de Spits 2</th></tr>
<tr><td>Project period</td><td>Jan–Oct 2013</td><td>Jan 2013–Feb 2014</td><td>Apr 2013–Feb 2014</td><td>Mar–Dec 2014</td><td>Mar–Dec 2015</td></tr>
<tr><td>No. of rewarding days</td><td>226</td><td>237</td><td>360</td><td>155</td><td>155</td></tr>
<tr><td>Project costs (incl. VAT)</td><td>€ 480,000</td><td>€ 4,200,000</td><td>€ 2,690,000</td><td>€ 5,362,000</td><td>€ 4,541,000</td></tr>
<tr><td>Type of registration</td><td>GPS (App)</td><td>ALPR cameras</td><td>ALPR/GPS (OBU and App)</td><td>ALPR cameras</td><td>ALPR cameras</td></tr>
<tr><td>Rewarding per avoidance</td><td>Lottery tickets</td><td>€ 3.00–6.00</td><td>€ 2.00</td><td>€ 3.00–4.00</td><td>€ 2.50–3.00</td></tr>
<tr><th>Project</th><th>Slim uit de spits 1 and 2</th><th>Spitsmijden Gale copperbrug 2A</th><th>Spitsmijden Gale copperbrug 2B</th><th>Slim uit de spits 3A</th><th>Slim uit de spits 3B</th></tr>
<tr><td>Project period</td><td>Jun 2013–Dec 2014</td><td>Aug–Dec 2014</td><td>Apr–Sep 2015</td><td>Sep–Oct 2015</td><td>Oct–Nov 2015</td></tr>
<tr><td>No. of rewarding days</td><td>338</td><td>90</td><td>66</td><td>30</td><td>30</td></tr>
<tr><td>Project costs (incl. VAT)</td><td>€ 9,475,000</td><td>€ 5,017,825</td><td>€ 4,084,175</td><td colspan="2">€ 1,267,601</td></tr>
<tr><td>Type of registration</td><td>ALPR cameras and GPS (App)</td><td>ALPR cameras</td><td>ALPR cameras</td><td>ALPR cameras</td><td>ALPR cameras</td></tr>
<tr><td>Rewarding per avoidance</td><td>€ 1.00–4.00</td><td>€ 4.00</td><td>€ 3.00</td><td>€ 4.00</td><td>€ 4.00</td></tr>
</table>

using tailored packages of measures that focus on the specific needs and behaviour of travellers and transport operators.

- Using a variety of measures. It is more effective to use a combination of complementary measures in an area instead of one blanket measure. For example, in Utrecht, we combined the provision of more sheltered cycle parking at stations along with improved way-finding and real-time information for passengers and free tickets to encourage increased usage of public transport. In total, there are more than 300 measures available.

Using the 300 measures, 12 rush-hour avoidance projects were created with a goal to change travel behaviour for the long-term.

20.4 A guide to implementing 'Spitsmijden'

20.4.1 Recruiting participants

The Spitsmijden projects were based on voluntary participation, and each project required the recruitment of its participants. Three strategies were used, which are as follows:

1. Monitored and registered licence plates by ALPR cameras along the route under study and sending an invitation letter to road users in the mail.
2. Engaged employers to reach out to their employees.
3. Outreach through mass media marketing and communication [e.g. billboards, variable message signs (VMS) messages, flyer teams at fuel stations, Facebook campaigns and advertisements in newspapers].

Recruiting participants via licence plate recognition and letter was by far the most effective approach. MuConsult's study revealed that 90% of participants in the pilot projects were recruited by letter, and overall there was a 25%–35% response rate after receiving personal letters [11]. As mentioned, only license-plate holders who consistently use a stretch of road or area are contacted, and the minimum frequency to participate can be determined for each project.

There may be some concerns over privacy. There is a risk that some people might not like their vehicles being 'watched' by the government or receiving an unsolicited offer to take part in a project. However, the number of complaints received per project was less than 1%, and the government has a public law duty to capture licence plates that legitimises this approach.

Engaging with companies and organisations to recruit their employees proved less effective than the letter/number plate approach. Prospective participants work for many different companies, making outreach expensive and complicated.

Social media campaigns were also deployed through Facebook and Twitter, along with local newspaper ads. Exact numbers are not known unfortunately.

Also, in most projects advertising signs or VMS were installed along roads to inform drivers of the rush-hour avoidance project. Unfortunately, either not many people notice the signs or remembered to sign up.

20.4.2 Types of rewards

During the last decade, the Dutch experimented with several ways to encourage participants to change their travel behaviour, which are as follows:

- cash;
- awarding points that are transferrable to products and gifts in online shops;
- intrinsic motivation (more free time, healthiness, no stress, etc.);
- gaming elements;
- lottery: option to win things;
- combination of options listed above.

Because monetary rewards are relatively expensive, research was undertaken to explore other options. Awarding points (that are transferrable to discounted products or services online) was one such option. Products included navigation devices, gift cards, car wash treatments, anti-skid courses and tickets to the zoo or theme parks. Not all the participants exchanged their points, so there this is a better business model than cash rewards.

Behavioural insights have also shown that people adapt their behaviour on several non-monetary motivations. Instead of using money or points ('external motivations'), behavioural scientists tried to arouse 'intrinsic motivations'. With the participants, they discussed quality of life benefits, such as more free time, healthiness, less stress, etc. In one project (Spitsmijden in Brabant), behavioural scientists also tested the theory of 'self-persuasion', meaning to get the participants to play an active role in persuading themselves to change their behaviour. Each week, some participants were asked to fill in a personal plan where they wrote how many times they planned to avoid rush hour by car ('implementation of intentions'). This specific group of participants made 27% more avoidances than the control group. The control group still made avoidances but 27% less.

Another idea to reduce project costs was to try a lottery system, called 'Winnen van de File', or 'beat the traffic'. Every time a participant avoided commuting during peak hours, he or she was automatically registered for a lottery ticket in the system. The more someone drove outside of peak hours, the higher their chance would be of winning the gift. Each month, a winner was chosen. The largest prize was a gift card for a weekend stay in a hotel, and the lowest prize was a gift card for an apple pie. This project wasn't much of a success. The results were not measurable because of too few participants and data.

20.4.3 Available technology

Technology used for these schemes is similar to those used for congestion charging. The first Spitsmijden experiment tested an OBU which was placed at the window of a car and communicated with the Electronic Vehicle Identification beacons which were placed above the A12 motorway. It also tested ALPR technology.

There is still some resistance towards getting people to put a device in their vehicles. Some people have issues with privacy and are uncomfortable with the fact that the car's location is always being tracked. Others do not like the inconvenience of having to make an appointment to get someone to install the device into the car. There are also people who are worried about damage to their dashboards and windshields.

Most of the Spitsmijden projects used ALPR-cameras to find participants who frequently drive during peak hours. In the Netherlands, this technique works well because there are not many alternative routes. There were some Spitsmijden projects where a lot of ALPR-cameras were used to ensure that participants would not get rewarded for driving alternative but equally undesirable routes. These projects were relatively expensive but the project leaders wanted to ensure that congestion was not displaced to other routes.

ALPR-cameras are proven to be at least 95% accurate, as long as the appropriate requirements are met for correct vehicle detection and a reliable recognition of a licence plate. It's important to contract a specialised ALPR company to get this minimum quality.

GPS tracking is another technology that can be used. GPS trackers can be installed onto an application (app) of a smartphone, on a device powered by the 12-V socket or by using the OBD port (on board diagnostics) in a car.

At the beginning of the Spitsmijden projects (between 2009 and 2012), not many people had smartphones. So participants in these projects could gain their own smartphone as a reward over time by avoiding peak hours. ALPR-cameras were used as a backup to the smartphones, because there was always a chance that people would leave their smartphone behind and be falsely rewarded. Smartphone ownership and usage rates grew rapidly after 2012, and people became much more devoted to their smartphones. By delivering a GPS track, participants show their travel behaviour and get rewarded if they travel off peak. When there is no GPS track, they don't get rewarded.

Every technology has its advantages and disadvantages, which have been summarised in Table 20.2 by MuConsult.

Table 20.2 Qualitative comparison of automatic registration systems used for Spitsmijden [12]

Aspect	**ALPR cameras**	**OBUs**	**Smartphones**
Costs	Average	High	Low
Chance of technological failure	High	Average	Low
Effort participants	Low	High	Average
Privacy participants	High	Low	Low
Privacy other road users (those who do not participate)	Low	High	High
Susceptibility to fraud	Low	Average	High
Combination with value-added services	Low	Average	High

20.4.4 Fraud prevention

One of the main criticisms of the Spitsmijden concept is its susceptibility to fraud. For example, participants could easily swap their vehicles with other people or buy a new one. Someone could also switch workplace location or home address without informing the project. Because of these concerns, the Dutch invented several control mechanisms to reduce the possibility of fraud, which are as follows:

- Monthly checks on the accuracy of licence plate information by the vehicle authority (when ALPR-technology is used). In the Netherlands, each vehicle has its own licence plate so if a car is sold, the licence plate stays with the car and the vehicle authority will pick up on this.
- Monthly check of the deviation of the peak-hour avoidances of an individual: when there is a high deviation, this could indicate someone moved away or changed their job.

If there is an indication that a participant's information is not correct, payment is immediately suspended. After an investigation, the information will either be corrected or the participant will be removed from the project.

About 7% of all the participants end up being excluded from the project. This doesn't mean 7% of the participants committed fraud, but a large group of participants are lax in responding to requests for information. In 2016, some members of the House of Chambers asked for explanation about the alleged fraud sensitivity because of media attention surrounding the issue. This led to the tightening of controls. It also led to the Dutch Minister of Infrastructure and the Environment promising to reclaim any unjust rewards and to impose a penalty in future Spitsmijden projects.

20.5 The results

20.5.1 Effects in rush-hour avoidances

Table 20.3 shows the average peak-hour avoidances per person per day during the rewarding period. It also shows that participants deliver 0.39 peak-hour avoidances on average each day. The absolute numbers vary by project. The most avoidances per participant per day were made by Spitsmijden Galecopperbrug 2A (0.52). The least avoidances per participant per day were made by Slim uit de Spits 1 and 2 (0.21).

The statistical analysis from MuConsult examined which factors most affect the rate of peak period avoidance. From this analysis, the following headline findings emerged [12]:

- **The amount of the compensation:** the higher the fee, the greater the average avoidance rates per person.
- **Length of the peak period:** If the peak period is longer, there are fewer opportunities to travel outside rush hour and consequently fewer avoidances are made.

Table 20.3 Reference behaviour of the participants before the rewarding period and avoidances during the rewarding period per week [13]

Project	Rewarding days	Participants	Avoidances			
			Per day	Maximum per day	Total	Per participant per day
Winnen van de File	226	1,192	–	–	–	–
Spitsmijden 010	238	5,049	1,464	2,101	348,700	0.29
Spitsvrij 2	159	3,288	865	998	137,500	0.26
Wild! van de Spits 1	155	10,500	4,924	6,822	763,200	0.47
Wild! van de Spits 2	155	12,500	4,623	5,813	716,600	0.37
Slim uit de Spits 1+2	390	21,056	4,456	6,967	1,735,200	0.21
Galecopperbrug 2a	90	11,134	5,745	–	517,050	0.52
Galecopperbrug 2b	66	7,619	3,557	–	234,762	0.47
Slim uit de Spits 3 (a+b)	55	3,250	802	890	48,100	0.25
Total	1,308	74,396	26,436	–	4,501,112	–
Average	164	9,300	3,662	4,540	562,639	0.39

- **Both peak hours:** if a project only rewards for peak avoidance in the morning, the average rate of overall avoidances during the day is lower.
- **Corridor or cordon:** if a project includes a cordon (e.g. the area of a city centre), there will be fewer avoidances made during rush hour than if it only concerns a corridor (e.g. a point location of a bridge or segment of a road). In a corridor, for example, there will more often be an alternative route. If you can't come into a city area, there is no chance to drive another route.
- **Commuting distance:** participants who live less than 15 km from their work make fewer avoidances than participants with greater commuting distances.
- **Residential area inside a cordon:** people who live inside the bypass cordon make fewer avoidances. For each ride to or from home during rush hour(s), they won't get rewarded, so for example, go shopping or taking children to school. Also, avoiding the area by selecting a different route here is not an option.
- **Work location inside a cordon:** when work location is inside the bypass cordon people make fewer avoidances. This seems logical, for each ride to or from the working site they won't get rewarded, including for example business trips. Also, avoiding the area by selecting a different route here is not an option.

- **Gender of the participants:** men make higher rates of avoidance than women.
- **Age of the participants:** participants over 55 make more avoidances than other age groups.
- **Leased vehicles:** participants with a leased car do not have a higher or lower rate of avoidance than those with owned cars.
- **Travel motive:** participants travelling for leisure do not have a higher or lower rate of avoidance than those commuting or working.
- **Telecommuting:** participants who have the possibility to work from home do not have a higher or lower rate of avoidance than those who do not.
- **Flexible hours:** participants with flexible working hours make more avoidances.
- **Fixed work address:** Those who normally work at a fixed address make more avoidances than those who do not.

20.5.2 Effects: behavioural change of participants

MuConsult also researched what participants did to avoid the rush hours (see Figure 20.2). The majority of participants still drove by car but managed to travel off peak (41%) or chose an alternative route (30%). In some projects (where there were not many alternatives), participants chose to work from home more often (14%). The use of alternative means of transport (public transport or bicycle/e-bike/moped/scooter) was done the least (5 resp. 10%) (Figure 20.2).

But what does peak avoidance bring us? As a rule of thumb, we know that a limited reduction of the traffic demand (3%–5%) has a relatively larger effect on the traffic flow (about 10%–15%). The effect is quite similar to how the traffic flow is affected during vacation periods. In the following section, we will elaborate on the congestion reduction of Spitsmijden.

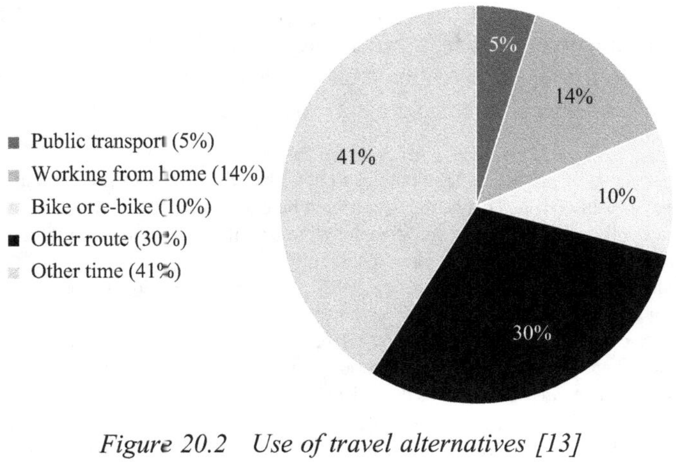

Figure 20.2 Use of travel alternatives [13]

20.5.3 Effects: congestion reduction on the road network

In the previous sections, we discussed the effects of rush-hour avoidance projects in terms of the absolute amount of rush-hour avoidances realised and the alternatives chosen by the participants. These effects can be labelled the 'output' of the projects. To determine the effects, in terms of 'outcome', meaning the contribution that is realised in improving accessibility, it is necessary to measure the congestion reduction in the designated project areas. In the evaluation study [13], a traffic model was used in order to measure the effects of the rush-hour avoidances which were realised by a rush-hour avoidance project, on the regional road network related to the specific project.

Because the projects had access to the origin and destination of each individual participant (based on survey information[3]), it was possible to determine the exact location of the rush-hour avoidance and to calculate the cumulative effect of this avoidance on the broader road network. For each rush-hour avoidance project, a separate run in the traffic model (NRM) was executed to isolate the effects of the rush-hour avoidance project.[4] The overall vehicle intensity in the road network is reduced by the number of rush-hour avoidances of each individual route of the participant. The assumption behind this method is that the number of cars during the rush hours has been reduced on the designated origin–destination–routes. The aggregated effect of the individual rush-hour avoidances is a reduction of the car intensity in specific parts of the road network.

The outcome of the model runs was presented two-fold (see Table 20.4): the gross effect presents the direct influence of the rush-hour avoidances on the road network, the net effect presents the effects in which the redistribution of traffic throughout the network is taken into account. To explain, due to the behaviour of the participants, traffic intensity is reduced and empty space is created in the network. This obviously leads to congestion reduction. However, part of this empty space is filled up by car users from other areas of the road network who continue driving during rush hours but redirect their route because it has become faster to do so. Due to this rebound effect, the net effect is smaller than the gross effect.[5] Table 20.4 shows that the gross effect of the investigated rush-hour avoidance projects on average resulted in a reduction of almost 2,000 vehicle loss hours (VLHs)[6] and over 100,000 car kilometres per working day. This means that, on average, each rush-hour avoidance project reduces car usage by 100,000 km, daily!

[3] From the perspective of privacy, it is important to note that this data does not contain personal and therefore privacy-sensitive data. That is, the data cannot be traced back to individual travellers, or their home and work addresses. This data is specific enough to be able to calculate the effect on congestion but also generic enough to safeguard the participants' privacy.

[4] A traffic model of the Dutch Highway Agency (NRM, 'Netherlands Regional Model') was used, with 2020 as the year of reference (version 2016).

[5] This effect only calculates immediate rebound effects of car traffic in the network. The medium and long-term latent demand, during which the congestion reduction causes travellers to reappraise car use at the possible cost of travel alternatives, was not calculated.

[6] Vehicle loss hour is a frequently used indicator for travel time loss in Dutch traffic and transport analyses. One vehicle loss hour signifies that a vehicle has suffered a delay of 1 h. With the help of the value of time index, this indicator makes it also possible to perform a societal cost–benefit analysis on measures influencing vehicle loss hours.

Table 20.4 Effects of rush-hour avoidance projects on congestion per day in the Netherlands

	Gross effect		**Net effect**		**Difference between gross and net effect**	
	VLH	**km**	**VLH**	**km**	**VLH**	**km**
1 Galecopperbrug (GCBA)	−2,218	−175,584	−1,087	−172,768	+51%	+2%
1 Galecopperbrug (GCBB)	−1,421	−106,880	−697	−103,808	−51%	−3%
2 Spitsvrij Utrecht (SVRIJ2)	−1,585	−76,928	−892	−77,056	−44%	0%
3 Spitsmijden 010 (S010)	−939	−52,960	−794	−51,456	−15%	−3%
4 Wild! van de Spits (WVDS) phase 1	−2,310	−81,152	−1,174	−81,184	−49%	0%
4 Wild! van de Spits (WVDS) phase 2	−2,310	−81,152	−1,174	−81,184	−49%	0%
5 Slim uit de Spits (SUDS1+2)	−3,031	−205,440	−2,408	−203,168	−21%	−1%
5 Slim uit de Spits 3 (SUDS3)	−1,571	−89,344	−2,055	−82,080	+31%	−8%
6 Winnen van de File (WVDF)	–	–	–	–	–	–
Average	**−1,923**	**−108,680**	**−1,285**	**−106,588**	**−33%**	**−2%**

Note: NRM West and NRM East version 2016, calculations by 4Cast.

Table 20.5 VLH reduction and kilometre reduction per rush-hour avoidance

Effect per rush-hour avoidance	**Gross effect**		**Net effect**	
	VLH	**km**	**VLH**	**km**
1 Spitsmijden Galecopperbrug (GCB2A)	0.4	34	0.2	33
1 Spitsmijden Galecopperbrug (GCB2B)	0.4	33	0.2	32
2 Spitsvrij Utrecht (SVRIJ2)	1.8	89	1.0	89
3 Spitsmijden 010 (SM010)	0.6	36	0.5	35
4 Wild! van de Spits (WVDS) fase 1	0.5	17	0.2	17
4 Wild! van de Spits (WVDS) fase 2	0.5	18	0.3	18
5 Slim uit de Spits (SUDS1+2)	0.7	46	0.5	46
5 Slim uit de Spits 3 (SUDS3)	1.8	102	2.3	94
6 Winnen van de File (WVDF)	–	–	–	–
Range	0.4–1.8	17–102	0.2–2.3	17–94
Weighted average	**0.6**	**34**	**0.4**	**33**

Note: NRM West and NRM East version 2016, calculations by 4Cast.

Furthermore, the table shows that, on average, one-third (−33%) of the reduction in travel time loss is negated due to the aforementioned shift in traffic flows. Interestingly, this negative effect does not occur with the indicator of car kilometres, making the differences in mileage between the gross and net effect negligibly small.

Another way of presenting the congestion reduction effects of these six projects is to look at the average effect of each rush-hour avoidance. While Table 20.4 shows the cumulative effect of a whole project, Table 20.5 presents the contribution of one single rush-hour avoidance within these projects.[7] The table shows that, on average, each rush-hour avoidance results in a reduction of 33 car kilometres and 0.4 VLH reduction. This makes clear that each individual rush-hour avoidance contributes significantly to the reduction of congestion in the road network.

The table reveals another remarkable outcome, namely that two specific rush-hour avoidance projects have an exceptionally positive outcome (a VLH reduction of 1.8, three times as high as the average). This in turn is caused by the high average travel distance of participants in this project. This outcome makes clear that one rush-hour avoidance is not the same as another, but that the specific project conditions matter. Projects with participants who are long-distance travellers have significantly higher impact on congestion reduction than projects comprising short-distance travellers.

20.5.4 Effects: long-term contribution of rush-hour avoidance projects to congestion reduction

An important question about the Spitsmijden projects is if the adapted behaviour of the participants is temporary or if it endures over a longer period after stopping rewarding participants within the project. The assumption is that participants adapted their behaviour and still avoid travelling during peak hours mostly, once the rewarding stopped.

Within three recent Spitsmijden projects, the behaviour of the participants has been measured after the period of rewarding. ALPR cameras were used for this. The first month afterwards the behaviour of the participants almost doesn't change (Table 20.6). Only in Utrecht at the Spitsmijden Galecopperbrug project (GCB2A and GCB2B), the conservation is less: only 70%–79% of the former behaviour of the group participants endures. Most likely, this is because of the application of Spitsmijden only during roadworks. For example, at the end of 2014, during the rewarding period of Spitsmijden Galecopperbrug (GCB2A), the participants made between 4,440 and 4,893 avoidances on average each day. In the beginning of 2015, when the rewarding was stopped, the former participants still showed avoidances, but these were less than in 2014: between 3,176 and 3,404 (minus 28%–30%).

The assumption is that people are inclined to return to their old behaviour when the period of roadworks has been terminated. This assumption is confirmed by a behavioural analysis among former participants of older projects [15]. During a survey, they were asked about their travel behaviour during their participation and their behaviour on average 36 months after participation. During participation, they reduced their peak hour travel by 34%. After 36 months, they still travel off peak 30% less. On average, the reduction is only 6%. There are big differences between

[7] The effect of a single rush-hour avoidance is calculated by dividing the total effect of a project by each rush-hour avoidance achieved. For example, the project SUDS3 resulted in 48,000 avoidance divided by 55 project days = 873 avoidances per day (figures presented in Table 20.3). The gross effect of a single avoidance in VLH per day is 1,571 (figure presented in Table 20.4) divided by 873 = 1.8.

Table 20.6 Average of working day avoidances during the last months rewarding and period after rewarding [13]

Month	Number of working days[a]	WvdS1	SUDS1+2	GCB2A	GCB2B
Avoidances during rewarding					
Aug 2014	10			4,893	
Sep 2014	22	6,342	6,103	4,819	
Oct 2014	18	5,377	6,187	4,701	
Nov 2014	20	5,407	6,348	4,556	
Dec 2014	15	4,938	6,806	4,440	
Apr 2015	8				2,780
May 2015	13				2,840
Jun–Jul 2015	30				2,828
Aug–Sep 2015 (till 11/9)	15				3,109
Avoidances after rewarding					
Jan 2015	20	5,066	6,605	3,176	
Feb 2015	15	5,817	6,645	3,404	
Mar 2015	22	6,205	–	3,300	
Apr 2015	7	6,287	–	–	
Sept 2015 (from 14/9)	13				2,302
Oct 2015	17				2,207
Nov 2015 (till 27/11)	20				2,196

[a]This indicates the number of days in the month that the Spitsmijden measures were in operation.

the participants where nothing has changed in working conditions and where something has changed. After 36 months, the increase in peak hour rides is higher among former participants who had changes in commuting or other working conditions. This indicates that the long-term effect on the former participants is not entirely due to the participation in the project but is influenced by other circumstances.

A behavioural analysis among former participants of the Wild! van de Spits projects in Rotterdam showed other results [15]. During this survey, the former participants were also asked about their travel behaviour during their participation and their behaviour 12–24 months after participation in Rotterdam. After 1 year, the reduction of the former peak hour drives is 49%. After 2 years, it becomes 67%. These percentages strongly differ from the 6% decrease (after 3 years of participation) measured in the behavioural analysis above. Differences in methodology seem to be the most logical explanation: during the survey [15], respondents were asked to fill in their behaviour before and after participation, while respondents from the Wild! van de Spits project only filled in their current behaviour. Their 'old' behaviour was already known. The results of Wild! van de Spits seemed to be more reliable because the results are measured more objectively rather than self-reported.

Table 20.7 Results of SCBA total costs and benefits of the six recent Spitsmijden projects (in millions euros) [13]

All projects (except Winnen van de File)	Participation period		3-year effect		10-year effect	
	Gross	Net	Gross	Net	Gross	Net
Benefits	**€ 65**	**€ 52**	**€ 97**	**€ 75**	**€ 132**	**€ 101**
Travel time benefits[a]	€ 49	€ 38	€ 76	€ 59	€ 98	€ 73
Distance benefits	€ 5	€ 5	€ 3	€ 3	€ 12	€ 12
Reliability[b]	€ 12	€ 10	€ 19	€ 15	€ 25	€ 18
Air quality	€ 0.3	€ 0.3	€ 0.5	€ 0.5	€ 0.8	€ 0.8
Climate	€ 0.3	€ 0.3	€ 0.5	€ 0.5	€ 0.8	€ 0.8
Noise	€ 0.3	€ 0.3	€ 0.5	€ 0.5	€ 0.7	€ 0.7
Road safety	€ 0.8	€ 0.8	€ 1.4	€ 1.3	€ 2.0	€ 2.1
Rewards participants[c]	–	–	–	–	–	–
Costs and benefits modal shift[d]	–	–	–	–	–	–
Loss of tax revenues	€ −2.9	€ −2.8	€ −4.7	€ −4.6	€ −7.0	€ −7.3
Costs	€ 37	€ 37	€ 37	€ 37	€ 37	€ 37
Project costs	€ 37	€ 37	€ 37	€ 37	€ 37	€ 37
Benefits/costs balance	€ 28	€ 15	€ 60	€ 38	€ 96	€ 64
Benefits/costs ratio	**1.8**	**1.4**	**2.6**	**2.0**	**3.6**	**2.8**

[a]Travel time benefits because of avoided VLHs. Other travel time benefits aren't considered. No multiplication factor has been applied for occupancy rates of vehicles.
[b]Increment of 25% to the post-travel time benefits.
[c]The rewards paid to the participants are included in the costs of the projects. Therefore, they should be counted as benefits as well. However, there was insufficient information to process this.
[d]The modal shift from car to (e-)bike or public transportation as well delivers costs and benefits. For example passenger revenues in public transport and costs. However, there was insufficient information to process this.

20.5.5 Costs–benefits

A common question about the Spitsmijden projects is how cost effective they are. Mobility policy is often based on (societal) cost–benefit analysis [(S)CBA]. The methods have different levels of detail and accuracy degrees. The SCBA is a widely accepted tool for determining cost effectiveness of projects. A benefit/cost ratio greater than 1.0 means that the social benefits of a project are higher than the social costs.

Table 20.7 shows the aggregated results of the SCBA of six recent Spitsmijden projects MuConsult evaluated, commissioned by the Ministry of Infrastructure and the Environment in 2017 [13]. There is a difference between the gross and net effects. Gross effect means that there will be no effect of the Spitsmijden schemes on the road usage by other motorists. Net effect means that there will be an effect of the Spitsmijden projects on route choice and departure times of other motorists in the network. Three variants are calculated: during the participation period, and a 3- and 10-year period without rewarding added to the participation period.

The aggregated results of the SCBA show that these projects are beneficial during the participation period already. Spitsmijden projects are especially cost

Table 20.8 Cost effectiveness per project

All projects	Participation period		With 3-year effect		With 10-year effect	
	Gross	Net	Gross	Net	Gross	Net
GCBA	0.8	0.5	1.5	0.8	2.7	1.6
GCBB	0.5	0.3	1.1	0.6	2.0	1.1
SV2	1.6	1.0	2.5	1.4	3.7	2.9
S010	0.8	0.7	1.3	1.1	1.6	0.9
WvdS1	1.1	0.6	1.9	1.0	2.6	1.9
WvdS2	1.2	0.7	2.1	1.1	2.9	2.1
SUDS1+2	3.9	3.5	4.9	4.4	5.9	5.3
SUDS3	1.8	2.2	5.4	6.5	8.6	5.3
WVDF	–	–	–	–	–	–
Unweighted average	**1.8**	**1.4**	**2.6**	**2.0**	**3.6**	**2.7**

Table 20.9 Development SCBA scores and costs Spitsmijden projects

	2013 Evaluation study		2017 Evaluation study	
	Projects	Average	Projects	Average
SCBA participation period	0.1–0.8	0.5	0.2–3.5	1.4
SCBA included long-term effects	0.3–1.5	1.1	1.0–5.3	2.6
Costs per avoidance		€ 22.00		€ 8.20
Cost per avoidances (including long-term effects)		€ 7.00		€ 3.50

effective when effects after rewarding are also counted during a period of 3–10 years. This takes account of project- and alternative-specific relapse each year and a discount rate of 4% per year.

Table 20.8 shows the cost effectiveness for each recent Spitsmijden project. Most projects are beneficial within the participation period, for example, Spitsvrij (SV2) and Slim uit de Spits (SUDS1+2 and SUDS3). There is an explanation for this. The participants of the SV2 and SUDS3 projects have relatively long-distance rides, so the effects of these rides are more significant. The SUDS1+2 projects show a relatively high cost effectiveness because of the relatively long participation period. Those participants achieved a high number of peak-hour avoidances during this project.

A comparison between the evaluated Spitsmijden projects in 2013 and 2017 shows an improvement of the cost effectiveness of the projects over time (Table 20.9). This can be explained by two main developments:

1. The lead time of the projects is shortened, but the projects recruited more participants. This results in less overhead expenses. Also, the projects could use the previously gained knowledge and technical build systems, so there was less investment needed.

2. The methodology to determine the SCBA scores has improved in recent years. In the 2013 evaluation study, most of the project-specific results have been extrapolated. In the 2017 evaluation study, each project had its own detailed calculation.

20.6 Public and political acceptance of rush-hour avoidance

In the previous sections, we described the (positive) effects of rush-hour avoidance projects in terms of behavioural change, congestion reduction and cost effectiveness. However, despite the overall positive contribution to accessibility, rush-hour avoidance projects are not without controversy. As is the case with congestion charging, there are both supporters and opponents to rush-hour avoidance projects. The public and political discord with rush-hour avoidance focuses on three main points of interest: financial rewarding, privacy and fraud.

20.6.1 Financial rewarding

Opponents are dissatisfied about the fact that motorist who already adapted their travel behaviour in the past cannot participate in the projects. The rush-hour avoidances achieved by this group of travellers are therefore not financially rewarded. The reason for this is that the motorists who already travel outside the rush hour are not registered anymore during peak hours by cameras. So, in the public eyes, only the travellers who still perform 'bad behaviour' qualify for participation and earn money. This can lead to the feeling that travellers who do not travel by car during rush hours feel themselves 'robbed' of the financial rewards when a rush-hour project is started in their region. Behaviourally, this can undermine the intrinsic motivation of travellers who have already chosen alternatives to car travel during the peak-hour period.

Despite these reservations, the overall public support for these projects is high. Based on a representative survey, amongst 15,000 Dutch travellers, we know that approximately 60% of the Dutch population supports the use of financial incentives to stimulate car users to travel outside the rush hours [15]. Only 10% think that using financial rewards is unacceptable. The interesting part of this outcome is that both participants and non-participants are very positive about this measure.

20.6.2 Privacy

More substantial concern and controversy surrounding rush-hour avoidance projects is related to privacy (protection of personal data). Depending on the exact technical and system choices, personal data will be processed. This starts with the reference measurement. Most projects use ALPR cameras to define which car owners drive frequently during rush hour. Not only the licence plates of participants but also the licence plates of other car owners will be registered and retained. This is without consent of the data subject. In the Netherlands, road authorities have public law duties to do so. The data of non-participants is deleted after 10 weeks. All other processing of personal data is with consent of the subject in terms of law.

Of the group of 15,000 respondents from the above-mentioned travel survey, 26% would not participate in rush-hour avoidance projects because they do not want their travel behaviour to be monitored [15]. This is much higher than the outcome of the survey held in 2013 [12], suggesting that privacy is an increasing factor of importance for road users. Obviously, former participants of rush-hour avoidance projects are less concerned with the monitoring of their travel behaviour. They even have a strong preference for ALPR technology as a means of registration, when compared with OBU or apps. This is probably because ALPR requires the least effort from the participants as the rush-hour avoidance are automatically measured with the use of cameras.

The outcome of the survey reveals that privacy surrounding Spitsmijden projects is not a concern for the majority of the Dutch population. Nevertheless, the group who do not want their travel behaviour monitored is very substantial, which indicates that privacy issues are a serious concern for these projects. Furthermore, politically privacy is also a strong concern, and there is strong opposition against the widespread use of ALPR cameras in the Netherlands. Recently, the Dutch parliament restricted the use of ALPR for rush-hour avoidance projects. Exactly because of the privacy issues involved, ALPR usage during Spitsmijden projects is now restricted to road construction works only.

20.6.3 Potential fraud

Politically, the strongest issue of Spitsmijden projects is potential fraud and fraud prevention. As discussed in Section 20.4, only 7% of all the participants end up being excluded from rush-hour avoidance projects. The vast majority of this group is not actually accused of committing fraud but are only lax in responding to requests for information. Nevertheless, despite a broad range of fraud prevention measures, fraud by rush-hour avoidance participants can never be fully ruled out.

In 2016, some members of the House of Chambers asked for explanation about the alleged fraud sensitivity because of media attention surrounding the issue. This led to the tightening of controls. It also led to the Dutch Minister of Infrastructure and the Environment promising to reclaim any unjust rewards and to impose a severe penalty in future Spitsmijden projects.

20.7 Conclusions

Both road pricing and rush-hour avoidance projects have the potential to be successful congestion reduction strategies. Making use of economic incentives can be a fruitful behavioural intervention approach to change existing travel routines of frequent car users. The financial rewarding and charging schemes provide a persuasive external impetus that makes travellers rethink their existing habits. They provide the necessary window of opportunity for travellers to alter existing behavioural routines and make them change their car use during peak hours.

Also from the point of societal issues involved, Spitsmijden projects and road pricing show remarkable similarities. The political and public debate on privacy, fraud and fraud prevention and the required technologies of registration are comparable. However, it is probably safe to state that, from the perspective of public acceptance, rush-hour avoidance feeds less controversy than road pricing. Not surprisingly, road users are more favourable to a rewarding than a charging scheme. Second, Spitsmijden projects are based on voluntary participation which greatly reduces the resistance. Furthermore, Spitsmijden projects are much easier and faster to implement.

The main aim of this contribution, however, is not to conduct an extensive comparison between the two approaches of charging and rewarding. Our main purpose was to show that, over the last 10 years, Spitsmijden projects have evolved to an effective congestion reduction measure which deserves serious investigation from other countries than the Netherlands. As far as known, no other countries have experimented with this concept. In this chapter, we examined the potential contribution of Spitsmijden projects and revealed that, in terms of behavioural change, congestion reduction and cost effectiveness, these projects show remarkable positive results in the Netherlands. Therefore, we encourage road authorities from other countries to start experimenting with their own version of Spitsmijden projects.

Especially, when large road construction works are involved, Spitsmijden projects can be highly successful in mitigating the negative effects of the capacity reduction and increased travel time loss. And, when the project is implemented correctly, the positive behavioural effects may long outweigh the period of the construction works. In addition to a short-term measure during roadworks, rush-hour avoidance can also be implemented to alleviate recurrent congestion at certain points and times on the road network, as opposed to congestion due to accidents or roadworks. In that sense, it can be implemented as an alternative to, or in advance of, road-charging schemes.

References

[1] Road congestion pricing, Santos, G. and E. Verhoef, 2011. In: Palma *et al.* (eds). A handbook of transport economics, 561–585. Cheltenham: Edward Elgar Publishing.

[2] Verse wortelen, sociale en morele vormen van belonen als aanvulling op, of alternatief voor, monetair belonen, J. Berveling, and KiM Den Haag, 2013.

[3] Buying people: the persuasive power of money, Bolderdijk, J. W. (2010). Groningen: Groningen University.

[4] Road pricing vs. peak-avoidance rewards: a comparison of two Dutch studies, Tillema, T., E. Ben-Elia, and D. Ettema, 2010. Paper submitted for the 12th World Conference on Transportation Research, Lisbon, 2010.

[5] The effects of rewards, Spitsmijden summary, Spitsmijden Group, April 2007.

[6] Interventions to break and create consumer habits, Verplanken, B. and W. Wood, 2006. Journal of Public Policy and Marketing, 12: 90–103.

[7] Beyond frequency: habit as mental construct, Verplanken, 2006. British Journal of Social Psychology, 45: 639–656.
[8] Predicting behaviour from actions in the past: repeated decision-making or a matter of habit? Aarts, H., B. Verplanken, and A. Van Knippenberg, 1998. Journal of Applied Social Psychology, 28: 1355–1374.
[9] Consuming mobility A practice approach to sustainable mobility transitions, J.O. Nijhuis, Wageningen: Wageningen Academic Publishers, 2013.
[10] Experimental design and modelling, Spitsmijden, Jasper Knockaert (ed.), Michiel Bliemer, Dick Ettema, Dusica Joksimovic, Albert Mulder, Jan Rouwendal, Dirk van Amelsfort, May 2007.
[11] Resultaten mobiliteitsprojecten, Ministerie van Infrastructuur en Milieu, 14 June 2011.
[12] Eindevaluatie Mobiliteitsprojecten, eindrapportage, MuConsult, 2 November 2013.
[13] Meta-evaluatie Spitsmijdenprojecten, eindrapportage, MuConsult, January 2017.
[14] http://www.beterbenutten.nl/en.
[15] Gedragsmeting Beter Benutten, I and O Research, September 2016, I and O Research commissioned by the Ministry of Infrastructure and the Environment.

Chapter 21

Summary and future prospects for road pricing: open research areas, future work and conclusions

John Walker and Andrew Pickford***

21.1 Introduction

According to Taylor [1] there are five main ways to manage road traffic congestion:

- Increase road capacity – which is expensive.
- Increase public transport (which may be expensive) and encourage 'active travel' (walking and cycling).
- Better management of existing capacity: coordinated traffic signal timing, coordinated public transport, better highway management.
- Technology: satnavs, shared transport services, autonomous vehicles.
- Road pricing to balance supply and demand; not liked by politicians and the public but favoured by transport experts because
  - Congested traffic means reduced throughput – inefficient and irritating;
  - Delay is non-linear – a few cars removed from the road significantly increases traffic speed and throughput;
  - Despite popular opinion to the contrary, some drivers can change mode, time of travel or route, at least some of the time;
  - Road pricing is no more regressive or inequitable than other motoring charges or taxes; and
  - Road-pricing works (as demonstrated in many chapters of this book).

As we saw in Chapter 1, the first two of these approaches, whilst often good investments, will not remove congestion on a sustainable basis, if implemented independently of other measures. The third and fourth certainly have their benefits, though the jury is still out on the impact of autonomous vehicles, but none of them are a complete solution. So the real issue is how to achieve the public (and hence political) acceptability of road pricing and create conditions that are favourable for its implementation.

*Transportation Research Group, University of Southampton, UK
**MVA Hong Kong Ltd., Hong Kong

21.2 Lessons learned from previous chapters

This section summarises and synthesises the lessons learned from the preceding chapters – often using the same words. Early research in the PRoGreSS project in 2004 [2] presented some emerging views from studies and pilots of road pricing, foretelling lessons learned over a decade later as presented in the preceding chapters.

21.2.1 Why road pricing and congestion charging?

Traffic congestion is a growing problem in our cities, imposing significant costs on citizens: wasted time for commuters and freight, contributing to a less efficient economy, increasing air pollution and impaired health. Revenues from vehicle taxation are declining. The UK Institute for Fiscal Studies projects annual decreases from £38Bn (2010) to £25Bn (2029), due to improved fuel efficiency and adoption of alternative fuels in the United Kingdom [3]. Similar concerns in the United States provide the background to the trials described in Chapter 18.

Previous chapters have demonstrated that road pricing and congestion charges do reduce congestion and local air pollution, particularly as part of a broader transportation policy package, as well as being a fair and reasonable substitute for other motoring taxes and charges. These should be the objectives for introducing charges – specifying whether the scheme is addressing pollution and/or congestion and/or fuel duty; and if the former, lower charges for cleaner vehicles.[1] Congestion pricing is an essential – but missing – piece of smart transportation policy. When designed well, a congestion charging scheme reduces traffic congestion and creates net economic benefits for the economy and for individual drivers.

More public transit, roads and cycling infrastructure provide drivers with alternatives, making it easier for them to respond to the congestion price by changing their behaviour. They are essential complements to congestion pricing. But without addressing the fundamental issue of misaligned incentives around free access to roads, traffic congestion will only get worse. Unlike water, gas and electricity, roads are not currently regarded as a utility; if they were, the implementation of 'user pays' policies for their provision and upkeep would be easier.

The prerequisites for a congestion charging scheme are

- There needs to be a well-defined, visible and local congestion problem to be solved.
- Strong political leadership, supported by mainstream political and public acceptance of the policy.

[1] Examples of road pricing schemes that have been implemented show that there is usually one dominant policy; either reduction of congestion or reduction in harmful emissions, so the management of expectations is invariably focused on one policy even if both are apparent. For example, lowering charges for clean vehicles could lead to an increase in the total vehicle population and a commensurate increase in usage rather than a direct substitution of older, polluting vehicles for cleaner vehicles. Policy flexibility, such as the progressive reduction in allowable thresholds of permitted harmful emissions, as used in London, can help maintain incentives for cleaner vehicles without resulting in increases in congestion.

- There will always be opposition; some stakeholders will prefer to use their cars.
- There needs to be a focused, well-managed design and implementation team in place.
- A complementary and coherent package of measures of which congestion charging is a part.

21.2.2 Public support

There will be a need to win over public and political support: the introduction of road pricing is a sociopolitical as well as a techno-economic process, its impact is far-reaching and the stakeholder engagement process needs to reflect this. Framing of the charges as a green policy to reduce congestion and pollution, or to replace fuel duty, is important. But it can take a long time to build up public acceptance and political confidence.

Behavioural psychology studies suggest that losses are twice as powerful as gains. When a congestion pricing project is proposed, even if it is a small part of a package of complementary measures, the public focus is on the potential tolls (loss), triggering the loss aversion reaction. And the potential complexity of congestion pricing results in confusion or doubts about the primary gains of time savings, and who will benefit. Media coverage concentrates on potential losses.

But after a project is implemented, and congestion has been reduced, the loss aversion effect switches in favour of congestion pricing – people do not want to lose the time savings benefit, and the tolls are clearly understood and of less concern.

There are also concerns about 'double taxation' – people think they had already paid for the highway through fuel taxes – ignoring the need for the ongoing costs of road maintenance; they also think that pricing is about raising revenue, not managing traffic. So, ideally, schemes should be revenue neutral, at least initially; but if there is surplus revenue, it should be spent on transport-related projects, including public transport and improved pedestrianisation. It is important to say what any revenues will be used for.

Lessons learned from Minnesota's outreach and education efforts, and from Milan's experiences, are

- Seeing is believing – show, don't just tell; citizens will accept and vote for pricing if they have actually observed it being used successfully.
- Community task forces are an excellent way to facilitate a detailed discussion of the issue amongst key leaders; 'grass-tops' support is a priority; top-level champions are key.
- Public outreach and meaningful engagement improves project design.
- Coalition requires constant maintenance – advocates of road pricing may have diverse and sometimes opposing views on other issues.
- Preparation must precede promotion – including simple visuals to explain the road-pricing concepts (such as Figure 5.4 and the UCLA Infographics on road pricing [4], which indicate how people are affected by road tolling, and how

equitable it is), preliminary project designs, surveys showing that citizens were open to the idea, answers to frequently asked questions and diverse and credible community champions.

- Preliminary technical details must be available – in Manchester and Edinburgh, people did not fully understood the plans and voted against them; few people realised that investment in public transport was the main aim and had to be implemented before the congestion charge could begin. It is also important to stress that the technology for congestion charging is available and proven, and that privacy concerns can be addressed (Section 21.2.4).
- No question should go unanswered; otherwise, it can quickly become an accusation that is believed.
- Flexibility and opportunism are important; unforeseen opportunities may arise; and customised messaging is needed for each stakeholder group.
- Accentuating the positive pays off – using terms such as 'express lanes' and branding such as 'MnPASS'.
- Choice sells – there should ideally be alternative and accessible non-charged routes, and solo drivers should be reminded that they have the option of paying a fee to use an uncongested express lane 'when they are in a hurry' – to link the benefits of an uncongested route (such as time savings) with the cost to drivers of using it.
- Non-governmental facilitators can be useful – such as the University of Minnesota's Humphrey School of Public Affairs – where people may have a mistrust of 'government'.

In Singapore, a key factor in getting the acceptance of the motoring public was that the electronic road pricing (ERP) system was completely reliable and easy to use. Publicity was also an important part of the programme, with road shows, talk shows on the radio and TV, and publicity material sent to all vehicle owners to explain how the ERP system worked and how it would benefit them. It was also stressed that there was no adverse health impact from the electromagnetic waves which were less powerful (and more localised) than mobile phones. An off-road trial, that was accessible for public visits, helped.

It is important to understand each city's own configuration and its place in a national context, and the important actors, arguments and agendas that shape a debate on this approach. This also implies that car dependency cannot be understood as a purely local phenomenon.

- The congestion charge should be placed in the context of a wider strategic transport goal for the area, which needs a proactive approach to stakeholder communication.
- Ideally, realistic alternatives should be considered and presented.
- Plans and claims about impacts must be consistent, and the impacts must be clear and explicit rather than implied.
- The scheme is not about the 'driver' vs. the rest of society but about how a diverse range of people access the services they need, breathe clean air and are protected from the impacts of climate change.

Branding is important; a scheme should have an attractive name, otherwise opponents will impose their own media-friendly branding with terms like 'C Charge' and 'Toll tax', reinforcing the focus on the charging element. Alternatives like 'clean air charge' or 'accessibility charge' may help to move the debate to the positive impacts of such a charge.

Transport policy should be developed strategically over time with education and awareness raising through public engagement. This is challenging since it pertains to societal change in how the transport system is seen.

As regards public support, one question in Sweden is why the referendum in Stockholm resulted in a majority in favour of the charges whereas in Gothenburg it did not. The reasons seem to be that

- Support was always lower in Gothenburg, because of lower usage of public transport and lower congestion levels, so more people were negatively affected by charges, and fewer drivers received significant travel time reductions.
- In Gothenburg, the charges were justified as a tax needed to finance infrastructure investment, but the low social returns of this investment did not help to build public support.

These points reinforce the necessity of tailoring schemes to local conditions; one size does not fit all. There is also a tension between simplicity and effectiveness. The planned charging regime in Edinburgh was more sophisticated than the London charge (only when crossing a cordon, only at certain times, only in the direction of peak traffic, and a double cordon); so it was more precisely targeted, with more options to avoid the charge. But that made it less easy to understand. So schemes should be kept simple, at least initially.

Conversely, in London, following a consultation, the Mayor proposed to reduce the charge for Heavy Goods Vehicles (HGVs) from £15 to £5, the same as for cars, simplifying (and reducing the cost of) the scheme because vehicle classes did not need to be identified. Also, a 100% discount was proposed for Blue Badge holders (mobility-impaired individuals), and a 90% discount for residents of the charging area, thereby increasing public acceptability.

As regards HGVs, the British Freight Transport Association has stated that its members are 'open to ideas about how new road infrastructure should be charged for in the future', under certain conditions which are summarised in their 'Commercial Vehicle Operators' Road Toll Charter' [5] – we need to remember that road haulage is important to the economy and to consumers, and that charges must be acceptable to hauliers as well as to the general public.

21.2.3 Consultation

The process of consultation implies interaction with relevant stakeholders; as various scenarios are developed, policy objectives and operating strategies need to be refined – to demonstrate that the consultation process is meaningful, able to resolve inconsistencies and to develop and maintain political and public acceptance. Perspectives of stakeholders need to be understood – 'what's in it for me?' They will have different perspectives on the impact (including costs and benefits), so consultation should be suitably targeted.

Stakeholder consultation and the means of communication will face one or more of the following:

- Visible transport-related problems: declining air quality, worsening congestion, increasing delays, increasing journey time variability, decreasing revenues from motoring taxes;
- Demand for new capacity and perhaps a lack of public funds to invest accordingly;
- Locally relevant precedents, such as Singapore's Area Licensing System that enabled the evolution to its current ERP policy and its planned 'ERP2' policy, and many of Europe's electronic truck tolling programmes that have evolved from earlier paper-based vignette schemes;
- The existence of proven enforcement processes, particularly those relating to moving traffic violations where compliance checking is largely automated and image-based evidence is acceptable to the courts;
- The availability (or lack) of alternative routes for through traffic, workforce flexibility to change time of travel; and high-quality alternatives such as bus rapid transit, light rail or mass transit services; and
- The stages of the economic cycle where there may be varying appetites for investments in new transport policies.

21.2.3.1 Types of consultation, levels of engagement and setting expectations

Consultation should be a process of dialogue with selected stakeholders, either 'internal' (within the administration) or 'external' (other public authorities, the public and business). Each phase of a consultation process should

- have a specified start and end date;
- inform decisions on the objectives, scope and implementation of the scheme;
- encourage greater involvement from stakeholders;
- help to refine the policy objectives and scheme operations strategy;
- include a range of formal and informal mechanisms; and
- continue after implementation to measure ongoing support and inform about potential changes to policy or operations.

Informal consultation helps to shape policy through dialogue with a limited range of stakeholders (mostly internal); formal consultation helps to refine emergent policies and gain support – or possibly uncover sufficient resistance that could render a scheme infeasible.

The complex relationship from communications through to measurement of attitudes is shown in Figure 8.3. Attitude measurement via social media channels is particularly beneficial, because insights are provided in near real time, allowing prompt responses to misunderstandings or errors.

The relatively long (2.5 years) planning and stakeholder consultation period of Berlin's Air Quality Plan (aimed at encouraging cleaner vehicle use) ensured sufficient levels of acceptance, reflected the complexities of a larger metropolitan population and allowed time to secure additional funding from the national government. The natural replacement rate and migration towards cleaner vehicles meant that tighter emissions targets could be progressively introduced.

21.2.3.2 Consultation in the United States

The efforts to implement a distance-based charging regime in Oregon, California and Washington were all preceded by extensive strategic engagement processes with lawmakers and the public. To convince them of the viability and appropriateness of distance charging to fund road systems, specific steps were needed, in two stages.

- The first stage involved *exploration* including policy development and technical research and development:
  - The government explains why exploring distance charging is a worthy endeavour; this entails communicating the nature of the problem being addressed. For distance charging, the increased fuel efficiency reduces revenue from fuel tax. But more particularly in Oregon, the public perception is the growing inequity of the fuel tax – some drivers paying a lot for road use, others paying very little.
  - The government takes into account particular sensitivities of their citizens – a natural pathway to resolution of issues identified by the public.
  - The government implements a demonstration or pilot trial of the distance charge – people need to experience what is proposed.
- For the second stage – *achieving a mandate* – a more extensive public outreach and education process is needed. Trials are one thing; a legislated mandate is on another scale, and lawmakers must feel confident that the public will consent. There needs to be a public information campaign to dispel myths and inform the voting public of the proposed action and its rationale. Legal action by Congress and the Oregon and California legislatures indicates that this is happening in the United States.

21.2.4 Privacy

In Singapore, the privacy issue was critical; the government announced that all valid ERP transactions would be deleted from the computer system once payments have been settled with the banks, typically within a day.

Oregon addressed privacy in three ways:

1. Motorists can choose whether government or private-sector organisations manage their data collection;
2. Motorists can choose the mileage reporting method and technology with at least one option not depending on knowledge of a vehicle's location;
3. Extensive statutory provisions require the government and the private sector account managers to ensure all personally identifiable information is protected from disclosure. Only evidence of contraventions is retained.

Consequently, privacy-protection issues have been substantially calmed in Oregon.

21.2.5 Equity

Current methods of paying for roads are not necessarily fair and progressive – a message that needs to be publicised, including to politicians and drivers' associations. According to Schweitzer and Taylor [6], 'Most of the equity concerns about road pricing stem from the fact that it is regressive; that is, poorer people spend a larger share of their incomes on tolls than do wealthier people. But in the US, road systems are financed primarily through fuel taxes, vehicle registration fees, property taxes, and, increasingly, sales taxes – all of which are also regressive. Thus the relevant question is not simply whether road pricing is regressive, or even if it will burden the poor. The relevant question is whether road pricing will burden the poor more than other ways of paying for roads'.

Schweitzer and Taylor go on to say, 'we should not subsidise all drivers (and charge all consumers) to help the small number of poor travellers who use congested freeways in the peak hours and peak directions. Rather we should help those who are less fortunate, and see to it that the rest of us pay our own way on the roads'.

Nonetheless, acceptability may be increased if there are mitigation measures such as help for people on low incomes, and reduction of public transport fares through subsidies. According to Chapter 14, level of car dependency among low-income citizens is very important for the resulting equity effects.

If congestion charges are part of a larger investment package, the social impact of the investments must not only be beneficial but must be perceived as being beneficial.

21.2.6 Political support

It is generally accepted that road pricing is politically a difficult thing to do and is perceived to restrain established freedoms. But a scheme needs a political champion.

The most powerful approach in getting public acceptance is 'seeing is believing'. For transportation and political leaders, this means that if they take a political risk and implement an effective and efficient road-pricing project, perhaps on a demonstration basis, they can have confidence that public scepticism will dissipate after residents have experienced it in operation. This was the case in Minnesota, London, Stockholm and Milan.

Before 2006, support was low among political parties in both Stockholm and Gothenburg; but in less than 10 years, congestion charges became uncontroversial and accepted by all political parties in Stockholm and Gothenburg, mainly because congestion charging plays a key role in getting national grants for transport investments. The power over the revenues, the system design and the role in the negotiations on national infrastructure grants seems to be critical for political support.

The rejected Manchester proposals were seen as a response to national priorities, and an opportunistic grab of national funds by local government. A more local approach, as part of an ongoing long-term transport strategy, might have worked better. National policy can be counterproductive in expecting cities to develop quick and competitive responses to national calls. Governance structures can also contribute to inertia in decision-making. Conurbations must be proactive in developing transport policy strategically, allowing time and democratic space for the involvement of businesses, citizens' groups and political representatives; this is not an appeasement process, but a collaborative development of transport policy.

It is important to realise that the viewpoints expressed on either 'side' of any debate are campaign messages, and not necessarily an accurate representation of public opinion. Claims have to be recognised as much tactical – i.e. what would help them achieve their goal – as representative. Furthermore, arguments presented by 'pros' and 'antis' are often two sides of the same coin:

- whilst a charge might 'keep the roads free for the rich', it would permit improvements giving low-income commuters more options not to drive;
- whilst the plans seemed to be saying 'on yer bike' (i.e. 'go away') to poorer commuters, facilitating active travel can be seen as a positive way of boosting health and social inclusion;
- whilst loan elements may be a risk, the funding is an opportunity;
- whilst businesses may incur charges, the scheme will reduce congestion, helping them make their deliveries on time.

The congestion charge must be presented in the context of a wider strategic transport goal for the area. This is not a simple task and implies the need for a proactive approach to communication, based on full knowledge and anticipation of a backlash against the plans, but also to avoid being on the defensive. Strong leadership and political consensus will instil a greater sense of public confidence in the proposed scheme.

In Manchester, the Local Authority felt unable to campaign for their own proposals – they thought they should be neutral, only co-ordinating the consultation and disseminating neutral information about the plans. This is a major and surely an unreasonable handicap, which at the very least could give the impression that they were not really in favour of their own proposals. Minnesota took a much more proactive approach.

Transition from a fuel tax to a distance-based charge for road funding has been a conundrum for US thought leaders, influenced by a long-term decline in federal income from that tax. Moving from an entirely government-operated system to an open commercial market facilitates transition; but legislatures must apply a distance charge mandate to a sizable number of motorists to enable the market to form and prosper. A strategic plan for full implementation must indicate the end state and how transition will occur, including recommendations for which road taxes will be replaced (e.g. fuel tax, vehicle registrations fees, licence fees), which vehicle types will be subject to the charge (and when) and the rate structure.

As the public and policymakers in the United States gain more intimate knowledge of distance charging models, political support should improve (provided the public continues to accept the long-standing 'user pays' policy for road funding). Widespread, mandated distance charging in the United States is not that far away.

21.2.7 Referendum – or not?

A referendum is a very limited way to make a complex decision, provides no guarantee of the most sustainable or fair outcome and limits the potential to revisit road pricing in the future. This is not to discount democracy – as opponents of referenda are commonly accused of – but to point out the benefits of persevering through existing democratic structures. In Manchester, the referendum led to a simplified representation of a complex issue; this suggests that other cities should not hold a referendum.

On the other hand, referenda in Stockholm and Milan gave positive outcomes, because people were shown the benefits in advance, and in Milan, in particular, the experience of the Ecopass low emission zone showed individuals exactly how much (or how little in most cases) the succeeding Area C congestion charge affected them.

In London, the congestion charging scheme was part of the Mayor's transport strategy, as outlined in his manifesto; since he was elected on this manifesto in 2000, he had a mandate for congestion charging without the need for a referendum.

21.2.8 Get your legislation in first

As indicated in Chapter 8, and inspired by London, the Mayor of New York attempted to implement a congestion pricing scheme in 2007, with wide support from local politicians, the press, business and environmental groups, academics, elected officials and city residents. But the scheme needed authorisation under state law; this was blocked, without a vote, in the state assembly, and the plan was abandoned. Schaller [7] points out that 'a relatively small group of auto users believed that congestion pricing was against their best interests... The intensive interests of one group were thus able to overcome widespread public support'.

Conversely, the Greater London Act (1999) included the powers to implement a congestion charging scheme, as did the Transport Act 2000 for the wider United Kingdom. Furthermore, the groundwork for the London scheme had been done in the Review of Charging Options for London (ROCOL) report [8], put together by a number of interested parties for the benefit of the Mayor and other politicians.

21.2.9 Design

21.2.9.1 General considerations for design

The design details of congestion-pricing policy are important. Congestion pricing is not a one-size-fits-all policy solution. Different cities face different types of congestion problems, and tailoring policies to local circumstances is critical for success.

The design of a congestion charging system is difficult and depends on local conditions, such as bottlenecks and route choices. There is a high risk of moving congestion around in the network or causing new congestion. A carefully calibrated local transport model should be used in the design process to reject bad designs.

In both Stockholm and Gothenburg, some suggested systems were rejected because the transport models indicated that they actually created problems. In Stockholm, with extensive spillback queues, a dynamic assignment model was necessary for capturing the full travel time reductions of congestion charges. In Gothenburg, where spillback queues are not a problem, a static assignment model accurately predicted travel time reductions.

Standardising approaches to charging within regional economies means that road users are more likely to understand how to pay for the use of roads, and face road-pricing policies that are familiar. Being able to use the same method to identify user accounts, such as tag or On-Board Unit (OBU) that works anywhere, has been one of the drivers for interoperability particularly in the United States, where interstate interoperability based on ANPR is also being implemented (Chapter 12). The European Electronic Toll Service (EETS) takes this further for EU member states by not only narrowing the choice of technologies for charging but also separates the organisational roles for charging, account management, governance and users, identifying the interfaces between them [9]. This also enables a competitive network of account management service providers and, in principle, means that a road-pricing operator does not need to invest in developing account management services since this resource would already exist from another service provider. Ireland provides a good example of how a competitive network of service providers may be established through role separation as described by the EETS.

There are lessons in the United States from the ongoing transition from regional to national interoperability. Transaction and valid account acceptance are the responsibility of the 'away' agency (the toll facility used away from home). Each 'home' agency maintains a tag status file and a customer licence plate file and distributes them to all other members. By presentation of these lists of transponders and licence plates, the home agency guarantees payment to the away agency. Away toll agencies in turn must determine if an identified transponder or licence plate is 'valid', i.e. associated with an account that a home agency will recognise for payment.

The key to successful inter-agency account acceptance in the United States has not been in the details of the technology, but rather the early adoption of a business rule to require prepaid toll account balances. American toll accounts are similar to debit cards – the money is already in the bank and use of the toll account just draws down on that money. The prepaid account balance allowed independent toll agencies to honour each other's accounts because they already had the customer's money, and each away agency could allow a valid vehicle to pass because they knew they could count on the home agency to pay. Establishing system-wide trust has proven to be a major hurdle in other countries where post-payment toll accounts are the accepted standard; toll operators have struggled with interoperability and the willingness to guarantee payments to other toll operators.

Having said that, American toll agencies today are considering various exceptions to the requirement for prepayment of tolls, so as to provide their customers with new options to increase toll programme attractiveness. In general, standards for the interface between tags and roadside readers are established on a de facto basis and apply regionally, such as E-ZPass in the North East of the United States. However, to prevent road users requiring multiple tags to identify applicable accounts, there were increasing needs for tags that would be interoperable with different readers and/or readers that could interrogate tags meeting different standards. Efforts in the United States towards developing and implementing national standards have been inconclusive although ISO 18000-6C has emerged, currently with limited use in the United States but has been adopted elsewhere such as in Malaysia and Taiwan as the basis of national tolling schemes. Other public domain standards include those developed by CEN TC278 in use throughout Europe, Australia, Chile and other countries, and in China where its national standard GB/T20851-2007 is being employed extensively for plaza-based tolling.

In 2012, the International Bridge Tunnel and Turnpike Association established a technical subcommittee to develop a process to identify a preferred [RFID-based] transponder/reader technology to [comply with] a 'national interoperability protocol' from among the seven different technologies in use in North America. A back-office subcommittee was created to come up with a uniform set of business rules that would work across all agencies.

In summary, we can contrast the top-down approach to EETS with the largely technology and systems-led approach in the United States. In both cases, interoperability is more than simply complying with regulations or standards but fairly allocating operating risks, setting and meeting the expectations for road users and on the whole establishing a reliable, trusted, accurate charging and enforcement regime. This could be considered one of the prerequisites for the political and public acceptability on which the pricing of our roads depends. In general though, each scheme faces its own challenges and defines its own benefits, and technologies are no more than an enabler as we have stressed above.

For example, the London Congestion Charging scheme depends on the collection of evidential quality images to check eligibility of vehicles to be used with the charged area. This was pre-EETS and the ROCOL report [8] suggested that a camera-based method would be most practical compared with a tag-based solution despite standards being available for the latter. The main conclusions drawn from the success of the London scheme are

- For the scheme to be fair, there must be checks and balances;
- Using licence plate recognition, the standard process must deal with the risk of duplicate registrations (clones or other countries using the same registration format);
- Robustness of evidence: images need to be supported by timing and location detail that is acceptable to the courts;
- Image detail has to allow comparison of features such as location on the street and context and the features on the vehicle or its number plate;
- The enforcement process must allow a customer to see the evidence and respond, to reduce risk of incorrect enforcement;
- Human involvement in enforcement is needed; automated processes cannot always distinguish image differences or anomalies that arise;
- Gathering and storage of the evidence and protection from tampering has to be auditable.

The US states of Oregon, California and Washington started with policy development for distance charging at the direction of each state's legislature. They adopted the common principles of user pay; open system; user choices; private-sector administration option; transparency; cost effectiveness and efficiency; system flexibility, adaptability and scalability; reliability, security and technological feasibility; simplicity of compliance and ease of use; ease of administration, enforceability; privacy protection; accuracy and system performance; fairness and equity; minimal non-governmental burden and integrated with other state policies. In addition, California and Washington specified accountable oversight and interoperability.

All three states piloted an open system design and a choice of commercially provided mileage reporting methods and account managers; Oregon and California also offered state account management. As a result, the key issues of complexity of implementation, difficulty of operations, simplicity of compliance, scalability, flexibility for policy adaptations and evolution of technologies and business systems are all essentially resolved. Security of data is continually managed. Interoperability among state systems and non-resident driving is being addressed and resolved. Finally, issues of equity by income, geography and vehicle type, enforcement, cost of start-up and administration and any transition process will be determined by the state legislatures.

21.2.9.2 Design of GNSS-based tolling systems

There is much experience with Global Navigation Satellite System (GNSS)-based tolling systems throughout Europe for HGVs (as well as in the United States for pilot distance-based schemes), yet each new tender has so far reinvented the approach towards defining and implementing a system. What lessons can we learn from the existing systems?

1. Requirements have varied widely so much of the technology had to be redeveloped, increasing costs and lead times, and increasing the risks of delay and failure.
2. Technology vendors can build an adaptable solution to mitigate these effects; but this can be challenging given the short time typically between contract award and system launch.
3. Most existing systems were devised to start operation with a large road user base and wide road network coverage. It would be better to gradually increase system size and complexity, starting initially with a few vehicles. Since the development overhead for the GNSS technology and Information Technology (IT) infrastructure is basically the same for both small and large numbers of users, the contract between the tendering authority and the vendor could specify a guaranteed initial number of users, with subsequent increases, so that the capital expenditures are distributed among a larger number of OBUs.
4. The risks of jamming and spoofing have so far been minor since the achievable economic gain is relatively small, while the perpetrator risks large fines. Galileo will introduce signal authentication services from 2018, to ensure that Galileo receivers will only accept direct signals from Galileo satellites and not from other potentially interfering sources.

5. At least for truck tolling, a dedicated OBU is difficult to replace; the necessity for reliable operation far outweighs the potential cost savings of smartphones (as described in Chapters 6 and 16, and Section 21.3.2). Hauliers are primarily interested in the smooth and reliable fulfilment of their tolling obligations – not having to worry that their OBUs are working properly.
6. Although most of the systems implemented so far favour a 'fat-client' solution (i.e. sophisticated OBUs with largely independent functionality, including maps, to calculate charges), there is a trend towards 'thin-client' architectures. With the improved coverage and bandwidth of mobile communication services, the advantages of making frequent changes in map information centrally can outweigh the higher data payload of transmitting position data from the OBU. Also, communication availability and round-trip delays in mobile networks have considerably improved.

The flexibility of GNSS technology potentially allows for all roads to be tolled, addressing traffic diversion and toll evasion, with a tariff structure that discourages vehicles from using specific roads or entering certain areas, and reduces vehicle emissions in sensitive areas, including in real time (e.g. a smog alarm). As an enabling technology, GNSS is flexible and notwithstanding compliance with EETS amongst EU member states and mandates emerging within the United States on road users, other approaches may afford reduced policy flexibility but may meet local needs better. Singapore's planned implementation of its GNSS-based ERP2 is likely to show the way ahead for all-roads charging but as long as there are other needs for charging, at toll plazas, cordons and for car parks, we may be heading for a multi-technology future where geolocation as well as short-range communication will be necessary to meet different policy demands on a typical journey, plus the occasional need to support the compliance checking process that may require localised data exchange with on-board equipment.

21.2.10 Exemptions

Exemptions have a large effect on the equity and efficiency of the charging system and should be carefully analysed in advance. Exemptions might also increase operational costs of the scheme, particularly if there are many categories or non-compliant behaviour is difficult to detect, so there should be as few as possible – though as we have seen, they may also increase acceptability of a scheme.

21.2.11 Lessons learned: operating costs

The system will have a significant cost to implement and operate; most scheme designs start with optimistically low cost estimates. For an ANPR-based scheme, the quality of vehicle registration and other data is key to successful operation; data processing should be automated to keep down operating costs but must allow manual intervention to resolve anomalies and unintended outcomes; a well-designed scheme should be scalable to operate in a larger area or serve multiple schemes; a shared services approach based on interoperability and standardised interfaces will maximise efficiency and cost effectiveness so that more net revenue can be invested in transport – rather than every new scheme implementing its own back office and customer payment channels; the impact of road-pricing schemes cannot always be predicted (though modelling will help) – we must allow for flexibility in outcome, including in supply chain and contracting – adaptability is key. Motorists are cost sensitive, depending on journey purpose, so some trips may be changed or abandoned, while others will not be.

There is some disagreement about scheme overheads – costs of operation (important for public acceptability). The authors of Chapter 13 are sceptical that it is possible to reach ambitiously low targets such as 5% of revenue collected – though this might be due to the particular circumstances of implementation of the London scheme.[2] But according to Chapter 18, the most recent research on the costs for a large-scale distance-based charging system indicates, unsurprisingly, that economies of scale apply; the more payers and charges that are paid, the lower the operational overhead. With 250,000 payers, it is estimated at 16%–18% of revenue; for 1 million payers, it is 8%–10%, and with 6 million, it falls to 4%–6%. These figures are in line with cost estimates for a Dutch scheme, which suggested that running costs could be as low as 3% – see Walker 2011 [16]; although this scheme was not implemented, the figures are 10-years old and equipment costs in particular have fallen significantly since then, implying that this low overhead cost could be achievable.

21.2.12 Short-term and long-term effects

In the short term, congestion levels are of key importance for there to be travel time gains. In the long term, city density and public transport market share are important factors for the success of congestion charging. Furthermore, as Eliasson [10] has pointed out, there are longer term processes, such as change of job or residence, which affect travel patterns.

21.2.13 Complementary measures

Experience in London indicates that complementary measures are needed alongside the implementation of a congestion charging scheme, including

- ‘Soft’ measures to promote usage of alternative modes of transport;
- Provision of additional public transport capacity prior to commencement of a congestion charging scheme (important for public acceptability reasons even if it doesn’t result in mode shift);
- Implementation of controlled parking zones (CPZs) to prevent drivers leaving their cars just outside the zone or swamping local parking near stations;

[2] The running costs of the London congestion charging scheme were very high at 42% of the revenue (see Walker [16] Table 13). But this was because the London scheme was totally new, with little comparable experience elsewhere. And as a pathfinder, the scheme had to be designed to be proof against legal challenge, which meant using relatively expensive technology.

- Ensuring that residents and businesses located just outside the zone boundary derive some benefits from the scheme, including CPZs;
- Traffic management measures to discourage traffic rerouting to evade the charge, thereby creating more congestion or other disbenefits;
- Complementary measures further from the zone boundary are important; drivers may make rerouting decisions a great distance away from the congestion charging zone;
- Improvements to the road network, to reduce congestion;
- Ensuring that there are no major roadworks within or near the zone boundary prior to 'go-live' date;
- Improvements to road signage, especially approaching the charging zone;
- Implementation or upgrading of the Traffic Control Centre;
- Additional journey time information gathering along the key radial routes.

21.2.14 Lessons learned: implementation and expansion

London's experience is that a coordinated and well-managed implementation can be delivered in a short timescale, such as a politician's term of office. Many activities can occur in parallel including preparing legislation, designing the operating model and the cost and revenue model, stakeholder consultation, technology trials, on-street surveys of infrastructure and the implementation plan – though many of these are interdependent. The programme must be treated as a series of phased milestones; and having a focused 'project manager for each major milestone' is more likely to result in success.

After several years of Area C in Milan, trips were down 22% and did not shift to other times of day; so trips were avoided or moved to public transport or non-motorised modes, in line with policy goals. Most vehicles entering the area are public (buses, taxies), authorised (police, ambulances, disabled) or non-polluting. There are peaks of vehicles entering just before and after the priced period, but they are no higher than during the previous Ecopass period and do not generate congestion because they are early or because they are counterflow traffic. Emissions are locally reduced, but only marginally in the larger metropolis. There was a less than 2% decrease in house prices and less than 1% increase in rents.

Other good practices are

- multiple payment channels to facilitate payment of the toll;
- no season tickets which would encourage extra unnecessary trips;
- transparency in policy goals and reporting of results.

The congestion charging zone and its technology can also serve other purposes such as low-emission zones. The most surprising fact of Milan Area C is not that it reduced the trips in the priced area but that the policy was accepted by the population. Some of this is in line with previous findings but others are less intuitive and would be good practice for other cities.

- The referendum took place after a several-year trial period in which the population saw how much they were individually affected, positively or negatively. As indicated in Section 21.2, 'seeing is believing' – this direct experience is worth much more than any communication campaign.

- The acceptability is different between occasional and regular travellers; the latter were already walking or using public transport to enter the city centre. Cars are typically used for infrequent occasions, such as a hospital visit or shopping, resulting in small annual payments.
- There are concessions to those who must drive regularly, such as residents and logistics operators, neither of whom are price sensitive.

However, there were other helpful factors in Milan which might not apply elsewhere:

- Most of the technological infrastructure (gates and control centre) already existed and was paid for. It may be hard to justify to the public the investment needed for a trial (though of course that is what happened in Stockholm).
- Area C is small, central, well served by public transport, with few residents and with a predominance of offices and shops. If it had been larger (like Stockholm) or less well served by public transport (like Rome), resistance to the toll would probably have been higher.

Other lessons are that

- The effects (price elasticities) of increased charges and extensions of the system cannot be expected to be as large as the effects at introduction.
- The unit costs per transaction can be reduced through a shared services approach to back office, payment channels and enforcement.

21.2.15 Post-event benefits

Singapore's experience in enhancing the ERP scheme is that there is always room for improvement. The motoring public asked for more features for their convenience e.g. payment of car-park charges at off-street car parks. With accurate position data, other value-added services can be provided; even if they do not generate additional revenue, they create added benefits that can help to increase user acceptance.

The congestion charging schemes are socially beneficial in both Stockholm and Gothenburg; cost–benefit analyses showed that the benefits (primarily travel time savings, but also safety, noise, pollution and CO_2 reductions) exceed the governmental costs (operating costs and fuel tax decreases).

Using the core modules of the HU-GO HGV tolling system in Hungary, additional transportation and customer account management services are being developed, including a national mobile payment service, a national electronic ticketing platform for public transport, a public road trade control system to combat fraud and tax evasion and a weigh-in-motion scheme.

21.3 The future

It would appear that GNSS technology will be central to the future of road pricing and congestion charging, because of its flexibility, continuing cost reductions and as a platform for other services. It is also the case that GNSS functional capability is being embedded within new vehicles anyway. New GNSS tolling systems are under investigation in Europe (Sweden, Switzerland, Bulgaria, Poland and the Czech Republic), South America, Singapore and the United States for charging vehicles for their road usage. We are witnessing a paradigm change, where not just road networks but also groups of vehicles (trucks, electric cars) can be charged distance-based fees. This will lead to a large increase in the installation base of GNSS OBUs, reducing system costs due to economies of scale. Associated technologies based on short-range communication will also play a part, particularly for enforcement, and to enable the continuation of smaller local schemes such as cordon-based congestion charging systems. In practice, enforcement of a GNSS system will also depend on DSRC to permit localised interrogation of GNSS OBUs, as shown for example in some GNSS-based heavy goods vehicle tolling schemes in Europe. The future of technologies for charging will therefore depend on hybrid equipment and, of course, the use of image-based enforcement will continue to depend on ANPR.

Nonetheless, we need more research and guidance – not so much in technology but in public acceptability, including behavioural economics and psychology [11–13]. We also need more implementations – there is no substitute for experience. Pathfinder projects such as *Rekening Rijden* and *Anders Betalen* in the Netherlands can help show the way, particularly if we regard road pricing as contributing to behaviour change.

21.3.1 Spitsmijden

'Spitsmijden' ('avoiding peak traffic') projects are an effective and cost-effective congestion reduction measure which deserve serious investigation from countries other than the Netherlands. The evidence is that Spitsmijden projects can significantly mitigate the capacity reduction and increased travel time during roadworks, and the effects may persist long afterwards. They can also be used to alleviate recurrent congestion on roads, either as an alternative to, or in advance of, road charging schemes.

One Dutch paper [14] has made a preliminary comparison of the effectiveness of road pricing vs. Spitsmijden on commuter behaviour, based on two different Dutch studies. The road-pricing study was 'stated preference' (in other words, drivers were asked what they would do), whereas the Spitsmijden study was based on actual behaviour, and the travel distances involved were different, so the results should be seen as indicative rather than definitive. But in line with psychological theory [15], which suggests that people are more motivated to change behaviour when rewarded rather than punished, rewarding (Spitsmijden) seems to be somewhat more effective in diverting commuting trips from the peak period than road pricing. In both cases, the most popular alternative to peak driving was off-peak driving, followed by using public transport; carpooling, cycling and working at home were significantly less popular options.

As the authors state, it would be interesting to have a joint study in which both rewarding and road-pricing measures were included simultaneously, allowing a more definitive comparison of effectiveness and acceptance – though in a real situation (as opposed to an experiment), one issue would be who pays the rewards (tax payer, employer, the average car driver?). They tentatively suggest that Spitsmijden is more appropriate at a local level, whereas road-pricing may be more viable at regional or nationwide scale.

Certainly further research is needed.

21.3.2 Smartphones

There are some small differences of opinion on the acceptability of smartphones as on-board units for charging. They are not acceptable in Singapore (Chapter 6) because of concerns about drivers forgetting to take the smartphone with them, or not activating the congestion charging application, as well as the potential for tampering and fraud – so the on-board device has to be a customised one that guarantees compliance. A similar opinion is found in Chapter 16. On the other hand, they are an option for drivers participating in the West Coast Distance Charge trials (Chapter 18); it will be interesting to see if they are found to be acceptable.

21.4 Conclusions and recommendations

21.4.1 Conclusions

The experience of the schemes described in this book have been positive, and the lessons learned are applicable to other schemes worldwide; they include

- To succeed needs substantial political commitment, leadership and a focused delivery team;
- If transportation and political leaders implement an effective and efficient road-pricing project, public scepticism will dissipate after residents have experienced it in operation;
- Complementary transport and highway measures are needed to make a scheme and the road network easier to use;
- Softer measures are as important as 'hard' measures to achieve success;
- There needs to be buy-in from the scheme proponents, road users and other stakeholders even if there is also significant opposition;
- Scheme costs can be optimised over time;
- Small schemes may be relatively expensive to implement and operate – but they can still be successful;
- The key to making schemes cost effective is to share common elements between cities, toll roads and crossings.

21.4.2 Recommendations

Our recommendations are

1. Cities should implement road-pricing pilot projects, customised to their local context, meeting the needs of local stakeholders and delivering local and visible benefits;

2. Provincial and regional governments should initiate, enable and facilitate road-pricing policies;
3. National governments should help fund pilot projects. They should carefully evaluate the performance of these projects, communicate the results broadly and incorporate lessons learned into future mobility policies, including at national as well as regional and local levels; and
4. Given how highways are currently funded, the shift towards use of electric and hybrid vehicles, and falling revenues from fuel and other motoring taxes, the arguments for national road pricing are compelling – though this should be implemented in an evolutionary fashion rather than by a 'big bang' approach.

Most importantly, in order that cities may grow without compromising their social, economic and environmental sustainability, the transportation policies need to be equally flexible and adaptable in the long term. In practice, this says less about vehicles but rather emphasises the importance of maintaining the movement of goods and people – in which road pricing has an indispensable role to play.

References

[1] Taylor, BD (2017) "THE ACCESS ALMANAC: Traffic Congestion Is Counter-Intuitive, and Fixable", ACCESS Magazine, Spring 2017, University of California, Los Angeles.

[2] PROGRESS Project 2000-CM.10390, Pricing Road use for Greater Responsibility, Efficiency and Sustainability in cities. (Bristol, Copenhagen, Edinburgh, Genoa, Gothenburg, Helsinki, Rome, Trondheim). Main Project Report, Version 1.0, July 2004. See also "Congestion Charging – The International Perspective", Dr Nick Ayland, Transport and Travel Research Ltd. (Undated).

[3] Johnson P, Leicester A, and Stoye G (2012) "Fuel for Thought: The What, Why and How of Motoring Taxation", London: Institute for Fiscal Studies and RAC Foundation, May 2012.

[4] UCLA (2017) Infographics on Road Pricing – http://www.its.ucla.edu/info-graphic-road-pricing/ [Accessed 21 Jan 2017].

[5] Freight Transport Association (undated), "Commercial Vehicle Operators' Road Toll Charter". http://www.fta.co.uk/policy_and_compliance/road/road_network/road_pricing.html [Accessed 8 July 2017]. But note that items 2 (Compensatory reduction in fuel duty) and 7 (Toll revenue to be invested in the roads to which they apply) appear to be inconsistent.

[6] Schweitzer L and Taylor BD (2010). Just road pricing. ACCESS (Magazine of the University of California Transportation Center), 36: 2–7.

[7] Schaller, Bruce, New York City's Congestion Pricing Experience and Implications for Road Pricing Acceptance in the United States, pub Transport Policy 17 (2010) pp. 266–273.

[8] ROCOL (1999). Road Charging Options for London: A Technical Assessment. London: HMSO.

[9] Directorate General for Mobility and Transport (2011) "Guide for the Application of the Directive on the Interoperability of Electronic Toll Systems" Sec 3.1. Available at https://ec.europa.eu/transport/sites/transport/files/media/publications/doc/2011-eets-european-electronic-toll-service_en.pdf [Accessed on 30 July 2017].

[10] Eliasson, J. (2010). So You're Considering Introducing Congestion Charging? Here's What You Need to Know: An FAQ Based on Stockholm's Experiences. OECD/ITF Joint Transport Research Centre Discussion Paper No. 2010/4, International Transport Forum/OECD Round Table on Implementing Congestion Charging.

[11] Sunstein CR and Thaler RH (2009) "Nudge: Improving Decisions About Health, Wealth and Happiness", New York and London: Penguin Books.

[12] Kahneman, D (2012) "Thinking, Fast and Slow", London and New York: Penguin.

[13] Halpern D (2015) "Inside the Nudge unit", London: Penguin Random House.

[14] Tillema T, Ben-Elia E, and Ettema D (2010), "Road pricing vs. peak-avoidance rewards: a comparison of two Dutch studies", 12th World Conference on Transportation Research, Lisbon 2010.

[15] Kahneman, D, and Tversky, A (1984), "Choices values and frames." American psychologist, 39 (4), 341–350. See also Kahneman, D., and Tversky, A. (1979), "Prospect theory – an analysis of decision under risk." Econometrica, 47 (2), 263–291.

[16] Walker, J. The Acceptability of Road Pricing. RAC Foundation; 2011. Available at http://www.racfoundation.org/research/economics/road-pricing-acceptability [Accessed 13 October 2017].

Chapter 22
Afterword

John Walker and Andrew Pickford***

The last 25 years have witnessed more changes in transport policy than we have seen over the last 100 years. We have the spectre of the decline in revenues from fuel duty, alternative-fuel cars, growth in application of 'user pays' principles, the 'big shift' of transport agencies from asset managers to service providers, ubiquitous mobile technologies, the advent of automation, new satellite constellations bringing accurate location determination – and of course, we continue to face the ugly realities of increasing congestion and the contribution of harmful emissions to climate change.

It is over 50 years since the UK government published its seminal work 'Traffic in Towns', known to professionals as 'The Buchanan Report' [1] which predicted the massive growth of road transport (at a time when the average level of car ownership in Europe was just over 100 cars per thousand inhabitants). Nevertheless, the report indicated that urban design and new technologies may have a role to play. The Smeed Report on the Technical and Economic Possibilities of Road Pricing was published by the same Government Ministry a year later in 1964 [2]. Many of the innovations in road user charging, described in this book and Pickford and Blythe [3], show that we are still facing challenges, in terms of efficiently managing traffic, internalising some (or all) of the external costs of road use and dealing with the disbenefits of travel, congestion and pollution. There is also the pressing need to mitigate the effects of greenhouse gases and their contribution to climate change – along with possible future energy shortages, two of the most pressing problems (both with a transport component) that the world must face over the coming decades. Pricing has a key role to play in weaning us off our love affair with the car in the West, or at least in mitigating its effects, and in helping the emerging economies to deal with traffic demand (which seems to be inextricably linked to economic growth). This is a challenge in which those of us dealing with road pricing and innovative demand management schemes have a clear and leading role to play.

*Transportation Research Group, University of Southampton, UK
**MVA Hong Kong Ltd., Hong Kong

The challenge will be met, and hopefully this book will help aspiring champions to come up with implementations of road user charging. However, as the foreword to the Buchanan Report so eloquently put it, 55 years ago:

> *We are nourishing a monster of great potential destructiveness. And yet we love him dearly ... the motor car is clearly a menace that can spoil our civilisation. But translated into terms of the particular vehicle that stands outside the door, we regard it as one of our most treasured possessions or dearest ambitions, an immense convenience, an expander of the dimensions of life, an instrument of emancipation, a symbol of the modern age. To refuse to accept the challenge it presents would be an act of defeatism .... we must meet it without confusion of purpose, without timidity over means, and above all without delay.* Sir David Crowther [1].

References

[1] Buchanan Report (1963). *Traffic in Towns*. UK Ministry of Transport, HMSO, London.

[2] Smeed R (1964). *Road Pricing: The Economic and Technical Possibilities*. UK Ministry of Transport, HMSO, London.

[3] Pickford A, and Blythe P (2006). *Road User Charging and Electronic Toll Collection*. Artech House, Boston, London.

Appendix A

Glossary of acronyms and technical terms

3-DES	triple DES encryption algorithm
3G	third generation mobile telephony e.g. W-CDMA
4G	fourth generation mobile telephony e.g. LTE
AASHTO	American Association of State Highway and Transportation Officials
ABD	Association of British Drivers
ABvM	Anders Betalen voor Mobiliteit – Different Payment for Mobility – a Dutch road pricing proposal
ACM	automatic coin machine, a machine for automatically processing and counting cash payments paid directly by road users on toll plazas
ACPO	the UK Association of Chief Police Officers (replaced in 2015 by a new body, the National Police Chiefs' Council)
ADEPT	Automatic Debiting and Electronic Payment for Transport: a European Research Project on advanced communications for road user charging (1991–1995)
ADU	application data unit
AES 128	Advanced Encryption Standard, an encryption standard based on 128 bit keys
AET	all-electronic toll collection
AETIS	Association of Electronic Toll and Interoperable Services.
AFC	Automatic Fee Collection (synonym for EFC)
AGMA	Association of Greater Manchester Transport Authorities (UK)
AID	application interface definition
AISCAT	Associazione Italiana Società Concessionarie Autostrade e Trafori. The association of Italian Toll Chargers.
ALPR	automatic license plate recognition (US usage – for enforcement of electronic tolling systems)
ALS	area licensing scheme – the Singapore paper licensing system used from 1975 to 1999

AMAT	Agenzia Mobilità Ambiente e Territorio (Agency for mobility, environment and land use) of the city of Milan
ANPR	automatic number plate recognition (UK usage – for enforcement of electronic tolling systems)
ANSI	American National Standards Institute
ANWB	Algemene Nederlandse Wielrijders Bond – the Royal Dutch Touring Club, the Dutch National Breakdown Service
APDU	application protocol data unit
Area C	the Milan congestion charge area
ASECAP	Association Européenne des Concessionnaires d'Autoroutes et d'Ouvrages à Péage (European Association of Companies with Concessions for Motorway, Bridge and Tunnel Tolls) www.asecap.com
ASETA	Asociación de Sociedades Españolas Concesionarias de Autopistas, Túneles, Puentes y Vías de Peaje – the association representing Spanish Toll Chargers)
ASFA	Association des Sociétés Francaises d'Autoroutes – the French Toll Charger's association
ASFINAG	Autobahnen und Schnellstrassen-Finanzierungs-Aktien Gesellschaft, the operator of the Austrian Truck Tolling scheme
ASN.1	abstract syntax notation one
ATI	the Alliance for Toll Interoperability
ATS	American Traffic Solutions – a provider of toll services for the rental car industry in the United States
AutoPASS	brand name for the Norwegian national road user charging system
Autopay	payment based on an automatic account set up by the driver (for the London Congestion Charging Scheme)
AVC	automatic vehicle classification
AVI	automatic vehicle identification
AVL	automatic vehicle location
BC	black carbon
Beidou	the Chinese satellite-based location system
BRT	bus rapid transit
BSP	bound service provider (in the Hungarian tolling system)
C2X	car to car and car to infrastructure communication
CAFE	Corporate Average Fuel Economy standards in the United States
CALM	continuous air interface for long and medium range initiatives in DSRC

CAM	commercial account manager
capex	capital expenditure
CARD-ME	Concerted Action for Research on Demand Management in Europe – a European project
CAV	connected and autonomous vehicles
CBA	cost–benefit analysis
CBD	central business district
CC	congestion charging
CCC	compliance check communication
CCTV	closed circuit television
CCZ	congestion charging zone
CEN	European Committee for Standardization (Comité Européen de Normalisation, www.cenorm.be)
CEN/CLC/TC 5	The CEN task committee TC 5 deals with topics related to space technology. Working Group 1, CEN/CLC/TC 5/WG 1, deals with navigation and positioning receivers for road applications.
CEN/TC 278	the CEN committee for Intelligent Transport Systems
CENELEC	Comité Européen de Normalisation Electrotechnique – the European Standards organisation
CEO	chief executive officer
CEPT	Conférence Européenne des administrations des Postes et des Télécommunications – the European Committee of PTTs
CESARE	Common Electronic Fee Collection System for a Road Tolling European Service – a series of projects set up by ASECAP, the Stockholm Group of European Road Public Administrations and the European Commission
CN	cellular networks
CAN bus	controller area network bus – a vehicle standard that allows microcontrollers and devices to communicate with each other
CO_2	carbon dioxide – a gas contributing to global warming
Congestion charging	a road user charging scheme in which the charge varies depending on the level of congestion (or, as a proxy, the time of day or day of the week)
COST	European Cooperation in Science and Technology
CPZ	controlled parking zone
CRCPP	The California Road Charge Pilot Program
CRM	customer relationship management

CS	central system – a generic term for the road user charging system's back office that encompasses data gathering, customer relations, operations and maintenance, enforcement record processing and reporting
CSC	customer service centre
CVO	commercial vehicle operations
DAB	digital audio broadcast
DBFO	design, build, finance and operate – a public sector initiative to inject private capital into road-building
DC	The District of Columbia, United States
Dead reckoning	more or less approximate method of measuring distance travelled, often by wheel rotation
DES	data encryption standard (a US government standard)
DfT	The UK Department for Transport
DG TREN	Directorate General for Transport and Energy, a European Commission Directorate
DGPS	differential GPS
DIRECTS	demonstration of interoperable road-user end-to-end charging and telematics systems – a UK DfT-funded project
DLR	The Docklands Light Railway in London
DMV	Department of Motor Vehicles – the US local government agencies that register vehicle and ownership details
DOT	Department of Transport (generic) or US Department of Transportation
DR	disaster recovery
DRG	dynamic route guidance
DSRC	dedicated short-range communication. A standard for microwave and infra-red-based vehicle to roadside communications
DVLA	The UK Driver and Vehicle Licensing Agency
EasyGo	a joint venture between Toll Chargers in Scandinavia and Austria, including the Norwegian and Swedish Road Administrations and publically owned companies ASFINAG (Austria), Storebæltbridge (Denmark) and Øresund-Bridge (Denmark). Its aim is to provide an interoperable ETC service to users throughout Norway, Sweden, Denmark and Austria
EC	European Community/European Commission
EETS	European Electronic Toll Service
EFC	Electronic Fee Collection
EGIS	a French engineering and transport systems company
EGNOS	European Geostationary Navigation Overlay System

eGo®	the brand name for a microwave tag supplied by TransCore. A communications protocol compliant with the ANSI National Committee for Information Technology Standards (NCITS) 256-2001 Part 4.2 and the ISO 18000-6B standards
EIRP	effective isotropic radiated power (i.e. radiated power relative to an isotropic source)
EMC	electromagnetic compatibility
Emovis	the new name for the ITS division of SANEF
EN	European Norm (Standard)
ENP	electronic number plate
ENV	European pre-standard
EPCglobal®	an initiative to innovate and develop industry-driven standards for the Electronic Product Code™ (EPC) to support the use of Radio Frequency Identification (RFID)
Ephemerides	accurate data about the GNSS satellite orbits. The position of each satellite can be calculated from this
ERC	European Radiocommunications Committee
ERM	ETSI EMC and radio spectrum matters TG37
ERP	electronic road pricing (as used in Singapore particularly)
ERTICO	ITS Europe – a partnership of companies and institutions involved in the production of Intelligent Transport Systems
ESA	European Space Agency
ETC	Electronic Toll Collection
ETR 407	Express Toll Route 407 (Canada) (www.etr407.com on 27 Dec 2016)
ETSI	European Telecommunications Standards Institute (www.etsi.org on 27 Dec 2016)
ETTM	electronic toll and traffic management
EU	European Union
EUR	the Euro currency
EVI	electronic vehicle identification
express lanes	an alternative term for high occupancy toll (HOT) lanes, but emphasising the consumer benefit
E-ZPass	The E-ZPass Group is an association of 38 toll agencies in 16 states, primarily in northern and eastern United States (http://www.e-zpassiag.com/ on 27 Dec 2016)
False positive	an undetected error such as a high confidence but incorrect ANPR record
FAST Act	Fixing America's Surface Transportation Act
fat client	Also known as 'thick client' or 'intelligent client'
FCC	Federal Communication Commission (US)

FDOT	Florida Department of Transport (www.dot.state.fl.us on 27 Dec 2016)
FHWA	US Federal Highway Administration
FL	The US state of Florida
FRAND	fair, reasonable and non-discriminatory approach
FTA	The British Freight Transport Association – which represents hauliers in the United Kingdom
GAGAN	GPS-Aided Geo Augmented Navigation – a satellite-based augmentation system for Indian Airspace
Galileo	the European satellite-based location system
GBP	British pounds sterling
GDF	geographic data file format
GDP	gross domestic product
geo-fence	a virtual geographic cordon implemented using GNSS technology
GFIP	Gauteng Freeway Improvement Project in South Africa
GHz	a frequency of 1,000,000,000 Hz
GIS	geographical information system
GLA	The Greater London Authority
GLC	The Greater London Council, formed in1965, abolished in 1986.
GLONASS	GLobal Orbiting Navigation Satellite System (Russia)
GM	Greater Manchester, United Kingdom
GMMG	Greater Manchester Momentum Group
GMPTA	Greater Manchester Passenger Transport Authority
GMPTE	Greater Manchester Passenger Transport Executive (now renamed Transport for Greater Manchester – TfGM)
GMTF	Greater Manchester Transport Fund
GMTIF	The bid of the Association of Greater Manchester Authorities to the UK government's Transport Innovation Fund
GMWDA	Greater Manchester Waste Disposal Authority
GNSS	Global Navigation Satellite System – the generic term for a satellite-based location system
GPRS	general packet radio service (GSM)
GPS	the US-operated Global Positioning System
'grasstops' leaders	influential advocates or officials with a high professional or public profile – as opposed to 'grassroots'
ground truth	reference data for corroboration of GNSS positioning
GSM	global system for Mobiles (originally 'Groupe Speciale Mobile') – the, European Cellular Mobile Phone Standard

GSS	global specification for short range communication – originally established by three European companies to achieve interoperability of Electronic Toll Collection equipment
GTA	Greater Toronto Area, Canada
GVA	gross value added
HDR	high data rate
HELP	heavy vehicle electronic license plate program. A US AVI system for trucks for interstate-border toll-payments
HGV	heavy goods vehicle
HMSO	Her Majesty's Stationery Office (United Kingdom) – the source of official Government publications
HOCAST	UK Home Office Centre for Applied Science and Technology
hot-listed	a vehicle which has been identified as having previous infringements
HTA	highway toll administration – a provider of toll services for the rental car industry in the United States
HMI	human machine interface
HOT	high occupancy & toll lane – the same as HOV but can also be used by single-occupancy vehicles on payment of a toll. Also known as managed lanes, or priced express lanes.
HOV	high-occupancy vehicle lane – a traffic lane which can be used only by vehicles with a certain minimum number of occupants
HTA	highway toll administration – a major US provider of toll services for the rental car industry
HUF	Hungarian Forint. There are currently (June 2017) 308, 274 and 353 forints per EUR, USD and GBP, respectively.
HU-GO	the Hungarian truck-tolling scheme
Hypothecation	the ring-fencing of funds collected by a scheme to support particular operations or transport services
Hz	Hertz – the unit of frequency
IAG	the E-ZPass InterAgency Group
IAP	interoperability application profile
IBTTA	International Bridge, Tunnel and Turnpike Association (www.ibtta.org on 27 Dec 2016)
ICT	information and communications technologies
ID	identification
IDUQ	Institut de développement urbain du Québec, Canada
IEEE	The US Institute of Electrical and Electronics Engineers
IFTA	International Fuel Tax Agreement

IM	interoperability management
INS	inertial navigation systems
Intelligent client	also known as 'thick client' – A GNSS OBU that can measure its location and match its position to an on-board digital roadmap
IP	internet protocol – or alternatively, intellectual property
IR	infra-red
IrDA	infrared data association
ISM	Industrial Scientific and Medical (radio band)
ISO	International Organization for Standardization (www.iso.ch 27 Dec 2016)
ISO-18000	a family of air-interface standards for RFID at various frequency ranges for tolling and other applications
ISTEA	Intermodal Surface Transportation Efficiency Act
ITS	Intelligent Transport Systems – IT and electronic communications applied to transportation
ITS AMERICA	Intelligent Transportation Systems Society of the United States
ITS(UK)	Intelligent Transport Systems (United Kingdom)
ITU	International Telecommunications Union – the international body for telecommunications standards
IU	in-vehicle unit
IVE	in-vehicle equipment
IVR	interactive voice response
IVU	in-vehicle unit
JPO	ITS Joint Program Office of the FHWA
JRC	EC Joint Research Centre in Ispra (Italy)
KPH	kilometres per hour
KPI	key performance indicator
LAC	location augmentation communication
LED	light-emitting diode
LEZ	Low Emission Zone, e.g. Greater London
LGV	light goods vehicle
LKW Maut	LastKraftWagen Maut (literally 'HGV Toll') – the German truck tolling scheme
LLEZ	London Low Emission Zone
LPR	license plate recognition
LRUC	lorry road user charge – a proposed and over-complicated UK HGV charging scheme abandoned before implementation

LSVA	Leistungsabhängige Schwerverkehrsabgabe – 'performance-related heavy vehicle fee' – the Swiss truck-tolling scheme
LTA	land transport authority (Singapore) www.lta.gov.sg 27 Dec 2016). The agency responsible for the ALS and road pricing schemes in Singapore
LTE	long term evolution – a 4G cellular radio standard
LTCC	London Traffic Control Centre
LTP	local transport plan
MA	the US state of Massachusetts
MaaS	mobility-as-a-service – integrated mobility services – public, private and shared – for on-demand, door-to-door, transport options for a monthly subscription fee, as an alternative to privately owned vehicle travel
managed lanes	see HOT
Map-matching	used in satellite-based navigation and charging systems, to correct errors or uncertainties in a vehicle's position as determined by GNSS by comparing to a digital map
MART	Manchester Against Road Tolls
MBUF	mileage-based user fee
meaconing	rebroadcasting a GNSS signal maliciously, accidentally or to improve reception but causing misreporting of a position
MEC	marginal external cost
MEMS	micro-electro-mechanical systems – miniaturised mechanical and electro-mechanical elements
MEN	*Manchester Evening News* – a UK newspaper
MHz	a frequency of 1 million Hz
MLFF	multi-lane free-flow – barrier-free (electronic) tolling
MMI	man machine interface
MnDOT	Minnesota Department of Transportation
MnPASS	the Minnesota HOT lanes 'brand'
MNO	mobile network operator
MOT	a new UK Ministry of Transport (MoT) Test inspection procedure was introduced so that only approved registered garages could issue MOT certificates – to combat cloning of number-plates
MoU	memorandum of understanding
mS	milliseconds – thousandths of a second
MTAB&T	Metropolitan Transportation Authority Bridges and Tunnels (New York)
NAFCP	The Hungarian National Automated Fare Collection Platform

NAI	UK National ANPR Infrastructure
NCDOT	North Carolina Department of Transportation
NCITS	ANSI National Committee for Information Technology Standards
NDS	The Slovak Road Authority
NFC	near-field communication
NGO	Non-Government Organisation
NHS	The UK National Health Service
NIOP	'National interoperability protocol' currently being developed by an IBTTA technical sub-committee from among the seven different tolling technologies in use in North America, to identify a preferred transponder/reader technology
nm	nano-metre – 10^{-9} of a metre
NO_x, NO_2	nitrogen oxides, nitrogen dioxide – harmful emissions especially from diesel engines
NPCC	The UK National Police Chiefs' Council
NPR	number plate recognition
NPRA	Norwegian Public Roads Administration
NRM	Netherlands Regional Model – A traffic model of the Dutch Highway Agency
NRP	National Road Pricing
NTPS	The Hungarian National Toll Payment Services PLC
NTSC	National Television System Committee – US television format
OBD port	On Board Diagnostics port – a socket into which a 'dongle' or other device can be fitted, giving access to a vehicle's internal electronic information
OBE	on-board equipment – subsystems located in the vehicle that interact with roadside charging and enforcement equipment
OBU	on-board unit
OCR	optical character recognition
OECD	Organisation for Economic Cooperation and Development
OEM	Original Equipment Manufacturer e.g. a car manufacturer
OMISS	Open Minimum Interoperability Specification Suite (UK DfT 2006)
opex	operational expenditure
OReGO	The branding of Oregon's Road Usage Charge Program
ORT	open road tolling (typically a US usage)
P&R	park-and-ride schemes – where drivers are encouraged to park in (usually purpose-built) car-parks outside the city and take a bus to the centre

PAL	phase alternation line – European television format
ParkPlus	a dynamic parking pricing scheme in Calgary
PAYD	pay as you drive (vehicle insurance)
PCN	penalty charge notice – a fine for vehicle owners not paying a charge or toll
PFI	private finance initiative
PLATON	'pay per ton' – Russia's truck tolling system
PM10	particulate matter less than 10 μm in diameter
PM2.5	particulate matter less than 2.5 μm in diameter
PNR	private non-residential parking
PNT	position, navigation and timing
PPP	public private partnership
PSDB	UK Home Office Police Scientific Development Branch
PTE	public transport executive
PTP	personal travel planning
PTT	post, telegraph and telephone – the generic term for the national telecommunications organisations
QoS	quality of service
QZSS	Quasi-Zenith Satellite System – a satellite-based augmentation system for GPS covering Japan
Rat running	drivers using secondary roads or residential streets instead of main roads, usually in urban or suburban areas, to avoid congestion or find a faster route
RCI	road charging interoperability – a European project
REETS	Regional European Electronic Toll Service
Retroreflective	reflecting light from an illuminator back into the source of the light or into a co-located camera or other detector including the eye
Revealed preference	studies in which consumer or traveller behaviour is shown via the actual choices they make. See also 'stated preference'
revenue-neutral	the revenues raised by a road pricing scheme are compensated by reductions in other motoring charges and taxes, so that on average motorists don't pay any more to travel on roads – though there may be individual winners and losers
RF	radio frequency
RFID	radio frequency identification
RFP	request for proposal
Rijkswaterstaat	Dutch Highway Agency, part of the Ministry of Infrastructure and the Environment, responsible for design, construction, management and maintenance of the main infrastructure facilities in the Netherlands, including the main road and waterway networks and water systems

ROCOL	Road Charging Options for London – a report that considered the feasibility of congestion charging in London, published in 2000
RRL	road research laboratory – the former name of the UK Transport Research Laboratory (TRL)
RSE	road-side equipment – the tolling equipment located at the road-side
RTTT	road transport and traffic telematics
RUC	road user charging (or sometimes Road Use Charging)
RUC West	a research partnership among various US state departments of transportation (14 members as of June 2017). See also WRUCC
RUFTF	Road User Fee Task Force in Oregon
SAM	security access module
SAM	the state account management option in the California RUC trials
SANEF	the French motorway operator (Société de Autoroutes du Nord Est de la France)
SANEF ITS	the ITS division of SANEF – now called EMOVIS
SaPPART	Satellite Positioning Performance Assessment for Road Transport – a COST Action which brings together experts in GNSS, ITS and mobility for standardisation and certification initiatives
SAR	Hong Kong Special Administrative Region
SBAS	satellite-based augmentation systems
SCBA	societal cost–benefit analysis
SDCM	System for Differential Correction and Monitoring (of GLONASS)
SeGo®	a brand name for a microwave tag supplied by TransCore, a high data rate (2 × eGo data rate) read/write tag protocol. It is compliant with ANSI NCITS 256-2001 Part 4.2, ISO 18000-6B, and ISO 10374 standards
SEMMMS	South East Manchester Multi-modal Strategy
SET	System Evaluation Test of GNSS for road charging in Singapore
SFMTA	San Francisco Municipal Transportation Agency
SFpark	a dynamic parking pricing pilot scheme in San Francisco
Shadow tolling	payments by public bodies to agencies operating roads – often based on traffic volumes and service levels
shoulder pricing	charges that are higher than at non-peak times but lower than at the peak
SHT	Société Habilité Télépéage – French toll service providers

SLA	service level agreement
SLPP	University of Minnesota Humphrey School of Public Affairs State and Local Policy Program
SMMC	the Hungarian State Motorway Management Company
SMS	short message service – a GSM data service
SOV	single occupancy vehicle
SP	service provider
spamming	the generation of signals, intentionally or otherwise, that result in a deterioration or failure of GNSS reception; GNSS signals are relatively weak and easy to jam
spillback queue	where a queue on a downstream road link obstructs an upstream link
spitsmijden	'avoiding peak traffic' – in Dutch
spoofing	the intentional generation of false, misleading or "forged" GNSS signals. It is illegal in most countries
stated preference	studies in which consumer or traveller behaviour is measured or predicted via the choices they say that they would make. See also 'revealed preference'
Stockholm Group	The Public Road Administrations of Finland, France, Germany, the Netherlands, Sweden, Switzerland and United Kingdom
STSFA	Surface Transportation System Funding Alternatives – part of the US FAST Act
SUC	The Santiago Urban Concessions tolling regime in Chile
SunPass®	the toll system operated in Florida. SunPass® is a registered trademark of the Florida Department of Transportation
TAC	California's Road Charge Technical Advisory Committee
Tag	also called OBE or OBU. A small transponder unit located in the vehicle that interacts with roadside charging and enforcement equipment
TC	Technical Committee (within CEN and ISO)
TC	toll charger
TC278	the CEN standard for microwave tags for tolling
TCA	transportation corridor agencies operating the toll system in southern California
TCP/IP	transmission control protocol/internet protocol
TDM	travel demand management
TDP	time distance place charging combines distance measurement with time-of-day and position
TDO	toll declaration operator
Telepass	the Italian toll service provider

TEN	trans-European road network
TfGM	Transport for Greater Manchester
TfL	Transport for London
TH	trunk highway
Thick client	see *intelligent client*
Thin client	A GNSS OBU that measures and reports its position to a central system for matching with a map database
TIF	transport innovation fund – a former UK government initiative to encourage Local Authorities to introduce congestion charging along with other transport improvements
TIS	Télépeage Inter-Société – the French interoperable electronic tolling system of motorway operators. There are two versions – TIS-Liber-t service for light vehicles, and TIS-PL (TIS Poids Lourds) for heavy vehicles
Toll charger	the organisation representing the road owner in terms of toll collection. A public or private organisation which levies tolls for the circulation of vehicles in an EETS domain
Toll service provider	the organisation responsible for contract and relations with the road user, e.g. billing, OBU distribution and dispute settlement. The toll service provider grants access to the service to the user
Toll2Go	the interoperable toll charge service between Germany and Austria
Trafikverket	The Swedish Transport Administration, http://www.trafikverket.se
TRANSUMO	Transition to Sustainable Mobility – a Dutch foundation
Transceiver	transmitter/receiver
TRL	The UK Transport Research Laboratory
TSP	toll service provider
TTCN	testing and test control notation
TTFF	time to first fix (GNSS OBUs)
TTI	traffic & travel information
TTP	trusted third party – an organisation issuing encryption keys or digital signatures
Tx	The US state of Texas
ULEZ	ultra-low emission zone, as proposed for London in 2019/20
UOBU	universal on-board unit – an OBU containing both DSRC and GNSS/CN technology
UHF	ultra-high frequency – frequency band from 300 MHz to 3 GHz

UMTS	universal mobile telecommunications service – the ETSI 3G mobile radio standard
UNI	the Italian standard for microwave tags for tolling
UPA	Minnesota's Urban Partnership Agreement
'urban canyon'	a phenomenon where satellite positioning signals are screened by tall buildings and are not visible to OBUs
US DOT	United States Department of Transportation
USD	US dollar
V5C	a UK new vehicle owner registration form (known as V5C) was introduced to require re-registration of all vehicles because of concerns about cloning of number-plates to evade the London congestion charge
Value pricing	another term for road user charging, road pricing etc.
VAS	value-added services
VAT	value added tax
VED	vehicle excise duty – the UK annual car tax
VEP	vehicle entry permit smart-card – a stored value card issued to each foreign car or motorcycle entering Singapore
VIA-T	the Spanish standard for microwave tolling on motorways
ViaVerde	The Portuguese Toll Service Provider
Video tolling	using (digital) cameras – usually with ANPR – to identify and charge vehicles
VIN	vehicle identification number
VKT	vehicle kilometres travelled
VLH	vehicle loss hour is a frequently used indicator for travel time loss in Dutch traffic and transport analyses. One vehicle loss hour signifies that a vehicle has suffered a delay of one hour.
VMS	variable message sign
VMT	vehicle miles travelled
VP	value pricing – a US term for road user charging
VRM	vehicle registration mark – the licence plate number
VTT	value of travel time
WAAS	US Wide Area Augmentation System (for satellite positioning)
Washington, DC	Washington, in the District of Columbia, is the capital of the United States
WAVE	wireless access for vehicular environments – a US standard for interoperable 5.9-GHz DSRC systems
WEZ	Western Extension Zone of the London Congestion Charging Scheme

White list	a list of valid/acceptable but possibly unequipped users of a toll system, such as public transport or emergency service vehicles – as opposed to a 'black list' of offenders
WG	working group
WiM	weigh in motion
WPL	workplace parking levy
WRI	World Resources Institute, http://www.wri.org/
WRUCC	Western Road Usage Charge Consortium, branded RUC West, in the United States
ZIP	a ZIP code or postal code is a code of letters and digits added to a postal address to aid in sorting mail for delivery
ZTL	Zona a Traffico Limitato in Milan

Appendix B

References and further reading

D'Artagnan Consulting (2015). "Briefing Book Compilation For The California Road Charge Technical Advisory Committee: February – August 2015". Reference Material for TAC Meeting #9. Prepared by D'Artagnan Consulting. September 16, 2015.

Eddington, R. (2006). The Eddington Transport Study. The Case for Action: Sir Rod Eddington's Advice to Government. Retrieved 25 February 2011 from http://www.thepep.org/ClearingHouse/docfiles/Eddington.Transport.Study%20-%20Rod.pdf.

Eliasson, J. (2010). So You're Considering Introducing Congestion Charging? Here's What You Need to Know: An FAQ Based on Stockholm's Experiences. OECD/ITF Joint Transport Research Centre Discussion Paper No. 2010/4, International Transport Forum/OECD Round Table on Implementing Congestion Charging.

Gullberg, A. and Isaksson, K. (eds.) (2009). "Congestion Taxes in City Traffic: Lessons Learnt from the Stockholm Trial". Stockholm: Nordic Academic Press.

Move NY "The Move NY Fair Pan", February 2015, iheartmoveny.org/wp-content/uploads/2015/02/Move-NY-Fair-Plan-150217v1.pdf. Accessed 25 June 2017.

Oehry, B. (2010). Critical Success Factors for Implementing Road Charging Systems. OECD/ITF Joint Transport Research Centre Discussion Paper No. 2010/3, International Transport Forum/OECD Round Table on Implementing Congestion Charging.

Pickford, A.T.W. and Blythe, P.T. (2006). "Road User Charging and Electronic Toll Collection", Artech House (Boston & London).

Richards, M.G. (2006). Congestion Charging in London: The Policy and the Politics. Basingstoke, UK, and New York. Palgrave Macmillan.

Smeed, R.J. (1964). Road Pricing: The Economic and Technical Possibilities. HMSO.

Walker, J. (2011). "The Acceptability of Road Pricing". http://www.racfoundation.org/research/economics/road-pricing-acceptability.

Walker, J. and Preston, J.M., (2017). "We can pay for better, safer, more reliable roads in a way that is fair to road users and good for the economy and the environment", ITS Review, November 2017, pp. 30–33.

Behavioural economics and psychology

Halpern, D. (2015). "Inside the Nudge Unit", London: Penguin Random House.

Kahneman, D. (2012) "Thinking, Fast and Slow", London & New York: Penguin Books.

Sunstein, C.R. and Thaler, R.H. (2009) "Nudge: Improving Decisions About Health, Wealth and Happiness", New York and London: Penguin Books.

Thaler, R.H. (2016) "Misbehaving: The Making of Behavioural Economics", New York and London: Penguin Books.

Tversky, A. and Kahneman, D. (1981). The Framing of Decisions and the Psychology of Choice. Science, New Series, Vol. 211, No. 4481, pp. 453–458.

Index

www.ingramcontent.com/pod-product-compliance
Lightning Source LLC
LaVergne TN
LVHW011854200926
848484LV00002B/454
* 9 7 8 1 7 8 5 6 1 2 0 5 3 *